# Appropriate Wisdom, Technology, and Management toward Environmental Sustainability for Development

# Appropriate Wisdom, Technology, and Management toward Environmental Sustainability for Development

Editor

**Shervin Hashemi**

MDPI • Basel • Beijing • Wuhan • Barcelona • Belgrade • Manchester • Tokyo • Cluj • Tianjin

*Editor*
Shervin Hashemi
Institute for Environmental
Research, Yonsei University
College of Medicine
Korea

*Editorial Office*
MDPI
St. Alban-Anlage 66
4052 Basel, Switzerland

This is a reprint of articles from the Special Issue published online in the open access journal *Sustainability* (ISSN 2071-1050) (available at: https://www.mdpi.com/journal/sustainability/special_issues/wisdom_technology_management_environmental_sustainability).

For citation purposes, cite each article independently as indicated on the article page online and as indicated below:

LastName, A.A.; LastName, B.B.; LastName, C.C. Article Title. *Journal Name* **Year**, *Volume Number*, Page Range.

**ISBN 978-3-0365-5189-0 (Hbk)**
**ISBN 978-3-0365-5190-6 (PDF)**

Cover image courtesy of Singkham

# Contents

# About the Editor

**Shervin Hashemi**

Along with being a senior researcher at the Institute for Environmental Research at Yonsei University College of Medicine, Seoul, Republic of Korea, Dr. Shervin Hashemi is also an honorary research associate at the Durban University of Technology, Faculty of Management Sciences, South Africa. His primary research expertise pertains to the provision of water and sanitation across remote areas. His remarkable invention of the Resource Circulated Sanitation system won him accolades such as the Leaving No One Behind Innovation Award 2019 and Energy Globe Republic of Korea National Award 2019. Furthermore, he also developed the Sanitation Sustainability Index to evaluate the technical, economic, and social sustainability of local sanitation systems. Since 2017, he has been an active member of the management committee of the International Water Association (IWA) Resource-Oriented Sanitation Specialist Group. In 2021, he was nominated for the IWA Global Water Award, which recognizes individuals who, through innovative leadership and practice, have made exceptional contributions to the mindful and wise use of water globally. He also serves as a member of the editorial board of the journal *Sustainability* (ISSN 2071-1050).

# Preface to "Appropriate Wisdom, Technology, and Management toward Environmental Sustainability for Development"

The protection and maintenance of environmental resources for future generations require responsible interaction between humans and the environment in order to avoid wasting natural resources. According to an ancient Native American proverb, "We do not inherit the Earth from our ancestors; we borrow it from our children." This indigenous wisdom has the potential to play a significant role in defining environmental sustainability.

Recent technological advances could sustain humankind and allow for comfortable living. However, not all of these advancements have the potential to protect the environment for future generations. Developing societies and maintaining the sustainability of the ecosystem require appropriate wisdom, technology, and management collaboration.

This book is a collection of 19 important articles (15 research articles, 3 review papers, and 1 editorial) that were published in the Special Issue of the journal *Sustainability* entitled "Appropriate Wisdom, Technology, and Management toward Environmental Sustainability for Development" during 2021–2022. The articles are covering the following subjects:

1. Appropriate Wisdom, including the following:

- Resource Valorization: Valuing Water, Wastes, and Other Resources;
- Development of Mathematical, Theoretical, and Computational Models;
- Scientific Hypothetical Approaches;
- Analyzing Historical Approaches and Ancient Wisdom.

2. Appropriate Technology, including the following:

- Nature-Based Solutions;
- Clean Technologies;
- Resource Circulation: Conservation and Recycling.

3. Appropriate Management, including the following:

- Human and Nature Interactions: Simulation and Optimization;
- Economic Management: Life Cycle Assessment (LCA) and Circular Economy;
- Social Awareness and Human Rights: Concepts of Leaving No One Behind;
- Sanitation, Hygiene, and Public Health;
- Risk Assessment;
- New Policies.

This book addresses the policymakers and decision-makers who are willing to develop societies that practice environmental sustainability, by collecting the most recent contributions on the appropriate wisdom, technology, and management regarding the different aspects of a community that can retain environmental sustainability.

I would like to acknowledge the journal *Sustainability* and the publisher, MDPI, for their generous support. In addition, I would like to thank the authors who contributed by submitting their valuable work to the Special Issue, as well as the reviewers and editorial staff at the journal

*Sustainability* for their time and effort in reviewing, proofreading, and publishing the articles included in this collection. I am grateful to Eng. Farid Hashemi for his kind support and encouragement.

This book and the Special Issue were edited during the peak of the COVID-19 pandemic. I would like to express my deepest sympathies to the people worldwide who have suffered losses as a result of the pandemic. I also wish to write in the memory of Prof. Parviz Tajdari and Dr. Shahram Soudagaran to acknowledge their important role in my life. I owe them so much and wish them to rest in peace.

**Shervin Hashemi**
*Editor*

*sustainability*

MDPI

*Editorial*

# Borrowing the Earth from the Next Generation: Appropriate Wisdom, Technology, and Management toward Environmental Sustainability for Development

**Shervin Hashemi** [1,2]

[1] Institute for Environmental Research, Yonsei University College of Medicine, 50-1 Yonsei-ro, Seodae-mun-gu, Seoul 03722, Korea; shervin@yuhs.ac

[2] Honorary Research Associate, Faculty of Management Science, Durban University of Technology, Durban 4000, South Africa

**Citation:** Hashemi, S. Borrowing the Earth from the Next Generation: Appropriate Wisdom, Technology, and Management toward Environmental Sustainability for Development. *Sustainability* **2022**, *14*, 8803. https://doi.org/10.3390/su14148803

Received: 14 July 2022
Accepted: 15 July 2022
Published: 19 July 2022



## 1. Introduction

Technological advancements, combined with rapid population growth, have led to an increase in environmental resource consumption. Accordingly, there are concerns regarding managing environmental interventions in social developments to protect and maintain resources for future generations.

Environmental sustainability can be defined as a sense of responsibility in consuming environmental resources, so that the next generation can have access to them. As a result, safe consumption and the protection of natural resources in a way that supports the ecosystems and the prosperity of the next generation, to ensure that they have at least an equal quality of life in the future, should be considered the foundation of developing communities [1–3]. Regardless of how a community approaches environmental sustainability, the goal remains the same: to provide responsible interactions with the environment through the appropriate wisdom, technology, and management [4].

This Special Issue of the journal *Sustainability* (ISSN 2071–1050), entitled "Appropriate Wisdom, Technology, and Management toward Environmental Sustainability for Development", was proposed to collect the most recent contributions regarding the appropriate wisdom, technology, and management in developing different aspects of a community that can retain environmental sustainability during the period 2022–2023.

Accordingly, the articles in the Special Issue cover at least one of the following subjects:

1. Appropriate wisdom, including:
    - Resource valorization: valuing water, wastes, and other resources;
    - Development of mathematical, theoretical, and computational models;
    - Scientific hypothetical approaches;
    - Analyzing historical approaches and ancient wisdom.
2. Appropriate technology, including:
    - Nature-based solutions;
    - Clean technologies;
    - Resource circulation: conservation and recycling.
3. Appropriate Management, including:
    - Human and nature interactions: simulation and optimization;
    - Economic management: life cycle assessment (LCA) and the circular economy;
    - Social awareness and human rights: concepts of leaving no one behind;
    - Sanitation, hygiene, and public health;
    - Risk assessment;
    - New policies.

## 2. List of Contributions

Up to 31 December 2021, 30 contributions were submitted for possible publication in this Special Issue. Contributions of a higher quality were subjected to the journal's precise review process, following an initial evaluation by the editor and the editorial office. Eighteen papers (15 research articles and 3 review papers) were finally accepted for publication and inclusion in the Special Issue. These papers' references are sorted in alphabetical order according to the first author.

1. Aderemi, B.A.; Olwal, T.O.; Ndambuki, J.M.; Rwanga, S.S. A Review of Groundwater Management Models with a Focus on IoT-Based Systems. *Sustainability* **2022**, 14, 148. https://doi.org/10.3390/su14010148
2. Alaloul, W.S.; Musarat, M.A.; Haruna, S.; Law, K.; Tayeh, B.A.; Rafiq, W.; Ayub, S. Mechanical Properties of Silica Fume Modified High-Volume Fly Ash Rubberized Self-Compacting Concrete. *Sustainability* **2021**, 13, 5571. https://doi.org/10.3390/su13105571
3. Alaloul, W.S.; Musarat, M.A.; Rabbani, M.B.A.; Iqbal, Q.; Maqsoom, A.; Farooq, W. Construction Sector Contribution to Economic Stability: Malaysian GDP Distribution. *Sustainability* **2021**, 13, 5012. https://doi.org/10.3390/su13095012
4. Aysan, A.F.; Bergigui, F.; Disli, M. Using Blockchain-Enabled Solutions as SDG Accelerators in the International Development Space. *Sustainability* **2021**, 13, 4025. https://doi.org/10.3390/su13074025
5. Eshete, Y.; Alamirew, B.; Bishaw, Z. Yield and Cost Effects of Plot-Level Wheat Seed Rates and Seed Recycling Practices in the East Gojam Zone, Amhara Region, Ethiopia: Application of the Dose–Response Model. *Sustainability* **2021**, 13, 3793. https://doi.org/10.3390/su13073793
6. González-Arqueros, M.L.; Domínguez-Vázquez, G.; Alfaro-Cuevas-Villanueva, R.; Israde-Alcántara, I.; Buenrostro-Delgado, O. Hazardous Solid Waste Confined in Closed Dump of Morelia: An Urgent Environmental Liability to Attend in Developing Countries. *Sustainability* **2021**, 13, 2557. https://doi.org/10.3390/su13052557
7. Kang, C.-W.; Kim, M.K.; Jang, E.-S. An Experimental Study on the Performance of Corrugated Cardboard as a Sustainable Sound-Absorbing and Insulating Material. *Sustainability* **2021**, 13, 5546. https://doi.org/10.3390/su13105546
8. Khan, S.A.R.; Ponce, P.; Thomas, G.; Yu, Z.; Al-Ahmadi, M.S.; Tanveer, M. Digital Technologies, Circular Economy Practices and Environmental Policies in the Era of COVID-19. *Sustainability* **2021**, 13, 12790. https://doi.org/10.3390/su132212790
9. Kumar, A.; Kumar, S.; Komal; Ramchiary, N.; Singh, P. Role of Traditional Ethnobotanical Knowledge and Indigenous Communities in Achieving Sustainable Development Goals. *Sustainability* **2021**, 13, 3062. https://doi.org/10.3390/su13063062
10. Mohammed, M.; Shafiq, N.; Elmansoury, A.; Al-Mekhlafi, A.-B.A.; Rached, E.F.; Zawawi, N.A.; Haruna, A.; Rafindadi, A.D.; Ibrahim, M.B. Modeling of 3R (Reduce, Reuse and Recycle) for Sustainable Construction Waste Reduction: A Partial Least Squares Structural Equation Modeling (PLS-SEM). *Sustainability* **2021**, 13, 10660. https://doi.org/10.3390/su131910660
11. Quiliche, R.; Rentería-Ramos, R.; de Brito Junior, I.; Luna, A.; Chong, M. Using Spatial Patterns of COVID-19 to Build a Framework for Economic Reactivation. *Sustainability* **2021**, 13, 10092. https://doi.org/10.3390/su131810092
12. Qv, D.; Duan, X.; Wang, J.; Hou, C.; Wang, G.; Zhou, F.; Li, S. Issues and Potential Solutions to the Clean Heating Project in Rural Gansu. *Sustainability* **2021**, 13, 8397. https://doi.org/10.3390/su13158397
13. Shehata, A.M. Current Trends in Urban Heritage Conservation: Medieval Historic Arab City Centers. *Sustainability* **2022**, 14, 607. https://doi.org/10.3390/su14020607
14. Song, N.V.; Ha, T.V.; Thuan, T.D.; Hanh, N.V.; Tien, D.V.; Tiep, N.C.; Phuong, N.T.M.; Tu, P.A.; Uan, T.B. Development of Rice Husk Power Plants Based on Clean Development Mechanism: A Case Study in Mekong River Delta, Vietnam. *Sustainability* **2021**, 13, 6950. https://doi.org/10.3390/su13126950

15. Sunny, A.R.; Mithun, M.H.; Prodhan, S.H.; Ashrafuzzaman, M.; Rahman, S.M.A.; Billah, M.M.; Hussain, M.; Ahmed, K.J.; Sazzad, S.A.; Alam, M.T.; Rashid, A.; Hossain, M.M. Fisheries in the Context of Attaining Sustainable Development Goals (SDGs) in Bangladesh: COVID-19 Impacts and Future Prospects. *Sustainability* **2021**, 13, 9912. https://doi.org/10.3390/su13179912
16. Supriyanto, E.; Sentanuhady, J.; Dwiputra, A.; Permana, A.; Muflikhun, M.A. The Recent Progress of Natural Sources and Manufacturing Process of Biodiesel: A Review. *Sustainability* **2021**, 13, 5599. https://doi.org/10.3390/su13105599
17. Tanveer, M.; Ahmad, A.-R.; Mahmood, H.; Haq, I.U. Role of Ethical Marketing in Driving Consumer Brand Relationships and Brand Loyalty: A Sustainable Marketing Approach. *Sustainability* **2021**, 13, 6839. https://doi.org/10.3390/su13126839
18. Tao, Y.-J.; Lin, Y.-S.; Lee, H.-S.; Gan, G.-Y.; Tu, C.-S. Using a Product Life Cycle Cost Model to Solve Supplier Selection Problems in a Sustainable, Resilient Supply Chain. *Sustainability* **2022**, 14, 2423. https://doi.org/10.3390/su14042423

## 3. Content and Significance of Contributions

### 3.1. Contributions to the Appropriate Wisdom Topic

There were six contributions related to the appropriate wisdom topic, including contributions 4, 9, 11–13, and 17.

In contribution 4, the authors reported on the application of blockchain-enabled solutions to achieving the Sustainable Development Goals (SDGs). In addition, the authors considered the knowledge gaps and introduced requirements for the provision of an appropriate blockchain technology that aims to accelerate global actions, particularly after the COVID-19 pandemic.

Based on an extensive and systematic literature review, the authors of contribution 9 explained that, among the 17 SDG goals, at least 7 of them, including SDGs 1–3, 12, 13, 15, and 17, are associated with traditional ethnobotanical knowledge (TEK). In order to achieve these goals, it is necessary to have a sufficient level of understanding of the indigenous wisdom of TEK.

In contribution 11, the authors proposed using humanitarian logistics theory to suggest an appropriate framework for economic reactivation in vulnerable communities during COVID-19. To accomplish this, the authors identified and analyzed factors related to COVID-19 mortality in Peru.

In contribution 12, the challenges were introduced regarding the applicability of the Rural Clean Heating Project (RCHP) in Gansu, China. The RCHP objects to making rural energy systems more flexible, enhancing the integration and distribution of green energy, and controlling the environmental impacts. The authors discussed appropriate policy solutions to major technical, economic, and ecological concerns regarding the application of the RCHP in Gansu.

In contribution 13, the current trends in Urban Heritage Conservation (UHC) in historical Arab cities were investigated, and the conservation parameters were identified. The author used these parameters to investigate the forthcoming changes in urban conservation, and provided useful suggestions for decision-makers and conservation authorities.

In contribution 17, a study was conducted on the impact of ethical marketing practices, looking at the sustainability of the value-adding products and customer–brand relationships in Pakistan. The authors also observed a positive association between brand loyalty, the sustainability of the value-added product and the customer value–brand relationship.

### 3.2. Contributions to the Appropriate Technology Topic

There were five contributions related to the appropriate technology topic, including contributions 2, 7, 8, 14, and 16.

In contribution 2, the authors reported results of improvements in the mechanical properties of self-compacting concrete by adding fly ash, crumb rubber, and silica fume.

In contribution 7, the applicability of corrugated cardboard for noise reduction was investigated. Based on the results, the authors suggested using perforated corrugated cardboard with multi-frequency resonators (PCCM) as an eco-friendly and sustainable noise-reduction material.

In contribution 8, the authors discussed the correlation among industry 4.0 technologies, COVID-19 pandemic, environmental regulation policies, and circular economy practices. Based on this study, while industry 4.0 technologies and environmental regulation policies have a significant impact on circular economy practices, even during the COVID-19 breakout, there is no evidence that COVID-19 was a major obstacle to the adoption of the circular economy.

In contribution 14, the authors demonstrated the Clean Development Mechanism (CDM) project for the Vietnamese electricity and energy sectors, with a focus on the use of rice husks for power generation in the Mekong Delta. The results demonstrated the possibility of using rice husk power to develop energy plants of up to 11 megawatts, along with other environmental benefits, such as a reduction in $CO_2$ emissions.

In contribution 16, a review of the recent advances in natural resources and manufacturing processes of biodiesels is presented. According to the authors, transesterification is currently the most sustainable biodiesel production method.

*3.3. Contributions to the Appropriate Management Topic*

There were seven contributions related to the appropriate technology topic, including contributions 1, 3, 5, 6, 10, 15, and 18.

Contribution 1 provides a review of the current groundwater management models, along with the systems that are enhanced by utilizing the Internet of Things (IoT).

In contribution 3, the authors used the data collected from the Department of Statistics Malaysia and the World Bank during 1970–2019 to determine the Malaysian construction sector's relationship with other sectors through complex linkages with significant contributions to the economy and gross domestic product (GDP). Based on the results, the authors proposed a sustainable conceptual framework for global sustainable development, which includes major factors affecting the growth in the construction industry.

Contribution 5 presented a case study, applying the dose–response model to evaluate the effects of seed rates and seed-recycling practices on wheat yield in Ethiopia using 450 sample respondents. The authors provided a cost–benefit analysis to prove that the farmers can cover seed costs through seed recycling. However, farmers' yields and net income can be significantly improved by utilizing unrecycled certified bread wheat seeds (CBWS).

In contribution 6, the authors analyzed solid samples from eight opencast wells to determine the presence of heavy metals and arsenic in the closed dump of Morelia. The results demonstrated the presence of heavy metals, including Pb, Cu, Ni, Zn, Cr, and Fe, as well as a high concentration of As. These findings can be used to support the requirement of essential improvements in territorial plans to enhance public and environmental health.

In contribution 10, the authors investigated the major challenges regarding construction waste in Malaysia. They suggested a model for the sustainable reduction in construction waste based on reducing, reusing, and recycling principles after identifying significant improvements and policy-related factors. The results showed a significant correlation between the improving factors, policy-related factors, and the amount of generated and reduced waste.

Contribution 15 reported on primary fieldwork and provided a secondary data analysis to obtain a clear overview of the performance and challenges in the fisheries sector in Bangladesh. In this paper, the authors evaluated the impact of the COVID-19 pandemic, as well as the effect of the fisheries sector in Bangladesh, on achieving SDGs.

Contribution 18 presents a report on the development of a product life-cycle cost model to support Taiwanese light-emitting diode (LED) manufacturers in capacity planning for sustainable and resilient supply chain management. In addition, the authors presented their decision-making tool, which was developed to assess sustainable procurement management in high-tech Taiwanese LED companies. The decision-makers can use the developed model to minimize the product life-cycle costs.

## 4. Conclusions

The Special Issue "Appropriate Wisdom, Technology, and Management toward Environmental Sustainability for Development" aimed to be a multidisciplinary one and collected relevant studies on approaches to environmental sustainability. It appealed for original and novel studies on the provision of responsible consumption, reasonable recycling, and sustainable development of natural resources in different environments. As a result, it can be regarded as a useful source, allowing for decision-makers and policy-makers in different environmental fields to be introduced to the latest advances and approaches to developing communities with environmental sustainability.

**Funding:** This research received no external funding.

**Acknowledgments:** The author would like to acknowledge the journal *Sustainability* and the publisher MDPI for their generous support. The author would also like to thank those who contributed papers to this Special Issue, as well as the reviewers and the authorial staff at the journal *Sustainability* for their time and effort in reviewing, improving, and publishing the included articles. The author is grateful to Eng. Farid Hashemi for his kind support and encouragement. The Special Issue was edited during the peak of the COVID-19 pandemic. The author would like to express his deepest sympathies to the people worldwide who suffered losses as a result of the pandemic.

**Conflicts of Interest:** The author declares no conflict of interest.

## References

1. Thangavel, P.; Sridevi, G. *Environmental Sustainability*; Thangavel, P., Sridevi, G., Eds.; Springer: New Delhi, India, 2015. [CrossRef]
2. Hashemi, S. Perspectives on Sustainable Integrated Clean Environment for Human and Nature. *Sustainability* **2021**, *13*, 4150. [CrossRef]
3. Hashemi, S. *Sustainable Integrated Clean Environment for Human & Nature*; Hashemi, S., Ed.; MDPI: Basel, Switzerland, 2021. [CrossRef]
4. Goodland, R. The Concept of Environmental Sustainability. *Annu. Rev. Ecol. Syst.* **1995**, *26*, 1–24. [CrossRef]

*sustainability*

MDPI

*Article*

# Using Blockchain-Enabled Solutions as SDG Accelerators in the International Development Space

**Ahmet Faruk Aysan [1,*], Fouad Bergigui [2] and Mustafa Disli [1]**

1 College of Islamic Studies, Hamad Bin Khalifa University, Doha 34110, Qatar; mdisli@hbku.edu.qa
2 SDI Global Consult, Dover, DE 19901, USA; f.bergigui@sdi.world
* Correspondence: aaysan@hbku.edu.qa

**Abstract:** As the world is striving to recover from the shockwaves triggered by the COVID-19 crisis, all hands are needed on deck to transition towards green recovery and make peace with nature as prerequisites of a global sustainable development pathway. In this paper, we examine the blockchain hype, the gaps in the knowledge, and the tools needed to build promising use cases for blockchain technology to accelerate global efforts in this decade of action towards achieving the SDGs. We attempt to break the "hype cycle" portraying blockchain's superiority by navigating a rational blockchain use case development approach. By prototyping an SDG Acceleration Scorecard to use blockchain-enabled solutions as SDG accelerators, we aim to provide useful insights towards developing an integrated approach that is fit-for-purpose to guide organizations and practitioners in their quest to make informed decisions to design and implement blockchain-backed solutions as SDG accelerators. Acknowledging the limitations in prototyping such tools, we believe these are minimally viable products and should be considered as living tools that can further evolve as the blockchain technology matures, its pace of adoption increases, lessons are learned, and good practices and standards are widely shared and internalized by teams and organizations working on innovation for development.

**Keywords:** blockchain; SDGs; innovation; COVID-19; green recovery; scorecard

**JEL Classification:** O16; O19; O35; P16; Q01; Q56

**Citation:** Aysan, A.F.; Bergigui, F.; Disli, M. Using Blockchain-Enabled Solutions as SDG Accelerators in the International Development Space. *Sustainability* **2021**, *13*, 4025. https://doi.org/10.3390/su13074025

Academic Editor: Shervin Hashemi

Received: 2 March 2021
Accepted: 29 March 2021
Published: 5 April 2021



## 1. Introduction

As we enter the decade of action to achieve the SDGs by 2030, the international community is facing unprecedented challenges to accelerate the pace towards meeting national sustainable development targets, at a time where the very development gains won over the last decades continue to be reversed by the aftershocks of the COVID-19 crisis [1]. If new technologies are going to contribute to the necessary transformation, there have to be adequate tools, methodologies, and standards to navigate the blockchain "hype cycle" and to move from a generalized "let's blockchain it" approach towards a rational narrative that is evidence-based.

Despite blockchain's promising potential, the reality can be challenging, there are not enough data, blockchain-backed applications for social impact are under-studied, and claims that blockchain-backed solutions can yield superior results when compared to other alternatives are yet to be supported by evidence [2]. While we can agree that blockchain has the potential to trigger disruptive innovations, we can also agree that the technology is not yet mature and that there is still a gap in terms of approaches and tools needed to develop blockchain use cases, evaluate blockchain applications, monitor experiments, mitigate associated risks, and manage organizational changes to galvanize innovation-readiness within organizations considering adopting blockchain technology and running use case experiments [3]. It is only by filling the existing gaps that we can make a stronger case for using blockchain as an SDG accelerator.

We can argue that the comparative advantages of blockchain can be explained by its ability to address the issue of trust within the global financial system in the aftermath of the 2008 global financial crisis [4]. Looking into the current practice among the key players in both public and permissioned blockchains, there is a costly race among early adopters towards becoming the standard and prototyping blockchain applications to do things other technologies cannot. Against this background, we can look at blockchain as a package of technologies and approaches that can be used to open up new opportunities for users to manage transactions, exchange values, and maintain digital trust [5].

In one forecast, the business value of blockchain was predicted to climb up to $176 billion by 2025, before skyrocketing to $3.1 trillion by 2030 [5]. Nevertheless, blockchain is yet to match the hype, despite its potential. The maturity of blockchain technology is expected to facilitate its wider adoption, widen the scope for its utilization, and consequently foster blockchain's ability to stimulate successful disruptions. This is in line with the need to stimulate business processes, enable regulatory frameworks, and trigger the necessary cultural shift and the underlying organizational structures.

In this paper, we examine how various blockchain solutions should be ideated, designed, and applied, and how related implementation choices can be made in an attempt to define a structured approach to develop effective use cases for blockchain applications in the fields of sustainable development and green recovery. We then identify potential gaps in terms of monitoring and evaluation, risks, and ethical considerations in order to propose tools to assess the impacts of blockchain-powered applications and manage the associated potential risks and ethical considerations. We also explore the extent to which the use of blockchain influences business processes and identify structured ways to manage transformational change within organizations. Individuals and organizations designing and experimenting blockchain-based solutions for sustainable development may benefit from the tools we have prototyped in this study. In this regard, we propose the SDG Acceleration Scorecard and the integrated approach to using blockchain-enabled solutions as SDG accelerators.

In doing so, we aim to stimulate a discourse around effectively harnessing the disruptive potential of blockchain as an SDG accelerator. This work will support development workers and organizations experimenting or willing to experiment with blockchain-enabled applications to address various challenges related to the achievement of the sustainable development goals by providing a practical toolbox. Such a toolbox will (i) facilitate the deployment of structured approaches and meaningful tools to build promising use cases for blockchain applications; (ii) monitor and evaluate the results of blockchain-enabled development interventions; (iii) assess and mitigate the risks associated with blockchain technology; (iv) manage organizational change and nurture a culture of change that will promote early and rational adoption of disruptive innovations.

## 2. Blockchain and Token Solutions

### 2.1. Blockchain Technology

Blockchain is an open-source technology that excludes the traditional third parties by relying on collective verification, thus offering a great alternative in terms of costs, traceability, security, and speed. When two financial entities such as banks receive a request to transfer money from one account to another, they have to update the balances of their respective customers. This costly and time-consuming coordination and synchronization exercise can be simplified on a blockchain by using a single ledger of transactions reflecting a single version of records instead of two different databases [6]. Blockchain applications go beyond finance and are growing to encompass a myriad of use cases [7,8].

Blockchains can be designed either as private or public; while decentralization remains a common denominator to both forms, there is a key difference in the level of access granted to participants [9]. In the case of a public blockchain, participants are typically encouraged to join the network through an incentivizing mechanism, such as in the case of Bitcoin [10]; anyone can join the network and decentralization is pushed to the fullest extent [9]. On the other hand, private or permissioned blockchains are closed networks where participants face restrictions in terms of who can write data and who can read it. Hence, while public blockchains maximize the anonymity, permissioned blockchains know the identities of their participants and determine which information they should or should not have access to [11].

While perceptions suggest that public and permissioned blockchains are competing with each other, they have different offerings and could be rather complementary in terms of the solutions they offer [11]. Public blockchains offer high security, an open environment, anonymity, and no restrictions, whereas private blockchains prioritize privacy, high efficiency, and stability. We can argue that permissionless blockchains empower the user by pushing transparency and decentralization to their full extents, while permissioned blockchains empower enterprises instead of individual employees [11,12]. The convergence of public and private blockchains is expected to pave the way for virtual ecosystems where a wide range of players can collaborate in a secure and auditable way [7].

The question remains of which blockchain is better for which applications? Indeed, public and private blockchains have distinct use cases. In general terms, public blockchains address business-to-consumer scenarios, while private blockchains are more applicable to business-to-business relationships, with some shared infrastructure between businesses [12]. The transparency and security features of public blockchains make them more suitable for developing blockchain-enabled solutions serving larger communities where trust is a key concern [9]. They are a viable option in situations where all users should be treated equally and when the protection of users' anonymity brings added value to the solution [12]. There are, however, some concerns about whether confidential data should be recorded on a public blockchain, assuming that the encryption could be hacked one day [9].

In the world of private blockchains, there are quite opposing concerns, since the players are reluctant to publicly share their business data. This is more appealing to financial institutions and corporations so they can know and predetermine who has access to what [9]. The downside though is that trust comes down to the credibility of the authorized nodes, as well as a relatively higher vulnerability to malicious attacks [9]. As blockchain technology keeps evolving, hybrid solutions could perhaps offer the best of both options by bringing together trust and security alongside efficiency and speed [9]. Given the current momentum in adopting blockchain applications across a large spectrum of industries, blockchain technology can only increase in popularity as the world enters the uncharted territories of the "new normal" in the post-COVID-19 era, where technologies are poised to play an extremely important role in redefining "business as usual".

### 2.2. Tokenization

Blockchain technology offers a myriad of value through a frictionless process of immutable and transparent records and through converting assets into digital tokens (i.e., tokenization) with smart contracts. These features offer solutions that are particularly suited to addressing challenges in the implementation of SDGs. Special importance is given to "impact tokens", which represent a group of tokens designed to unlock investments for projects with positive social and environmental impacts [13]. The deployment of these impact tokens in blockchain offers new mechanisms to improve ESG ratings, as it offers proof that a particular investment has delivered a positive impact [14]. A key advantage of tokenization is that it offers traceability across the supply chain ecosystem. In this regard, token-based incentive schemes should further broaden the accessibility of blockchain-based solutions for sustainable development, in alignment with UN SDGs. Furthermore, as suggested by Uzsoki and Guerdat [13], the UN or other international bodies could facilitate the adoption of impact tokens by setting common standards that set out both their characteristics and achievement of SDGs.

Supported by UN World Food Program, Fishcoin is a blockchain-based data-sharing platform that incentivizes catch registration and data-sharing across the seafood supply chain [15]. Data contribution to the blockchain platform is rewarded by tokens, which can be exchanged for mobile phone credit. Amply, a pilot project in South Africa that has been funded by UNICEF Innovation Fund and Innovation Edge, tracks school attendance by providing children with self-sovereign digital identities on the blockchain. When a teacher confirms attendance on a mobile application, a token is generated that the school can redeem for further subsidies. Moeda is a blockchain-based cooperative banking system that leverages on a fiat-pegged digital token. The Moeda initiative facilitates access to finance to unbanked and underbanked entrepreneurs, whereas impact investors are able to keep track of their investments.

## 3. Methodology

To prototype an integrated approach for this research, we will go through several stages. First, we will define the minimum viable product for the integrated approach by examining the available literature using four thematic clusters: building blockchain use cases, monitoring and evaluation of blockchain-based applications, risks and ethical considerations associated with blockchain-powered experiments, and change management in organizations deploying blockchain-based solutions. Second, the integrated approach built on the initial findings from the literature review will be further improved based on our observations and the inputs shared by resource persons approached to conduct this research.

During the last stage, we will finally propose an integrated approach together with the SDG Acceleration Scorecard to assess the potential for SDG acceleration. While we recognize the limitations of this work and the need to conduct further field investigations to explore the proof of concept by reaching out to a larger number of respondents to gather and analyze critical data, we believe that the proposed package composed of the integrated approach and the SDG Acceleration Scorecard is a necessary first step. Such tools should be considered as "living tools" that can be further improved as blockchain technology matures, based on the feedback from end-users as they domesticate and internalize these tools to develop meaningful blockchain solutions.

We consider four thematic clusters according to which we will test the following research questions and assumptions in an attempt to verify their validity (Table 1).

### *3.1. Building a Use Case*

The low experimentation costs of blockchain solutions driven by platforms made available via service providers known as blockchain as a service (BaaS) offerings [16], combined with the increasingly complex nature of delivering development and humanitarian interventions in a time where both developing and developed countries are pursuing their sustainable development targets, has stimulated many investments in use cases for blockchain solutions in a wide array of contexts. Here, we are looking beyond the hype at the fundamental question of what can be and what cannot be solved by blockchain? We can also agree that given the scarcity of finances available for development, the increasing demand and challenges faced by organizations and practitioners in today's world to deliver effective yet lasting results, the return on investment of innovation experiments, including blockchain, should not be limited to the proof of concept but should also consider impacts at scale.

While blockchain can be the appropriate choice in some cases, other conventional technologies might be more appropriate in other situations [17]. Given the blockchain hype, we consider that the critical starting point to make a use case for a given blockchain application is to avoid at all costs developing a blockchain solution when there is no problem to be solved with it. The specific problems that need to be fixed and the feasibility of the blockchain solution should be rigorously examined while building a promising use case for blockchain [18]. Other criteria to consider could include the realities on the ground in line with the specific development or humanitarian contexts for which a blockchain solution is sought, the capabilities of the staff involved in the blockchain experiment, the flexibility of the applicable business processes, and the underlying managerial structures.

**Table 1.** The thematic clusters proposed to prototype the integrated approach.

| Thematic Cluster | Research Questions and Assumptions |
| --- | --- |
| Building a use case | What can be solved and what cannot be solved by blockchain? Which structured approach should be followed to understand the potential of blockchain in a given context and design specific and promising use cases? We assume that the solution should not come before the problem. To make a meaningful use case for blockchain there are many boxes to tick. |
| Risks and ethical consideration | Are blockchain applications in sustainable development risk-free? What are the potential risks associated with the utilization of this technology and which ethical reflections should be taken into consideration? We assume that there is a need to explore ethical considerations around the use of blockchain to determine potential risks for development professionals to make informed decisions and come up with appropriate mitigation measures. |
| Monitoring and Evaluation | Which structured approach and tools can be used to monitor and evaluate blockchain applications? Looking beyond pilots, can blockchain solutions deliver impacts at scale? How can we measure such impacts? We assume that while blockchain applications are contributing to solve some of the challenges around the implementation of the SDGs, this has been mostly demonstrated and reported in the available literature through pilot applications that were not replicated at a larger scale. To better assess blockchain's impact on the SDGs, to examine whether or not a given result can be fully attributed to the blockchain, or if blockchain rather partially contributed to such a result, we assume that there is a need to develop adequate tools to support monitoring and evaluation of different blockchain experiments. Such tools should not only track progress but should also provide a standardized platform to ensure comparability and assess criteria of relevance, effectiveness, efficiency, sustainability, and impact, in addition to gender equality. |
| Change management | What about change management in organizations adopting blockchain? How does embracing innovation, in the case of blockchain applications, affect the way development and humanitarian organizations do business? We assume that change management is a dynamic process that needs to be understood to enable organizations to develop an evolutionary adaptation approach to navigate the transition, address resistance to change, and nurture innovation. |

Another vital factor is the coordination among stakeholders involved in the design and implementation of a given blockchain solution. To design viable blockchain solutions, due consideration should also be given to the ability to manage technology constraints through trade-offs, to convince various stakeholders of the relevance of the blockchain use case, to cooperate with other players, and to comply with the applicable regulations and standards on a case-by-case basis.

For this research, we have adapted the stages defined in the blockchain use case development method (BUD) proposed by Fridgen et al. [19] using a design thinking process [20]. We argue that these stages will evolve to reflect the latest developments in blockchain as the technology matures and reaches a stage of producing impacts at scale.

Step 1—Understand blockchain

We assume that the focus here should not be on understating the technology itself, but on what can be done with it. This also takes into consideration the capabilities organizations should have to design and implement blockchain solutions.

Step 2—Define the problem you are trying to solve with blockchain

Here we need to clearly define the challenge that we are trying to overcome with blockchain. Individuals and organizations need to consider feasibility aspects, while various organizational implications with regard to business processes and change management should also be taken into consideration. Guiding questions can be provided, such as:

- What is the exact problem you are trying to solve?
- How does it affect your sustainable development context?

- Which capabilities do you need to develop a blockchain solution to address this problem?

Step 3—Pitch your ideas

We assume that while teams can consider existing blockchain use cases in their fields of interest, they should only use such cases for inspiration and come up with their innovative ideas by brainstorming and working together in cross-functional teams. Guiding questions can be provided, such as [21]:

- Are transparency and traceability important in your context?
- Do you need a centralized or decentralized solution?
- Which transactions require trusted partners or processes?
- Which intermediaries can be eliminated?
- Will you reduce transaction time for end-to-end processes with blockchain?
- Which costs can be reduced?
- Which data are most sensitive and which data will be stored?
- How much control do you need for stored data?
- Who are your potential end-users?
- What transformational change can result from using blockchain in your organization?

Step 4—Build Consensus

We assume that while consensus protocols make blockchain platforms immutable, achieving such consensus may require huge computational power and large amounts of energy, while design choices can have implications on how much time is needed for consensus and the degree of decentralization.

- To what extent are timing and decentralization critical factors in your context?
- How much computational power is needed to achieve consensus? Can it be reduced?
- Which consensus protocols can be used?

Step 5—Prototype the solution

We assume that issues of efficiency, scalability, performance, and impact should be factored-in while defining the parameters for any given blockchain prototype. Guiding questions may include:

- What is the cost of developing this blockchain solution?
- What would be the return on investment?
- Which blockchain platform should be used in your case (public, private, or hybrid)?

Step 6—Test and iterate

We assume that blockchain solutions should be tested with potential end-users to identify residual issues that will be taken back to the prototyping phase through an iterative process.

Figure 1 summarizes the steps in the blockchain use case development method.

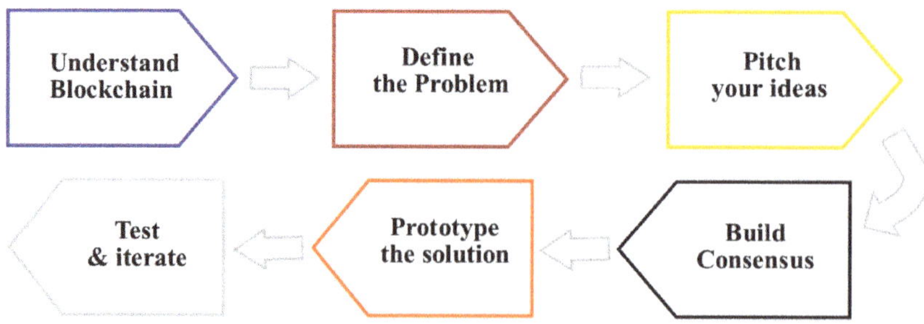


**Figure 1.** Presentation of the blockchain use case development method. Source: The author based on Fridgen et al. [19].

### 3.2. Risks and Ethical Considerations

We can praise blockchain for its disruptive potential and all the benefits it can bring into a given use case. Nevertheless, we can also agree that there is no return on investment without risks. Organizations and practitioners must avoid at all costs creating more problems than the ones a blockchain application is designed to solve. Given the complex nature of blockchain technology, it is not an easy task to assess the underlying risks, especially in the absence of standards and regulations [22]. For this research, we adapted the frameworks for risk analysis and management [23–25]:

- Analyze (identify core risks and assess their potential impact and likelihood);
- Manage (develop proper management responses to mitigate risks);
- Anticipate (detect new risks that are arising from various ethical considerations).

#### 3.2.1. Risk Analysis

At the design stage, there are no blockchain standards, as the technology is not mature yet. Lessons learned and good practices should be coded into standards to minimize risks and pave the way towards enhanced cooperation. We assume that the interoperability between blockchain platforms will remain a key issue, and thus represents a key risk. Other technical risks may potentially stem from design-related choices, for instance whether forking another version of the ledger is allowed or whether automated actions are built-in on a blockchain infrastructure.

Implementation comes with its own set of risks. Consensus-building algorithms can cause serious delays given the nature of participatory processes. High maintenance costs can cause projects to fail and scalability might not be achieved through economies of scale. Cybersecurity risks and fraud are key challenges for shared technological infrastructure that can originate from the outside or from within. While the likelihood of outside attacks appears to be minimal, risks stem from the irreversible nature of internally produced mistakes, such as sharing an erroneous code across the network or building consensus on a false entry. Updates and maintenance managed by third parties may qualify as potential risks as well, especially if the skills required and the availability of service providers are in low supply.

Other risks in general terms are linked to the constantly evolving blockchain regulations in various jurisdictions, the cost of technology updates, competition from other disruptive innovations, and the ecological footprint of blockchain given its energy consumption. In general terms, the more we understand blockchain and the more we understand the problem we are trying to solve with blockchain, the higher our chance of coming up with the most appropriate risk mitigation measures.

#### 3.2.2. Risk Management

We consider three risk categories that can be analyzed at different stages before, during, and after the implementation of the blockchain-backed solution.

##### Preventable Risks

Here, we are looking at internal and avoidable risks arising from within organizations and teams. With blockchain, the need for human interventions can be reduced when using automated processes and smart contracts, which in return can decrease the likelihood of alteration of records or corruption of people involved. However, risks may stem from the irreversible nature of internally produced mistakes, such as sharing an erroneous code across a network or building consensus on a false entry. We assume that active prevention through a rules-based control model is the most suitable risk management response to risks listed under this category, which can be achieved by monitoring business processes and by guiding human-made decisions related to blockchain applications.

Strategic Risks

These are risks organizations and teams are willing to take to achieve high targets. Here, the managerial response is rather towards reducing the likelihood for risks to occur and containing such risks in the case they materialize. While blockchain can minimize risks related to compliance and litigation, early investors, driven by an appetite for innovation and to be pioneers their fields of practice, may see their projects seriously delayed by consensus-building algorithms given the nature of the participatory processes, or their projects may even fail due to poor buy-in from their stakeholders and high maintenance and scalability costs.

External Risks

These are consequences of events happening in the outside world that cannot be accurately forecasted, and over which organizations and individuals have no control whatsoever. Proper mechanisms should be set in place to track such risks and mitigate their impacts to the extent possible using appropriate managerial responses. Although the likelihood of outside attacks appears to be minimal, cybersecurity risks are a key challenge for shared technological infrastructure. Updates and maintenance managed by third parties may qualify as potential risks as well, especially if the skills required and the availability of service providers are in low supply. Other risks in general terms are linked to the legal–technology mismatch due to the constantly evolving blockchain regulations in various jurisdictions, in addition to the cost of technology updates and competition from other disruptive innovations.

#### 3.2.3. Ethical Considerations

We can argue that ethics made a late appearance in the blockchain sphere because it is seemingly hard to understand how the technology works. Several ethical considerations need to be addressed. Due to massive energy consumption, public blockchains based on mining can potentially have a significant impact on the environment. The annual total electricity consumption of bitcoin assuming constant power usage over one year is estimated at 77.38 TWH [26]. This number should be understood as an educated estimate, as it is not possible to accurately calculate the electricity consumption of bitcoin processes.

The hyper-efficiency that is indeed a valuable comparative advantage of blockchain applications can disrupt the jobs of many people involved, for example those working in data maintenance [27]. Additionally, blockchain can over-promise and fail to meet expectations or inadvertently facilitate crime-related transactions and oppressive conduct [28]. Blockchain can be used for money laundering in illicit business activities, such as weapons and drugs or other related transactions [29].

Equally important is the way we design blockchain applications based on trade-offs, which can have far-reaching consequences [30]. Such trade-offs can codify biases and exacerbate social dynamics within a given community, while the transparent nature of blockchain information can put persons at risk because of their ethnicity, religious background, or sexual orientation [30]. The same applies to the immutability of digital identity records if a person is threatened and requires anonymity or a new identity, or if erroneous data are recorded. If there is no way of retrieving a lost private key, this might simply mean that a person loses control over their digitally recorded assets. When the encryption algorithms used are outdated, sensitive information might be exposed and transactions can be forged [30].

Due consideration should also be given to data privacy. In the absence of natural or legal persons to fulfill data privacy requirements as data controllers, given the decentralized nature of blockchain platforms, one of the key challenges for blockchain-powered solutions is how to define controllership over data. Another key challenge is the underlying assumption that data can be erased should such a need arise based on data privacy requirements [31].

There is a steep compliance curve ahead for blockchain practitioners to comply with the requirements stipulated in General Data Protection Regulations (GDPR). In order to build privacy-compliant use cases of blockchain-backed solutions using GDPR lenses, silos should be broken down through proactive dialogue involving technology disruptors, developers, and regulators to address key challenges related to defining data controllers or deleting personal data [32]. By doing so, more privacy-friendly operational procedures can be codified into the process of building blockchain use cases.

### 3.3. Monitoring and Evaluation (M&E)

#### 3.3.1. Monitoring

We argue that the core ingredients to successfully monitor blockchain interventions would be for organizations to have clear monitoring policies in place, for individuals to have clearly defined roles and responsibilities in the monitoring process, and to have a first-hand understanding of the operational context in which a given blockchain-powered application is being deployed. Appropriate M&E frameworks should be developed to track specific project outputs, especially related to the annual targets that blockchain projects expect to achieve in line with their result matrices. Monitoring actions should capture and analyze data to report on the progress towards outcomes, impeding factors, partners' roles in achieving the expected results, and lessons learned to develop and disseminate knowledge products. While blockchain experiments for social good remain largely understudied, such knowledge products are undoubtedly needed by the global community given the scarcity of data available to identify good practices. Additionally, monitoring data and reports are key resources that can be used to inform decision-making and future experiments.

For this research, we propose a three-step approach for monitoring blockchain projects, namely data gathering, analysis, and reporting:

- Collecting data through available tools. Data can be collected based on field investigations and through different reporting mechanisms, then triangulated for validation and discussed via participatory processes, such as during coordination meetings with partners;
- Analyzing the data gathered to extract useful information, identify patterns, and detect bottlenecks. Data interpretation in terms of what needs to be done may be used to generate useful and user-friendly insights to guide decision-makers;
- Reporting to inform future decisions. Stakeholders and partners involved in blockchain projects can be guided to make informed decisions to ensure the project is on track to deliver their expected results.

#### 3.3.2. Evaluation

Since their adoption in 1991 [33], the five OECD DAC criteria are the most referenced and used standards for evaluating international development interventions. For many years, evaluation professionals have considered relevance, effectiveness, efficiency, impact, and sustainability as the standards to assess development interventions. The main issue here seems to be how these criteria should be used, as one mainstream mistake is to start using such criteria in a mostly mechanical way by "ticking boxes" instead of using them to support a non-linear critical analysis [34]. There are ongoing discussions regarding the extent to which the DAC criteria should be revised to capture the new paradigms of sustainable development. We can argue that not every evaluation should cover all 5 requirements; even if this is the case, not all criteria should be analyzed in the same depth, and most importantly these criteria may not be entirely suitable for the needs of today [34]. We are looking here into disruptive innovations and specifically blockchain-enabled applications for sustainable development to extract key guiding criteria, which evaluation professionals could consider when evaluating blockchain-backed projects.

One common critique of blockchain applications is the lack of evidence to support claims for disruptive results. This can be partially explained by the fact that the technology is not mature yet, or that lessons learned from blockchain experiments are poorly documented. As a result, we can argue that the hype has been driven to some extent by a "fear of missing out" bubble based on proof-of-concept experiments design to make headlines [35]. With that being said, there is a promising potential for blockchain use cases to have a transformative business impact by improving productivity and quality, increasing transparency, and reinventing products and processes [36], hence the need to fill in the gaps by developing proper M&E evaluation frameworks for blockchain applications for international development.

After developing a promising use case for blockchain, securing the support of various partners, unlocking the necessary funding, and before kicking-off the implementation of a given blockchain-powered solution, there is a need to set up robust M&E frameworks. Organizations might consider hiring full-time M&E practitioners or investing in diffuse M&E skills throughout their workforce assigned to design and implement blockchain-based applications. However, various considerations make it difficult to perform M&E functions for sustainable development interventions driven by innovative processes. For instance, the uncertainty associated with innovation-based solutions, the speculative nature of the results one should expect down the road, and the hard-to-measure nature of innovation-driven benefits, such as stimulating an appetite for a diffuse "business as unusual" culture, are all vital aspects to consider while monitoring and evaluating blockchain-centered experiments.

Before implementation, certain criteria need to be considered for blockchain practitioners to come up with the right decisions [37]. In terms of governance, for instance, it is vital to identify who controls access to the blockchain application, who makes decisions, and who enforces such decisions to alter the blockchain solution as the needs evolve in a given development or humanitarian context. One of the main aims for developers of blockchain applications in sustainable development contexts is to minimize their power consumption and ecological footprint [26]. In this regard, choosing the right consensus scheme and offering the right incentives can enhance the acceptability, and thus the success, of blockchain-enabled solutions. To decide on the right type of blockchain platform to deploy, organizations need to take into consideration whether the data are meant to be accessible publicly, in a restricted circle, or as a combination of both.

For this research, we have adapted the evaluation framework that was developed by Fridgen et al. [38] to evaluate the applicability of blockchain technology in the public sector. We argue that this framework is not only suitable for the evaluation of blockchain use cases, but also for the provision of a set of evaluation criteria to consider while conducting mid-term or final evaluations of blockchain projects for sustainable development. Here, we consider 3 sets of criteria, namely technical, functional, and legal criteria.

Technical Criteria

In this set, we have included 4 parameters, namely performance, scalability, security, and usability.

1. *Performance*

Due consideration should be given to how fast transactions can be operated within a blockchain network. Furthermore, duration and latency need to be verified using available metrics, such as the number of blocks and the size of transactions.

2. *Scalability*

This is a key element given that blockchain experiments are usually designed to be tested in small pilot cases. Once the candidate blockchain applications deliver proof of the concept and graduate from the experimentation phase, they are expected to be applicable on a larger scale to address international development. It is, thus, critical to factor-in scalability parameters such as the network size, number of possible nodes, and transactions needed down the road on a larger scale.

Blockchain-powered solutions leverage the network effects of the investment universe by unlocking funds from small investors. Network effects through blockchain technology are achieved without the disruptive consequences the presence of an intermediary brings with it. In this regard, Catalini and Gans [39] point out that decentralized networks reduce transaction costs and enable network effects without being subject to monopoly pricing and control. Blockchain has the power to enable an accelerated realization of SDGs, as it helps to connect and integrate different databases and information flows [40]. The decentralized principle of value creation through blockchain-based architectures will benefit sustainable development by facilitating simultaneous collaboration and competition among partners within the loop [41]. The network effects result in a positive feedback loop, whereby the services offered become strengthened with the adoption of the network [42]. It is evident that blockchain network effects will manifest at an increasing pace when the ecosystem becomes more valuable to its participants as more people partake in it.

3. *Security*

We can assume that a fair number of development solutions have been designed as blockchain applications given the enhanced security features of blockchain technology. This points out the need to carefully consider the security features provided by a given blockchain application and assess whether it answers specific contextual requirements related to the security of data and transactions.

4. *Usability*

User-friendly applications tend to be more accepted. The ease of use is a critical parameter to consider, which comes down to ensuring that everyone can benefit from the blockchain solution and that no one is left behind. Important details to factor-in can range from internet traffic to the ability to fork new features, the available tools and languages, and the number of possible users.

Functional Criteria

This set includes criteria related to the R&D costs required to build a blockchain-backed solution, implement it, and ensure its operationalization and maintenance, in addition to other criteria related to the flexibility and transparency of the process.

1. Costs

Some blockchain applications may require heavy transactions and storage of data, implying high costs. This should be duly considered while budgeting to include not only the prototyping and deployment of the blockchain application, but also its operational and maintenance costs.

2. Flexibility

Another important feature in international development setups is the ability of blockchain-powered pilots to be replicated or altered in a way to serve new needs or address new challenges. For instance, ff a blockchain platform is only meant to be deployed in a restricted area, such limitations are to be considered before any investments are made upfront.

3. Transparency

It is important to find the right trade-off between transparency and other outputs. Confidentially can be increased by storing encrypted data, however it will reduce transparency and performance. Transparency can be in conflict with commercial confidentiality, such as in cases where competitors exist [17].

Legal Criteria

This set includes criteria related to data privacy, legal procurement frameworks, employee protection rights, and other legal regulations. Here, we highlight various legal considerations applicable in the jurisdictions and to the entities involved in blockchain experiments. These include requirements for data privacy, contractual modalities specified in procurement laws, employment protection rights, and any other legal instruments that are potentially applicable to blockchain within a specific jurisdiction.

### 3.4. *Change Management*

We can argue that blockchain technology stimulates a progressive evolution in the ways business is conducted. Development organizations and practitioners running various innovation experiments and aspiring to harness the disruptive potential of blockchain are facing the challenge of managing the change towards a "new normal". New avenues and meaningful methods are being explored to nurture a culture of innovation, keep risks at acceptable levels, and address pockets of resistance regarding both organizations and individuals being able to follow the trajectory of evolutionary adaptation and achieve transformational change in terms of doing business [5]. Blockchain should not be considered as a technology option but should be understood in terms of its ability to address specific issues and deliver precise solutions. Organizations and individuals should also consider experiments where they can explore decentralized business processes and allocate staff who have both technological and technical capabilities [5].

We also argue that organizations and individuals embracing the disruption wave brought about by blockchain technology need to be change-adept. When an organization invests in a blockchain application, it has already identified a problem and came up with a use case for blockchain as a solution. In this respect, managers will need to set up innovation-prone environments to ensure effectiveness and efficiency, reinforce collaboration, address siloed-behaviors, provide leadership and coaching, and avail resources needed for the blockchain-enabled solution to deliver on its promises. Here, the focus is on the capability of the organizations to implement a blockchain-backed solution. In fact, without a structured approach to implementation, an innovative solution is likely to fail. We assume that change management is a dynamic process that needs to be understood in order to enable organizations to develop an evolutionary adaptation approach to navigate the transition, address resistance to change, and nurture innovation.

For this research, we will focus on change at the individual level as the cornerstone for change to happen at the organizational level. The following change management approach was inspired by the ADKAR (awareness, desire, knowledge, ability and reinforcement) and Kurt Lewin change management models [43,44]:

1. Unfreeze

At this initial stage, the focus should be on raising awareness beyond the hype to visualize potential opportunities to adopt blockchain-based solutions and seize the most promising ones.

2. Change

This is an intermediary stage where the spotlight is on motivating individuals and teams to design blockchain-powered solutions. This includes coaching, managing resistance, fostering readiness to become a change agent, as well as investing in core capabilities through learning and practice for individuals and teams to acquire a wide spectrum of skills deemed necessary to implement blockchain projects.

3. Refreeze the change

This is the final stage, where the focus in on re-calibration through corrective actions whenever applicable, but also on measuring achievements and celebrating success.

## 4. Results

### 4.1. *SDG Acceleration Scorecard*

Given the current hype, one can argue that in general terms, development practitioners and organizations tend to act prematurely to address a giving problem without necessarily being aware of the alternatives and without running proper assessments that would or would not prioritize blockchain as the best way forward. Even after an assessment indicates that a blockchain solution is the best option, one can still do more to harness not only the direct benefits to address the initial problem for which a blockchain solution is needed, but equally important are the wider sustainable development co-benefits to which such blockchain solutions can contribute. Given the interlinked nature of the SDGs, one blockchain-backed solution can contribute to more than just one SDG at a time. In other terms, one should think beyond the simple issue to be solved by a blockchain application. While such an application can solve a specific problem, it can also contribute to achieving other development targets if properly designed and implemented, hence the SDG acceleration effect. That is why we truly believe that for blockchain-powered applications to deliver wider development co-benefits, it is critical to assess their SDG acceleration potential by taking into consideration an initial set of criteria and indicators related to use case development, monitoring and evaluation, risks and ethical considerations, and change management. We refer to Table 2, which showcases examples through which dimensions of SDG targets can be accelerated.

**Table 2.** Sustainable development goal (SDG) acceleration matrix.

| Acceleration Area | Multiplier Effect (# of SDG Targets) | Balance across the 3 Dimensions | | | Score |
|---|---|---|---|---|---|
| | | Social | Economic | Environmental | |
| Example: Female empowerment | 15 | Yes | Yes | Yes | 15 \| 3 |
| Example: Impact investing | 10 | Yes | Yes | Yes | 10 \| 3 |
| Example: Health records | 8 | Yes | Yes | No | 8 \| 2 |

Source: Adapted from the SDG Accelerator and Bottleneck Assessment Tool [45]; the acceleration areas, multipliers, balances, and scores are examples and will depend on the specific context applicable to a given development intervention.

Our attempt here is to provide a simplified scorecard that will help development practitioners and organizations running blockchain experiments understand the bigger picture, pilot their interventions in a way to accelerate the achievement of the SDGs, and fulfill the sustainable development aspirations of individuals, communities, cities, regions, and governments worldwide. Inspired by the use of scorecards in the development practice, one of the experimental tools we have proposed in this work is the SDG Acceleration Scorecard (SAS). We argue that scorecards can be extremely useful in helping development practitioners monitor progress and measure results. To monitor capacity development, scorecards can be used as tools to quantify qualitative processes and measure the change in capacities using a set of relevant indicators to which specific ratings are assigned.

This was, for example, the case with the Capacity Development Scorecard developed to monitor the progress made towards achieving global environmental benefits for projects funded by the Global Environment Facility [46]. This scorecard was adapted to measure increases in capacity in the case of the implementation of the Access and Benefit Sharing (ABS) mechanism of the Nagoya Protocol, which governs the utilization of genetic resources and their associated traditional knowledge [47]. Simplified scorecard systems were also used to gather general information and conduct assessments. The WWF–World Bank Marine Protected Area (MPA) Scorecard, for instance, was designed for protected marine areas. It mainly relies on the available literature, in addition to the opinions of site managers and independent assessors [48]. Such a scorecard can be rapidly deployed, is less costly to implement, and broadly covers the relevant issues. Nevertheless, the depth of analysis provided by such tools remains low [49].

In our attempt to develop a scorecard to assess the SDG acceleration potential of blockchain-enabled solutions, we applied the core criteria defined in the thematic clusters to propose acceleration areas and tried to identify possible drivers of acceleration by asking specific questions to generate useful information that can be used for two main objectives. Under the first scenario, the scorecard can be used to assess the multiplier effect of a given blockchain-enabled solution in terms of how many SDGs are expected to benefit from such an intervention. The information generated by the scorecard can also be used to run a quick assessment to check whether or not the gaps in design and implementation have been properly addressed. In the second scenario, gathering data using the scorecard can enable a comparison between different blockchain-backed candidate solutions, in terms of their rational; risk category; the existence of proper mechanisms to generate data, track progress, assess outcomes, and manage change; and their overall ability to deliver tangible and holistic development outcomes.

The proposed SDG Acceleration Scorecard consists of acceleration areas, drivers, and a scoring system. To characterize the acceleration effect of a given blockchain-enabled solution, the scorecard provides descriptive sentences for each driver in the acceleration areas, each corresponding to a numerical rating from zero (0) to three (3) based on a 33-point scoring system. In the following paragraphs, we provide practical explanations related to the acceleration areas, the drivers within each area, and explain how the scores are assigned using descriptive sentences to describe each driver.

A. Acceleration area 1: Use Case Development

When use cases for blockchain-enabled solutions are properly designed by taking into consideration the interlinked nature of the SDGs, they can have a multiplier effect on the development dividends expected down the road. This in turn will boost the ongoing efforts to achieve the SDGs; hence, the acceleration effect.

Driver A1

This driver generates useful information to explain whether or not there is a strong justification for the design choice of a blockchain-backed solution. We propose asking the following or other alternative questions: Why would you consider blockchain as a solution to the problem encountered in your case?

*Scorecard rating:* We assigned scores of 0 to 3 to the responses provided:

0—No response or not sure;

1—Blockchain is an emerging innovation in international development;

2—Blockchain has the potential to solve certain development challenges we face today;

3—Blockchain is the best alternative to address the specific challenges we are facing.

Driver A2

This driver generates useful information to explain whether or not a given blockchain-backed solution was compared to other alternatives. We propose asking the following or other alternative questions: Did you compare blockchain to other alternative solutions in terms of efficiency, scalability, performance, impact, and cost-effectiveness?

*Scorecard rating:* We assigned scores of 0 to 1 to the responses provided:

0—No/not sure;

1—Yes.

Driver A3

This driver generates useful information to understand whether or not end-users were involved in testing and iterating a given blockchain-backed solution during the design, which is crucial in order to strengthen ownership and for consensus-building. We propose asking the following or other alternative questions: How many end-users were or will be approached during the design of your blockchain experiment?

*Scorecard rating:* We assigned scores of 0 to 3 to the responses provided:

0—None or not sure;

1—Less than 2;

2—2 to 5;

3—More than 5.

Driver A4

This driver generates useful information to assess the multiplier effect of a given blockchain-powered solution in terms of its ability to generate wider development co-benefits by contributing to multiple SDGs. We propose asking the following or other alternative questions: How many SDGs do you expect your blockchain project to contribute to?

*Scorecard rating:* We assigned scores of 0 to 3 to the responses provided:

0—None or not sure;

1—Only 1 SDG;

2—From 2 to 3 SDGs;

3—Multiplier effect across more than 3 SDGs.

B. Acceleration area 2: Risks and Ethical Considerations

This part of the scorecard can generate risk-related information to reveal whether risks and ethical considerations related to blockchain technology are well understood and to ensure that a given blockchain experiment does not create more problems than the ones it was designed to solve.

Driver B1

This driver generates useful information to assess whether or not a risk analysis was undertaken for a given blockchain-powered solution. We propose asking the following or other alternative questions: Have you conducted a risk analysis to identify and assess the core risks associated with your blockchain experiment?

*Scorecard rating:* We assigned scores of 0 to 1 to the responses provided:

0—No or not sure;

1—Yes.

Driver B2

This driver generates useful information to assess the ability to identify and categorize risks in a structured way to enable a proper risk management response. As a follow-up to the question under B1, we propose asking the following or other alternative questions: What were the risks you have identified?

*Scorecard rating:* We assigned scores of 0 to 3 to the responses provided:

0—None or not sure

1—Mostly external risks (high risk);

2—Mostly strategic risks (moderate risk);

3—Mostly preventable risks (low risk).

Driver B3

This driver generates useful information about whether or not there is a risk management mechanism for a given blockchain-powered solution to mitigate different types of risks. We propose asking the following or other alternative questions: Did you come up with a proper management response to mitigate the risks identified?

*Scorecard rating*: We assigned scores of 0 to 1 to the responses provided:

0—No or not sure;

1—Yes.

Driver B4

This driver generates useful information on whether or not there is a mechanism in place to update risk logs and detect emerging risks as blockchain-backed solutions are prototyped and deployed in a rapidly evolving legal, institutional, and technological landscape. We propose asking the following or other alternative questions: Did you set up a mechanism to detect and anticipate new emerging risks and address ethical considerations?

*Scorecard rating:* We assigned scores of 0 to 1 to the responses provided:

0—No or not sure;

1—Yes.

C. Acceleration area 3: Monitoring and Evaluation

This part of the scorecard can generate M&E-related information to explain how activities will be monitored, how progress will be tracked, and which criteria will be used to evaluate the applicability and the outcomes of blockchain experiments.

Driver C1

This driver generates useful information to assess the extent to which data are being collected for a given blockchain-powered solution. We propose asking the following or other alternative questions: Do you collect any type of data on your blockchain experiment?

*Scorecard rating:* We assigned scores of 0 to 3 to the responses provided:

0—There is essentially no data collection;

1—There is some sort of data collection;

2—Data are collected and analyzed;

3–Data are systematically collected, analyzed, and publicly shared to inform peers and decision-making.

Driver C2

This driver generates useful information to assess whether or not an M&E mechanism is in place for a given blockchain-powered solution. We propose asking the following or other alternative questions: Do you have an M&E mechanism in place to ensure proper monitoring of your blockchain experiment?

*Scorecard rating:* We assigned scores of 0 to 3 to the responses provided:

0—There is essentially no mechanism;

1—There is some sort of mechanism;

2—There is a relatively good mechanism;

3—There is a well-functioning mechanism.

Driver C3

This driver generates useful information to explain the set of criteria used in assessing the applicability of blockchain experiments and evaluate their outcomes. We propose asking the following or other alternative questions: Which of the following criteria would you consider in evaluating your blockchain project? Performance, scalability, usability, security, cost, flexibility, transparency, legal regulations, environmental footprint, social impact, and SDG acceleration:

*Scorecard rating:* We assigned scores of 0 to 3 to the responses provided:

0—None or not sure;

1—At least 1 criterion;

2—At least 3 criteria;

3—At least 5 criteria including the one on SDG acceleration.

D. Acceleration area 4: Innovation Management

The last part of the scorecard can be used to generate information to better understand the change dynamics in a given development setup and trigger adequate change management responses towards nurturing a culture of innovation and addressing pockets of resistance.

Driver D1

This driver generates useful information to assess whether or not there is some sort of awareness about change management. We propose asking the following or other alternative questions: Would you conduct any awareness-raising activities within your organization or team to visualize the potential of blockchain-enabled applications and co-create promising use cases?

*Scorecard rating:* We assigned scores of 0 to 1 to the responses provided:

0—No or not sure;

1—Yes.

Driver D2

This driver generates useful information to assess the willingness of managers to understand the root causes of resistance to change, and more specifically to blockchain-driven innovations. We propose asking the following or other alternative questions: Are you planning to explore the reasons behind any resistance that blockchain-backed solutions may face within your organization or team?

*Scorecard rating:* We assigned scores of 0 to 3 to the responses provided:

0—No or not sure;

1—Yes.

Driver D3

This driver generates useful information to assess whether or not there are proper mechanisms in place to enable an innovation-centered environment. We propose asking the following or other alternative questions: Would you consider motivating your staff, partners, and end-users involved in your blockchain experiment?

*Scorecard rating*: We assigned scores of 0 to 3 to the responses provided:

0—No or not sure;

1—Yes, by allocating appropriate time and resources for the team members to innovate and prototype promising solutions;

2—Yes, by investing in core capabilities and skills to implement blockchain projects through learning and practice;

3—Yes, through coaching for readiness to become a change agent.

Driver D4

This driver generates useful information to assess whether or not there is some sort of policy or strategy to maintain an innovation-prone set up in a given context. We propose asking the following or other alternative questions: Do you have a strategy to maintain an innovation-friendly environment within your organization?

*Scorecard rating:* We assigned scores of 0 to 3 to the responses provided:

0—No or not sure;

1—Yes, by recommending corrective actions for improvement;

2—Yes, by measuring achievements;

3—Yes, by celebrating success.

Table 3 presents the SDG Acceleration Scorecard. We emphasize that the scoring system proposed here is not exclusive and that the scores assigned to each descriptive sentence can be revised and weighted accordingly on a case-by-case basis to better reflect the context within which a blockchain-backed solution is deployed. The scorecard can be used as a benchmark at the beginning of the design phase, but also to generate useful insights as the implementation of blockchain experiments is moving forward and to stay on track towards reaching the established targets. The SDG Acceleration Scorecard is a flexible tool and can be adapted and further improved to reflect the differentiated contexts underpinning blockchain experiments.

**Table 3.** SDG Acceleration Scorecard.

| Acceleration Areas | Drivers | Scorecard | Score |
|---|---|---|---|
| Use Case Development | Why would you consider blockchain as a solution to the problem encountered in your case? | 0—No Response/Not sure<br>1—Blockchain is an emerging innovation in international development<br>2—Blockchain has the potential to solve some development challenges we face today<br>3—Blockchain is the best alternative to address the specific challenges we are facing | |
| | Did you compare blockchain to other alternative solutions in terms of efficiency, scalability, performance, impact, and cost-effectiveness? | 0—No/Not sure<br>1—Yes | |
| | How many end-users were/will be approached during the design of your blockchain experiment? | 0—None/Not sure<br>1—Less than 2<br>2—2 to 5<br>3—More than 5 | |
| | How many SDGs do you expect your blockchain project to contribute to? | 0—None/Not sure<br>1—Only 1 SDG<br>2—From 2 to 3 SDGs<br>3—Multiplier effect across more than 3 SDGs | |
| Risks & Ethical Considerations | Have you conducted a risk analysis to identify and assess the core risks associated with your blockchain experiment? | 0—No/Not sure<br>1—Yes | |
| | If yes, what were the risks you have identified? | 0—None/Not sure<br>1—Mostly external risks (High risk)<br>2—Mostly strategic risks (Moderate risk)<br>3—Mostly preventable risks (Low risk) | |
| | Did you come up with a proper management response to mitigate the risks identified? | 0—No/Not sure<br>1—Yes | |
| | Did you set up a mechanism in place to detect and anticipate new emerging risks and address ethical considerations? | 0—No/Not sure<br>1—Yes | |

**Table 3.** *Cont.*

| Acceleration Areas | Drivers | Scorecard | Score |
|---|---|---|---|
| Monitoring & Evaluation | Do you collect any type of data on your blockchain experiment? | 0—There is essentially no data collection<br>1—There is some sort of data collection<br>2—Data are collected and analyzed<br>3—Data are systematically collected, analyzed, and publicly shared to inform peers and decision-making | |
| | Do you have an M&E mechanism in place to ensure proper monitoring of your blockchain experiment? | 0—There is essentially no mechanism<br>1—There is some sort of mechanism<br>2—There is a relatively good mechanism<br>3—There is a well-functioning mechanism | |
| | Which of the following criteria would you consider in evaluating your blockchain project? Performance, scalability, usability, security, cost, flexibility, transparency, legal regulations, environmental footprint, social impact, and SDG acceleration. | 0—None/Not sure<br>1—At least 1 criterion<br>2—At least 3 criteria<br>3—At least 5 criteria including the one on SDG acceleration | |
| Innovation Management | Would you conduct any awareness-raising activities within your organization or team to visualize the potential of blockchain-enabled applications and co-create promising ones? | 0—No/Not sure<br>1—Yes | |
| | Are you planning to explore the reasons behind any resistance that blockchain-backed solutions may face within your organization or team? | 0—No/Not sure<br>1—Yes | |
| | Would you consider motivating your staff, partners, and end-users involved in your blockchainexperiment? | 0—No/Not sure<br>1—Yes, by allocating appropriate time and resources for the team members to innovate and prototype promising solutions<br>2—Yes, by investing in core capabilities and skills to implement blockchain projects through learning and practice<br>3—Yes, through coaching for readiness to become a change agent | |
| | Do you have a strategy to maintain an innovation-friendly environment within your organization? | 0—No/Not sure<br>1—Yes, by recommending corrective actions for improvement<br>2—Yes, by measuring achievements<br>3—Yes, by celebrating success | |

Source: Authors.

We fully recognize the limitations of using an SDG Acceleration Scorecard in the context of blockchain experiments. While it was designed as a low-cost and rapidly deployable tool, it should not be seen as a replacement for academically proven methodologies. The concept of scoring can entail risks of distortion, which leave ample room to improve accuracy by weighting the scores [38]. In this case, we assume that the weights assigned to each question in the SDG Acceleration Scorecard reflect some sort of rationale, which might not be necessarily the case. We considered a simple scoring system while recognizing its limitations.

Again, while recognizing the limitations of the proposed acceleration scorecard as a minimum viable product, which should not replace academically tested methodologies, the scorecard can help individuals and organizations to generate useful insights to guide their efforts throughout the journey of co-creating blockchain-powered solutions in the field of international development, and more specifically to meet the 2030 deadline for achieving the SDGs.

### 4.2. Integrated Approach

Intentional design is the recipe for developing promising blockchain applications. Unlike with other technologies, it is not easy to fix flows, alter records, or change contracts because of the immutable and distributed nature of data. By intentional design, here we mean an integrated approach that goes beyond identifying the problem faced and implementing an intervention to achieve expected outcomes; that is, going the extra mile to understand risks, ethical considerations, and technology choices that will usually make the difference.

As can be observed from Table 4, the prototype for the integrated approach proposed at this stage is a living tool that can be further improved as blockchain technology matures over the years to come, based on feedback from end-users as they domesticate and internalize the tool in their quest to adopt the technology and run experiments on blockchain-backed solutions in the international development space.

**Table 4.** Integrated approach to using blockchain-enabled solutions as SDG accelerators.

| 1. Build a Use Case | 2. Factor-In Risks and Ethical Considerations | 3. Setup M&E Frameworks | 4. Enable Evolutionary Adaptation |
|---|---|---|---|
| We propose the following stages to the blockchain use case development method:<br>• Understand blockchain<br>• Define the problem to be solved with blockchain<br>• Pitch your ideas<br>• Build consensus<br>• Prototype the solution<br>• Test and iterate | Risks<br>We propose a three-step approach for risk management:<br>• Analyze (identify and assess core risks)<br>• Manage (develop proper responses)<br>• Analyze (detect new risks)<br>Ethical considerations<br>There are several ethical considerations with regards to blockchain applications that need to be properly addressed, such as environmental impacts, job disruptions, or their potential illicit utilization (criminality, weapons, drugs, money laundering, etc.). | Monitoring<br>We propose a basic three-strep approach for monitoring blockchain projects:<br>• Collect data<br>• Analyze data<br>• Report findings<br>Evaluation<br>We provide sets of evaluation criteria to set up robust M&E frameworks for blockchain projects and support a critical analysis while conducting evaluations:<br>• Technical criteria<br>• Functional criteria<br>• Legal criteria | Change management<br>Organizations embracing blockchain need to be change-adept. In the following change management approach, the focus is on change at the individual level as the corner stone for change to happen at the organizational level:<br>• Unfreeze<br>• Change<br>• Refreeze the change |

Source: Authors.

## 5. Discussion and Conclusions

Can blockchain be a game-changer as an SDG accelerator? Or is it hype that is driven by high expectations for its applications? We argue that the power of blockchain can be effectively harnessed to deliver significant progress towards achieving SDGs. Nevertheless, we also argue that there are limitations to what can be done, hence the need to stress that development professionals cannot solve all issues around SDGs with blockchain and that implementation choices in designing and implementing blockchain-enabled solutions should be rigorously justified to demonstrate tangible added value compared to other alternative solutions, which might be less costly and less technical.

While we recognize the limitations in developing the SDG Acceleration Scorecard and the integrated approach to using blockchain-enabled solutions as SDG accelerators, these are meant to be minimally viable products and serve as living tools, making room for further improvements and finetuning as the technology matures and lessons learned from various experiments are widely shared across the international development space.

"Don't over-hype" and "don't over-promise" indeed remain key messages for individuals and organizations experimenting with blockchain-powered solutions in international development. We believe this work offers good food for thought to further stimulate the discussions within the research community, but also among innovation and development practitioners working at the frontlines to raise the double challenge of adapting to the new normal post-COVID19 and meeting the SDG targets by 2030.

**Author Contributions:** Conceptualization, A.F.A., F.B. and M.D.; methodology, F.B.; software, F.B. and M.D.; validation, A.F.A., F.B. and M.D.; formal analysis, F.B.; investigation, A.F.A., F.B. and M.D.; resources, F.B. and M.D.; data curation, F.B. and M.D.; writing—original draft preparation, F.B.; writing—review and editing, A.F.A. and M.D.; visualization, F.B. and M.D.; supervision, A.F.A.; project administration, A.F.A. and M.D. All authors have read and agreed to the published version of the manuscript.

**Funding:** This research received no external funding.

**Institutional Review Board Statement:** Not applicable.

**Informed Consent Statement:** Not applicable.

**Data Availability Statement:** Not applicable.

**Conflicts of Interest:** The authors declare no conflict of interest.

## References

1. Barbier, E.B.; Burgess, J.C. Sustainability and development after COVID-19. *World Dev.* **2020**, *135*, 105082. [CrossRef] [PubMed]
2. Upadhyay, N. Demystifying blockchain: A critical analysis of challenges, applications and opportunities. *Int. J. Inf. Manag.* **2020**, *54*, 102120. [CrossRef]
3. Zhu, S.; Song, M.; Lim, M.K.; Wang, J.; Zhao, J. The development of energy blockchain and its implications for China's energy sector. *Resour. Policy* **2020**, *66*, 101595. [CrossRef]
4. Jiang, L. The Age of Trust—The Problem Blockchain Solves That Others Cannot. 13 December 2018. Available online: https://medium.com/swlh/the-age-of-trust-the-problem-blockchain-solves-that-others-cannot-6024ebf47cad (accessed on 24 September 2019).
5. Gartner. Digital Disruption Profile: Blockchain's Radical Promise Spans Business and Society. 2018. Available online: https://www.gartner.com/en/doc/3855708-digital-disruption-profile-blockchains-radical-promise-spans-business-and-society (accessed on 29 February 2020).
6. Mougayar, W. If You Understand Google Docs, You Can Understand Blockchain. Coindesk. 2016. Available online: https://www.coindesk.com/understand-google-docs-can-understand-blockchain (accessed on 18 November 2019).
7. Deloitte. *Blockchain: A True Disruptor for the Energy Industry, Use Cases and Strategic Questions*; Deloitte: New York, NY, USA, 2018.
8. Dalla Palma, S.; Pareschi, R.; Zappone, F. What is your Distributed (Hyper) Ledger? In Proceedings of the 4th International Workshop on Emerging Trends in Software Engineering for Blockchain (WETSEB'21) at ICSE 2021, Madrid, Spain, 23–29 May 2021.
9. SelfKey. Understanding Public vs. Private Blockchain. HYPERLINK. 2020. Available online: https://selfkey.org/understanding-public-vs-private-blockchain/ (accessed on 4 December 2019).
10. Jayachandran, P. The Difference between Public and Private Blockchain. 2017. Available online: IBM:Https://www.ibm.com/blogs/blockchain/2017/05/the-difference-between-public-and-private-blockchain/ (accessed on 20 November 2019).

11. Garriga, M.; Dalla Palma, S.; Arias, M.; de Renzis, A.; Pareschi, R.; Andrew Tamburri, D. Blockchain and cryptocurrencies: A classification and comparison of architecture drivers. *Concurr. Comput. Pract. Exp.* **2020**, e5992. [CrossRef]
12. Massessi, D. Public vs. Private Blockchain in a Nutshell. 2018. Available online: Medium:Https://medium.com/coinmonks/public-vs-private-blockchain-in-a-nutshell-c9fe284fa39f (accessed on 15 March 2021).
13. Uzsoki, D.; Guerdat, G. Impact Tokens: A Blockchain-Based Solution for Impact Investing. International Institute for Sustainable Development. 2019. Available online: https://www.iisd.org/publications/impact-tokens-blockchain-based-solution-impact-investing (accessed on 12 March 2021).
14. Sincock, C.; Lewis, R. ESG in Financial Services: Today and in the Future. Capco Intelligence. 2021. Available online: https://www.capco.com/Intelligence/Capco-Intelligence/ESG-In-Financial-Services (accessed on 5 January 2020).
15. Fishcoin. A Blockchain Based Data Ecosystem for the Global Seafood Industry. 2018. Available online: https://fishcoin.co/files/fishcoin.pdf (accessed on 24 November 2019).
16. DeCleene, J. How Blockchain-as-a-Service (Baas) Platforms are Speeding up Blockchain Adoption. Data Driven Investor. Available online: https://www.datadriveninvestor.com/2018/02/17/how-blockchain-as-a-service-baas-platforms-are-speeding-up-blockchain-adoption/# (accessed on 30 May 2018).
17. Sin Kuang, L.; Xiwei, X.; Yin Kia, C.; Qinghua, L. Evaluating Suitability of Applying Blockchain. In Proceedings of the 2017 International Conference on Engineering of Complex Computer Systems, Fukuoka, Japan, 6–8 November 2017. [CrossRef]
18. Mckinsey. Blockchain beyond the Hype: What Is the Strategic Business Value? June 2018. Available online: https://www.mckinsey.com/business-functions/digital-mckinsey/our-insights/blockchain-beyond-the-hype-what-is-the-strategic-business-value# (accessed on 24 September 2019).
19. Fridgen, G.; Lockl, J.; Radszuwill, S.; Schweizer, A.; Urbach, N. A Solution in Search of a Problem: A Method for the Development of Blockchain Use. In Proceedings of the 24th Americas Conference on Information Systems (AMCIS), New Orleans, LA, USA, 16–18 August 2018.
20. Dam, R.; Siang, T. 5 Stages in the Design Thinking Process. The Interaction Design Foundation. 2019. Available online: https://www.interaction-design.org/literature/article/5-stages-in-the-design-thinking-process (accessed on 15 October 2019).
21. Wachal, M. Factors to Evaluate When Using Blockchain Technology. 2019. Available online: http://scrypt.media/2019/08/08/factors-to-evaluate-when-using-blockchain-technology/ (accessed on 17 February 2020).
22. Panetta, K. Blockchain Combines Innovation with Risk. *Gartner*, 16 October 2016. Available online: https://www.gartner.com/smarterwithgartner/blockchain-combines-innovation-with-risk/(accessed on 5 December 2019).
23. Kaplan, S.R.; Mikes, A. Managing Risks: A New Framework. Harvard Business Review. June 2012. Available online: https://hbr.org/2012/06/managing-risks-a-new-framework (accessed on 14 November 2019).
24. Tasca, P. Managing Risk under the Blockchain Paradigm. Linkedin. Available online: https://www.linkedin.com/pulse/managing-risk-under-blockchain-paradigm-paolo-tasca (accessed on 3 March 2017).
25. Thivaios, P. Managing the Risks of Blockchain. 2018. Available online: https://knect365.com/riskminds/article/c515ed63-da12-4d4e-874a-eb83eefb3e66/managing-the-risks-of-blockchain (accessed on 1 October 2019).
26. CBECI. Cambridge Bitcoin Electricity Consumption Index. 2019. Available online: https://www.cbeci.org/ (accessed on 3 October 2019).
27. Houlding, D. Blockchain: 6 Key Ethical Considerations. Available online: Lifeboat:Https://lifeboat.com/blog/2019/01/blockchain-6-key-ethical-considerations (accessed on 18 January 2019).
28. Longstaff, S. Blockchain: Some Ethical Considerations. The Ethics Center. Available online: https://ethics.org.au/blockchain-some-considerations/ (accessed on 16 March 2019).
29. Smith, J. Beyond Crypto—Blockchain Ethics. Hackernoon. Available online: https://hackernoon.com/beyond-crypto-blockchain-ethics-eabd8df6faf5 (accessed on 4 February 2019).
30. Apointe, C.; Fishbane, L. The Blockchain Ethical Design Framework. Beeck Center for Social Impact + Innovation—Georgetown University. 2018. Available online: https://www.mitpressjournals.org/doi/pdf/10.1162/inov_a_00275 (accessed on 14 December 2019).
31. EPRS. Blockchain and the General Data Protection Regulation Can Distributed Ledgers Be Squared with European Data Protection Law? European Parliamentary Research Service. 2019. Available online: https://www.europarl.europa.eu/RegData/etudes/STUD/2019/634445/EPRS_STU(2019)634445_EN.pdf (accessed on 17 March 2021).
32. Schwerin, S. Blockchain and privacy protection in the case of the european general data protection regulation (GDPR): A delphi study. *J. Br. Blockchain Assoc.* **2018**, *1*, 3554. [CrossRef]
33. OECD. DAC Principles for Evaluation of Development Assistance. Organisation for Economic Co-Operation and Development. 1991. Available online: https://www.oecd.org/dac/evaluation/2755284.pdf (accessed on 16 October 2019).
34. Pasanen, T. 2018: Time to Update the DAC Evaluation Criteria? 2018. Available online: https://www.odi.org/blogs/10594-2018-time-update-dac-evaluation-criteria (accessed on 4 October 2019).
35. Jiang, L. Breaking Blockchain—A Framework to Evaluate Blockchain Use Cases. 2018. Available online: https://hackernoon.com/breaking-blockchain-a-framework-to-evaluate-blockchain-use-cases-9efbc30a3fa7 (accessed on 5 December 2019).
36. WEF. *Building Value with Blockchain Technology: How to Evaluate Blockchain's Benefits*; World Economic Forum: Cologny, Switzerland, 2019.
37. Vangala, N. Evaluation Criteria for Blockchain. *Medium*, 27 June 2018. Available online: https://medium.com/nhct-nanohealth-care-token/evaluation-criteria-for-blockchain-e33a54b477aa(accessed on 23 September 2019).

38. Fridgen, G.; Guggenmos, F.; Lockl, J.; Rieger, A.; Schweizer, A.; Urbach, N. Developing an Evaluation Framework for Blockchain in the Public Sector: The Example of the German Asylum Process. Amsterdam: Proceedings of the 1st ERCIM Blockchain Workshop 2018, Reports of the European Society for Socially Embedded Technologies (ISSN 2510-2591). 2018. Available online: https://www.fim-rc.de/Paperbibliothek/Veroeffentlicht/756/wi-756.pdf (accessed on 30 May 2020).
39. Catalini, C.; Gans, J.S. *Some Simple Economics of the Blockchain (No. w22952)*; National Bureau of Economic Research: Cambridge, MA, USA, 2016.
40. Swan, M. *Blockchain: Blueprint for a New Economy*; O'Reilly Media, Inc.: Newton, MA, USA, 2015.
41. Narayan, R.; Tidström, A. Tokenizing coopetition in a blockchain for a transition to circular economy. *J. Clean. Prod.* **2020**, *263*, 121437. [CrossRef]
42. Abrahamson, E.; Rosenkopf, L. Social network effects on the extent of innovation diffusion: A computer simulation. *Organ. Sci.* **1997**, *8*, 289–309. [CrossRef]
43. Prosci. What Is the ADKAR Model? 2019. Available online: https://www.prosci.com/adkar/adkar-model (accessed on 3 October 2019).
44. Talib Hussain, S.; Shen, L.; Akram, T.; Jamal Haider, M.; Hadi Hussain, S.; Ali, M. Kurt Lewin's Change Model: A Critical Review of the Role of Leadership and Employee Involvement in Organizational Change. *J. Innov. Knowl.* **2016**. Available online: https://study.com/academy/lesson/lewins-3-stage-model-of-change-unfreezing-changing-refreezing.html (accessed on 21 November 2019).
45. UNDP. SDG Accelerator and Bottleneck Assessment. 2017. Available online: https://www.undp.org/content/dam/undp/library/SDGs/English/SDG_Accelerator_and_Bottleneck_Assessment_Tool.pdf (accessed on 29 March 2020).
46. Bellamy, J.-J.; Hill, K. Monitoring Guidelines of Capacity Development in Global Environment Facility Projects. Global Support Programme, Bureau for Development Policy, United Nations Development Programme. 2010. Available online: https://www.thegef.org/sites/default/files/publications/Monitoring_Guidelines_Report-final.pdf (accessed on 7 January 2020).
47. CBD. Overview of Access and Benefit-Sharing Capacity-Building Tools and Resources. Convention on Biological Diversity. 2018. Available online: https://www.cbd.int/doc/c/99fa/59d2/f895959bebd76275b1fbf283/abs-cbiac-2018-01-02-add2-en.pdf (accessed on 8 February 2020).
48. Leverington, F.; Hockings, M.; Lemos, K.; Courrau, J.; Pavese, H. *Management Effectiveness Evaluation in Protected Areas, a Global Study, Overview of Approaches and Methodologies*; Supplementary Report No.1, The University of Queensland: Gatton, Australia, 2008.
49. Staub, F.; Hatziolos, M. *Score Card to Assess Progress in Achieving Management Effectiveness Goals for Marine Protected Areas*, Revised Version; The World Bank: Washington, DC, USA, 2004.

sustainability

MDPI

*Review*

# Role of Traditional Ethnobotanical Knowledge and Indigenous Communities in Achieving Sustainable Development Goals

**Ajay Kumar [1,*], Sushil Kumar [2], Komal [3], Nirala Ramchiary [4] and Pardeep Singh [5]**

1 Department of Plant Science, School of Biological Sciences, Central University of Kerala, Periye, Kasaragod 671316, Kerala, India
2 Department of Botany, Government Degree College, Kishtwar, Kishtwar 182204, Jammu and Kashmir, India; sushilthakur863@gmail.com
3 Department of English, Government Degree College, Bani, Kathua 184206, Jammu and Kashmir, India; komalkanthalia108@gmail.com
4 Translational and Evolutionary Genomics Laboratory, School of Life Sciences, Jawaharlal Nehru University, New Delhi 110067, India; nramchiary@jnu.ac.in
5 Department of Environmental Science, PGDAV College, University of Delhi, New Delhi 110065, India; psingh@pgdav.du.ac.in
* Correspondence: ajay@cukerala.ac.in

**Citation:** Kumar, A.; Kumar, S.; Komal; Ramchiary, N.; Singh, P. Role of Traditional Ethnobotanical Knowledge and Indigenous Communities in Achieving Sustainable Development Goals. *Sustainability* **2021**, *13*, 3062. https://doi.org/10.3390/su13063062

Academic Editor: Emanuele Radicetti

Received: 19 January 2021
Accepted: 4 March 2021
Published: 11 March 2021

**Publisher's Note:** MDPI stays neutral with regard to jurisdictional claims in published maps and institutional affiliations.



**Abstract:** The sustainable development goals (SDGs) are a set of 17 goals with 169 targets. The Agenda 2030 of the United Nations envisages a holistic approach to achieve these goals by focusing on humankind and the planet. In this review, we analyzed the scientific literature and technical reports of international bodies such as the United Nations and Food and Agriculture Organization relating to traditional ethnobotanical knowledge (TEK). The literature on TEK was mapped with the targets of the SDGs to determine the role of traditional knowledge in the realization of selected goals and targets. Our extensive and systematic reviewing of available literatures suggests that, of the 17 goals, at least seven goals are associated with TEK. To achieve these seven goals, a thorough understanding is required to disentangle the intricacies involving TEK, indigenous people holding TEK, and their future role in achieving the SDGs. Our review points towards the role of TEK in achieving goals linked to poverty, health and wellbeing, responsible consumption and production, climate action, life on land, and partnerships. In summary, we argue that achieving the intended outcomes of the SDGs and the targets requires concerted efforts of all relevant stakeholders, including indigenous communities, common citizens, scientists, policy makers, and world leaders.

**Keywords:** sustainable development goals; ethnobotany; human health; poverty; traditional knowledge; sustainable agriculture

## 1. Introduction

The United Nations General Assembly in its 70th meeting on 25 September 2015 adopted a resolution "transforming our world: the 2030 agenda for sustainable development" [1]. Its member countries adopted 17 set of goals called the sustainable development goals (SDGs) to end poverty, protect the planet, and ensure prosperity for all. Each goal has certain targets to be achieved by 2030. Sustainable development goals are an extension of millennium development goals (MDGs) and part of a new sustainable development agenda to complete what MDGs did not achieve [2]. For the goals to be realized, everyone needs to do their part, including governments, the private sector, and civil society. The Agenda 21 of the Rio Earth Summit in 1992 [3] where the concept of sustainable development emerged, advocated a pivotal role for indigenous people and other local communities in the management of environment and sustainable development because of their traditional knowledge and associated practices [4]. The official UN document of 2015 (transforming our world: the 2030 agenda for sustainable development) did not explicitly explain the role of ethnobotanists and traditional ethnobotanical knowledge (TEK) in achieving these sustainable development goals. Apart from the role of different countries and various stakeholders mentioned in the document in achieving the goals and targets, we advocate and reiterate a similar role for indigenous communities by recognizing and supporting their identity, cultures, customs, practices, and interests and by enabling their effective and active participation in the realization of sustainable development goals, including the goals that were mentioned in Principle 22 of the 1992 Rio Declaration on the Environment and Development [3]. It was reaffirmed that traditional ethnobotanical knowledge, ethnobotanists, and people's participation can significantly contribute towards achieving the sustainable development goals by 2030 and beyond.

People have a long history of using plants for various purposes such as for food, medicine, shelter, decoration, construction, and clothing [5]. The usage of plants for various purposes by indigenous and local communities comprises traditional ethnobotanical knowledge, which is not well documented by indigenous people, but has been orally and vertically transmitted from generation to generation [6]. However, ethnobotanists have played an important role in unraveling and documenting these plant-people interactions and unlocked the knowledge by conducting various types of interviews and surveys [7,8]. Besides these, archeological, paleontological, and archaeogenetic evidence has also recently been used to determine plant-people interactions from prehistoric times [9–11]. Though the records on the consumption of plants by humans in the prehistoric times are scarce, recently it was discovered that people in South Africa used leaves of *Cryptocarya woodii* for the construction of bedding 77,000 years ago, and the identity of the plants used was established using modern archaeogenetic tools [12]. This plant is still used by the people living in the area of its recovery, and it has now been established that this plant is toxic to mosquitoes because of its larvicidal properties [12]. The recovery of starch granules from the surfaces of at least 105,000 years old stone tools from Mozambique suggests that early humans consumed grass seeds [9]. The discipline of ethnobotany works at the intersection of plants-people at one end and science at the other end; therefore, ethnobotanists can act as a bridge between them [13]. Tuxill and Nabhan [14] have suggested that ethnobotany can act as useful vehicle and process of development. The outcomes of the interactions between the plants and people as deciphered by ethnobotanists holds enormous potential to solve some of the issues that the world faces today. The common global challenges we face today are ending poverty, achieving zero hunger, improving the nutritional status of the people, promoting sustainable agriculture, improving the health status of the people, providing affordable health care services, and combating climate change [15,16]. Therefore, in the present review article, we identified seven sustainable development goals towards which the traditional ethnobotanical knowledge can contribute significantly. These are SDG (1) no poverty, (2) zero hunger, (3) good health and wellbeing, (12) responsible consumption and production, (13) climate action, (15) life on land, and (17) partnerships for the goals. In the following sections, a brief background of major challenges and the potential roles of ethnobotany in overcoming these challenges to achieve the SDG by 2030 are presented (Figure 1).

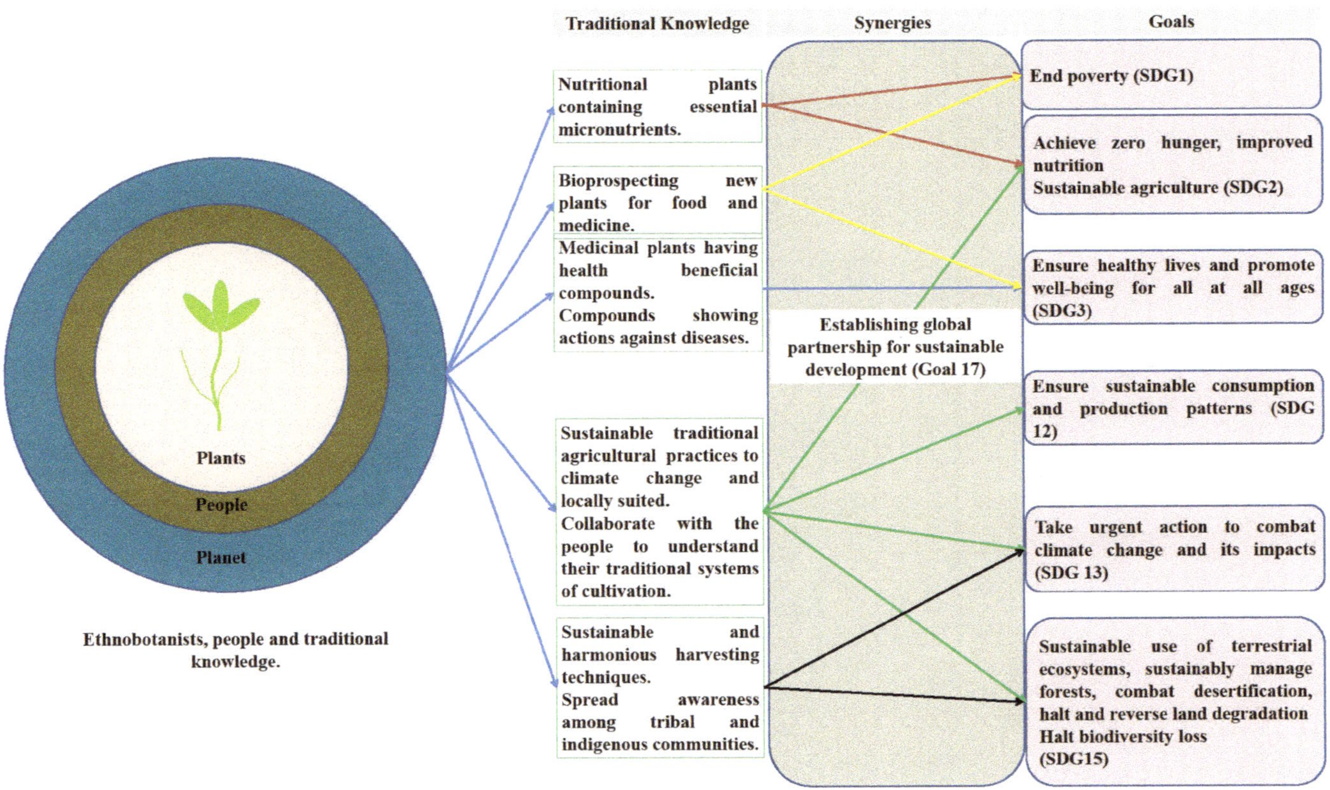


**Figure 1.** A schematic representation highlighting the potential roles of ethnobotanists towards achieving seven sustainable goals and targets.

## 2. Methodology

To study the role of ethnobotanical knowledge in the realization of sustainable development goals, we searched available literature and information available on the subject [17]. Various reports and documents of the international bodies such as the UN and FAO were studied to understand the sustainable development goals and the targets envisaged under it. We conducted an extensive literature survey to find out papers and reports related to traditional knowledge to link with the sustainable development goals. The research question in the present study was to know whether there were any goals and targets related to ethnobotanical knowledge and whether ethnobotanical knowledge could help in achieving them. After a thorough study, we have identified seven goals related to ethnobotanical knowledge which are SDG 1 (no poverty), 2 (zero hunger), 3 (food health and wellbeing), 12 (responsible consumption and production), 13 (climate action), 15 (life on land), and 17 (partnerships for the goals). The seven goals and targets thereunder are abridged into four broad areas (Table 1) as few goals have overlapping/common targets, and then the gaps or problems in those broad areas are identified. For example, SDG 2 is common for broad area 1 (poverty-hunger-malnutrition) and broad area 2 (desertification, land degradation, and sustainable agriculture). Similarly, SDG 15 is repeated for broad area 2 and 4 (ensuring sustainable consumption-production and reversal of biodiversity loss). The goal 17 (SDG 17) refers to establishing partnerships and synergies among various stakeholders. Therefore, the need of partnerships to achieve different goals or outcomes is discussed in the broad areas. Following this criteria, various papers related to ethnobotanical studies were analyzed to fill the gaps or to solve the problems. Furthermore, various terms relating to sustainable development goals such as "traditional knowledge", "ethnobotanical knowledge", "traditional agricultural knowledge", "traditional ecological knowledge", "food security", "abiotic stresses", "climate change", "end hunger", "food security", "medicinal plants", "sustainable agriculture", "climate resilience" and "ethnobotanical knowledge" and various combinations of key terms were searched from various databases such as PubMed, Google Scholar, Web of Science, AGRICOLA, and Scopus [17,18]. The terms were searched either individually or in multiple combinations. The electronic search queries generated more than 3000 papers from various databases. We manually analyzed and removed the duplicates and only 105 articles including reports and websites were included for writing this review. For the survey of various reports relating to sustainable development goals, a simple Google search was used or the website of the organizations were directly visited.

**Table 1.** The identified four broad areas and the sustainable development goals linked with them.

| Broad Area | Goals Linked to the Broad Area |
|---|---|
| Poverty-hunger-malnutrition | SDG 1: End poverty.<br>SDG 2: Achieve zero hunger and improve nutritional status of the public. |
| Desertification, land degradation, and sustainable agriculture | SDG 2: Achieve sustainable food production systems.<br>SDG 15: Protect, restore, and promote sustainable use of terrestrial ecosystems; sustainably manage forests; combat desertification; and halt and reverse land degradation and halt biodiversity loss. |
| Health and wellbeing of the people | SDG 3: Ensure healthy lives and promote well-being for all at all ages. |
| Ensuring sustainable consumption-production reversal of biodiversity loss | Goal 12: Ensure sustainable consumption and production patterns.<br>Goal 13: Take urgent action to combat climate change and its impacts.<br>SDG 15: Protect, restore, and promote sustainable use of terrestrial ecosystems; sustainably manage forests; combat desertification; and halt and reverse land degradation and halt biodiversity loss. |
| Partnerships (important for achieving all the goals) | SDG 17: Revitalize the global partnership for sustainable development. |

## 3. Results and Discussion

### 3.1. Poverty-Hunger-Malnutrition Conundrum and the Role of Wild Food Plants

The ever-increasing human population projections indicate an increase in the number of consumers and demand for goods and services [19]. The Agenda 2030 seeks to end poverty (SDG1), achieve zero hunger, and improve nutritional status of the public (SDG2). Poverty, hunger, and nutrition are three interlinked issues with cause and effect relationships which need to be addressed systematically in an integrated manner [20]. Targeting poverty alone may help to achieve zero hunger, but unless people have access to nutritious food, it may not deliver the expected result [21]. Therefore, the initiatives for alleviating poverty must be coupled with addressing the issues related to diet, malnutrition, and undernutrition. Approximately one billion of the poorest people of the world remain undernourished today, and the Food and Agriculture Organization (FAO) estimated that we will need to increase food production by 70% to feed the 9.1 billion people by 2050 [22]. These data suggest the gravity of challenges the world is facing to ensure food security for all.

As per the FAO [23], food security has four main dimensions: (1) physical availability of food (supply of food), (2) economic and physical access to food, (3) food utilization, and (4) the stability of the other three dimensions over time (Figure 2). The ethnobotanical consumption of wild vegetables and food plants by indigenous communities are in congruence with the above four dimensions as locally grown wild vegetables can help increase the supply of food thereby maintaining demand-supply equilibrium (as described in Figure 2). Since the wild food plants are collected and consumed locally by people, the change in the prices of the market foods may not affect the economic access to it. Besides this, the plants are locally grown, adapted to particular local environmental conditions, resilient to climate change, and are suitable to the local agro-climatic conditions [24]. Therefore, researchers have now emphasized the urgent need to diversify and expand the food basket of the people under unpredictable climatic conditions [25], and in this scenario wild food plants if adopted in mainstream diet are the best option to achieve this goal [26,27]. Several study groups also reported the importance and wide usage of wild food plants in various countries such as India [7,8,27,28], Ethiopia [29,30], China [31], Indonesia [32], Greece [33], and Italy [34,35], to name a few. Since the locally grown plants are rich in vitamins and micronutrients, they can help in reducing diet related malnutrition considerably [32,36]. Ethnobotanists have already generated immense amounts of data on the use of wild plants for food. A search on Google Scholar titled "wild plants as food" shows 1,220,000 results (within the years 2000–2020). All these results suggest that many wild plants under consumption by various indigenous communities have already been documented. Therefore, the present challenges of food insecurity necessitate the mainstreaming of wild food plants, their formal addition into the existing food basket of the public by scientifically validating the nutritional composition, the study and development of the mode of their multiplication/reproduction, and the life cycle-related requirements for their germination and cultivation. To achieve this, multidisciplinary approaches and experts in agronomy, plant science, phytochemistry, dietetics, and modern analytical approaches for nutritional profiling need to be employed. Furthermore, potential health effects also need to be understood as some plants may have the presence of harmful compounds which may exert negative side effects on humans. Combining multidisciplinary approaches along with the clues obtained from ethnobotanical data on wild plants for consumption as fruits and vegetables can provide alternative options and aid in diversifying the food basket of the people [25]. The issue of food security can be catered by popularizing and mass propagation of the locally grown plants especially in the developing countries in Africa and Asia. This can further reduce the dependency of people on only a few crop plants available for their nutritional purposes. In the long run, apart from the role of wild food plants in the diversification of food basket, they can also act as valuable genetic resources for the crop improvement programs as some of the food plants consumed indigenously are the wild relatives of the domesticated crops plants currently grown/consumed.

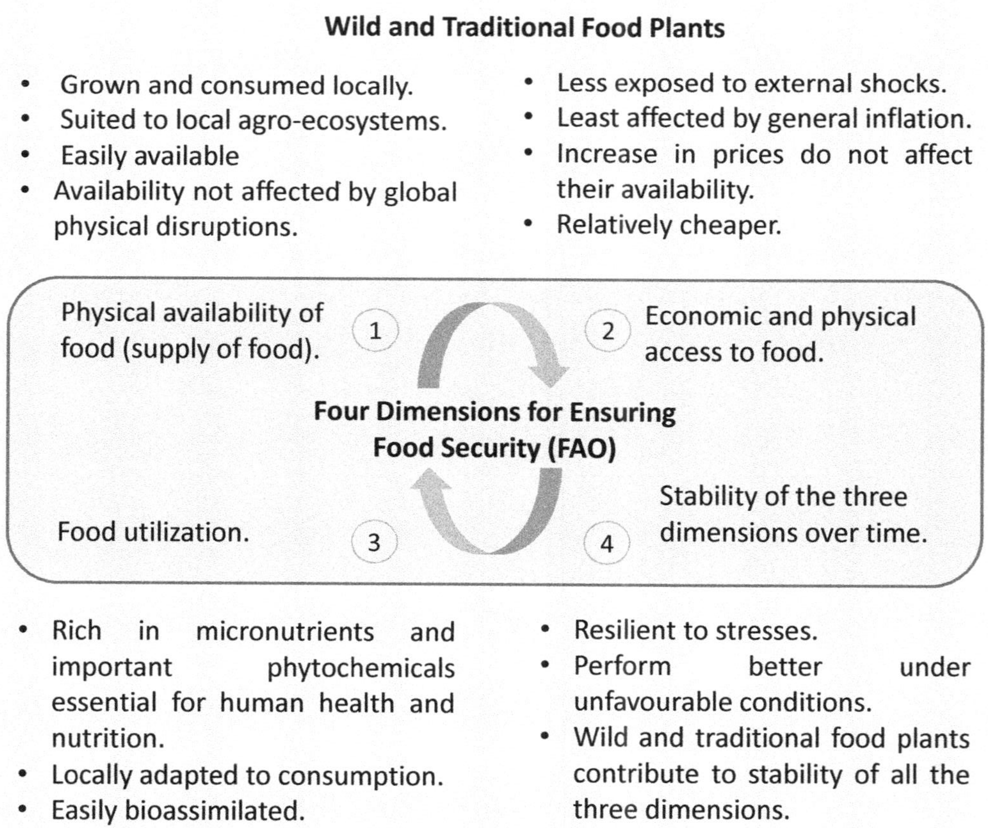


**Figure 2.** The role of Wild and Traditional Food Plants in FAO's four dimensions for ensuring global food security.

### *3.2. Overuse of Chemicals in Agriculture, Desertification, Land Degradation and the Need for Sustainable Agriculture*

The increased demand for nutritious, safe, and healthy food for the burgeoning human population and the promise to maintain biodiversity and other natural resources are posing a major challenge to agriculture that is already threatened by climate change and excessive use of chemical fertilizers and pesticides [25]. The excessive use of fertilizers has rendered soils unfit for agricultural purposes. Its long-term use has increased the salinity of the soils and reduced the total land available for cultivation of crops [37]. This problem of soil degradation is severe in tropics and subtropics, and it has led to decrease in the soil ecosystem services by 60% in the 60 years from 1950 to 2010 [38]. It has been reported that nearly 500 million hectares (Mha) of the land is affected due to the soil degradation in the tropics [39], and globally 33% of the total land is affected due to land degradation [40,41]. Besides affecting agronomic production, soil degradation can also slow down the economic growth of developing countries which are majorly dependent on agriculture [42]. The use of chemical pesticides has a deteriorating effect on the non-target beneficial insects such as honey bees [43–45]. Pesticides can affect bee populations directly by causing mortality and by altering their behaviour through sublethal effects. Sublethal pesticides also interfere with brood development and shorten the life cycle of adults [46]. An unusual phenomenon of the disappearance of bees from beehives was observed in the US in 2006 which was termed as colony collapse disorder (CCD) [43,47]. It has been established that many factors might contribute to CCD but pesticides play a synergistic role in this disorder [43,48,49]. Interestingly, organic beekeepers did not face such CCD-like situations [50]. Considering the importance of bees in pollination of the food crops, the increased use of pesticides can lead to reduced food production endangering our food security [50]. The fertilizer and pesticide runoff have also adversely affected quality of surface and groundwater [51]. Thus total land available for cultivation is reduced due to its degradation caused by various activities such as water and wind erosion, salinity, sodicity, alkalinity, reduction in soil fertility, and urban expansion [52]. The increase in productivity of land available using sustainable practices, conservation of the remaining land resources, and reclamation of the degraded land are some of the challenges in land use and sustainable agriculture [53–55].

Besides achieving zero hunger and improved nutrition, SDG2 also seeks to achieve sustainable food production systems (sustainable agriculture) through the implementation of resilient agricultural practices that increase productivity by maintaining the health of the ecosystems [56]. Through the intimate association with the plants and agriculture related activities, people have gained tremendous knowledge on the various aspects of agriculture and crops plants [57]. The knowledge ranges from the soil types, season of sowing a particular crop, the water and nutrient requirements, and other conditions which promote or limit crop productivity [58,59]. The knowledge also pertains to the diseases caused by pathogens and attack by pests and their management strategies [60,61]. The ethnobotanical knowledge is vast, not uniformly distributed, and locally suited as per the environmental conditions and the availability of particular crops [62,63]. In addition to the lack of proper documentation, researchers have suggested that increased modernization may have led to the loss of this knowledge in many places [64]. Turner and Turner [65] have ascribed some other reasons to this loss which include dynamism and changing knowledge systems, loss of indigenous languages, lack of time and opportunities for cultural practices, urbanization of indigenous people, globalization, and industrialization. These challenges in the world agriculture calls for the immediate attention not only to protect the knowledge base of the communities involved in ethnobotany in active sense but also to revive where it is dormant [24]. Rockstrom et al. [66] have stressed the necessity of identifying environmental conditions that enable prosperous human development and set limits for the planet to remain in that state. The study further suggests that the Holocene epoch provides a reference point when the naturally occurring environmental change was within the limits of earth's regulatory capacity and it helped our ancestors to develop agriculture and modern societies to prosper [66]. Therefore, efforts must be taken to understand, preserve, and promote traditional agricultural knowledge locked with the indigenous communities encompassing wide range of domains of agriculture ranging from types of soils, diseases, environmental conditions, to the management of diseases, interventions required in the soils for disease prevention, irrigation, the types of genotypes, and their selection for a particular soil type [63]. Literature on linking of various domains on traditional agricultural knowledge is scarce [67], and it is advisable to take up studies that consider linking across domains of traditional agricultural knowledge. The knowledge of traditional and sustainable agricultural practices must be strengthened in the interest of the public and the planet; therefore, a greater role of traditional knowledge experts and the communities holding this knowledge is suggested [68–70]. Greater public participation, more funding, and scientific research must be promoted. Sustainable agricultural practices need to be followed which involve the minimal use of chemical fertilizers and pesticides. The reduced use of chemicals and promotion of sustainable agriculture will help combat climate change (SDG 13). Sustainable agriculture and climate resilient cultivation practices will further arrest and reverse land degradation caused by the rampant use of chemical fertilizers. This would also halt biodiversity loss especially the bees and other pollinator species, which is very important for crop diversification (SDG 15).

### 3.3. *Health and Wellbeing of the People: Blend of Traditional Knowledge and Modern Science*

Of the various uses of plants, medicinal plants used for human health and well-being are the most important because of the presence of specialized metabolites in those plants (SM's) [71]. Specialized metabolites with medicinal properties are non-uniformly distributed in certain families of plants, and these plants act as natural chemical factories for the production of SM's [72]. The SM's have promising health promoting effects and are an important source of many of the present-day drugs [73]. It has been suggested that medicinal plants play a major role in the primary health care of nearly 80 percent of people living in the developing countries [74]. Not only in the developing countries, but natural products and medications derived from them also contribute significantly to the health care systems of the remaining 20 percent of the people residing in developed countries [75]. Interestingly, of the 25 best-selling pharmaceutical drugs, 12 are natural products derived from the plants [76,77]. Ethnobotanists have significantly contributed to the discovery of numerous medicines such as artemisinin (from *Artemisia annua* L.), aspirin (from *Filipendula ulmaria* (L.) Maxim.), ephedrine (*Ephedra* spp.) codeine and papaverine (from *Papaver somniferum* L.), colchicine (from *Colchicum autumnale* L.), taxol (from *Taxus brevifolia* Nutt.), digoxin and digitoxin (from *Digitalis purpurea* L.), capsaicin (from *Capsicum* spp.), and tetrahydrocannabinol and cannabidiol (from *Cannabis sativa* L.), to name a few [71,78–80]. The importance of the discovery of Artemisinin was recognized when Tu Youyou, a Chinese scientist, was awarded Nobel prize in 2015 [81]. Further, a search on Google Scholar titled "wild plants as medicine" shows 216,000 results (between the years 2000–2020).

Indigenous and tribal communities use thousands of plants for medicinal purposes, many of them are not even botanically named, and many drugs of the plant origin are waiting to be discovered by modern science [82–84]. The famous anthropologist David Maybury-Lewis had emphasized the role of tribal communities in assisting the discovery of more and more medicinal plants which are used by them for medicinal purposes [85]. Ethnobotanists can expedite the identification process of probable medicinally valuable plants, and it is suggested that instead of undertaking random screening expeditions, clues and leads can be derived from the ethnobotanical knowledge that can ease the task of bioprospecting of the plants [86]. Garnatje et al. [6,71] suggested the term "ethnobotanical convergence" for the similar uses of plants included in the same node of a phylogeny. Although the term "ethnobotanical convergence" was criticized by Hawkins and Teixidor-Toneu [87], it is nevertheless believed that linking of new technologies with the traditional ethnobotanical knowledge can expedite the process of target-based drug discovery. Linking ethnobotany with other disciplines such as phytochemistry, pharmacology, pharmacognosy, and molecular biology can aid in the identification and screening of important plants for their promising role in treating diseases [88]. Furthermore, approaches such as genomics and omics can also be employed to identify the genes underlying the (specialized) metabolites present in the plants characterized by high throughput metabolomics approaches such as gas chromatography-mass spectrometry (GCMS), liquid chromatography-mass spectrometry (LCMS) and nuclear magnetic resonance spectroscopy (NMR) [89,90]. The proper identification, utilization, and conservation of medicinal plants can assist in providing better alternative healthcare services in rural areas, especially in developing countries [91,92]. Furthermore, a number of wild medicinal and aromatic plants are highly valuable, and a significant proportion of the people consume them for medicinal purposes [93]. The scarcity of better health care systems ensuring healthy lives and promoting wellbeing of the people at all ages (SDG 3) in the developing countries of Asia and Africa underpins the importance of ethnomedicinal plants [91,92]. The upward trend of dependency on plant-based medicines, especially in the past few decades, suggests that the role of plant-based drugs will continue to grow in the coming years which may put pressure on the available medicinal plant resources. With the growing burden of diseases coupled with issues such as population growth and climate change, the discovery of plant-based medicines needs to be hastened using leads from indigenous communities in collaboration with experts from multiple disciplines.

### 3.4. Ensuring Sustainable Consumption-Production Patterns Would Halt Biodiversity Loss

Harvesting practices of wild plants are generally invasive and are destructive to the naturally occurring wild vegetables and medicinal plants which may pose serious threat to these important plants and if kept unchecked sometimes may lead to the extinction of some of the important species [94]. According to the World Wildlife Fund and International Union for Conservation of Nature, approx. 50,000 to 80,000 species of flowering plants are currently being used for medicinal purposes worldwide [95]. Nearly 15,000 of them are threatened with extinction due to excessive exploitation and habitat destruction [96]. The issues related to harvesting coupled with over-exploitation, over usage, and climate change necessitates the need for change in collection and consumption patterns. The collection and consumption patterns need to be congruous with the self-regeneration potential of wild genetic resources; therefore, ethnobotanists can play important roles in the conservation of genetic resources by providing feedback to the communities relying on collection and consumption of medicinal plants [97]. Several researchers have demonstrated the importance of ethnobotany in the conservation and management of natural resources. For example, Phillips and Gentry [98,99] have shown that the number of uses of a plant and its popularity can be used to indirectly access the harvesting pressure on a species and the roles of the communities involved in its usage. Bussmann [75] has also highlighted the role of ethnobotany in the conservation of biodiversity [75]. Ethnobotany can help conserve biodiversity by evolving achievable models for natural resource use and effective management that can be integrated into decision and policy making [100]. A case study from Southern Ecuador has stressed the need to include interdisciplinary approaches for the conservation of ethnomedicinal plants to prevent them from over harvesting. To reduce the pressure on wild plants, alternative methods can be adopted to propagate them. For example, the most popular herbal tea of Southern Ecuador is "Horchata" which consists of more than 30 medicinal herbs, and the harvesting of them directly from the wild may endanger their survival. Nowadays, these 30 ingredients are organically produced by local farmers in small managed gardens instead of directly collecting from the wild, thereby reducing the pressure on the wild populations [75]. Peters, Alexiades, and Laird [101] have suggested the need of imparting skills to the indigenous communities for the better management of the tropical forests. This would reduce the dependency on the external inputs and create a skilled workforce of local communities. Experts from various disciplines such as forestry, ecology, ethnobotany, economics, and anthropology can intervene and play a crucial role in managing the forests and other resources. Skills and knowledge of the best practices such as how to grow, harvest, and consume wild vegetables and other resources in a sustainable manner (SDG 12) would also play a synergistic role in strengthening the community management of the valuable resources. Awareness programs at larger scale must also be integrated while training selected members of local indigenous communities to highlight negative effects of destructive harvesting practices. The sustainable consumption production patterns would further halt biodiversity loss (SDG15). Besides this, the conservation initiatives can be undertaken along with the help of conservation scientists, geneticists, and people's participation. Integration of traditional knowledge into ecological research for biodiversity conservation involving local communities holds the potential towards sustainable development [102], and it must be recognized and promoted [103,104].

## 4. Conclusions

The role of ethnobotanical knowledge, indigenous communities, and ethnobotanists has to be recognized on an urgent basis in realizing sustainable development goals. An international collaboration consortium deriving people from various countries and various fields can be established to reap the benefits of traditional ethnobotanical knowledge to alleviate poverty, end hunger, provide better healthcare facilities, combat climate change, and conserve biodiversity and solve biodiversity related issues. Digitization and the creation of universal databases of the usage of plants for various purposes as a global common can be initiated to disseminate information with regard to ethnobotanically important plants and the knowledge associated with it. The modern scientists can use these clues to further establish scientific reasoning, for example, to investigate which compound may be responsible for treating a particular disease; what the nutritional profile of a plant is; whether it can be recommended as a source of nutrition, and if yes, how much is sufficient. Thus, we call for strengthening ethnobotanical studies, and sufficient funding needs to be channeled for promoting research in this field in order to meet SDGs. This can be concluded with a quote by Dr. Margaret Chan (former Director General of WHO), "The two systems of traditional and Western medicine need not clash. Within the context of primary health care, they can blend together in a beneficial harmony, using the best features of each system, and compensating for certain weaknesses in each. This is not something that will happen all by itself. Deliberate policy decisions have to be made. But it can be done successfully" [105]. This holds true not only for the traditional medicines but also for other domains of traditional knowledge such as food plants, sustainable agriculture, biodiversity conservation, and climate change. Traditional knowledge can be supplemented with the modern advancements in science. This integrated approach involving a blend of traditional knowledge and modern advancements in science can contribute to achieving the SDGs if planned and implemented properly. These integrated approaches are in consonance with the SDG 17 (revitalize the global partnership for sustainable development) which emphasizes the importance of global partnership for achieving the rest of the 16 goals. We believe that ethnobotany research groups and societies from various parts of the world must initiate collaborations and partnerships among themselves and with other fields in a cross-disciplinary manner for realizing the sustainable development goals in the greater interest of people and the planet.

**Author Contributions:** A.K., S.K. and N.R.: conceptualization, methodology, writing—original draft preparation, review and editing, visualization; K., and P.S.: writing—original draft preparation, review and editing. All authors have read and agreed to the published version of the manuscript.

**Funding:** This research received no external funding.

**Institutional Review Board Statement:** Not applicable.

**Informed Consent Statement:** Not applicable.

**Data Availability Statement:** All data generated or analyzed during this study are included in this publication.

**Acknowledgments:** We gratefully acknowledge the continuing support by Central University of Kerala and Jawaharlal Nehru University, New Delhi.

**Conflicts of Interest:** The authors declare that they have no competing interests.

## References


1. UN General Assembly. Transforming our World: The 2030 Agenda for Sustainable Development; Report No. A/RES/70/1. Available online: http://www.un.org/ga/search/view_doc.asp?symbol=A/RES/70/1&Lang=E (accessed on 16 May 2020).
2. International Council for Science and International Social Science Council. Review of Targets for the Sustainable Development Goals: The Science Perspective. Available online: https://www.icsu.org/publications/review-of-targets-for-the-sustainable-development-goals-the-science-perspective-2015 (accessed on 13 April 2020).
3. United Nations General Assembly. Report of the United Nations Conference on Environment and Development (Rio de Janeiro, 3–14 June 1992). 1992. Available online: http://www.un.org/documents/ga/conf151/aconf15126-1annex1.htm (accessed on 17 May 2020).
4. Kimerling, J. Rio+10: Indigenous Peoples, Transnational Corporations, and Sustainable Development in Amazonia; CUNY Academic Works. 2002. Available online: https://academicworks.cuny.edu/cl_pubs/252 (accessed on 28 June 2020).
5. Balick, M.J. Transforming Ethnobotany for the New Millennium. *Ann. Mo. Bot. Gard.* **1996**, *83*, 58. [CrossRef]
6. Garnatje, T.; Peñuelas, J.; Vallès, J. Reaffirming 'Ethnobotanical Convergence'. *Trends Plant Sci.* **2017**, *22*, 640–641. [CrossRef]
7. Saha, D.; Sundriyal, M.; Sundriyal, R.C. Diversity of Food Composition and Nutritive Analysis of Edible Wild Plants in a Multi-Ethnic Tribal Land, Northeast India: An Important Facet for Food Supply. *Indian J. Tradit. Knowl.* **2014**, *13*, 698–705.
8. Chauhan, S.H.; Yadav, S.; Takahashi, T.; Łuczaj, Ł.; D'Cruz, L.; Okada, K. Consumption Patterns of Wild Edibles by the Vasavas: A Case Study from Gujarat, India. *J. Ethnobiol. Ethnomed.* **2018**, *14*, 57. [CrossRef] [PubMed]
9. Mercader, J. Mozambican Grass Seed Consumption during the Middle Stone Age. *Science* **2009**, *326*, 1680–1683. [CrossRef] [PubMed]
10. Liu, L.; Bestel, S.; Shi, J.; Song, Y.; Chen, X. Paleolithic Human Exploitation of Plant Foods during the Last Glacial Maximum in North China. *Proc. Natl. Acad. Sci. USA* **2013**, *110*, 5380–5385. [CrossRef]
11. Allaby, R.G.; Kistler, L.; Gutaker, R.M.; Ware, R.; Kitchen, J.L.; Smith, O.; Clarke, A.C. Archaeogenomic Insights into the Adaptation of Plants to the Human Environment: Pushing Plant–Hominin Co-Evolution Back to the Pliocene. *J. Hum. Evol.* **2015**, *79*, 150–157. [CrossRef]
12. Wadley, L.; Sievers, C.; Bamford, M.; Goldberg, P.; Berna, F.; Miller, C. Middle Stone Age Bedding Construction and Settlement Patterns at Sibudu, South Africa. *Science* **2011**, *334*, 1388–1391. [CrossRef] [PubMed]
13. Alexiades, M. Ethnobotany in the Third Millennium: Expectations and Unresolved Issues. *Delpinoa* **2003**, *45*, 15–28.
14. Tuxill, J.D.; Nabhan, G.P. *People, Plants, and Protected Areas: A Guide to in Situ Management*; Earthscan: London, UK, 2001; ISBN 9781853837821.
15. Food and Agriculture Organization of the United Nations. Food and Agriculture: Key to Achieving the 2030 Agenda for Sustainable Development. 2009. Available online: http://www.fao.org/policy-support/tools-and-publications/resources-details/en/c/422261/ (accessed on 17 February 2021).
16. The Sustainable Development Goals Report. Available online: https://unstats.un.org/sdgs/report/2017/ (accessed on 17 February 2021).
17. Snyder, H. Literature Review as a Research Methodology: An Overview and Guidelines. *J. Bus. Res.* **2019**, *104*, 333–339. [CrossRef]
18. Kugley, S.; Wade, A.; Thomas, J.; Mahood, Q.; Jørgensen, A.K.; Hammerstrøm, K.; Sathe, N. Searching for Studies: A Guide to Information Retrieval for Campbell Systematic Reviews. *Campbell Syst. Rev.* **2017**, *13*, 1–73. [CrossRef]
19. Newson, R.S.; Lion, R.; Crawford, R.J.; Curtis, V.; Elmadfa, I.; Feunekes, G.I.; Hicks, C.; van Liere, M.; Lowe, C.; Meijer, G.W.; et al. Behaviour Change for Better Health: Nutrition, Hygiene and Sustainability. *BMC Public Health* **2013**, *13*, S1. [CrossRef] [PubMed]
20. Siddiqui, F.; Salam, R.A.; Lassi, Z.S.; Das, J.K. The Intertwined Relationship between Malnutrition and Poverty. *Front. Public Health* **2020**, *8*, 453. [CrossRef]
21. One Health, Zero Hunger. Available online: https://www.globalhungerindex.org/issues-in-focus/2020.html (accessed on 16 February 2021).
22. How to Feed the World in 2050—An Issue Brief Prepared for the High Level Expert Forum. 2009. Available online: http://www.fao.org/wsfs/forum2050/wsfs-forum/en/ (accessed on 17 February 2021).
23. Food and Agriculture Organization of the United Nations. An Introduction to the Basic Concepts of Food Security (FAO, Rome). Available online: http://www.fao.org/documents/card/en/c/2357d07c-b359-55d8-930a-13060cedd3e3/ (accessed on 17 February 2021).
24. Talberth, J.; Susan, L. Reviving Dormant Ethnobotany: The role of women and plant knowledge in a food secure world. In Proceedings of the 36th Annual Conference of the Society of Ethnobiology in University of North Texas, Denton, TX, USA, 15–18 May 2013. Available online: https://sustainable-economy.org/wp-content/uploads/2013/06/Reviving-Dormant-Ethnobotany.pdf (accessed on 10 March 2021).
25. Dwivedi, S.L.; Lammerts van Bueren, E.T.; Ceccarelli, S.; Grando, S.; Upadhyaya, H.D.; Ortiz, R. Diversifying Food Systems in the Pursuit of Sustainable Food Production and Healthy Diets. *Trends Plant Sci.* **2017**, *22*, 842–856. [CrossRef] [PubMed]
26. Shaheen, S.; Ahmad, M.; Haroon, N. Diversity of Edible Wild Plants: Global Perspectives. In *Edible Wild Plants: An Alternative Approach to Food Security*; Springer International Publishing: Cham, Germany, 2017; pp. 59–64; ISBN 9783319630366.

27. Ray, A.; Ray, R.; Sreevidya, E.A. How Many Wild Edible Plants Do We Eat—Their Diversity, Use, and Implications for Sustainable Food System: An Exploratory Analysis in India. *Front. Sustain. Food Syst.* **2020**, *4*, 56. [CrossRef]
28. Thakur, A.; Singh, S.; Puri, S. Exploration of Wild Edible Plants Used as Food by Gaddis-A Tribal Community of the Western Himalaya. *Sci. World J.* **2020**, *2020*, 1–6. [CrossRef] [PubMed]
29. Ashagre, M.; Asfaw, Z.; Kelbessa, E. Ethnobotanical Study of Wild Edible Plants in Burji District, Segan Area Zone of Southern Nations, Nationalities and Peoples Region (SNNPR), Ethiopia. *J. Ethnobiol. Ethnomed.* **2016**, *12*, 32. [CrossRef]
30. Duguma, H.T. Wild Edible Plant Nutritional Contribution and Consumer Perception in Ethiopia. *Int. J. Food Sci.* **2020**, *2020*, 1–16. [CrossRef]
31. Zhang, L.; Chai, Z.; Zhang, Y.; Geng, Y.; Wang, Y. Ethnobotanical Study of Traditional Edible Plants Used by the Naxi People during Droughts. *J. Ethnobiol. Ethnomed.* **2016**, *12*, 39. [CrossRef]
32. Pawera, L.; Khomsan, A.; Zuhud, E.A.; Hunter, D.; Ickowitz, A.; Polesny, Z. Wild Food Plants and Trends in Their Use: From Knowledge and Perceptions to Drivers of Change in West Sumatra, Indonesia. *Foods* **2020**, *9*, 1240. [CrossRef] [PubMed]
33. Psaroudaki, A.; Nikoloudakis, N.; Skaracis, G.; Katsiotis, A. Genetic Structure and Population Diversity of Eleven Edible Herbs of Eastern Crete. *J. Biol. Res. Thessalon.* **2015**, *22*, 7. [CrossRef]
34. Pieroni, A.; Nebel, S.; Santoro, R.F.; Heinrich, M. Food for Two Seasons: Culinary Uses of Non-Cultivated Local Vegetables and Mushrooms in a South Italian Village. *Int. J. Food Sci. Nutr.* **2005**, *56*, 245–272. [CrossRef]
35. Ghirardini, M.; Carli, M.; del Vecchio, N.; Rovati, A.; Cova, O.; Valigi, F.; Agnetti, G.; Macconi, M.; Adamo, D.; Traina, M.; et al. The Importance of a Taste. A Comparative Study on Wild Food Plant Consumption in Twenty-One Local Communities in Italy. *J. Ethnobiol. Ethnomed.* **2007**, *3*, 22. [CrossRef]
36. Pinela, J.; Carvalho, A.M.; Ferreira, I.C.F.R. Wild Edible Plants: Nutritional and Toxicological Characteristics, Retrieval Strategies and Importance for Today's Society. *Food Chem. Toxicol.* **2017**, *110*, 165–188. [CrossRef]
37. Osorio, N.W. *Soil Nutrient Management in the Tropics*; Universidad Nacional de Colombia: Medellin, Colombia, 2012.
38. León, J.D.; Osorio, N.W. Role of Litter Turnover in Soil Quality in Tropical Degraded Lands of Colombia. *Sci. World J.* **2014**, *2014*, 693981. [CrossRef]
39. Lamb, D.; Erskine, P.D.; Parrotta, J.A. Restoration of Degraded Tropical Forest Landscapes. *Science* **2005**, *310*, 1628–1632. [CrossRef] [PubMed]
40. Bini, C. Soil: A Precious Natural Resource. In *Conservation of Natural Resources*; Kudrow, N.J., Ed.; Nova Science Publishers: Hauppauge, NY, USA, 2009; pp. 1–48. ISBN 9781607411789.
41. Lal, R. Restoring Soil Quality to Mitigate Soil Degradation. *Sustainability* **2015**, *7*, 5875–5895. [CrossRef]
42. Scherr, S.J. The Future Food Security and Economic Consequences of Soil Degradation in the Developing World. In *Response to Land Degradation*; Bridges, M.E., Hannam, I.D., Oldeman, R.L., Penning de Vries, F.W.T., Scherr, S.J., Sombatpanit, S., Eds.; Broken Sound Plaza: Seattle, WA, USA, 2009; pp. 153–170.
43. Quarles, W. Pesticides and Honey Bee Colony Collapse Disorder. *IPM Pract.* **2008**, *30*, 1–10.
44. Quarles, W. Protecting Native Bees and Other Pollinators. *Common Sense Pest Control Q.* **2008**, *24*, 4–14.
45. Stokstad, E. Field Research on Bees Raises Concern about Low-Dose Pesticides. *Science* **2012**, *335*, 1555. [CrossRef]
46. Quarles, W. Pesticides and Honey Bee Death and Decline. *IPM Pract.* **2011**, *33*, 1–8.
47. Pollinator Health and Colony Collapse Disorder. Hearing to Review the Status of Pollinator Health Including Colony Collapse Disorder Hearing before the Subcommittee on Horticulture and Organic Agriculture of the Committee on Agriculture House of Representatives; Serial No. 110-39. 26 June 2008. Available online: https://www.govinfo.gov/content/pkg/CHRG-110hhrg50679/html/CHRG-110hhrg50679.html (accessed on 13 August 2020).
48. Spivak, M.; Mader, E.; Vaughan, M.; Euliss, N.H. The Plight of the Bees. *Environ. Sci. Technol.* **2011**, *45*, 34–38. [CrossRef] [PubMed]
49. Colony Collapse Disorder and Pollinator Decline. Hearing to Review the Status of Pollinator Health Including Colony Collapse Disorder Hearing before the Subcommittee on Horticulture and Organic Agriculture of the Committee on Agriculture House of Representatives. 29 March 2007. Available online: https://www.govinfo.gov/content/pkg/CHRG-110hhrg36465/html/CHRG-110hhrg36465.html (accessed on 12 August 2020).
50. Schacker, M. *A Spring without Bees: How Colony Collapse Disorder Has Endangered Our Food Supply*; Lyons Press: Guilford, NC, USA, 2008; ISBN 9781599216003.
51. Aktar, W.; Sengupta, D.; Chowdhury, A. Impact of Pesticides Use in Agriculture: Their Benefits and Hazards. *Interdiscip. Toxicol.* **2009**, *2*, 1–12. [CrossRef] [PubMed]
52. The Conservation of Lands in Asia and the Pacific. *CLASP: A Framework for Action*; FAO: Rome, Italy, 1995; Available online: http://www.fao.org/docrep/v9909e/v9909e02.html (accessed on 12 August 2020).
53. Bhan, S. Land Degradation and Integrated Watershed Management in India. *Int. Soil Water Conserv. Res.* **2013**, *1*, 49–57. [CrossRef]
54. Spiertz, H. Challenges for Crop Production Research in Improving Land Use, Productivity and Sustainability. *Sustainability* **2013**, *5*, 1632–1644. [CrossRef]
55. Winterbottom, R.; Reij, C.; Garrity, D.; Glover, J.; Hellums, D.; McGahuey, M.; Scherr, S. *Improving Land and Water Management. Creating a Sustainable Food Future*; Installment Four, Washington, Working Paper; World Resources Institute: Washington, DC, USA, 2013. Available online: https://www.wri.org/publication/improving-land-and-water-management (accessed on 6 August 2020).

56. Gil, J.D.B.; Reidsma, P.; Giller, K.; Todman, L.; Whitmore, A.; van Ittersum, M. Sustainable Development Goal 2: Improved Targets and Indicators for Agriculture and Food Security. *Ambio* **2019**, *48*, 685–698. [CrossRef] [PubMed]
57. Altieri, M.A. Linking Ecologists and Traditional Farmers in the Search for Sustainable Agriculture. *Front. Ecol. Environ.* **2004**, *2*, 35–42. [CrossRef]
58. Barrera-Bassols, N.; Zinck, J.A. Ethnopedology: A Worldwide View on the Soil Knowledge of Local People. *Geoderma* **2003**, *111*, 171–195. [CrossRef]
59. Mikkelsen, J.H.; Langohr, R. Indigenous knowledge about soils and a sustainable crop production, a case study from the Guinea Woodland Savannah (Northern Region, Ghana). *Geogr. Tidsskr. Dan. J. Geogr.* **2004**, *104*, 13–26. [CrossRef]
60. Grzywacz, D.; Stevenson, P.C.; Mushobozi, W.L.; Belmain, S.; Wilson, K. The Use of Indigenous Ecological Resources for Pest Control in Africa. *Food Secur.* **2014**, *6*, 71–86. [CrossRef]
61. Bentley, J.W.; Thiele, G. Bibliography: Farmer Knowledge and Management of Crop Disease. *Agric. Hum. Values* **1999**, *16*, 75–81. [CrossRef]
62. Brush, S.B. Ethnoecology, Biodiversity, and Modernization in Andean Potato Agriculture. *J. Ethnobiol.* **1992**, *12*, 161–185.
63. Brush, S.B. *Farmers' Rights and the Protection of Traditional Agricultural Knowledge*; CAPRi WORKING PAPER NO. 36, 1–41; International Food Policy Research Institute: Washington, DC, USA, 2005. Available online: https://www.ifpri.org/publication/farmers-rights-andprotection-traditional-agricultural-knowledge (accessed on 10 March 2021).
64. Wolf, P.; Medin, D.L. Measuring the evolution and devolution of folk-biological knowledge. In *On Biocultural Diversity: Linking Language, Knowledge and the Environment*; Maffi, L., Ed.; Smithsonian Institution Press: Washington, DC, USA, 2001; pp. 212–227.
65. Turner, N.J.; Turner, K.L. "Where Our Women Used to Get the Food": Cumulative Effects and Loss of Ethnobotanical Knowledge and Practice; Case Study from Coastal British Columbia This Paper Was Submitted for the Special Issue on Ethnobotany, Inspired by the Ethnobotany Symposium Organized by Alain Cuerrier, Montreal Botanical Garden, and Held in Montreal at the 2006 Annual Meeting of the Canadian Botanical Association. *Botany* **2008**, *86*, 103–115. [CrossRef]
66. Rockström, J.; Steffen, W.; Noone, K.; Persson, Å.; Chapin, F.S.; Lambin, E.F.; Lenton, T.M.; Scheffer, M.; Folke, C.; Schellnhuber, H.J.; et al. A Safe Operating Space for Humanity. *Nature* **2009**, *461*, 472–475. [CrossRef] [PubMed]
67. Dove, M.R. The life-cycle of indigenous knowledge, and the case of natural rubber production. In *Indigenous Environmental Knowledge and Its Transformations*; Bicker, A., Ellen, R., Parkes, P., Eds.; Harwood Academic Publishers: Amsterdam, The Netherlands, 2000; pp. 213–251.
68. Brush, S.B. Farmers' Rights and Protection of Traditional Agricultural Knowledge. *World Dev.* **2007**, *35*, 1499–1514. [CrossRef]
69. Maldonado, J.; Bennett, T.M.B.; Chief, K.; Cochran, P.; Cozzetto, K.; Gough, B.; Redsteer, M.H.; Lynn, K.; Maynard, N.; Voggesser, G. Engagement with Indigenous Peoples and Honoring Traditional Knowledge Systems. *Clim. Chang.* **2016**, *135*, 111–126. [CrossRef]
70. Subramanian, S.M.; Pisupati, B. *Traditional Knowledge in Policy and Practice: Approaches to Development and Human Well-Being*; United Nations University Press: Tokyo, Japan, 2010; ISBN 9789280811919.
71. Garnatje, T.; Peñuelas, J.; Vallès, J. Ethnobotany, Phylogeny, and 'Omics' for Human Health and Food Security. *Trends Plant Sci.* **2017**, *22*, 187–191. [CrossRef]
72. Zhu, F.; Qin, C.; Tao, L.; Liu, X.; Shi, Z.; Ma, X.; Jia, J.; Tan, Y.; Cui, C.; Lin, J.; et al. Clustered Patterns of Species Origins of Nature-Derived Drugs and Clues for Future Bioprospecting. *Proc. Natl. Acad. Sci. USA* **2011**, *108*, 12943–12948. [CrossRef]
73. Hussain, M.d.S.; Rahman, M.d.A.; Fareed, S.; Ansari, S.; Ahmad, I.; Mohd, S. Current Approaches toward Production of Secondary Plant Metabolites. *J. Pharm. Bioallied Sci.* **2012**, *4*, 10. [CrossRef]
74. Ekor, M. The Growing Use of Herbal Medicines: Issues Relating to Adverse Reactions and Challenges in Monitoring Safety. *Front. Pharmacol.* **2014**, *4*, 177. [CrossRef] [PubMed]
75. Bussmann, R.W. Ethnobotany and Biodiversity Conservation. In *Modern Trends in Applied Terrestrial Ecology*; Ambasht, N.K., Ambasht, R.S., Eds.; Springer: Boston, MA, USA, 2002; pp. 343–360. [CrossRef]
76. Baker, J.T.; Borris, R.P.; Carté, B.; Cordell, G.A.; Soejarto, D.D.; Cragg, G.M.; Gupta, M.P.; Iwu, M.M.; Madulid, D.R.; Tyler, V.E. Natural Product Drug Discovery and Development: New Perspectives on International Collaboration. *J. Nat. Prod.* **1995**, *58*, 1325–1357. [CrossRef]
77. Farnsworth, N.R.; Akerele, O.; Bingel, A.S.; Soejarto, D.D.; Guo, Z. Medicinal plants in therapy. *Bull. World Health Organ.* **1985**, *63*, 965–981. [CrossRef]
78. Rao, K.V. Taxol and Related Taxanes. I. Taxanes of Taxus Brevifolia Bark. *Pharm. Res.* **1993**, *10*, 521–524. [CrossRef]
79. Abourashed, E.A.; El-Alfy, A.T.; Khan, I.A.; Walker, L. Ephedra in Perspective—A Current Review. *Phytother. Res.* **2003**, *17*, 703–712. [CrossRef] [PubMed]
80. Sarpras, M.; Gaur, R.; Sharma, V.; Chhapekar, S.S.; Das, J.; Kumar, A.; Yadava, S.K.; Nitin, M.; Brahma, V.; Abraham, S.K.; et al. Comparative Analysis of Fruit Metabolites and Pungency Candidate Genes Expression between Bhut Jolokia and Other *Capsicum* Species. *PLoS ONE* **2016**, *11*, e0167791. [CrossRef]
81. Su, X.Z.; Miller, L.H. The Discovery of Artemisinin and the Nobel Prize in Physiology or Medicine. *Sci. China Life Sci.* **2015**, *58*, 1175–1179. [CrossRef] [PubMed]
82. Alves, R.R.; Rosa, I.M. Biodiversity, Traditional Medicine and Public Health: Where Do They Meet? *J. Ethnobiol. Ethnomed.* **2007**, *3*, 14. [CrossRef] [PubMed]

83. Uniyal, S.K.; Singh, K.; Jamwal, P.; Lal, B. Traditional Use of Medicinal Plants among the Tribal Communities of Chhota Bhangal, Western Himalaya. *J. Ethnobiol. Ethnomed.* **2006**, *2*, 14. [CrossRef] [PubMed]
84. Pan, S.Y.; Litscher, G.; Gao, S.H.; Zhou, S.F.; Yu, Z.-L.; Chen, H.Q.; Zhang, S.F.; Tang, M.-K.; Sun, J.-N.; Ko, K.-M. Historical Perspective of Traditional Indigenous Medical Practices: The Current Renaissance and Conservation of Herbal Resources. *Evid. Based Complementary Altern. Med.* **2014**, *2014*, 525340. [CrossRef]
85. Wright, R.M.; Kapfhammer, W.; Wiik, F.B. The clash of cosmographies: Indigenous societies and project collaboration—Three ethnographic cases (Kaingang, Sateré-Mawé, Baniwa). *Vibrant Virtual Braz. Anthropol.* **2012**, *9*, 382–450. [CrossRef]
86. Saslis-Lagoudakis, C.H.; Savolainen, V.; Williamson, E.M.; Forest, F.; Wagstaff, S.J.; Baral, S.R.; Watson, M.F.; Pendry, C.A.; Hawkins, J.A. Phylogenies Reveal Predictive Power of Traditional Medicine in Bioprospecting. *Proc. Natl. Acad. Sci. USA* **2012**, *109*, 15835–15840. [CrossRef] [PubMed]
87. Hawkins, J.A.; Teixidor-Toneu, I. Defining Ethnobotanical Convergence. *Trends Plant Sci.* **2017**, *22*, 639–640. [CrossRef] [PubMed]
88. Obakiro, S.B.; Kiprop, A.; Kowino, I.; Kigondu, E.; Odero, M.P.; Omara, T.; Bunalema, L. Ethnobotany, Ethnopharmacology, and Phytochemistry of Traditional Medicinal Plants Used in the Management of Symptoms of Tuberculosis in East Africa: A Systematic Review. *Trop. Med. Health* **2020**, *48*, 68. [CrossRef] [PubMed]
89. Schilmiller, A.L.; Pichersky, E.; Last, R.L. Taming the Hydra of Specialized Metabolism: How Systems Biology and Comparative Approaches Are Revolutionizing Plant Biochemistry. *Curr. Opin. Plant Biol.* **2012**, *15*, 338–344. [CrossRef]
90. Kumar, A.; Kumar, S.; Thomas, T.D.; Ramchiary, N.; Swamy, M.K.; Ahmad, I. Linking Omics Approaches to Medicinal Plants and Human Health. In *Natural Bio-Active Compounds: Volume 3: Biotechnology, Bioengineering, and Molecular Approaches*; Akhtar, M.S., Swamy, M.K., Eds.; Springer: Singapore, 2019; pp. 31–57. ISBN 9789811374388.
91. Sofowora, A.; Ogunbodede, E.; Onayade, A. The Role and Place of Medicinal Plants in the Strategies for Disease Prevention. *Afr. J. Tradit. Complementary Altern. Med.* **2013**, *10*, 210–229. [CrossRef]
92. Karunamoorthi, K.; Jegajeevanram, K.; Vijayalakshmi, J.; Mengistie, E. Traditional Medicinal Plants: A Source of Phytotherapeutic Modality in Resource-Constrained Health Care Settings. *J. Evid. Based Complementary Altern. Med.* **2013**, *18*, 67–74. [CrossRef]
93. Plotkin, M.J. *Tales of a Shaman's Apprentice: An Ethnobotanist Searches for New Medicines in the Amazon Rain Forest*; Viking Penguin: New York, NY, USA, 1993; p. 318.
94. Schipmann, U.; Leaman, D.J.; Cunningham, A.B.; Walter, S. Impact of Cultivation and Gathering of Medicinal Plants on Biodiversity: Global Trends and Issues. *Acta Hortic.* **2005**, 31–44. [CrossRef]
95. Miththapala, S. (Ed.) *Conserving Medicinal Species: Securing a Healthy Future; Ecosystems and Livelihoods Group*; Asia Asian Regional Office The World Conservation Union (IUCN): Colombo, Sri Lanka, 2006; ISBN 9789558177419.
96. Bentley, R. *Medicinal Plants*; Domville-Fife Press: London, UK, 2010.
97. Cunningham, A.B. The role of ethnobotany and indigenous knowledge in conservation of plant genetic resources. *Dinteria* **1992**, *23*, 119–131.
98. Phillips, O.; Gentry, A.H. The Useful Plants of Tambopata, Peru: II. Additional Hypothesis Testing in Quantitative Ethnobotany. *Econ. Bot.* **1993**, *47*, 33–43. [CrossRef]
99. Phillips, O.; Gentry, A.H. The Useful Plants of Tambopata, Peru: I. Statistical Hypotheses Tests with a New Quantitative Technique. *Econ. Bot.* **1993**, *47*, 15–32. [CrossRef]
100. De Albuquerque, U.P.; de Sousa Araújo, T.A.; Ramos, M.A.; do Nascimento, V.T.; de Lucena, R.F.P.; Monteiro, J.M.; Alencar, N.L.; de Lima Araújo, E. How Ethnobotany Can Aid Biodiversity Conservation: Reflections on Investigations in the Semi-Arid Region of NE Brazil. *Biodivers. Conserv.* **2009**, *18*, 127–150. [CrossRef]
101. Peters, C.M.; Alexiades, M.; Laird, S.A. Indigenous Communities: Train Local Experts to Help Conserve Forests. *Nature* **2012**, *481*, 443. [CrossRef] [PubMed]
102. Peters, C.M. Economic Botany and Management Potential of Neotropical Seasonally Dry Forests. In *Seasonally Dry Tropical Forests: Ecology and Conservation*; Dirzo, R., Young, H.S., Mooney, H.A., Ceballos, G., Eds.; Island Press: Washington, DC, USA, 2011; pp. 239–257. [CrossRef]
103. Posey, D.E. Traditional knowledge, conservation and the Rain Forest Harvest. In *Sustainable Harvest and Marketing of Rainforest Products*; Plotkin, M.J., Famolare, L., Eds.; Island Press: Washington, DC, USA, 1992; pp. 46–50.
104. Pimbert, M.P.; Pretty, J.N. Parks, People and Professionals: Putting "Participation" into Protected Area Management"; UNRISD Discussion Paper 57. 1995. Available online: https://www.unrisd.org/unrisd/website/document.nsf/(httpPublications)/FD45EA7929FCAECC80256B67005B66D8?OpenDocument (accessed on 10 March 2021).
105. World Health Organisation (WHO). Address of Director-General (Dr Margaret Chan) of the World Health Organization at the WHO Congress on Traditional Medicine. Available online: http://www.who.int/dg/speeches/2008/20081107/en/ (accessed on 13 August 2020).

*sustainability*

MDPI

*Article*

# Using Spatial Patterns of COVID-19 to Build a Framework for Economic Reactivation

**Renato Quiliche [1], Rafael Rentería-Ramos [2,*], Irineu de Brito Junior [3,4], Ana Luna [1] and Mario Chong [1]**

1 Facultad de Ingeniería, Universidad del Pacífico, Lima 150113, Peru; r.quilichealtamirano@up.edu.pe (R.Q.); ae.lunaa@up.edu.pe (A.L.); m.chong@up.edu.pe (M.C.)
2 Escuela de Ciencias Básicas, Tecnologías e Ingenierías, Universidad Nacional Abierta y a Distancia, Bogotá 110311, Colombia
3 Environmental Engineering Department, Sao Paulo State University, Sao Jose dos Campos 12247-004, Brazil; irineu.brito@unesp.br
4 Graduate Program in Logistics Systems Engineering, Sao Paulo University, Sao Paulo 05508-060, Brazil
* Correspondence: rafael.renteria@unad.edu.co

**Abstract:** In this article, we propose an application of humanitarian logistics theory to build a supportive framework for economic reactivation and pandemic management based on province vulnerability against COVID-19. The main research question is which factors are related to COVID-19 mortality between Peruvian provinces? We conduct a spatial regression analysis to explore which factors determine the differences in COVID-19 cumulative mortality rates for 189 Peruvian provinces up to December 2020. The most vulnerable provinces are characterized by having low outcomes of long-run poverty and high population density. Low poverty means high economic activity, which leads to more deaths due to COVID-19. There is a lack of supply in the set of relief goods defined as Pandemic Response and Recovery Supportive Goods and Services (PRRSGS). These goods must be delivered in order to mitigate the risk associated with COVID-19. A supportive framework for economic reactivation can be built based on regression results and a delivery strategy can be discussed according to the spatial patterns that we found for mortality rates.

**Keywords:** humanitarian logistics; pandemic; economic reactivation; spatial modelling

**Citation:** Quiliche, R.; Rentería-Ramos, R.; de Brito Junior, I.; Luna, A.; Chong, M. Using Spatial Patterns of COVID-19 to Build a Framework for Economic Reactivation. *Sustainability* **2021**, *13*, 10092. https://doi.org/10.3390/su131810092

Academic Editor: Shervin Hashemi

Received: 26 July 2021
Accepted: 3 September 2021
Published: 9 September 2021



## 1. Introduction

Since 2020, the novel coronavirus has caused changes in the livelihoods of the people and the nature of work. These changes motivated a set of public policies that aims to reactivate economies while minimizing the total number of contagions to reduce mortality rates. For a case study of Peru, the government adopted a preventive framework including early lockdowns, curfews, prolonged mobility restrictions between regions, and prohibitions to agglomeration in closed spaces [1–5]. After these lockdowns, a set of subsidiary policies for the population and companies [1,2,6], and macroeconomic policies oriented toward reactivate consumption and production were implemented in order to mitigate the impacts of COVID-19 [7].

The COVID-19 pandemic had an impact both on the people directly affected by the virus, who used a lot of resources in their recovery, and on those indirectly affected, who suffered from the scarcity of income sources due to the short-run contractions of the economy, which includes the livelihoods of healthy people affected by lockdowns. Recent literature aimed to generate evidence on the direct and indirect impacts of COVID-19 on variables such as macroeconomic consumption, investment, production, microeconomic poverty, health, and labor, among others [8–10]. This branch of COVID-19 literature proposed policies oriented toward mitigating the impacts of COVID-19 such as the optimal duration of quarantines and labor-oriented policies, among others. However, what is missing in the general COVID-19 literature is a humanitarian logistics perspective, which is important because it aims to reduce human suffering in the context of a pandemic [11–13] and allows us to build a supportive framework for economic reactivation based on vulnerability assessment and disaster management principles [14].

Another branch of COVID-19 literature explores the factors affecting the rates of contagions or mortality by COVID-19 [15,16]. This branch uses statistical modeling and supporting machine learning methodologies in order to determine the relationship between a set of observable factors and the COVID-19 mortality rates [17,18]. This literature contributes to the empirical evidence using vulnerability assessment, while most of the COVID-19 related literature is based on calibrated methodologies. Empirical methodologies still represent a gap in the COVID-19 literature.

This work focuses on supporting economic reactivation and pandemic management in the post-lockdown context. A humanitarian logistics supportive framework is built based on provinces' vulnerability against COVID-19 [14]. This vulnerability according to the literature review could be directly related to economic activity [15–17], socio-demographic characteristics [18–20], the population's health [21–23], and health system performance [24–27]. The vulnerability against COVID-19 is directly related to the mortality rates, which is the variable of interest in this study since mortality focuses on symptomatic population, and it represents a better proxy for the demand for relief goods. The main research question is which factors are related to COVID-19 mortality between Peruvian provinces? Additionally, with this information, how can a supportive framework be implemented for underpinning the economic reactivation? We conduct a regression analysis to explore which factors determine the differences in COVID-19 cumulative mortality rates for 189 Peruvian provinces up to December 2020. In order to contribute to the empirical literature, in this work, we also test the significance of the spatial interaction effects of the COVID-19 vulnerability drivers extracted from the literature [15,16,24]. From a theoretical perspective, there are several reasons to suspect that these vulnerability drivers have spatial interaction effects, as socioeconomic variables such as wealth or employment tend to have spatial spillovers [28,29] and these could be correlated with COVID-19 mortality. Controlling these spatial effects helps reduce the bias on the coefficients, so a post-estimation analysis is more reliable. This is an important step because model predictions and vulnerability assessments are used as the basis of our supportive framework for economic reactivation.

After the regression analysis, we use the results to build support for economic reactivation, which is based on the delivery of essential goods and services to face the pandemic. The type of goods is drawn from literature and is defined as Pandemic Response and Recovery Supportive Goods and Services (PRRSGS). We classify the demand for these essential supplies according to quartiles low, medium, high, and very high demand, and this classification is used to assess economic reactivation based on the level of vulnerability. For the case of seven provinces with zero deaths, we use an elastic-net regression (ENR) to forecast their demand classification. This methodology covers the four phases of humanitarian logistics: risk mitigation, preparedness, response, and recovery [30]. Risk mitigation can be achieved by implementing policies based on the vulnerability drivers (seen in regression analysis), and disaster preparedness policies can be developed for the zero-death provinces, where there has been no infection yet. Response and recovery policies require defining PRRSGS and tracing a delivery strategy based on demand estimation.

The main contribution of this work is methodological, as a humanitarian logistics perspective is missing in COVID-19 literature. This is in the form of a proposal of an alternative framework for economic reactivation and pandemic management, which includes supportive policies that prioritize the vulnerable. This work also contributes to the empirical literature with the estimation of spatial dependence models; a statistical modeling approach [31]; and ENR, a conventional supervised machine learning technique. To the best of our knowledge, both applications are missing in the empirical literature.

The remainder of this study is structured as follows. Section 2 presents the relevant literature and the theoretical framework. Section 3 describes the data collection procedures and the data processing methods to estimate provinces' vulnerability to COVID-19 and the demand for PRRSGS. Section 4 outlines the main results and discusses the hypothesis. Section 5 addresses the implications of our results for policymakers. Finally, Section 6 presents the conclusions and recommendations for future research.

## 2. Literature Review

First, in this literature review, we show how humanitarian logistics serves disaster management, specifically in the case of a pandemic, considered as a type of disaster. Second, we review some work in which the sets of determinants for the cases and number of deaths resulting from COVID-19 are analyzed. Third, vulnerability to COVID-19 mortality is defined based on those factors, and a contribution to the definition of vulnerability based on the effects of spatial interaction is established.

### *2.1. A Humanitarian Logistics Framework for Pandemic Assessment*

The problem can be defined as follows: there are different outcomes of COVID-19 deaths between Peruvian provinces that are caused by a set of explanative factors; the main goal of this paper is to find the determinants of COVID-19 mortality rates for Peruvian provinces in order to guide disaster response and recovery at the national level. However, the number deaths due to COVID-19 deaths is in a strong relationship with the number of infections—causally, infection leads to mortality—but focusing our attention on the mortality rates is how we can assess the severity of a pandemic and guide response and recovery.

Governments around the world are applying social policies oriented toward mitigating the direct and indirect impacts of COVID-19 as an economic catastrophe [32,33]. Direct impacts include people infected by COVID-19, as they have to expend time and resources for their recovery. Indirect impacts arise from the changes in the operations, with the economies contracting themselves as a consequence of the pandemic; therefore, consumption, production, and thus the livelihoods of people are affected [8,34]. Human suffering rises [12] because, in most economies, supply could not respond to rising demands regarding health care services (hospitalization and intensive care units), key health care goods (oxygen and automatic respirators), and social aid (because with lockdowns, unemployment causes hunger and raises the incidence of poverty). All of the aforementioned lead to an increase in deprivation costs that need to be mitigated [12]. In this work, the concept of deprivation costs is emphasized.

The central aim of the humanitarian logistics discipline is to minimize the deprivation costs or the lack of essential goods to survive the aftermath of a disaster [11–13]. Then, four phases of disaster management are followed: risk mitigation, preparedness, response, and recovery [30]. This discipline also tries to prioritize cases where the costs of deprivation are the highest or where help is required immediately [35]. For our case, this happens when a community faces high COVID-19 mortality because mortality is a consequence of the scarcity of essential goods and services, and of essential supplies, in general, to recover from COVID-19 symptoms.

A logistics-based humanitarian framework is especially useful for pandemic outbreaks management due to the need to improve demand response and to prioritize vulnerable populations. However, for the Peruvian case, policymakers are not considering the implementation of this framework [1]. Instead, the government adopted a preventive framework, including early closures, curfews, prolonged restrictions on mobility between regions, and prohibitions on crowding in closed spaces [2–5]. Given the restrictive nature of the first measures, it is counterintuitive that Peru is one of the most affected countries by COVID-19 due to its number of cases and deaths.

Furthermore, the humanitarian logistics conceptual framework allows us to identify and prioritize a vulnerable population based on the level of deprivation costs, which are specially related to higher COVID-19 mortality rates. Higher mortality rates can be explained by high economic activity indicators such as employment or internal mobility. Especially in informal economies, the working population cannot work from home and all of people who maintain face-to-face interactions with other people are especially vulnerable [24]. COVID-19 deaths are more likely if a set of vulnerability characteristics are met by provinces (see Section 2.3 for more details). It is worth mentioning that we emphasize the demand for an analysis on the deaths rather than the contagions because we want to avoid the asymptomatic population and prioritize the resolution of human suffering.

The pandemic response and recovery framework that was applied in Peru until the end of 2020 included a set of subsidiary policies for the population and companies [3,4,6], and macroeconomic policies oriented toward reactivate consumption and production [7]. In addition to this, the humanitarian logistics proposal for response and recovery phases while economic reactivation is carried out is a supportive framework aimed at reactivating the economy by mitigating the impacts of COVID-19 (based on the idea that maintaining low mortality rates is required to carry out economic reactivation). This framework is based on the estimation of the demand for a set of PRRSGSs at the provincial level, which is based on province vulnerability, so that supply can respond to the demand. In Table 1, we list the PRRSGS and consider the goods from [36,37]:

**Table 1.** Pandemic recovery supportive goods and services.

| Types | Support for Response and Recovery from COVID-19 Pandemic [36] | COVID-19 Diagnosis and Treatment and Disinfection and Sterilization Medical Equipment [37] |
|---|---|---|
| Goods and services | Oxygen, automatic respirators, dexamethasone, prednisone, acetaminophen, antibiotics (azithromycin, levofloxacin), anti-clotting medication, KN95 masks, gloves, intensive care beds, alcohol, other disinfection products. Health care, emergency transport, hospitals, doctors. Funerary protocols (incineration) and transport. | Stethoscope, thermometer, sphygmomanometer, oxygen flowmeter, oxygen saturation monitor, air disinfection machine, crash cart, defibrillator, monitor, micro-injection pump, sputum elimination machine, noninvasive ventilator, invasive ventilator, continuous renal replacement therapy (CRRT), extracorporeal membrane oxygenation (ECMO), designated computed tomography, polymerase chain reaction (PCR) machine, nucleic acid detector, ultraviolet disinfection machine, anesthesia machine, ventilator circuit disinfection machine, infrared thermal imager, and forehead thermometer. |

Source: [36,37]

The responding supply has a significant impact on the deprivation costs and serves to mitigate injuries for the response phase and to accelerate the transition to a business-as-usual scheme regarding the recovery phase. In zero-contagion/death provinces, estimating its possible demand serves preparedness policies regarding the COVID-19 pandemic.

### 2.2. Previous Works on Determinants of COVID-19 Cases and Deaths

Since the beginning of the pandemic outbreak, a large amount of literature has been published. One branch explores the determinants of COVID-19 cases [15–21] and of deaths due to COVID-19 [22,23]. Ref. [15] found age, disability, language, race, and employment to be the determinants of COVID-19 cases for the United States. [16] discussed the importance of a disadvantaged socioeconomic position on COVID-19 disease and mortality, concluding that high-quality data on socioeconomic characteristics are needed to generate more evidence about this relationship. For Ceará, a state of Brazil, [17] found that the number of intensive care beds has spatial patterns, but they do not coincide with the ones for COVID-19 cases. Ref. [18] concluded that socioeconomic status is essential to staying healthy in the context of a pandemic; thus, poverty may lead to more contagions in the US. Ref. [19] found that per capita income, average household size, population density, and minority composition are significant predictors of COVID-19 cases in nursing homes in the US. Ref. [20] found that, in the US, states with little tolerance for deviance from enforced rules saw faster early epidemic growth and that population density is a significant predictor for COVID-19 cases. Ref. [21] focused on the significance of geographical factors, finding population density and topographic altitude as predictors of COVID-19 cases. Ref. [22] showed that pulmonary embolism was the main determinant of excess out-of-hospital deaths during COVID-19 in Paris, France. Finally, ref. [23] found that population density, the proportion of elderly residents, and percent population tested are key predictors for both COVID-19 cases and deaths.

Another branch of literature has suggested the adoption of data-driven machine learning (ML)-based frameworks for pandemic assessment, these frameworks focused on the estimation of the expected number of COVID-19 cases or deaths [24–27] and on prioritizing humanitarian operations for the vulnerable. However, the better/worse performance of ML methods concerning traditionally applied statistical analysis is still being discussed. Among the applied ML methods, some authors [24–26] applied unsupervised cluster-analysis algorithms, allowing them to explore the agglomeration patterns in the data to recover categories that explain the differences between reported COVID-19 cases or deaths. Inherent Risk of Contagion [24] was applied for municipalities, and a country-level cluster analysis was applied by [25]. Supervised approaches were discussed in [26] and [27], which are useful for making short-term forecasts of the number of COVID-19 cases or deaths. Supervised approaches included Random Forest (RF), Gradient Boosted Machine (GBM) [26], Linear Regression (LR), Support Vector Machine (SVM), Least Absolute Shrinkage and Selection Operator (LASSO), and Exponential Smoothing (ES) [27] for the country-level number of COVID-19 cases or deaths. According to the results, unsupervised approaches outperformed supervised ones. Among the most important predictors, economic activity, size of the population, the prevalence of chronic illness, and environmental pollution remained.

Spatial modeling applied to COVID-19 outcomes has been performed with scarce literature. Ref. [38,39] applied spatial modeling, finding that there are significant spatial interaction effects that reduce bias in the analysis of COVID-19 determinants. At this point, our paper contributes to the empirical literature. Spatial modeling is a special case of statistical modeling literature, and we want to test for significative spatial interaction effects between COVID-19 mortality and vulnerability drivers.

### 2.3. COVID-19 Pandemic in Peru: A Brief Review

The Peruvian response to the pandemic was an early lockdown with a set of prohibitions and mobility restrictions. There are several ways in which the pandemic affected Peruvians' livelihoods, mental health, and information technology development, among others [40–42]. According to [40], Peru reached a milestone of more than 1200 deaths per one million of inhabitants by February 2021. Possible explanations for these poor outcomes are the overwhelmed and fragmented public health system, gaps in the infrastructure, a lack of specialized personnel, and deficient leadership of political authorities. Following [41], there was an info-demic composed of excessive and unfounded information that hindered an appropriate public health response and harmed the mental health of the population. However, the authors argue that Peru could control the info-demic due the introduction of prison sentences to discourage creating and sharing fake news. Finally, following [42], there were good advances in digitalization that were caused by the pandemic; nevertheless, more efforts are needed to implement telemedicine and to offer access to the internet to achieve high-quality telemedicine to all vulnerable groups in Peru.

### 2.4. Vulnerability Sources of Peruvian Provinces Regarding COVID-19 Pandemic

A spatial dependence econometric model (SDM) is proposed to investigate the factors affecting the number of COVID-19 cases and deaths in 189 Peruvian provinces. This regression model allows us to test for spatial interaction effects while we conduct a regression analysis on COVID-19 vulnerability drivers. This model is used to estimate the demand for PRRSGS based on COVID-19 mortality rates. If spatial interaction effects are significant, the predictions of the demand is less biased compared with the case of a simple linear model. The spatial analysis contributes to the planning of PRRSGS distribution among the provinces at the country level. If there are strong COVID-19 mortality rate spatial patterns, it is possible to study the patterns and to use the information to trace a delivery strategy for PRRSGS.

Based on the concept of deprivation costs [12] and on the literature of social determinants of health [43–46], we can theoretically build up a definition of vulnerability at the province level, which serves to motivate the SDMs and to define the expected signs of the effects of the variables. Our goal is to test the theoretical framework for vulnerability against COVID-19; thus, some of the propositions can be empirically rejected by the evidence. This definition has seven blocks that are built using previous literature that explores COVID-19 determinants [15–21], ML applications [24–27], and social determinants [43–46]. This definition also contains the main hypothesis that is tested in this paper.

It is important to notice that, for some groups of variables, we could have two opposite effects according to the relationship of the variable with poverty, which is the main social determinant of health: a poverty condition causes more vulnerability to COVID-19 mortality due to the scarcity of resources to face illness, malnutrition, and other vulnerability conditions. Poverty in this paper is considered a variable that has a negative relationship with mortality rates, even if it could be positively associated with vulnerability. The reason behind this consideration is that low poverty is a proxy for high demand for help, as provinces with high economic activity tend to have more cases. This argument is defended by the machine learning literature [47], and we acknowledge that it is opposite to the social determinants argument.

Assuming that empirical poverty is negatively associated with COVID-19 mortality, the example of health insurance makes the interpretations more complex: a good coverage of health insurance may lead to lower mortality rates, but this implies high economic activity so it may lead to higher mortality rates at the same time. The literature that assesses social determinants of health has a different perspective of the effects of the following definition blocks. For this literature, poverty is a condition that exacerbates COVID-19 deaths, and all other factors related to poverty such as educational level, black skin color, low income, and bad system performance, among other variables [43–46]. In this paper, we expect a negative relationship because the context contains the data when pandemic was in its first stage, and thus, it is plausible that the relationship between COVID-19 and poverty is negative for the case of Peru. The following section contains the definition blocks and the perspective bridging theories from machine learning applications and social determinants of health:

- First: a province is vulnerable in the post-lockdown context if it has a lower incidence of poverty, unemployment, and other indicators of bad economic performance. Economic activity, which is inversely related to poverty, requires internal mobility and interpersonal interaction [47]. This does not mean that we should give less importance to poor provinces, even if they might have a lower demand for PRRSGS. The optimal policy must give the latter equal weight in the humanitarian objective function. We argue that this proposition is valid if we measure long-run poverty from households, and short-run measures may not prove this relationship.
- Second: provinces that have more proportion of the vulnerable population in terms of age (the older); sex (the males); skin color (black); and chronic diseases such as hypertension, diabetes, and obesity, among others. Deprivation costs are higher for people with a higher probability of dying due to COVID-19, including people with the aforementioned characteristics [43–45] reach the same signs of effects for the case of USA). Two types of overcrowding measures are accounted for regarding vulnerability: overcrowding in households and overcrowding in cities (proxied by population density). Provinces with higher overcrowding in households may have worse COVID-19 outcomes [44]. We expect the same for provinces with higher overcrowding in cities. Poverty is generally negatively associated with overcrowding in cities but positively associated with overcrowding in households [45].
- Third: A short-run households' income drastically reduced and then slowly recovered, so there is a shortage in the available resources to battle the contagion, which cannot

be covered. In the absence of high coverage of health insurance, this would lead to a greater number of deaths. However, we acknowledge that, in order to obtain high health insurance coverage, provinces must work, so the economic activity must be high for provinces with high health insurance coverage, and thus, this could lead to higher mortality rates. In consequence, we face two possible effects of health insurance: whether health insurance coverage positively or negatively affects the number of deaths due to COVID-19 is an empirical question.

- Fourth: health system performance indicators play a key role in the determination of COVID-19 outcomes, especially when the COVID-19 cases exceed the capacity of hospitals and other health facilities. Provinces where the health system performs bad are more vulnerable and likely to have a greater number of deaths [24]. However, as economic activity is a condition for good health system performance, the final effect is theoretically ambiguous [45]. Whether the effect is finally positive or negative regarding COVID-19 deaths is also an empirical question.
- Fifth: the COVID-19 outcome of a province regarding mortality rates may be affected by neighboring provinces with a high incidence of COVID-19 deaths. These are called endogenous interaction effects.
- Sixth: considering previous exploratory works on determinants of COVID-19 cases and deaths, and the previous vulnerability definition block, there is no reason not to suspect that the COVID-19 outcome in a province is affected by the exogenous outcomes of neighboring provinces (i.e., poverty, demographics, health systems, etc.). These are called exogenous interaction effects.
- Seventh: after considering all of the information above, it is possible that the spatial dependence model still has a residual that represents all of the variability of COVID-19 outcomes that could not be explained by the predictors. In this residual, there may be correlated effects, as is pointed out by [48], where the COVID-19 outcomes of a province are affected by unobserved similar characteristics of their neighbors.

The following Table 2 shows the final expected signs for groups of variables:

**Table 2.** Expected signs for variables.

| Variable | Expected Sign of Effect Regarding Mortality Rates |
| --- | --- |
| Poverty | − |
| Age | + |
| Sex | + |
| Black skin color | + |
| Chronic diseases | + |
| Overcrowding in cities | + |
| Overcrowding in households | + |
| Health insurance | ? |
| Health system performance | ? |
| Endogenous interaction effects | + |
| Exogenous interaction effects | + |
| Correlated effects | + |

− negative, + positive and ? undetermined. Authors' own elaboration based on reviewed literature.

Empirically, we test blocks fifth, sixth, and seventh, and we estimate seven models to visualize the different types of spatial interaction effects. These models or specifications passed a model selection phase where we selected the one that better fits and matches the data.

The main contribution of the framework proposed in this paper to the literature is that it configures pandemic management in the function of vulnerability regarding COVID-19 mortality, arguing that higher deprivation costs are found where there are greater mortality rates. Our proposal tries to build a supportive framework for economic reactivation-based on-demand assessment. This framework supports the policymakers in optimizing the available resources and improves the disaster management policies.

## 3. Materials and Methods

This section presents the data collection and data processing methods. First, the construction of variables is explained. Second, the spatial regression models are defined. Third, the model selection phase as well as the metrics used for model selection are explained. Fourth, the methodology for the assessment of demand is defined.

Two different software were used in this study: STATA 16 for spatial regression models and Python 3.6 for ENR and demand forecasting. Additionally, we used QGIS software for mapping of COVID-19 mortality rates.

### 3.1. Data Collection Methods

For this paper, the data listed in Table 2 were collected from different open-access sources such as Ministerio de Salud (MINSA by its Spanish acronym), the XII Population, VII Household, and III Indigenous Communities Census carried out in 2017, the Encuesta Demográfica y de Salud Familiar (ENDES by its Spanish acronym), the Encuesta Nacional a Hogares (ENAHO by its Spanish acronym), and the Sistema Nacional de Defunciones (SINADEF by its Spanish acronym). We rescaled the data at the province level to obtain indicators about poverty; employment; education; demographic characteristics; the prevalence of chronic illness such as diabetes, hypertension, and obesity; health insurance; and health system performance indicators. The number of COVID-19 deaths was recovered from the MINSA (2020) database (from 6 March 2020 to 12 December 2020). The deaths are reported by familiars and confirmed by autopsy. The quality of these data in representing the real differences between provinces is not the best; however, we show that the data are reliable for the analysis when interpreting the empirical results. Table 3 describes all of the variables included in this study.

**Table 3.** Description of variables.

| Name | Description | Source |
|---|---|---|
| Logdeaths1000 | The logarithm of the cumulative number of COVID-19 deaths per 1000 inhabitants until 12 December 2020 | MINSA (2020) |
| Composite_Poverty_Index | The average index of poverty estimated by the Multiple Correspondence Analysis (MCA) methods | Census * (2017) |
| Employed | People employed per 1000 inhabitants | S.A. [1] |
| Secondary_Educ | People having a complete secondary education per 1000 inhabitants. | S.A. |
| Vulnerable_pop | Elderly people (>65 years) per 1000 inhabitants | S.A. |
| LogPD_1000 | The logarithm of the population density measured as 1000 inhabitants per $km^2$ (urban area) | S.A. |
| Males_1000_Inhab | Males per 1000 inhabitants | S.A. |
| White_1000_Inhab | White people per 1000 inhabitants | S.A. |
| Black_1000_Inhab | Black people per 1000 inhabitants | S.A. |
| Assian_1000_Inhab | Asian people per 1000 inhabitants | S.A. |
| Overcrowding | The average proportion of households with more members than rooms | S.A. |
| Natural_Region1 | Province is located in the Coastal region | S.A. |
| Natural_Region2 | Province is located in the Highlands region | S.A. |
| Natural_Region3 | Province is located in the Jungle region | S.A. |
| Hypert1 | People with hypertension measured by results on differential pressure per 1000 inhabitants | S.A. |
| Hypert2 | People with diagnosed hypertension per 1000 inhabitants | S.A. |
| Diabetes | People with diagnosed diabetes per 1000 inhabitants | S.A. |
| Obesity | People with obesity by body mass index per 1000 inhabitants | S.A. |
| Chronic | People with a chronic illness (COPD, diabetes, hypertension, etc.) per 1000 inhabitants | ENAHO * (2019) |
| Health_Insurance | Number of people with health insurance per 100 inhabitants | Census * (2017) |
| Life_Expectancy | The average life between 2017 and 2019 before the COVID-19 pandemic | S.A. |
| Days_Till_Attended | The average days until medical attention | ENAHO * (2019) |
| SD_DTA | The standard deviation of days until medical attention | S.A. |
| Travel_Time | The average hours of travel time to health facility | S.A. |
| SD_TTtHFH | The standard deviation of hours of travel time to health facility | S.A. |
| Waiting_Time | The average hours of waiting time until attention in health facility | S.A. |
| SD_WT4AH | The standard deviation of hours of waiting time until attention at the health facility | S.A. |

[1] S.A.: same as above. Source: MINSA (2020), CENSO (2017), ENDES (2019), ENAHO (2019), and SINADEF (2019). * data recovered by National Institute of Statistics and Informatics (INEI in Spanish).

### 3.2. Data Processing Methods

The SDM to be used has to be selected from a family of cross sectional spatial dependence models described in [48,49]. This family of models arises from different ways in which spatial patterns can affect a dependent variable outcome: the endogenous interaction effects, the exogenous interaction effects, and the correlated effects. There are a total of seven models, including the single ordinary least squares model (OLS), which is a sub-case with none of the effects listed above (these models are estimated in the literature for spatial dependence models application). From the most general aspect to the particular one, these models are defined as follows:

$$SDM\ (1,1,1):\ Y_i = \beta X_{ik} + \theta W X_{ik} + \lambda W Y_i + v_i;\ v_i = \rho W v_i + \epsilon_i \text{ Manski model,} \tag{1}$$

$$SDM\ (1,0,1):\ Y_i = \beta X_{ik} + \lambda W Y_i + v_i;\ v_i = \rho W v_i + \epsilon_i \text{ Kelejian–Prucha model,} \tag{2}$$

$$SDM\ (1,1,0):\ Y_i = \beta X_{ik} + \theta W X_{ik} + \lambda W Y_i + v_i \text{ Spatial Durbin model,} \tag{3}$$

$$SDM\ (0,1,1):\ Y_i = \beta X_{ik} + \theta W X_{ik} + v_i;\ v_i = \rho W v_i + \epsilon_i \text{ Spatial Durbin Error model,} \tag{4}$$

$$SDM\ (1,0,0):\ Y_i = \beta X_{ik} + \lambda W Y_i + v_i \text{ Spatial lag model,} \tag{5}$$

$$SDM\ (0,0,1):\ Y_i = \beta X_{ik} + v_i;\ v_i = \rho W v_i + \epsilon_i \text{ Spatial error model} \tag{6}$$

$$SDM\ (0,0,0):\ Y_i = \beta X_{ik} + v_i \text{ OLS model,} \tag{7}$$

In Equations (1)–(7), $Y_i$ represents a vector of $N \times 1$ observations on the dependent variable, which is the number of deaths by COVID-19. First, on the right side, $X_{ik}$ is a matrix of $N$ observations for the $K$ exogenous variables of the model and $\beta$ is the $K \times 1$ vector of parameters related to the $K$ exogenous regressors. Second, $WX_{ik}$ is the matrix of spatially lagged exogenous variables pre-multiplied by the vector $\theta$ of $K \times 1$ parameters associated with these spatially lagged variables (to account for exogenous interaction effects). Third, $\lambda W Y_i$ represents the spatially lagged dependent variable, with the scalar $\lambda$ (called the autoregressive parameter) pre-multiplying the $WY_i$ term, which is a $N \times 1$ vector of observations on the spatially lagged dependent variable (to account for endogenous interaction effects). Finally, $v_i$ is an error, which can be correlated with its spatial lag as it is represented by Equations (1), (2), (4), and (6), and $\rho$ is a scalar parameter that accounts for the spatial correlation of the error term. Finally, $\epsilon_i$ is the idiosyncratic stochastic error, which is assumed to be white noise that captures all of the remaining variability of the dependent variable and is uncorrelated with other variables in the model.

The W matrix represents a spatial weighting matrix that is important to test if the spatial interaction effects are significant. This matrix provides information about the spatial units that, in this work, are the provinces. A spatial weights matrix, formally, is a $N \times N$ positive symmetric matrix for which the weights named $w_{ij}$ define the spatial relations among locations and, therefore, determine the spatial autocorrelation statistics [50].

The following Equation (8) has been used for defining spatial weight matrix W:

$$\mathrm{W} = \begin{pmatrix} w_{11} & \ldots & w_{1n} \\ \vdots & \ddots & \vdots \\ w_{n1} & \ldots & w_{nn} \end{pmatrix} \tag{8}$$

where $w_{ij} = \begin{cases} 1 & if\ i\ and\ j\ are\ contiguous \\ 0\, if\ i\ and\ j\ are\ not\ contiguous \end{cases}$.

Nevertheless, there are several ways of defining W matrix; we list three of them that are based on Euclidian distance: Queen's criteria (contiguity), Gabriel criteria, and K-nearest neighbors' criteria [38,39]. We based our selection of W on the maximization of the value of the Moran Statistic that is estimated with a W definition [39]. The Moran Statistic is a measure of spatial autocorrelation that lies in the interval [−1,1], but its significance determines the level of observable spatial autocorrelation among a variable. We found that Queen's criteria maximize the significance of Moran statistic, and thus, this is the optimal way to define the W matrix. We focus on the spatial patterns that can be measured with the data.

Equations (1)–(7) are estimated with maximum-likelihood methods assuming that $\epsilon_i \sim N(0, \sigma^2 I)$. This implies that the distribution of COVID-19 mortality is normal. The parameters are obtained, maximizing the following Equation (9):

$$lnL\left(Y\middle|\psi, \lambda, \rho, \sigma^2\right) = -\frac{N}{2}ln(2\pi) - \frac{N}{2}ln\left(\sigma^2\right) + ln||I - \lambda W|| + ln||I - \rho W|| + \frac{1}{2\sigma^2}(\epsilon'\epsilon) \quad (9)$$

where $\psi = (\beta, \theta)$ and errors $\epsilon = F(\psi, \lambda, \rho)$ are obtained from the reduced form of models from (1) to (7). The terms $ln||I - \lambda W||$ and $ln||I - \rho W||$ can appear or disappear in the expression according to the model formulation. Definition of errors and the abovementioned terms changes according to the model; for example, if we estimate Equation (6), which is the spatial error model, the errors will be $\epsilon = F(\psi, \rho)$ with $\psi = (\beta)$ and we only maintain the term $ln||I - \rho W||$ in the likelihood equation. Equation (9) is known as the unconcentrated log-likelihood function and is discussed in Lee (2004).

A disadvantage of SDMs is that they do not support out-of-sample predictions, which is the case for the seven provinces. Data are available for 189 provinces out of a total of 196 provinces in Peru. The missing provinces are considered as contributing zero cases or deaths. However, this hypothesis is questionable because, in some of these provinces, there are high indicators of economic activity (for example, Cañete with a value of 38.60 in the composite poverty index, which reflects low poverty below the average of 56.81). Therefore, for an estimation of the provinces with missing data, we use ENR, a machine learning tool that comes as close as possible to linear models of spatial dependency. The following Equation (10) shows the parameter estimation method and the objective function for ENR:

$$\beta = \min_{\beta}\left(||Y - X\beta||^{\hat{2}} + \lambda_2||\beta||^2 + \lambda_1||\beta||\right) \quad (10)$$

where $\lambda_1$ is the parameter for the term associated with LASSO and $\lambda_2$ to the Tikhonov regularization associated with ridge regression. The parameters $\lambda_1$ *and* $\lambda_2$ can take any value from zero to infinity, and they measure the weight of the penalization for irrelevant or correlated regressors, respectively [51]. The method sacrifices bias to get a lower variance estimator that has been proven to outperform low bias over-fitted estimators in the context of out-of-sample predictions. The selection of $\lambda_1$ and $\lambda_2$ penalization size and the ratio are made by repeated k-fold cross-validation (with $k = 4$), maximizing the out-of-sample predictive power measured with mean absolute error.

In order to qualitatively estimate the demand of PRRSGS, Equations (1)–(7) are estimated; then, we selected the model that better fits the data based upon the following metrics [51] (the less-biased model, considering only the 189 nonzero provinces that were part of the estimation of Equations (1)–(7)):

$$MF\ R^2 = 1 - \frac{LL_m}{LL_0} \quad (11)$$

$$R^2 = \frac{\sum_{i=1}^{N}\left(\hat{Y}_i - \overline{Y}_i\right)^2}{\sum_{i=1}^{N}\left(Y_i - \overline{Y}_i\right)^2} \quad (12)$$

$$AIC = -2LogL + 2q \quad (13)$$

$$BIC = 2Log(L) + qLog(N) \quad (14)$$

$$WT = (Rb - r)'\left(RVR'\right)^{-1}(Rb - r),\ Rb = \begin{pmatrix} \theta \\ \lambda \\ \rho \end{pmatrix}\ and\ r = \begin{pmatrix} 0 \\ 0 \\ 0 \end{pmatrix} \quad (15)$$

$$RMSE = \sqrt{\sum_{i=1}^{N}\frac{\left(\hat{Y}_i - \overline{Y}_i\right)^2}{N}} \quad (16)$$

This information is complemented with ENR outcomes; this model does not pass for model selection phase, but it passes through repeated k-fold cross-validation to maximize its predictive power.

*3.3. Demand Assessment*

Once the model selection phase is carried out, the one that better fitted the data was used to make predictions. Within the sample, predictions were used as proxies for the demand of PRRSGS and out-of-sample predictions used with ENR tool are employed as forecasts of demand for provinces that have zero contagions or deaths (Equation (10)). The data are visualized spatially, and we report the results for the best models.

Demand is assessed by quartiles of the number of predicted COVID-19 mortality rates. This implies that the demand assessment is based on vulnerability as the predicted mortality rates are a function of a set of vulnerability drives listed in Section 2.3. We can distinguish four groups. This categorization of demand helps to trace a delivery strategy and to identify vulnerability patterns in the space, as mortality is ranked from low to very high categories:

- $COVID19_{Q1}$: Low demand
- $COVID19_{Q2}$: Medium demand
- $COVID19_{Q3}$: High demand
- $COVID19_{Q4}$: Very high demand

In the end, every province has an established demand, so that a distribution strategy for the 196 provinces can be discussed. In the Results and Discussion section, we focus on spatial insights and their contribution to the humanitarian logistics framework. In managerial and research implications, we summarize and discuss the main results, and their utility at the moment of tracing a humanitarian logistics supportive framework strategy for pandemic management.

The methodology first does a regression analysis using the SDM in Equations (1)–(7) and (10); then, it passes the model selection phase using Equations (11)–(16). After the model selection phase, we use the best SDM specification to predict the level of demand of PRRSGS, and for provinces with zero deaths, we use estimates from Equation (10) to forecast demand. Finally, with the post estimation results and demand quartile classification, we discuss a delivery strategy at the national level using provinces as the unit of analysis.

## 4. Results

The COVID-19 mortality rates have been log-transformed to draw a normal distribution for it. Figure 1 shows a box plot with a swarm plot and a kernel density of the log-transformed mortality rates.

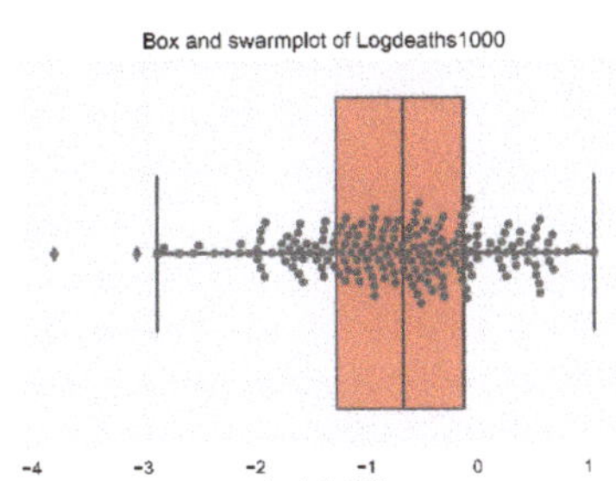


(a)

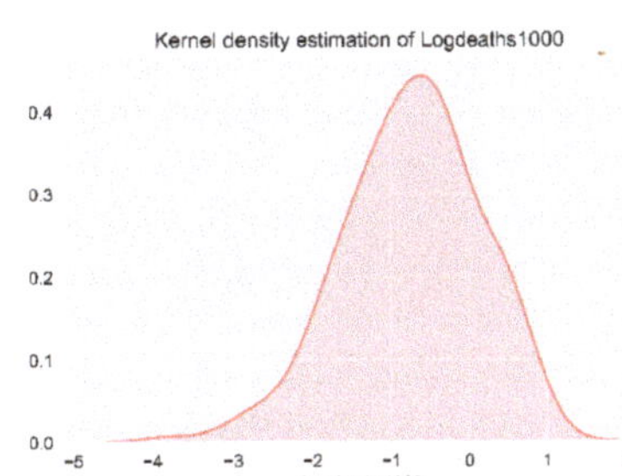


(b)

**Figure 1.** Distribution plots of Logdeaths1000. (**a**) Boxplot of Logdeaths1000. (**b**) Kernel density plot of Logdeaths1000. Authors' own elaboration.

We observe that, on average, provinces face less than one death per 1000 inhabitants. This is consistent with the mortality rates from international estimations for COVID-19. Although the mortality rates are low, they can vary a lot, and this is why it is important to understand the factors behind those differences. For this purpose, we propose a set of variables listed in Table 3 in the previous section. Figure 2 shows a plot of the correlation matrix between the variables that explain the differences between COVID-19 mortality rates.

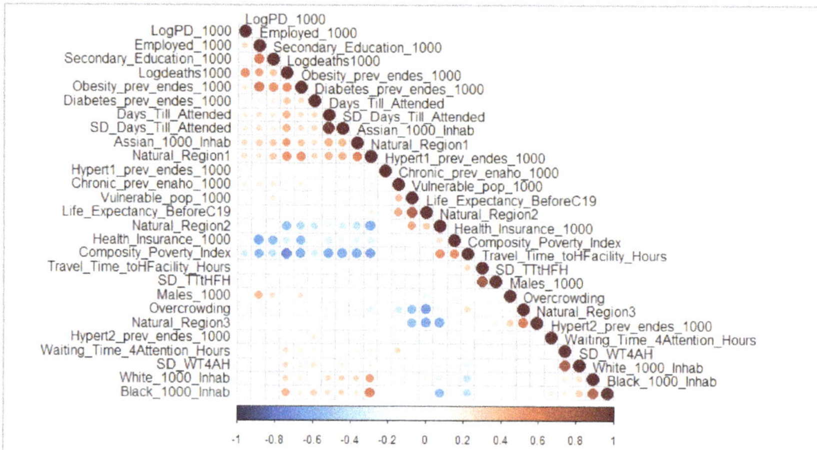


**Figure 2.** Correlation matrix of variables. Authors' own elaboration.

In Figure 2, blue circles represent the negative correlation and red ones represent the positive correlation. The size and transparency of the circles represent the magnitude of the correlation between the variables. The mortality rates (Logdeaths1000) are, on the one hand, negatively related to belonging to the highlands region (Natural_Region2), health insurance (Health_Insurance_1000), and poverty (Composite_Poverty_Index). On the other hand, they are positively related to belonging to the coastal region (Natural_Region1), presence of Asian population (Assian_1000_Inhab), presence of Black population (Black_1000_Inhab), the average days until a patient is attended (Days_Till_Attended), the inequity of the days for medical attention (SD_Days_Till_Attended), the average waiting time for attention in a health facility (Waiting_Time_4Attention_Hours), the inequity of waiting time for attention (SD_WT4AH), the prevalence of diabetes (Diabetes_prev_endes_1000), and the log population density (LogPD_1000).

The long-run poverty (Composite_Poverty_Index) variable is related to many other variables in the model. It is negatively related to the log population density (LogPD_1000), employment (Employed_1000), human capital with secondary education (Secondary_Education_1000), log of mortality rates (Logdeaths1000), obesity prevalence (Obesity_prev_endes_1000), the average days until a patient is attended in a health facility (Days_Till_Attended), the inequity in time for patient attention (SD_Days_Till_Attended), the Asian population (Assian_1000_Inhab), and the belonging to the coastal region (Natural_Region1), and it is positively related to belonging to highlands region (Natural_Region2) and being health insured (Health_Insurance_1000). According to these correlations, poor provinces are those with a low population density, employment, human capital, obesity prevalence, bad performance of health system, and low health insurance coverage.

This correlation could lead to a possible multicollinearity problem. Nevertheless, the mean-variance inflation factor across all variables after an OLS estimation is 4.96, which reflects low multicollinearity. The following Table 4 shows the descriptive statistics for all the variables considered in the paper:

**Table 4.** Descriptive statistics.

| Variable | Mean | S.D. | Min | Max | Median |
|---|---|---|---|---|---|
| Logdeaths1000 | −0.74 | 0.87 | −3.80 | 1.05 | −0.67 |
| Employed_100 | 371.12 | 72.03 | 140.31 | 589.99 | 379.94 |
| Males_1000_Inhab | 501.72 | 19.55 | 471.90 | 604.85 | 498.97 |
| Vulnerable_pop | 199.71 | 40.98 | 83.02 | 350.88 | 197.91 |
| Health_Insurance | 811.87 | 93.86 | 488.04 | 958.34 | 817.18 |
| Secondary_Educ | 313.98 | 52.78 | 186.49 | 436.08 | 314.60 |
| Life_Expectancy | 63.34 | 6.49 | 35.09 | 74.81 | 65.03 |
| Chronic | 364.96 | 96.99 | 117.03 | 630.40 | 362.31 |
| Composite_Poverty_Index | 56.62 | 12.17 | 25.71 | 78.74 | 60.63 |
| White | 32.24 | 25.05 | 0.71 | 119.25 | 27.29 |
| Asian_1000_Inhab | 0.18 | 0.40 | 0.00 | 3.32 | 0.04 |
| Black_1000_Inhab | 21.10 | 24.99 | 0.00 | 102.08 | 12.16 |
| Hypert1 | 434.69 | 80.36 | 96.14 | 724.68 | 430.87 |
| Hypert2 | 89.86 | 44.14 | 0.00 | 277.53 | 84.54 |
| Diabetes | 25.32 | 25.18 | 0.00 | 171.49 | 21.70 |
| Obesity | 184.44 | 94.21 | 0.00 | 514.62 | 173.56 |
| Days_Till_Attended | 1.06 | 1.79 | 0.01 | 12.97 | 0.34 |
| SD_DTA | 3.67 | 5.10 | 0.00 | 35.44 | 1.49 |
| Travel_Time | 0.65 | 0.45 | 0.20 | 4.37 | 0.53 |
| SD_TTtHFH | 1.61 | 2.84 | 0.16 | 30.70 | 0.82 |
| Waiting_Time | 0.11 | 0.05 | 0.02 | 0.28 | 0.11 |
| SD_WT4AH | 0.41 | 0.16 | 0.05 | 1.21 | 0.43 |
| logPD_1000 | −0.31 | 3.33 | −9.85 | 7.70 | −0.15 |
| Overcrowding | 17.24 | 7.85 | 7.54 | 53.62 | 14.71 |
| Natural_Region1 | 0.19 | 0.39 | 0.00 | 1.00 | 0.00 |
| Natural_Region2 | 0.62 | 0.49 | 0.00 | 1.00 | 1.00 |
| Natural_Region3 | 0.19 | 0.39 | 0.00 | 1.00 | 0.00 |

Authors' own elaboration.

After running an OLS regression, we test for spatial dependence in the residuals using a contiguity spatial weighting matrix using the residuals. The Moran's standard deviates statistic is 5.0581, with the lowest $p$-value between other spatial weighting matrices (see Table 5 for results with other spatial weighting matrices), rejecting the null hypothesis that states that the errors are independently and identically distributed with 99% of confidence and a $p$-value of 0.000, favoring the hypothesis of spatially correlated errors.

**Table 5.** Moran's standard deviate statistic and significance ($p$-values).

| Spatial Weighting Matrix | Moran's I Standard Deviate | Significance ($p$-Value) |
|---|---|---|
| Queen's contiguity criteria | 5.058 | $4.23 \times 10^{-7}$ |
| K-nearest neighbors (K = 1) | 2.864 | $4.18 \times 10^{-3}$ |
| K-nearest neighbors (K = 2) | 3.613 | $3.02 \times 10^{-4}$ |
| K-nearest neighbors (K = 3) | 3.648 | $2.64 \times 10^{-4}$ |
| K-nearest neighbors (K = 4) | 3.874 | $1.07 \times 10^{-4}$ |

Authors' own elaboration.

We proceed with the results of the SDMs, in Table 6, we summarize the main ones that contribute to model selection. We select the one with a lower bias for in-sample predictions. For out-of-sample prediction, we use an ENR, which minimizes the variance, sacrificing some bias to obtain lower variance predictions.

First, the variables used in OLS regression produced a McFadden pseudo-$R^2$ of 0.52 and an $R^2$ of 0.74. These results support the hypothesis of the positive relationship between vulnerability to COVID-19 and the mortality rates. Even though the lowest AIC and BIC outcomes are for the $SDM(1,1,0)$ and $SDM(0,1,1)$, they have a higher bias than OLS, as they have 0.275 and 0.273 outcomes in RMSE, respectively, versus 0.269. Furthermore, RMSE is reduced in $SDM(1,1,1)$ to 0.267. This means that $SDM(1,1,1)$ is the best candidate for the model that better fits the data. Although SDM(1, 1, 1), SDM(1, 1, 0), and SDM(0, 1, 1) are equivalent in McFadden pseudo-$R^2$, $R^2$, and RMSE, we must consider that the spatial dependence in residuals is strongly significant, so instead of basing the model selection decision on AIC BIC, it is better to select the model with higher significance on spatial terms; this is measured by the Wald Test for significance of spatial terms. In consequence, the best model that better fits the data is the SDM(1, 1, 1).

**Table 6.** Models for COVID-19 deaths (per 1000 inhabitants).

| Model | SDM (1,1,1) | SDM (1,0,1) | SDM (1,1,0) | SDM (0,1,1) | SDM (1,0,0) | SDM (0,0,1) | OLS (0,0,0) | ENR |
|---|---|---|---|---|---|---|---|---|
| MF $R^2$ | 0.67 | 0.53 | 0.67 | 0.67 | 0.53 | 0.53 | 0.52 | N.E. |
| $R^2$ | 0.75 | 0.72 | 0.75 | 0.75 | 0.73 | 0.72 | 0.74 | 0.68 |
| AIC | 443.80 | 283.20 | 267.30 | 266.80 | 281.90 | 281.90 | 287.00 | N.E. |
| BIC | 443.80 | 377.20 | 439.1 | 438.60 | 372.70 | 372.70 | 374.60 | N.E. |
| WT | 98.63 * | 11.95 * | 88.16 * | 99.55 * | 7.138 * | 18.75 * | N.E. | N.E. |
| RMSE | 0.27 | 0.28 | 0.27 | 0.27 | 0.28 | 0.28 | 0.27 | 0.30 |

Authors' own elaboration. N.E.: Not estimated. * For statistically significant WT statistics ($p < 0.01$).

We consider the possibility that residuals vary in a non-random way. Specifically, we believe that the dataset suffers from heteroscedasticity. In order to test it, we used the OLS model residuals. The Breusch–Pagan test chi-squared statistic, subject to one degree of freedom, is 16.69, which rejects the null hypothesis of homoscedasticity with 99% of confidence and a $p$-value of 0.000. The following Figure 3 plots the mortality rates versus the predictions of the SDM(1, 1, 1) model; it confirms the presence of heteroscedasticity:

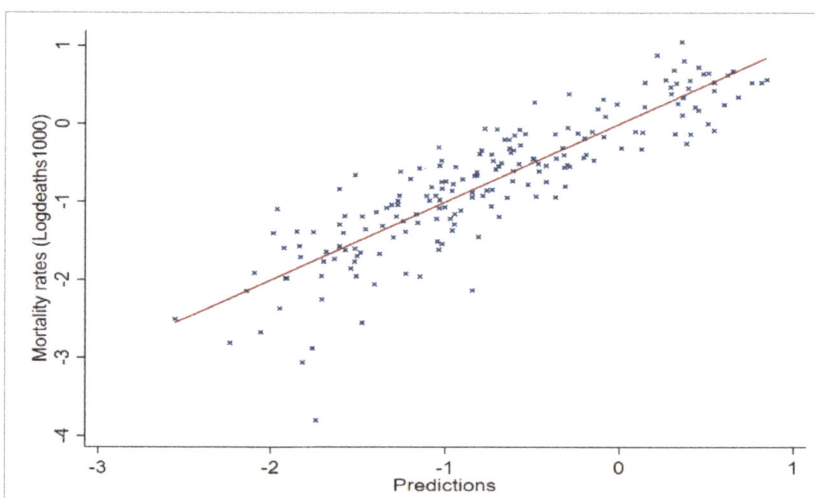


**Figure 3.** Linear fit plot of real mortality rates versus predictions. Authors' own elaboration.

We observed that provinces with lower mortality rates have a higher variance in the residuals than other provinces. Furthermore, the variance for other provinces seemed to be constant. For this reason, we report robust standard errors for the individual significance of coefficients in the SDM(1, 1, 1) model. The following Table 7 shows the regression results:

**Table 7.** Regression results (coefficients).

| | SDM(1,1,1) | | | | ENR |
|---|---|---|---|---|---|
| **VARIABLES** | **Logdeaths1000** | **S.E.** | **W** | **S.E.** | **Logdeaths1000** |
| Employed_100 | 0.001 | (0.001) | −0.005 * | (0.003) | 0.000 |
| Males_1000_Inhab | −0.001 | (0.003) | −0.002 | (0.005) | 0.002 |
| Vulnerable_pop | 0.001 | (0.002) | 0.003 | (0.005) | 0.004 |
| Health_Insurance | −0.001 | (0.001) | 0.001 | (0.002) | 0.001 |
| Secondary_Educ | 0.002 | (0.001) | 0.005 | (0.003) | 0.001 |
| Life_Expectancy | 0.018 | (0.012) | 0.006 | (0.035) | 0.000 |
| Chronic | −0.000 | (0.000) | −0.002 * | (0.001) | 0.000 |
| Composite_Poverty_Index | −0.031 *** | (0.009) | 0.009 | (0.024) | −0.048 |
| White_1000_Inhab | −0.006 * | (0.003) | 0.024 *** | (0.006) | −0.002 |
| Assian_1000_Inhab | −0.076 | (0.122) | −0.430 | (0.387) | 0.000 |
| Black_1000_Inhab | 0.004 | (0.003) | −0.020 ** | (0.008) | 0.007 |
| Hypert1 | −0.000 | (0.000) | 0.000 | (0.001) | 0.000 |
| Hypert2 | 0.000 | (0.001) | 0.002 | (0.003) | −0.001 |
| Diabetes | −0.000 | (0.002) | 0.007 | (0.004) | 0.000 |
| Obesity | 0.000 | (0.001) | 0.002 | (0.002) | 0.001 |
| Days_Till_Attended | −0.021 | (0.061) | −0.117 | (0.186) | 0.000 |
| SD_DTA | 0.008 | (0.020) | 0.059 | (0.066) | −0.003 |
| Travel_Time | −0.244 | (0.162) | −0.083 | (0.528) | 0.000 |
| SD_TTtHFH | 0.051 ** | (0.025) | 0.046 | (0.080) | 0.000 |
| Waiting_Time | 1.498 | (1.450) | −4.008 | (4.174) | 0.000 |
| SD_WT4AH | 0.219 | (0.391) | 1.434 | (1.083) | 0.000 |
| LogPD_1000 | 0.077 *** | (0.012) | 0.131 *** | (0.050) | 0.053 |
| Overcrowding | 0.019 ** | (0.008) | −0.043 *** | (0.015) | 0.019 |
| Natural_Region1 | (base) | | (base) | | 0.000 |
| Natural_Region2 | −0.056 | (0.187) | −0.483 | (0.460) | 0.000 |
| Natural_Region3 | 0.577 ** | (0.259) | −0.238 | (0.525) | 0.000 |
| Logdeaths1000 | | | 0.109 | (0.100) | N.E. |
| e.Logdeaths1000 | | | 0.305 ** | (0.151) | N.E. |
| var(e.Logdeaths1000) | | | 0.190 *** | (0.027) | N.E. |
| Constant | 1.468 | (1.524) | | | 0.403 |

Robust standard errors in parentheses: *** $p < 0.01$, ** $p < 0.05$, * $p < 0.1$. Authors' own elaboration.

The results can be summarized as follows:

Poverty (Composite_Poverty_Index): there is a strong negative relationship between long-run poverty and COVID-19 mortality rates. While economic reactivation is carried out, provinces that had a high level of economic activity, which is related to lower poverty and is captured in the index, are especially vulnerable to COVID-19 mortality. After the short-run shock, the economy tends to return to its long-run level; this is the main reason for the sign of this effect. Poor provinces are vulnerable to the indirect effects of a pandemic: unemployment and shortage in available resources, among others. This insight favors the proposal of an implementation of a reactivation framework rather than a lockdown: provinces are returning to their long-run level economic activity, and this must be reintroduced safely. There is no evidence that poverty produces exogenous interaction effects on mortality rates.

Overcrowding in cities or population density (LogPD_1000): there is a strong positive relationship between population density and COVID-19 mortality rates. Higher mortality is found in provinces with higher population density. In low population density provinces or rural provinces, the health system may be inexistent, but since the data used for this analysis represent the initial period of COVID-19 with limited spread between provinces, we do not observe the effects of bad health system performance, but instead, we observe the effect of population density or overcrowding in cities where infection spread earlier. According to empirical results, poor provinces are less overcrowded in cities (negative correlation of −0.50) but more overcrowded in households (correlation of 0.30). The effect of population density is on the same line as the effect of poverty: those provinces that are rich and have high population density (since both attributes are correlated) are more vulnerable to COVID-19 deaths. There is marginal evidence that population density produces positive exogenous interaction effects on mortality: increases in population density, in neighboring provinces, can lead to higher mortality rates, but the significance of this effect is not sufficient to reject the hypothesis of no effect at all. Future deeper research is needed to better explain this relationship.

Overcrowding in households (overcrowding) is intuitive in that there is a strong positive relationship between overcrowding in households and COVID-19 mortality rates. Overcrowding is a serious issue in less developed countries, and it is a consequence of the informal settlement and bad birth control policies, among other factors. Mortality is magnified by the fact that more individuals live in a household than normal given the number of bedrooms. In order to improve pandemic management, the government should minimize the contagion not only outside but also inside the household. Nevertheless, there is strong evidence that overcrowding produces negative exogenous interaction effects on mortality: increases in overcrowding, even in neighboring provinces, lead to lower mortality rates on a province. The indirect effect and the direct effect are counteracted, but the indirect effect is greater than the direct effect (see Table 4). The final effect of an increase in the proportion of overcrowded households is a small decrease in mortality rates. Policymakers should consider that an increase in overcrowding in households may lead to lower mortality rates, even though the size of the effect is small. This is another reason why poor provinces, which are more overcrowded in households, have lower mortality rates.

Population ethnic composition (White_1000_Inhab and Black_1000_Inhab) produces significant effects on mortality rates. These effects are spatial interaction effects. An increase in the white population in neighboring provinces would lead to higher mortality rates in a province; on the other hand, an increase in the black population in neighboring provinces leads to lower mortality rates in a province. These effects are contrary to the effects of ethnic groups found by [46] for the US. It is interesting to acknowledge that a white population has a higher probability of dying according to the results of the model for the case of Peru. We reject the hypothesis of positive effects in favor of negative effects for skin color and null effect of other variables.

ENR validates the effects for poverty, population density, overcrowding, vulnerable population, health insurance, white population, black population, hypertension2 (second way of measuring diagnosed hypertension), obesity, and inequity in health. However, the size of the effect for the vulnerable population, health insurance, white population, black population, hypertension2, obesity, and inequity in health is relatively small (i.e., $0.001 < \text{effect} < 0.01$). These results prove the robustness of the SDM$(1,1,1)$. The following Table 8 shows the direct and indirect (from spatial spillovers) impacts predicted by the model:

**Table 8.** Direct, indirect, and total effects predicted after SDM$(1,1,1)$.

| Logdeaths1000 | Direct | Indirect | Total |
|---|---|---|---|
| Employed_100 | 0.001 | −0.001 | 0.001 |
| Males_1000_Inhab | −0.001 | −0.007 | −0.007 |
| Vulnerable_pop | 0.001 | −0.001 | 0.001 |
| Health_Insurance | −0.001 | 0.003 * | 0.002 |
| Secondary_Educ | 0.000 | 0.003 | 0.003 |
| Life_Expectancy | 0.013 | 0.001 | 0.014 |
| Chronic | 0.000 | −0.001 | 0.000 |
| Composite_Poverty_Index | −0.029 *** | 0.008 | −0.021 ** |
| White_1000_Inhab | −0.007 *** | 0.011 * | 0.004 |
| Assian_1000_Inhab | 0.038 | 0.247 | 0.285 |
| Black_1000_Inhab | 0.005 * | −0.009 | −0.004 |

**Table 8.** *Cont.*

| Logdeaths1000 | Direct | Indirect | Total |
|---|---|---|---|
| Hypert1 | 0.000 | 0.001 | 0.001 |
| Hypert2 | 0.000 | −0.001 | 0.000 |
| Diabetes | 0.000 | 0.006 | 0.006 |
| Obesity | 0.000 | 0.003 | 0.003 |
| Days_Till_Attended | −0.048 | 0.104 | 0.055 |
| SD_DTA | 0.017 | −0.047 | −0.031 |
| Travel_Time | −0.218 | −0.693 | −0.911 |
| SD_TTtHFH | 0.045 * | 0.106 | 0.150 * |
| Waiting_Time | −0.108 | −1.800 | −1.908 |
| SD_WT4AH | 0.672 * | 0.511 | 1.183 |
| LogPD_1000 | 0.079 *** | 0.038 | 0.117 *** |
| Overcrowding | 0.019 *** | −0.031 ** | −0.012 |
| Natural_Region1 | −0.100 | −0.290 | −0.391 |
| Natural_Region2 | 0.337 | −0.206 | 0.131 |
| Natural_Region3 | 0.001 | −0.001 | 0.001 |

*** $p < 0.01$, ** $p < 0.05$, * $p < 0.1$. Authors' own elaboration.

First, keeping the remaining variables constant, the effect of an increase in one unit of Composite_Poverty_Index produces a significant direct decrease of 0.029% deaths per 1000 inhabitants, and the same increase in neighboring provinces produces an insignificant increase of 0.008% deaths per 1000 inhabitants. The total effect is a significant decrease of 0.021 in the deaths per 1000 inhabitants. Second, an increase of 1% in logPD_1000, is related to a direct increase of 0.079% in deaths per 1000 inhabitants, and an insignificant indirect increase of 0.038%. The total effect is insignificant. Third, an increase of one percentage unit in overcrowding would lead to a positive direct effect of 0.019% in deaths per 1000 inhabitants, and a negative indirect effect of 0.031%. Both effects are statistically significant, and the total effect is insignificant. After reviewing the marginal effects, it is clear that provinces vulnerable to COVID-19 mortality are those with higher economic activity, high population density, and a high proportion of overcrowded households. Most of the indirect effects are insignificant except for overcrowding. However, overcrowding's direct and indirect effects are counteracted, and then, policies linked to this variable may not produce the desired impact. We recommend that future research should evaluate the impact of overcrowding in households on mortality rates.

The following provinces listed in Table 9 have zero reported deaths (if there were zero COVID-19 cases, the deaths were replaced by zero):

**Table 9.** Province forecast of deaths.

| Department | Province | ENR COVID-19 Deaths Forecast | Demand Category |
|---|---|---|---|
| Ucayali | Purús | 0.35 | Medium |
| Ancash | Asunción | 0.28 | Medium |
| Ancash | Antonio Raymondi | 0.23 | Low |
| Huánuco | Marañón | 0.20 | Low |
| Lambayeque | Ferreñafe | 0.72 | High |
| Lima | Cañete | 1.28 | Very high |
| Loreto | D.Marañón | 0.31 | Medium |

Authors' own elaboration.

The ENR regression model was cross-validated using a repeated k-fold algorithm, it resulted in parameters $\lambda_1 = 0.07$ and $\lambda_2 = 0.93$. It predicts demand based on vulnerability characteristics against COVID-19 mortality. The predictions vary from 0.20 to 1.28 deaths per 1000 inhabitants. Additionally, we have prediction categories from low to very high demand. According to their observable characteristics, it is not very likely that these provinces have zero deaths, as is reported on the dataset. In Figure 4, we plot the spatial distribution of the mortality rates. In the first panel, we show the real mortality rates; in the second, the rates predicted by SDM(1, 1, 1), and in the third panel, we show the predictions for ENR model. We observe that the predictions of SDM(1, 1, 1) are higher in the north. We have high demand predictions in the coastal and jungle regions. In the highlands, the predominant category is high, and in central-south highlands, medium demand is predominant. Low demand is present in the south and central-north highlands.

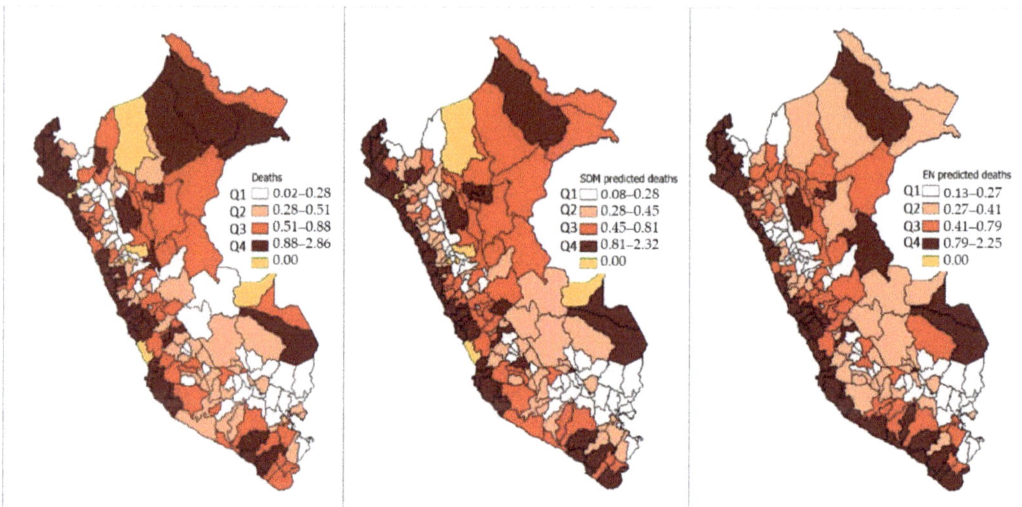


**Figure 4.** Number of deaths mapped by COVID-19 (per 1000 inhabitants) versus the predictions. First panel shows the actual mortality rates. Second panel shows the predicted mortality rates by SDM (1,1,1). Third panel shows the forecasts of ENR. Authors' own elaboration.

The government should deliver to very high demand provinces first; furthermore, the central warehouses for PRRSGS must be located in coastal provinces and jungle provinces with very high-demand categories. The second echelon of the delivery could be from very high-demand provinces to high-, medium-, and low-demand ones. The number of warehouses depends on the number of spatial clusters of very high-demand points. Across the coastal region, delivery must use the main road to reach all very high-demand provinces and, then, in a second echelon, reach the other provinces. In the second echelon, the delivery must be from coastal to highlands and from jungle to highlands. The low-demand provinces must be provided with PRRSGS at the end so that we can prioritize medium- and high-demand provinces. To follow this delivery strategy, policymakers should consider the predictions of the SDM(1, 1, 1) model complemented by the predictions of the ENR model. This must be performed in order to overcome data limitations and to establish a delivery strategy for PRRSGS. As pandemic outbreak evolves, more and more supplies will be needed, so establishing a national-level delivery strategy is very important for successful pandemic management. In Figure 4, we also observe spatial clusters in the coastal and jungle regions. The delivery of PRRSGS must be made first in the clusters of high COVID-19 mortality, and then, the supplies must be delivered to the medium-and low-demand quartiles, according to the spatial distribution. If we consider the forecasts of demand present in the third panel of Figure 4, we also find that, in coastal and jungle regions, we have spatial clusters with high COVID-19 mortality.

## 5. Discussion

Based on the results of this work, we summarize the managerial and research implications:

The highly vulnerable provinces are those with lower poverty, good economic performance, and high urban population density (Population density: total population divided by total urban area for each province). Economic activity is related to population density and agglomeration, and provinces that maintain a certain level of economic activity tend to have higher mortality rates; that is the reason why we need a supporting framework for economic reactivation. High economic activity provinces suffer from a lack of supply for PRRSGS; that is the reason why we have found this effect.

In order to mitigate risks, decision-makers must pay attention to economic performance, population density, and household overcrowding. High mortality should be expected in highly vulnerable provinces with the previously defined characteristics; thus, risk mitigation policies must consider these vulnerability drivers. Population overcrowding in households was a significant predictor, but its direct effect is counteracted by its indirect spatial spillover effect, so more research is needed to clarify its impact on COVID-19 mortality.

There are significant spatial patterns in COVID-19 mortality rates (based on Moran's statistic's significance). This allowed us to trace a delivery strategy based on spatial clusters. We must, first, deliver to very high-demand provinces; second, deliver PRRSGS to neighboring high-demand provinces; and third, deliver to medium- and low-demand provinces. This delivery is from the high-demand points to the low- and medium-demand points. Policymakers should treat forecasts as complementary information and include the zero-death provinces into the general delivery strategy.

The forecasts are made based on province-level observable characteristics and suggest that we can follow the same strategy of delivery from coastal and jungle regions to highlands.

Pandemic preparedness, response, and recovery success depend on the methodology used to estimate the demand. This methodology guarantees that we can used vulnerability to COVID-19 mortality as a proxy for demand. Deprivation costs are high where there is high mortality against COVID-19, so we propose that a demand assessment must follow this criterion.

## 6. Conclusions

In this paper, we argued that a supportive framework for economic reactivation can be built based on vulnerability against COVID-19 mortality. This vulnerability arises from deprivation costs that are high as a consequence of the pandemic. Following this idea, we decided to estimate SDMs, where the dependent variable was the mortality rates for COVID-19 at the provincial level. This regression model helped to understand, first, the difference between mortality rates as the difference in vulnerability against COVID-19. The rest of the methodological steps included a model selection and post-estimation analysis that helped to find the best model and to use their predictions to trace a delivery strategy for PRRSGS. The demand for PRRSGS was proxied by the predicted COVID-19 mortality rates, which determined vulnerability against COVID-19 mortality.

The principal result for the regression analysis is that $\mathrm{SDM}(1,1,1)$, the more general SDM, is the one that better fits the data, suggesting that there are statistically significant endogenous, exogenous, and correlated spatial interaction effects. This means that mortality rates exhibit spatial patterns. Two branches of literature were analyzed: the prediction-oriented machine learning literature and the social determinants of health literature. The evidence favors the hypotheses related to the machine learning literature for the early stages of the pandemic. Poverty, population density, and overcrowding remained the main predictors. However, overcrowding's direct and indirect (spatial spillovers) effects are counteracted, so further research is needed to clarify the relationship between overcrowding and mortality rates. Provinces are inevitably returning to their long-run level of economic activity, and high mortality rates exist because there is scarcity in essential supplies that are needed to save lives regarding infection with COVID-19. Population density is positively related to mortality rates, so for provinces with high population density (overcrowding in cities), policymakers should expect higher mortality rates, and thus, the delivery of PRRSGS must be prioritized. The rest of the vulnerability drivers related to demographics, health insurance, and health system performance were not significant, and thus, empirically, we reject the hypotheses regarding these variables.

The hypothesis of provinces' vulnerability to natural disasters (and all of its blocks) is partially accepted; there are important sources of vulnerability based on poverty, population density, and overcrowding in households. It is important to notice that poverty interacts with other variables theoretically and empirically, and in this case, it helps to define the effect of population density and overcrowding in households. Poor provinces are more overcrowded in households but less overcrowded in cities; they also have lower mortality rates, which means that overcrowding in households is negatively correlated with mortality rates. Overcrowding in cities or population density is positively correlated with COVID-19 mortality. These correlations include spatial interactions, which were important at the moment when the final effect on the mortality rates is determined. Regarding age, sex, skin-color, chronic disease prevalence, obesity, and health system performance, we have not arrived at statistically significant results, and thus, we cannot reject the null hypothesis of no relationship between mortality rates and these variables.

This paper contributes empirical methodologies, and statistical (spatial regression) and machine learning (ENR)-related methods applied to the field of humanitarian logistics. The evidence provided by this paper mean new insights for a specific case, and spatial interaction effects represent new mechanisms through which vulnerability drivers affect COVID-19 mortality. There is an increasing demand of empirical methodologies in this field because they help to integrate academia with practitioners. Consequently, the novelty in this paper regarding similar papers on the field is the application of humanitarian logistics principles to a data-driven analysis, using the insights obtained to build a framework for economic reactivation based on an analysis of the demand and identifying the goods needed for the delivery of humanitarian aid.

The economy must be reactivated considering humanitarian logistics principles, which requires an assessment of the demand for humanitarian aid, considering COVID-19 mortality as the main proxy for this demand. The applied methodology shows that data-driven techniques are especially useful in improving disaster management techniques and decision-making. This methodology can be replicated with other cases to optimize the policies for pandemic response and recovery phases. However, future research is still needed to overcome the limitations of data and to validate these first exploratory results. Further research in the field should be encouraged in order to link practitioners with academia; this can be performed by using empirical methods in humanitarian logistics research. This paper contributes to the empirical literature, but research in the field should be further expanded in order to overcome the limitations and other related studies.

**Author Contributions:** Conceptualization, R.R.-R., R.Q., I.d.B.J. and M.C.; methodology, R.Q., R.R.-R. and A.L.; software, R.Q.; validation, R.R.-R., A.L. and R.Q.; formal analysis, R.Q., R.R.-R. and A.L.; investigation, I.d.B.J., M.C. and R.Q.; resources, R.R.-R., I.d.B.J. and M.C.; data curation, R.Q.; writing—original draft preparation, R.Q., I.d.B.J. and M.C.; writing—review and editing, A.L., I.d.B.J., M.C., R.R.-R. and R.Q.; visualization, R.R.-R. and A.L. supervision, M.C.; project administration, M.C.; funding acquisition, M.C., R.R.-R. and I.d.B.J. All authors have read and agreed to the published version of the manuscript.

**Funding:** This research was funded by Vicerrectorado de Investigación (VRI) y Centro de Investigación de la Universidad del Pacífico (CIUP), Universidad Nacional Abierta y a Distancia and the Coordination for the Improvement of Higher Education Personnel-Brazil (CAPES), Procad Defesa 88887.387760/2019-00.

**Institutional Review Board Statement:** Not applicable.

**Informed Consent Statement:** Not applicable.

**Data Availability Statement:** The data as well as software program used to perform the estimation is attached in the following link: https://github.com/renatoquiliche/COVID19-socsi (accessed on 20 December 2020).

**Conflicts of Interest:** Authors declare no conflicts of interest.

## References


1. Presidencia del Consejo de Ministros. Decreto Supremo que precisa los alcances del artículo 8 del Decreto Supremo N 044-2020-PCM, que declara el estado de emergencia nacional por las graves circunstancias que afectan la vida de la nación a consecuencia del brote del COVID-19. In *Decreto Supremo N° 045-2020-PCM*; El Peruano: Lima, Peru, 2020.
2. Presidencia del Consejo de Ministros. Decreto Supremo que modifica el Decreto Supremo N° 116-2020-PCM, Decreto Supremo que establece las medidas que debe seguir la ciudadanía en la nueva convivencia social y prorroga el Estado de Emergencia Nacional por las graves circunstancias que afectan la vida de la Nación a consecuencia del COVID-19, modificado por los Decretos Supremos N° 129-2020-PCM, N° 135-2020-PCM, N° 139-2020-PCM, N° 146-2020-PCM, N° 151-2020- PCM, N° 156-2020-PCM, N° 162-2020-PCM y N° 165-2020-PCM. In *Decreto Supremo N° 170-2020-PCM*; El Peruano: Lima, Peru, 2020.
3. Poder Ejecutivo. Decreto de urgencia que establece el retiro extraordinario del fondo de pensiones en el sistema privado de pensiones como medida para mitigar efectos económicos del aislamiento social obligatorio y otras medidas. In *Decreto de Urgencia N° 034-2020*; El Peruano: Lima, Peru, 2020.
4. Poder Ejecutivo. Decreto de urgencia que establece medidas para reducir el impacto en la economía peruana, de las disposiciones de prevención establecidas en la declaratoria de estado de emergencia nacional ante los riesgos de propagación del COVID-19. In *Decreto de Urgencia N° 033-2020*; El Peruano: Lima, Peru, 2020.
5. Ministerio del Interior. Aprueban protocolo para la implementación de las medidas que garanticen el ejercicio excepcional del derecho a la libertad de tránsito en el marco del Estado de Emergencia Nacional declarado mediante D.S. N° 044-2020-PCM. In *Resolución Ministerial N° 304-2020-IN*; El Peruano: Lima, Peru, 2020.
6. Ministerio de Economía y Finanzas. Aprueban el Reglamento Operativo del Fondo de Apoyo Empresarial a la MYPE (FAE-MYPE). In *Resolución Ministerial N° 124-2020-EF/15*; El Peruano: Lima, Peru, 2020.
7. Montoro, C.; Pérez, F.; Herrada, R. Medidas del BCRP Frente a la Pandemia del Nuevo Coronavirus. *Rev. Moneda* **2020**, *182*, 10–18.
8. Brodeur, A.; Gray, D.; Islam, A.; Bhuiyan, S. A Literature Review of the Economics of Covid-19. IZA Discussion Paper No. 13411. 2020. Available online: https://ssrn.com/abstract=3636640 (accessed on 20 December 2020).
9. Acemoglu, D.; Chernozhukov, V.; Werning, I.; Whinsnton, M.D. *A Multi-Risk SIR Model with Optimally Targeted Lockdown Working Paper No. 27102*; Working Paper Series; National Bureau of Economic Research: Cambridge, MA, USA, 2020. [CrossRef]
10. Béland, L.-P.; Brodeur, A.; Mikola, D.; Wright, T. *The Short-Term Economic Consequences of Covid-19: Occupation Tasks and Mental Health in Canada*; Social Science Research Network: New York, NY, USA, 2020. Available online: https://papers.ssrn.com/abstract=3602430 (accessed on 20 December 2020).
11. Overstreet, R.E.; Hall, D.J.; Hanna, J.B.; Rainer, R.K. Research in humanitarian logistics. *J. Humanit. Logist. Supply Chain Manag.* **2011**, *1*, 114–131. [CrossRef]
12. Holguín-Veras, J.; Pérez, N.; Jaller, M.; Van Wassenhove, L.N.; Aros-Vera, F. On the appropriate objective function for post-disaster humanitarian logistics models. *J. Oper. Manag.* **2013**, *31*, 262–280. [CrossRef]
13. Leiras, A.; Junior, I.D.B.; Peres, E.Q.; Bertazzo, T.R.; Yoshizaki, H.T.Y. Literature review of humanitarian logistics research: Trends and challenges. *J. Humanit. Logist. Supply Chain Manag.* **2014**, *4*, 95–130. [CrossRef]
14. Renteria, R.; Chong, M.; de Brito Junior, I.; Luna, A.; Quiliche, R. An entropy-based approach for disaster risk assessment and humanitarian logistics operations planning in Colombia. *J. Humanit. Logist. Supply Chain Manag.* **2021**, *11*, 428–456. [CrossRef]
15. Andersen, L.M.; Harden, S.R.; Sugg, M.M.; Runkle, J.D.; Lundquist, T.E. Analyzing the spatial determinants of local Covid-19 transmission in the United States. *Sci. Total Environ.* **2021**, *754*, 142396. [CrossRef]
16. Khalatbari-Soltani, S.; Cumming, R.G.; Delpierre, C.; Kelly-Irving, M. Importance of collecting data on socioeconomic determinants from the early stage of the COVID-19 outbreak onwards. *J. Epidemiol. Community Health* **2020**, *74*, 620–623. [CrossRef] [PubMed]
17. Pedrosa, N.L.; de Albuquerque, N.L.S. Spatial analysis of COVID-19 cases and intensive care beds in the state of Ceará, Brazil. *Cienc. Saude Coletiva* **2020**, *25*, 2461–2468. [CrossRef]
18. Rollston, R.; Galea, S. COVID-19 and the Social Determinants of Health. *Am. J. Health Promot.* **2020**, *34*, 687–689. [CrossRef]
19. Sugg, M.M.; Spaulding, T.J.; Lane, S.J.; Runkle, J.D.; Harden, S.R.; Hege, A.; Iyer, L.S. Mapping community-level determinants of COVID-19 transmission in nursing homes: A multi-scale approach. *Sci. Total Environ.* **2020**, *752*, 141946. [CrossRef]
20. White, E.R.; Hébert-Dufresne, L. State-level variation of initial COVID-19 dynamics in the United States. *PLoS ONE* **2020**, *15*, e0240648. [CrossRef]
21. Gupta, A.; Banerjee, S.; Das, S. Significance of geographical factors to the COVID-19 outbreak in India. *Model. Earth Syst. Environ.* **2020**, *6*, 2645–2653. [CrossRef]
22. Benzakoun, J.; Hmeydia, G.; Delabarde, T.; Hamza, L.; Meder, J.; Ludes, B.; Mebazaa, A. Excess out-of-hospital deaths during the COVID -19 outbreak: Evidence of pulmonary embolism as a main determinant. *Eur. J. Heart Fail.* **2020**, *22*, 1046–1047. [CrossRef]
23. Zhang, C.H.; Schwartz, G.G. Spatial Disparities in Coronavirus Incidence and Mortality in the United States: An Ecological Analysis as of May 2020. *J. Rural. Health* **2020**, *36*, 433–445. [CrossRef]
24. Bruno, G.; Wemole, R.; Ronsivalle, G.B.; Foresti, L.; Poledda, G. *A Prototype Model of Georeferencing the Inherent Risk of Contagion from COVID-19*; ResearchGate preprint: Rome, Italy, 2020. [CrossRef]
25. Carrillo-Larco, R.M.; Castillo-Cara, M. Using country-level variables to classify countries according to the number of confirmed COVID-19 cases: An unsupervised machine learning approach. *Wellcome Open Res.* **2020**, *5*, 56. [CrossRef]

26. Chakraborti, S.; Maiti, A.; Pramanik, S.; Sannigrahi, S.; Pilla, F.; Banerjee, A.; Das, D.N. Evaluating the plausible application of advanced machine learnings in exploring determinant factors of present pandemic: A case for continent specific COVID-19 analysis. *Sci. Total Environ.* **2021**, *765*, 142723. [CrossRef]
27. Rustam, F.; Reshi, A.A.; Mehmood, A.; Ullah, S.; On, B.-W.; Aslam, W.; Choi, G.S. COVID-19 Future Forecasting Using Supervised Machine Learning Models. *IEEE Access* **2020**, *8*, 101489–101499. [CrossRef]
28. Crandall, M.S.; Weber, B.A. Local Social and Economic Conditions, Spatial Concentrations of Poverty, and Poverty Dynamics. *Am. J. Agric. Econ.* **2004**, *86*, 1276–1281. [CrossRef]
29. Wardhana, D.; Ihle, R.; Heijman, W. Agro-clusters and Rural Poverty: A Spatial Perspective for West Java. *Bull. Indones. Econ. Stud.* **2017**, *53*, 161–186. [CrossRef]
30. Alexander, D.E. *Principles of Emergency Planning and Management*; Oxford University Press: Oxford, UK, 2002.
31. Smith, C.D.; Mennis, J. Incorporating Geographic Information Science and Technology in Response to the COVID-19 Pandemic. *Prev. Chronic Dis.* **2020**, *17*, 246. [CrossRef] [PubMed]
32. Zeckhauser, R. The economics of catastrophes. *J. Risk Uncertain.* **1996**, *12*, 113–140. [CrossRef]
33. Pindyck, R.S.; Wang, N. The Economic and Policy Consequences of Catastrophes. *Am. Econ. J. Econ. Policy* **2013**, *5*, 306–339. [CrossRef]
34. Yoshizaki, H.; de Brito Junior, I.; Hino, C.; Aguiar, L.; Pinheiro, M. Relationship between Panic Buying and Per Capita Income during COVID-19. *Sustainability* **2020**, *12*, 9968. [CrossRef]
35. Gutjahr, W.J.; Fischer, S. Equity and deprivation costs in humanitarian logistics. *Eur. J. Oper. Res.* **2018**, *270*, 185–197. [CrossRef]
36. Alibaba. *COVID-19 Outbreak Hospital Response Strategy*; Zhejiang University School of Medicine: Hangzhou, China, 2020.
37. Alibaba. *Handbook of COVID-19 Prevention and Treatment*; Zhejiang University School of Medicine: Hangzhou, China, 2020.
38. Guliyev, H. Determining the spatial effects of COVID-19 using the spatial panel data model. *Spat. Stat.* **2020**, *38*, 100443. [CrossRef]
39. Sun, F.; Matthews, S.A.; Yang, T.-C.; Hu, M.-H. A spatial analysis of the COVID-19 period prevalence in U.S. counties through 28 June 2020: Where geography matters? *Ann. Epidemiol.* **2020**, *52*, 54–59.e1. [CrossRef]
40. Schwalb, A.; Seas, C. The COVID-19 Pandemic in Peru: What Went Wrong? *Am. J. Trop. Med. Hyg.* **2021**, *104*, 1176–1178. [CrossRef] [PubMed]
41. Alvarez-Risco, A.; Mejia, C.R.; Delgado-Zegarra, J.; Del-Aguila-Arcentales, S.; Arce-Esquivel, A.A.; Valladares-Garrido, M.J.; Del Portal, M.R.; Villegas, L.F.; Curioso, W.H.; Sekar, M.C.; et al. The Peru Approach against the COVID-19 Infodemic: Insights and Strategies. *Am. J. Trop. Med. Hyg.* **2020**, *103*, 583–586. [CrossRef] [PubMed]
42. Alvarez-Risco, A.; Del-Aguila-Arcentales, S.; Yáñez, J.A. Telemedicine in Peru as a Result of the COVID-19 Pandemic: Perspective from a Country with Limited Internet Access. *Am. J. Trop. Med. Hyg.* **2021**, *105*, 6–11. [CrossRef]
43. Turner-Musa, J.; Ajayi, O.; Kemp, L. Examining Social Determinants of Health, Stigma, and COVID-19 Disparities. *Health Care* **2020**, *8*, 168. [CrossRef]
44. Burström, B.; Tao, W. Social determinants of health and inequalities in COVID-19. *Eur. J. Public Health* **2020**, *30*, 617–618. [CrossRef] [PubMed]
45. Fielding-Miller, R.K.; Sundaram, M.E.; Brouwer, K. Social determinants of COVID-19 mortality at the county level. *PLoS ONE* **2020**, *15*, e0240151. [CrossRef] [PubMed]
46. Thakur, N.; Lovinsky-Desir, S.; Bime, C.; Wisnivesky, J.P.; Celedón, J.C. The Structural and Social Determinants of the Racial/Ethnic Disparities in the U.S. COVID-19 Pandemic. What's Our Role? *Am. J. Respir. Crit. Care Med.* **2020**, *202*, 943–949. [CrossRef] [PubMed]
47. Engle, S.; Stromme, J.; Zhou, A. *Staying at Home: Mobility Effects of COVID-19*; SSRN: New York, NY, USA, 2020.
48. Elhorst, J.P. Applied Spatial Econometrics: Raising the Bar. *Spat. Econ. Anal.* **2010**, *5*, 9–28. [CrossRef]
49. Kelejian, H.H.; Prucha, I.R. A Generalized Moments Estimator for the Autoregressive Parameter in a Spatial Model. *Int. Econ. Rev.* **1999**, *40*, 509–533. [CrossRef]
50. Zhou, X.; Lin, H. Spatial Weights Matrix. In *Encyclopedia of GIS*; Springer: Berlin/Heidelberg, Germany, 2008; p. 1113. [CrossRef]
51. O. Ogutu, J.; Schulz-Streeck, T.; Piepho, H.-P. Genomic selection using regularized linear regression models: Ridge regression, lasso, elastic net and their extensions. *BMC Proc.* **2012**, *6*, S10. [CrossRef]

*sustainability*

MDPI

*Article*

# Issues and Potential Solutions to the Clean Heating Project in Rural Gansu

**Dehu Qv [1,*], Xiangjie Duan [1], Jijin Wang [2], Caiqin Hou [1], Gang Wang [1], Fengxi Zhou [1,*] and Shaoyong Li [1,*]**

[1] Department of Building Environment and Energy Application Engineering, School of Civil Engineering, Lanzhou University of Technology, Lanzhou 730050, China; dxj2021001@163.com (X.D.); zjshcqzxy@163.com (C.H.); wang_g@lut.edu.cn (G.W.)

[2] School of Architecture, Harbin Institute of Technology, Key Laboratory of Cold Region Urban and Rural Human Settlement Environment Science and Technology, Ministry of Industry and Information Technology, Harbin 150090, China; wangjijin126@163.com

* Correspondence: QDH000@126.com (D.Q.); geolut@163.com (F.Z.); lishaoyong99@163.com (S.L.); Tel.: +86-931-2973715 (D.Q.)

**Abstract:** Rural clean heating project (RCHP) in China aims to increase flexibility in the rural energy system, enhance the integration of renewable energy and distributed generation, and reduce environmental impact. While RCHP-enabling routes have been studied from a technical perspective, the economic, ecological, regulatory, and policy dimensions of RCHP are yet to be analysed in depth, especially in the underdeveloped areas in China. This paper discusses RCHP in rural Gansu using a multi-dimensional approach. We first focus on the current issues and challenges of RCHP in rural Gansu. Then the RCHP-enabling areas are briefly zoned into six typical regions based on the resource distribution in Gansu Province, and a matching framework of RCHP is recommended. Then we focus on the economics and sustainability of RCHP-enabling technologies. Based on the medium-term assessment of RCHP in the demonstration provinces, various technical schemes and routes are analysed and compared in order to determine which should be adopted in rural Gansu. In addition to technical and economic effects of those schemes, the corresponding ecology, policy, finance, and market implications are also concerned. We briefly discuss how the national regulators incentivise the implementation of RCHP in rural Gansu. Major barriers to RCHP are identified as the sustainability of technology, economy, ecology, policy, finance, and market. Subsequently, some policy solutions to overcome these barriers are proposed.

**Keywords:** rural clean heating project; rural Gansu; sustainability; potential solutions; benchmarking

**Citation:** Qv, D.; Duan, X.; Wang, J.; Hou, C.; Wang, G.; Zhou, F.; Li, S. Issues and Potential Solutions to the Clean Heating Project in Rural Gansu. *Sustainability* **2021**, *13*, 8397. https://doi.org/10.3390/su13158397

Academic Editor: Shervin Hashemi

Received: 20 June 2021
Accepted: 15 July 2021
Published: 28 July 2021



## 1. Introduction

### 1.1. Building Conservation-Minded Society

After the first two-decade development in the 21st century, many countries in the world regard "harmonious development between human and nature" as the key of state construction in the next stage, especially after the COVID-19 disaster [1–5]. Particularly, the developed countries in Europe put "renewable energy evolution based new-age industrial revolution" into their national development programs [6–10].

In China, people are confronted with three challenges and/or opportunities: the scientific and technological revolution, the energy revolution, and the ecological revolution. Based on the "development-improvement" experience by western world, Chinese people adopt a sustainable development strategy in building a conservation-minded society (CMS) [11–15].

Energy saving is an important part of CMS, and building energy saving (BES) is the substantial portion [16–18]. Therefore, BES as a national strategic support is in line with the scientific development law.

Through long-term, active, and extensive participation of the whole society, China has made an extraordinary progress in three major sectors of urban residential energy saving, public building energy saving, and central heating system energy saving in the north. However, rural China, being the key rear area supporting China's reform and opening-up, faces many challenges from the clean heating project [17,18].

### 1.2. Rural Clean Heating Project

Since 2005, the characteristics of rural housing energy consumption and its composition have undergone profound changes with the promotion of New Rural Construction (NRC (The detailed information about NRC can be taken on http://qgxnc.org/, accessed on 3 February 2021)) and the improvement in peasant farmer's living standards. The most conspicuous variation is rural energy-use structure. Over the past 20 years, the consumption of commodity energy in rural China has been increasing year by year. Meanwhile the consumption of non-commodity energy (biomass energy, especially) has been decreasing [17,18].Furthermore, the total energy consumption in rural areas continues to rise, although the resident population (composed mainly of old people, young children, and school-age children (More information on rural China can be gotten on http://www.gov.cn/, accessed on 3 February 2021)) in rural China continues to decline in the meantime, which means that the rural energy-use intensity tends to grow all the time.

On the one hand, increased rural energy demand may indicate the improvement in peasant farmer's living conditions. On the other hand, the lower proportion of renewable energy in rural energy consumption is very worrying, since it works against the low-carbon development path advocated by the government. Therefore, the key to transform energy use in rural China lies in structural optimization of energy use and increase in renewable proportion [18].

Field research shows that the crucial domestic terminals of rural energy use consist of heating, cooking (including domestic hot water), and electricity consumption for household electric appliances and lighting. Among them, wastage from heating accounts for 53.6% and even more than 60% of waste in some areas in rural China [17,18]. Thus, the rural clean heating project (RCHP) is of great significance to the evolution of CMS in China.

### 1.3. Progress of RCHP in Rural Gansu

#### 1.3.1. Current State of Rural Gansu

Information on the current state of Gansu Province can be found on http://www.gansu.gov.cn/col/col10/index.html, accessed on 3 February 2021. Gansu Province (Figure 1 (Gansu Province map can also be found on http://www.gansu.gov.cn/col/col10/index.html, accessed on 3 February 2021.)) is the provincial administrative region of the People's Republic of China, and its capital is Lanzhou. It is located in the west of China, in the middle and upper reaches of the Yellow River, with a vast area (more than 1600 km winding from the east to the west, with a total area of 425,900 km$^2$, accounting for 4.72% of China's total area). There are complex terrains, crisscross mountains, and great differences in altitude. Meanwhile, there are rich mineral resources, wind resources, hydro-power resources, and solar energy resources. The province has a variety of climate types and abundant wildlife. As the birthplace of ancient Chinese civilisation, it has contributed extremely important history and culture to the world.

However, as obvious as the above advantages are, its natural ecology is fragile, the economic development is lagging behind, and the living standard of peasant farmers is low (e.g., the per capita GDP ranking of Gansu Province in China is 31/31 (http://www.gov.cn/shuju/chaxun/index.htm, accessed on 3 February 2021.)). Hence, determining how to carry out RCHP in the complex human and geographical environment, while taking into account the regional political and economic characteristics, is the core of RCHP in rural Gansu.

**Figure 1.** Gansu Province map. Note: This map is provided by http://www.gansu.gov.cn/col/col10/index.html, accessed on 3 February 2021.

#### 1.3.2. Living Energy-Use Reality in Rural Gansu

Table 1 details the annual consumptions of diverse energy resources in rural Gansu [17,18]. Figure 2 depicts the commercial energy consumption in a comparison with the north average [17,18]. It can be found in Table 1 that the total annual energy consumption in rural Gansu is lower than that in the north average, which implies that the peasant farmers' living standard in rural Gansu is lower than that in northern rural China. Fortunately, this living-standard gap has been narrowed since 2006.

**Table 1.** Annual consumptions of diverse energy resources in rural Gansu.

| Year | Name | Unit | Value | Average | Deviation |
|---|---|---|---|---|---|
| 2006 | Total building area | $10^8$ m$^2$ | 3.4 | 5.3 | −35.8% |
| | Coal | $10^4$ t | 779 | 1018.9 | −23.5% |
| | Liquefied gas | $10^4$ t | 1.2 | 13.5 | −91.1% |
| | Power | $10^8$ kWh | 15.7 | 29.4 | −46.6% |
| | Firewood | $10^4$ t | 55 | 287.6 | −80.9% |
| | Crop straw | $10^4$ t | 92 | 357.1 | −74.2% |
| | Commercial energy | $10^4$ tec | 610 | 849.4 | −28.2% |
| | Non−Com. energy | $10^4$ tec | 79 | 348.4 | −77.3% |
| | Sum | $10^4$ tec | 689 | 1200.5 | −42.6% |
| 2014 | Total building area | $10^8$ m$^2$ | 3.7 | 5.9 | −37.3% |
| | Coal | $10^4$ t | 890 | 1056.1 | −15.7% |
| | Liquefied gas | $10^4$ t | 3.5 | 14.8 | −76.4% |
| | Power | $10^8$ kWh | 16.3 | 46.4 | −64.9% |
| | Firewood | $10^4$ t | 45 | 329.7 | −86.4% |
| | Crop straw | $10^4$ t | 80 | 352.9 | −77.3% |
| | Commercial energy | $10^4$ tec | 695 | 928.1 | −25.1% |
| | Non−Com. energy | $10^4$ tec | 67 | 383.1 | −82.5% |
| | Sum | $10^4$ tec | 762 | 1311.1 | −41.9% |

**Table 1.** *Cont.*

| Year | Name | Unit | Value | Average | Deviation |
|---|---|---|---|---|---|
| 2018 | Total building area | $10^8$ m$^2$ | 3.4 | 5.3 | −35.8% |
| | Coal | $10^4$ t | 918.8 | 803.6 | 14.3% |
| | Liquefied gas | $10^4$ t | 4.3 | 13.5 | −68.1% |
| | Power | $10^8$ kWh | 20.7 | 58.3 | −64.5% |
| | Firewood | $10^4$ t | 46.5 | 265.5 | −82.5% |
| | Crop straw | $10^4$ t | 82.6 | 315.5 | −73.8% |
| | Commercial energy | $10^4$ tec | 723.6 | 818.5 | −11.6% |
| | Non−Com. energy | $10^4$ tec | 69 | 317.1 | −78.2% |
| | Sum | $10^4$ tec | 792.6 | 1135.5 | −30.2% |

Note: the commercial energy mentioned in this table includes coal, liquefied gas, and power electricity; the non-commercial energy (Non-Com.) includes firewood and crop straw. The source of data in Table 1 is Reference [17], National Bureau of Statistics of China. China Statistical Yearbook. http://www.stats.gov.cn/tjsj/ndsj/; accessed on 3 February 2021.

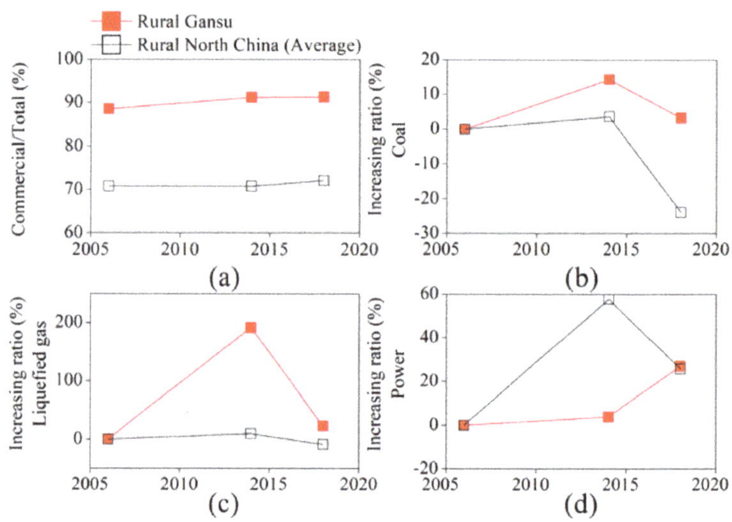


**Figure 2.** Commercial energy consumption in rural Gansu; (**a**) the percentage of commercial energy consumption to total energy consumption; (**b**) the increasing ratio of coal consumption; (**c**) the increasing ratio of liquid gas consumption; (**d**) the increasing ratio of power consumption; Note: The complete data can be found in References [17,18].

On the other hand, in terms of annual commercial energy consumption, its proportion in total energy use in rural Gansu is dramatically higher than that in north average, and this percentage still increases (Figure 2a). Although the coal consumption in the northern rural China has been dropped since 2014, the peasant farmers living in rural Gansu tend to consume more (Figure 2b). The liquefied-gas consumption shows a similar trend in past years as Figure 2c depicts. By contrast, the increasing rate of power consumption in rural Gansu is accelerating, which is contrary to the slowing-down tendency in the northern rural China (Figure 2d) [17,18].

Based on the other studies we can understand that there is a significant gap between energy-use demand and energy supply in rural Gansu, and there is a great divergence between its energy structure and the sustainability goals of China [11–16]. Furthermore, the contradictions existing in the peasant farmers' living conditions and their health targets are obvious [18]. In particular, the indoor air temperatures in rural Gansu are generally low with cold air infiltration, and the indoor air quality is poor caused by coal combustion [18].

## 2. Challenges

This section firstly summarises mutual issues of the clean heating project in northern rural China. Then, the specific issues in rural Gansu, which make the advance in RCHP in rural Gansu harder, are introduced from different aspects.

### 2.1. Mutual Issues in Northern Rural China

Mutual issues in northern rural China have been studied carefully in the reference [18].

(1) Different from the cities with concentrated population, the rural population is scattered and forms a small-scale settlement pattern with villages as the main administrative unit. Thus, rural building energy saving should be carried out by villages.

(2) A combination of production and living functions has made rural houses special. The farmhouses have large courtyards, mostly vacant or idle rooms, and are mainly single storey or low-rise buildings.

(3) The traditional lifestyles in the northern rural areas have a profound impact on the energy consumption of heating and cooking: (1) In the cold winter, peasant farmers spontaneously integrate some room functions to reduce the number of heating rooms and shorten the heating hours. (2) Peasant farmers' clothing level is based on the short-term outdoor work "not cold", and their clothing is thicker. In the meantime, peasant farmers do not adapt to the large indoor-outdoor temperature difference. Hence the indoor design temperature of rural residential buildings in winter is generally lower than 16 °C, conventionally ranged 10–15 °C. (3)The thrifty living habits of Chinese peasant farmers, together with the traditional ways of life and labour, jointly determine the heating characteristics of "part of the time and part of the space" in northern rural China, and thus generate the heating demand of "room regulation; use as you go; one key operation".

(4) The midterm assessment for the northern RCHP, which started in 2016, was completed. By the end of 2018, the clean heating rate of the regions participating in the project has exceeded the medium-term target [17,18]. During this period, a series of clean heating schemes suitable for rural areas in northern China have been verified, which include coal to electricity, coal to gas, coal to biomass, solar based multi energy complementation, and coal to central heating [17,18]. Each scheme has different applicable conditions, advantages, and disadvantages, which are detailed in Tables 2–5 (The corresponding data come from the reference [18]). In particular, the scheme of coal to central heating is only suitable for cogeneration and industrial surplus heat recovery, which require speasant farmers to live in concentrated dwellings, so that the transmission and distribution pipes will not be too long [19–25].

(5) Other problems exposed in practice include: (1) there is a lack of coordination among departments as well as there are conflicts among policies, whereas such conflicts are usually ignored because people tend to consider the disharmony as a symbol that each department performs its own tasks; (2) financial subsidies do not fully consider the differences (economic, environmental protection, sustainability, etc.) between various technical schemes; (3) the economic incentive policy is not clear enough, and the subsidy is insufficient; (4) the government needs stronger scientific guidance of planning and technical path, and the overall consideration of follow-up operation and maintenance; (5) part of the implementation plan is too rough, which directly affects the enthusiasm of peasant farmers' participation; (6) the operation mechanism of marketisation is unsound, which relies too much on government subsidies, and financing is difficult; (7) the sustainable development and long-term market mechanism have not yet formed, and it is difficult for peasant farmers to bear fully without the subsidy; (8) the lack of quantitative evaluation mechanism makes it difficult to revise the technical route in time [17,18].

**Table 2.** Assessment of coal to electricity.

| Name | Details |
| --- | --- |
| Optimum scheme | Low-temperature air source heat-pump air heater |
| Technical advantages | At present, China is in the leading position in the application fieldin the world.<br>The heat-pump products can provide reliable and efficient heating whilethe ambient temperature is higher than −30 °C, which covers most of climatic regions with heating demand in China. |
| Economical advantages | The power consumption of each household in a whole heating season is less than 2000 kWh.<br>Even if there is no electricity subsidy, the total cost is less than 1000 RMB. |
| Common advantages | The demand of external power capacity is small, and there is no need for special rural power grid upgrading. |
| Using advantages | One key operation; hot air output; without water pump energy consumption; without water loop leakage; without antifreeze or other issues. |
| Summary | It is anideal technical scheme widely suitable for RCHP in China. |

**Table 3.** Assessment of coal to gas.

| Name | Details |
| --- | --- |
| Economic disadvantages | The price of gas is higher than the psychological acceptance price of peasant farmers, which leads to the gas company spending a lot of money to build the gas pipeline network, but the actual effective utilisation rate is low, preventing obtainment of an investment return. |
| Safety disadvantages | There are common safety hazards, involving products, installations, constructions, operations, and daily safety awareness. |
| Common disadvantages | The pressure of gas resources is great, and gas companies generally suffer serious losses. |
| Financial disadvantages | The subsidy input is large and increasing year by year, resulting in a heavy financial burden on the local government. |
| Summary | This scheme is almost unacceptable in northern rural China |

**Table 4.** Assessment of coal to biomass.

| Name | Details |
| --- | --- |
| Optimum solutions | Plan A: Hot air type biomass pellet heating furnace<br>Plan B: Hot water type biomass granule heater |
| Technical advantages | Plan A: Compared with scheme of coal to gas, the transformation cost is reduced by 38% and the operation cost is reduced by 52%. In areas rich in biomass resources, it is worth promoting.<br>Plan B: In addition to heating load, it also takes into account the demand of cooking and domestic hot water, and closely fits with the traditional living habits of peasant farmers. At present, this technology has been fully automated, intelligent, has one key operation, and is convenient and safe. Even without government subsidy, each household consumes about 1.5 t biomass particles in a whole heating season, and the heating cost is about 1000 RMB, which is equivalent to the cost of burning bulk coal, but it is more convenient and environmentally friendly. |
| Technical defects | The current biomass particles are mainly wood, which cannot make full use of more abundant crop straw resources. |
| Market defects | Furnace products lack standard specifications. |
| Industrial defects | The market specialisation of biomass energy industry chain is low, and the complete system has not yet formed. |
| Common defects | The national and local governments have not given clear opinions on the environmental protection requirements of biomass energy, resulting in many regions still waiting. |

**Table 5.** Assessment of solar based multi energy complementation.

| Name | Details |
|---|---|
| Mainstream technologies | Integrate solar energy collection system with electric energy, natural gas, or other energy sources to provide hot water or hot air for peasant farmers. |
| Technical and economical disadvantages | This system is usually complex. In order to ensure the reliability of unattended, solar energy is often abandoned and auxiliary heat source is used instead, which is contrary to the original intention of clean heating. |
| Sustainability disadvantage | The system control is complex, and there is large amount of operation and maintenance, which is not suitable for rural areas. |

### 2.2. *Special Issues in Rural Gansu*

Special issues in rural Gansu can be found on http://www.gansu.gov.cn/col/col10/index.html, accessed on 3 February 2021.

#### 2.2.1. Resource Distribution

(1) Land. The unused land is concentrated in the north, and dry land is concentrated in the southeast. The woodlands are scattered along the southeast border, and pastures are concentrated in the southwest.
(2) Mineral resources. The nonmetal resources are mainly distributed in the south-central part of the province, and metal minerals are distributed in the whole province, but more in the south.
(3) Clean energy. It is mainly distributed in the west and north of Jiuquan and Jiayuguan line.
(4) Circular economy base planned by the government. Zhangye-Wuwei-Dingxiarea is thecircular economy base for processing characteristic agricultural and sideline products. Gannan-Linxia-Longnan area is the base of ecological circular economy. Other areas are heavy industry and general industrial bases.
(5) Human resources. There are few higher education institutions in the province, and they are concentrated in the provincial capital. Restricted by economy, human settlements, and other factors, all kinds of talent-introduction work lags far behind other provinces. Further, even the talent exchange with neighbouring Shaanxi Province is not frequent, and the development of information and technology is lagging behind.

#### 2.2.2. Population Distribution

They are mainly distributed in Gannan-Linxia-Longnan area, or around the provincial capital, and in various industrial areas. In the area west and north of the Jiuquan-Jiayuguan area, which is rich in clean energy, the natural conditions are hard and the population is sparse.

#### 2.2.3. Economic Distribution

Rural poverty is generally deep, showing a spatial pattern of "high in the southeast and low in the northwest". The spatial agglomeration characteristics of rural economy and social poverty are significant. The spatial pattern of environmental poverty is "high on both wings and low in the middle".

#### 2.2.4. History and Culture

With a long history and culture, traditional notions are deeply rooted in the hearts of the natives. On the one hand, the common customs of thrift and simplicity still exist. On the other hand, there is little internal motivation for new things and trends.

#### 2.2.5. Ecological Environment

(1) Water environment. Precipitation decreases from southeast to northwest, and water resources per capita is low. For example, the average annual precipitation of the

whole province is 398.5 mm, which is less than the national average (632 mm). It is one of the provinces with the least precipitation in China. About 70% of the area has annual precipitation less than 500 mm (http://www.gansu.gov.cn/col/col2085/index.html, accessed on 3 February 2021.).

(2) Atmospheric environment. Although the air quality has been improved, it is still disturbed by dust.

(3) Land environment. The effectively used area of land is small. Soil fertility remains weak, and soil erosion and desertification are still serious.

(4) Self repair function. The Yellow River Basin is dry and rainless all the year round, the river runoff is small, the self purification ability is feeble, and the ecological environment is still fragile.

## 3. Framework of RCHP in Rural Gansu

### *3.1. Rural Residential Insulation Transformation*

So far, the proportion of residential insulation in rural Gansu is about 2.6%, while it is 30.7% in Beijing [17,18]. Thus, there is greater improvement potential, which is the first step of RCHP.

#### 3.1.1. Existing Farmhouse

In accordance with the characteristics of rural residential heating demand, internal thermal insulation (ITIS) is more suitable for the external wall. Compared with external insulation, internal insulation is simpler, which supports the use of local rooms and reduces indoor continuous operation area. Thus, the construction is safer and more convenient, which is conducive to improve construction efficiency and reduce labour intensity. In addition, the construction is not affected by outdoor climate.

The overall implementation principles of ITIS include low-cost targeted thermal insulation and extra insulation that comprises of external doors and windows. The mentioned targeted thermal insulation means that the frequently used rooms are insulated first. Repairing roofs and insulating north facing external walls have priority over the others. In the meantime, sunlight in the south should be fully used, and interiorceiling can be increased to reduce indoor floor height. When the selected insulated materialispolymer resin, total investment of the whole household is about 2275 RMB, and the comprehensive energy saving ratio is about 30% [18]. If the budget is still sufficient, extra insulation should be considered. Here the total investment of the whole household will be increased by 1325 RMB, and the comprehensive energy saving rate will be about 35% [18].

#### 3.1.2. New Farmhouse

When the funds are limited, we can adopt the insulation route for existing rural houses. If the fund is surplus, by contrast, the overall insulation of the enclosure structure can be considered.

Straw board and brick insulation is preferred in wheat, pasture, and grass production areas. The thermal conductivity of dry grass leaves is about 0.1 W/(m·K), which is equivalent to that of cement-based perlite. The heat transfer coefficient of a 330 mm straw brick wall is about 0.3 W/($m^2$·K), which is about 1/3 of that of 370 mm red brick wall [18]. At present, the technology has been successfully applied in demonstration projects in Gansu and Heilongjiang provinces, and the overall cost per unit building area is 600–700 RMB/$m^2$, which is lower than that of conventional brick-and-tile houses [18]. In the absence of biomass raw materials, new insulation blocks can be considered as external wall insulation and foam cement roofs as roof insulation. This technology has been used for many years in the north of China. It is mature, reliable, and economical. In addition, PVC plastic window and glass fiber reinforced plastic window have good thermal insulation performance with relatively low price, which are proper insulation solutions for rural China.

#### 3.1.3. Relevant Issues

Relevant issues in rural Gansu can be found on http://www.gansu.gov.cn/col/col10/index.html, accessed on 3 February 2021.

(1) Gansu Province belongs to cold climates, so that the casement windows rather than sliding windows should be adopted.
(2) The fire-resistant rate of internal insulation materials is required to be B1 and above. In the meantime, safety awareness of peasant farmers must be improved.
(3) The condensation phenomenon on the inner surface of external wall is rare because indoor heating temperature of rural houses is low, absolute moisture content of air is small, and its dew point temperature is lower.
(4) There is an air layer between the insulation material and the inner surface of the external wall, and the upper part of the air layer is connected with the ceiling to ensure smooth exhaust of moisture.
(5) There is no risk of condensation in the overall external insulation structure as the previous research demonstrated [26].

### *3.2. Clean Heating Route*

(1) The basic principles include: (1) divide the implementation with the characteristics of resources distribution (including material, energy, population, economy, etc.); (2) take ecology as the fundamental, and efficiently use the regional resources; (3) streamline configurations and technologies.
(2) Reasonably exploit the biomass energy without wasting, nor impairing the development of wind power, hydropower, photovoltaic power, etc.
(3) The cost of energy transmission and distribution must be carefully calculated and compared among various RCHP schemes. The main stream technique is the distributed energy system technique.
(4) Pay attention to the technical economy and check the payback period and sustainability of the scheme.
(5) Under the guidance of the government, green development funds for poor and backward areas should be set up to support clean heating enterprises to issue green bonds and support the flow and exchange of talents.
(6) Draw lessons from advanced experience, establish clean heating standards and specifications led by the government, and introduce independent supervision of the third party.
(7) Support and guide peasant farmers' self-organized biomass cooperatives, and conduct the biomass-energy poverty alleviation and development project based on the rational use of tax lever.

## 4. Methods

Figure 3 draws the methodological map of this study, which also plots the underlying connections among various methods and tools.

### *4.1. Investigation and Survey*

The sources of data are statistical data by National Statistical Yearbook, micro survey results from individual or organ interview, modelling data stemmed from CBEM by Tsinghua University, and comprehensive summary of measured data which provided by a range of independent research.

### *4.2. Benchmarking*

In this section, we are concerned with demographic factors, natural environmental factors, cultural and historical environmental factors, human resource factors, natural resource factors, economic factors, and policy factors.

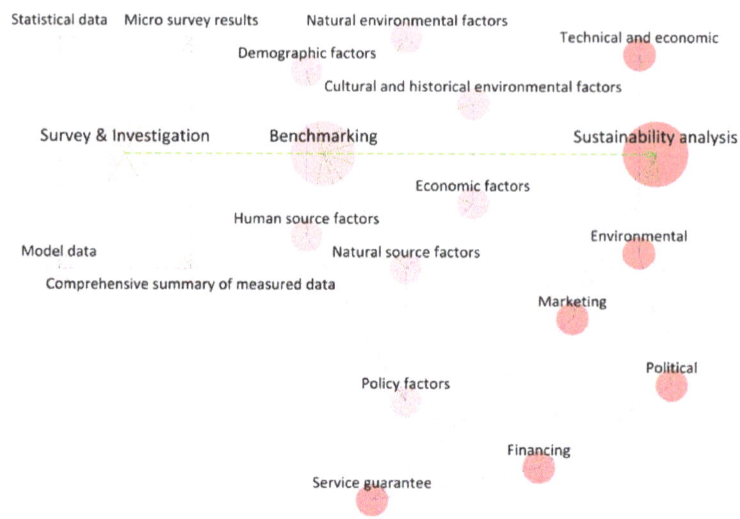


**Figure 3.** Methodological map.

### *4.3. Sustainability Analysis*

Sustainability analysis covers technical and economic, environmental, political, marketing, financing, and service-guarantee aspects. More specifically, the adopted technologies have to be reliable, renewable (avoid using high quality energy in a low-quality way), and economical (the investment is less and the payback period is short). Owing to an originally fragile ecology in Gansu Province, environment protection has to be the fundamental of development. Pollution-treatment route cannot be considered. Quantifying environmental impact indexes, assessing various technical solutions, and often revising technology roadmap is necessary and significant.

Additionally, policies must stand the test of time and the policy makers must abandon opportunism and concentrate on long-term development. Subsidy policies should be different in accordance with various technical and economic indicators of different schemes. Policy formulation, implementation, and supervision must be managed systematically. With respect to the market, the adopted scheme should be helpful to improve the market mechanism and establish a sustainable market order, which is commercially viable, even if preferential policies are absent. Meanwhile, the product quality must be strictly supervised and controlled, and a complete supporting industry chain that serves RCHP is necessary. Moreover, the government must lead the establishment of green development funds, and actively encourage the introduction of social capital to benefit the people. Step by step, a benign operation model of "enterprise oriented, government driven and residents affordable" can be set up. Ultimately, a circular ecology is formed based on long-term planning of the government, which carries out the biomass-based poverty alleviation project and supports (and guides) peasant farmers to build their own ecological home.

## 5. Results and Discussion

In this part, the information sources are taken on http://www.gansu.gov.cn/col/col10/index.html, accessed on 3 February 2021.

### *5.1. Longdong Zone with Grass Field Ecological Agriculture and Animal Husbandry Industrial Area*

#### 5.1.1. General Principles

The arid and semi-arid hilly areas in Eastern Gansu Province are grassland zonal vegetation. Based on the link between Leguminosae and Gramineae, the dry farming farmland will be rotated in the Loess Plateau to improve the soil fertility and develop the family breeding of cattle, pigs, sheep, and chickens with grass feed.

In order to prevent soil erosion and desertification, it is necessary to take returning farmland to forest and grassland as the main direction for the ridge, hill and gully land with serious soil and water loss. In the meanwhile, establish a farmland shelterbelt system with shrubs as the main component and trees, shrubs, and grasses combined, to improve the conditions of agricultural production. Take feed processing as the leader to increase the utilisation rate of feed and other plants and enhance the output of livestock products.

Based on the initial investment constraints and operating cost constraints, we should focus on the long-term development of this zone and form a household-scale biomass energy cogeneration model.

#### 5.1.2. Energy Saving Based on Thermal Insulation of Rural Buildings

The rural houses in this area are mainly rebuilt. Based on the full biomass resources in this zone, we adopt the overall insulation structure of straw board and brick. The cost of external wall is about 125 RMB/$m^2$, the cost of external window is about 320 RMB/$m^2$, and the cost of roof insulation is about 50 RMB/$m^2$. The overall cost of unit building area ranges from 600 to 700 RMB/$m^2$, which is slightly lower than that of conventional brick-and-tile houses. The average heating energy saving of each household is more than 30% [17,18].

#### 5.1.3. Construction of Biomass Industry Cycle

This project benchmark is the nearly zero carbon emission mode of biomass cogeneration with "gas-electricity-heat-fertiliser" cycle in Ping'an County, Hebei Province. This project takes households as the unit based on the water collection and storage facilities. A household-scale cogeneration pattern with efficient planting, house feeding and breeding, and biogas energy supply is set up to coordinate the contradictions among gas, electricity, heat, and fertiliser.

We take Pingliang and Qingyang as project demonstration areas, based on the ecological compensation mechanism, and make full use of clean coal and power resources in Pingliang, and water conservation in Qingyang, and gradually promote the dry farming circular ecological agriculture mode with unit of households (Figure 4 [27]).

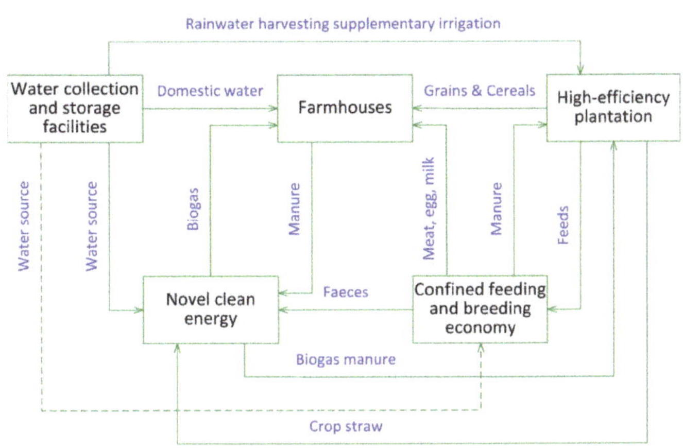


**Figure 4.** Schematic of dry farming circular ecological agriculture mode; Note: The detailed information about Figure 4 can be found in Reference [27]; S L Jin. Investigation on the model of dry farming circular ecological agriculture with farmer as unit in semi-arid rain fed agricultural area of Gansu Province. Communication of Agricultural Science and Technology; 2011, (2): 11–3.

The expected operation effects include: (1) the centralised treatment rate of sewage is more than 91%; (2) the treatment rate of domestic waste is more than 99%; (3) the recycling rate of wasted agricultural film is more than 83%; (4) the comprehensive utilisation rate of large-scale livestock and poultry breeding waste is more than 97%; (5) the comprehensive utilisation rate of crop straw is more than 84%; (6) the annual utilisation rate of waste is 12,000 t/MW; (7) the annual productivity of combustible gas is 3–40 million $m^3$/MW, replacing standard coal by 3000 t/MW; (8) reduction of 72 t/MW of sulfur dioxide, 22 t/MW of nitrogen oxide, and 7200 t/MW of carbon dioxide; (9) 2400 t/MW of solid carbon and 1800 t/MW of organic matter active extract obtained; (10) the income of peasant farmers is increased by 3100 RMB/capita [18].

### *5.2. Longxi Dry Farming and Rain Collecting Agricultural and Livestock Industrial Area*

#### 5.2.1. General Principles

The agricultural industrialisation mode of Rainfed irrigation in Longxi is developed by the local people in long-term agricultural practice. The key of this model is to collect water for irrigation and improve the utilisation rate of precipitation.

In order to solve the energy problem in rural areas, it is necessary to return farmland to forest and grass, expand the vegetation coverage, and combine the engineering measures to renovate the barren mountains and ditches, build fuel forest.

The development of animal husbandry in agricultural areas enhances organic fertiliser for crop production, improves soil fertility, and achieves positive interaction. In order to build a sound carbon cycle, we need to tap the regional woody biomass energy.

#### 5.2.2. Energy Saving Based on Thermal Insulation of Rural Buildings

The rural houses in this area are mainly rebuilt. In order to fully use the regional biomass resources, we adopt the economic targeted thermal insulation system. North facing exterior wall is internally insulated with 30 mm polymer resin of 35 RMB/$m^2$. Indoor ceiling is insulated by 30 mm polymer resin of 35 RMB/$m^2$. The total investment of the whole household is about 2275 RMB, and the comprehensive energy saving rate is about 30% [17,18].

As for households with good financial conditions, we can attach external window thermal curtain (60 RMB/$m^2$) and external door thermal curtain (60 RMB/$m^2$). The total investment of the whole household (including external wall and ceiling reconstruction costs) is about 3600 RMB, and the comprehensive energy saving rate is about 35% [18].

#### 5.2.3. Distributed Biomass Energy System

This project benchmark is intelligent biomass pellet fuel heating furnace in Beichuan County, Sichuan Province. It sets up biomass pellet fuel processing stations in villages to realise the pattern of self-production for self-use, and interconnecting villages and towns. We take Longxi as the demonstration area and build an organic agriculture integrated with biomass cycle model.

The expected operation effects include: (1) the processing capacity of each hundred square meters (workshop) is 30–40 t/month, and the charge standard for the substitute processing is about 400 RMB/t; (2) biomass fuel is local collection and production, bringing additional income of 100–200 RMB/t for peasant farmers; (3) the original living (and cooking) habits of peasant farmers are not changed, and the cleaning stove spontaneously realises cooking and heating; (4) the biomass stove is directly installed indoors and heated up fast and effectively; (5) the combustion efficiency of burner reaches more than 95%, the thermal efficiency of rated working condition reaches more than 80%, the thermal efficiency of cooking reaches more than 40%, and the waste heat recovery efficiency of flue gas reaches more than 10%; (6) the daily consumption of biomass fuel for each household is not more than 1 kg, and the annual fuel consumption is about 8–12 RMB/$m^2$, saving more than 80% of fuel consumption compared with conventional firewood stoves; (7) the pollutant emission is reduced by 90%and the indoor air quality is significantly improved; (8) the agricultural and livestock wastes such as dead branches and leaves, crop straw, livestock manure, and other biomass wastes are reasonably consumed, the heating expenditure is decreased, and the peasant farmers' income is increased in disguised form [18].

### *5.3. Longnan Mountainous 3D Ecological Characteristic Agricultural Area*

#### 5.3.1. General Principles

Basic farmland with high and stable yield is built on both sides of the river valley with good water and heat conditions and on the gentle low hillside to ensure food demand. Convert farmland to forests, plant economic forests and firewood forests, and develop diversified management in the areas close to the mountains and steep slopes.

In the area that is not suitable for planting economic forests, we need to plant trees, shrubs, and grasses to increase vegetation coverage, store soil and water, and improve ecological environment. Remote mountains are built as ecological forests. Briefly, in this zone, we should fully use the regional advantages of abundant woody biomass to build a unique model of forest economy.

#### 5.3.2. Energy Saving Based on Thermal Insulation of Rural Buildings

The rural houses in this area mainly focus on energy-saving transformation, and adopt economic targeted thermal insulation system, which is consistent with the technical route adopted in Longxi area. Meanwhile, we need to fully concern the humid climate of this region.

#### 5.3.3. Distributed Biomass Energy System

This project benchmark is intelligent biomass pellet fuel heating furnace in Raoyang County, Hebei Province. It fully uses the forest advantages to establish biomass particle fuel processing stations with the unit of villages and realises the pattern of self-production for self-use, and village interconnection.

Here we take Tianshui, Dingxi, and Longnan as demonstration areas, and establish a virtuous cycle of biomass fuel processing and consuming. The expected project operation effects can be seen by examining the case in the Longxi zone.

### *5.4. Gannan-Linxia Artificial Ecological Grass and Livestock Industry Area*

#### 5.4.1. General Principles

Here the climate is cold and humid with obvious difference in vegetation types. There are high-quality grassland and rich forests, so that we can boost the intensive and long-chain development of grass and livestock industry, based on the artificial and semi artificial grassland. In accordance with the constraints of initial investment and operating costs, the government has to focus on the long-term development of this zone, and gradually sets up a complete industrial biomass cycle.

#### 5.4.2. Energy Saving Based on Thermal Insulation of Rural Buildings

The rural houses in this zone are mainly rebuilt, which is consistent with the technical route adopted in Longdong area. Meanwhile, the cold and humid climate of this region should be carefully considered.

#### 5.4.3. Construction of Biomass Industry Cycle

This project benchmark is the nearly zero carbon emission mode of biomass cogeneration with "gas-electricity-heat-fertiliser" cycle in Ping'an County, Hebei Province. We make full use of the regional ecological advantages and abundant water resources in Linxia to build a new ecological agriculture model of efficient planting integrated with breeding and regional biogas ecological tourism. Subsequently, we promote a large-scale cogeneration of gas, electricity, heat, and fertiliser.

In this zone, 4958 administrative villages are involved, and the total investment in ecological construction is 19.41 billion RMB (The mentioned data come from http:

//www.gansu.gov.cn/col/col10/index.html, accessed on 3 February 2021.). Based on the technical route of biomass waste plus clean energy plus organic fertiliser, a complete cogeneration pattern is set up with multi-level utilisation of resources and material virtuous recycling (Figures 5 and 6 [28]). The expected project operation effects can be referred to the case in the Longdong area.

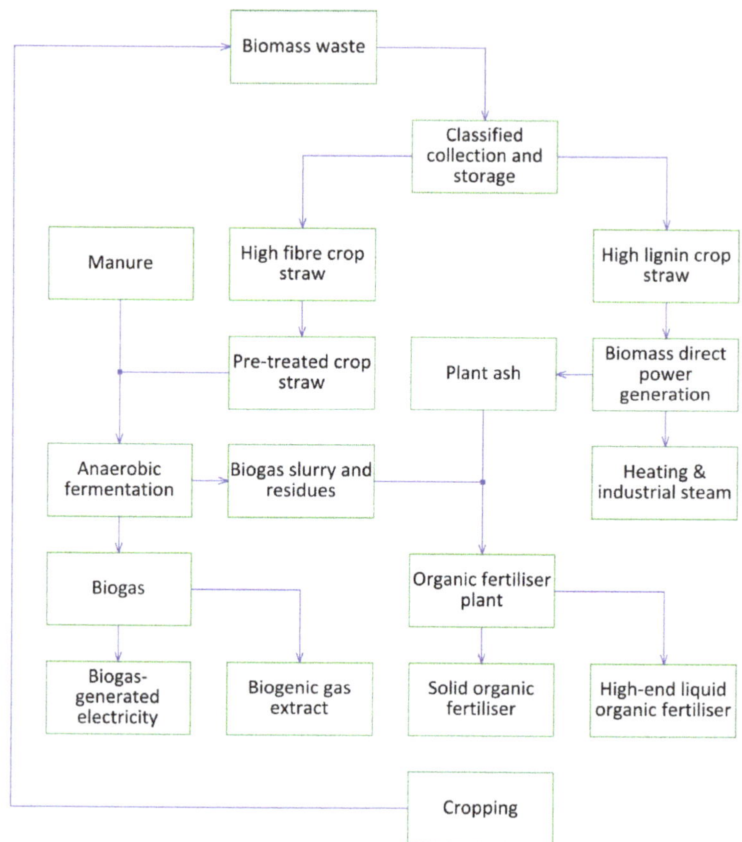


**Figure 5.** Gas-electricity-heat-fertiliser cogeneration mode.

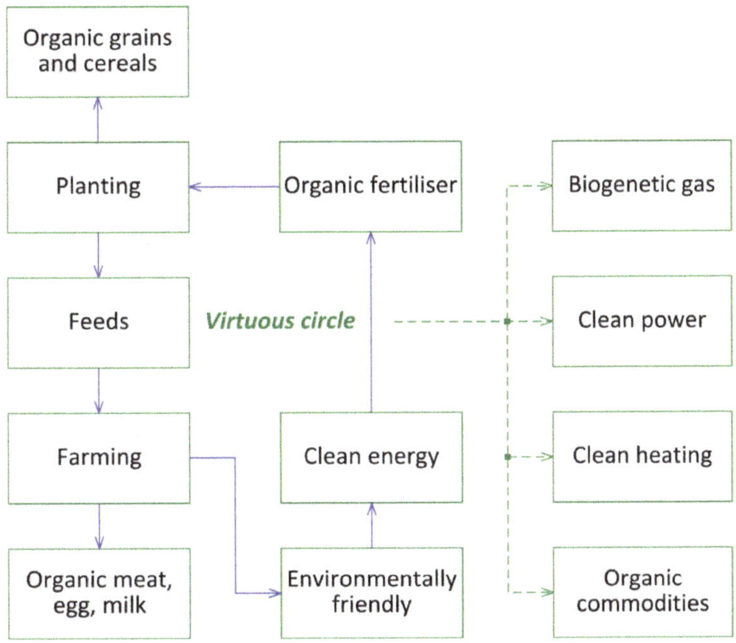


**Figure 6.** Schematic multi-level utilisation of resources and material virtuous recycling; Note: The detailed information about Figures 5 and 6 can be found in Reference [28]; W He. A preliminary study on the mode of comprehensive agricultural development for poverty alleviation and ecological environment improvement in the poverty stricken mountainous areas of central and southern Gansu Province. Communication of Agricultural Science and Technology; 2008, (16): 16–7.

### *5.5. Hexi Corridor Oasis Ecological Agriculture Area*

#### 5.5.1. General Principles

The key to the industrialisation of oasis ecological agriculture is water. Qilian Mountain is the water conservation and supply area of the river in Hexi Corridor. In the oasis, scientific irrigation and drainage system should be set up, water resources should be reasonably distributed, and water-saving technology should be popularised.

Grassland is a fragile ecosystem in arid desert area on the edge of oasis, and its carrying capacity is very limited. Thus, the rehabilitation of grassland is really essential. Based on the local water, soil, light and heat resources, afforestation, green barrier construction, plastic film and greenhouse agriculture have to be developed in the desert area around the oasis. According to the initial investment constraints and operating cost constraints, we intend to develop a regional solar thermal or photovoltaic system for the ecology evolution in this zone.

#### 5.5.2. Energy Saving Based on Thermal Insulation of Rural Buildings

The rural houses in this area mainly focus on energy-saving transformation, and adopt economic targeted thermal insulation system, which is consistent with the technical route adopted in Longxi area.

#### 5.5.3. Clean Heating Project

There are two plans for RCHP including household photovoltaic power generation integrated with photo-thermal heating, and passive solar house. As for Plan A, household photovoltaic power generation integrated with photo-thermal heating, both domestic power and heating demands (hot water is involved too) are satisfied simultaneously, and the comprehensive efficiency of solar utilisation is increased remarkably. Meanwhile, the electricity consumption of peasant farmers is reduced substantially, which promotes peasant farmers' income in disguised form since the power generated by this system is connected to the municipal grid. When it is in continuous rainy or snowy weather or at night, low-temperature ASHP unit is employed as an auxiliary heat source to provide heat for peasant farmers, which reduces commodity energy consumption dramatically. In practice, the system operates automatically, and peasant farmers can adjust indoor air temperature by regulating the indoor fan speed, based on their personal sensations. The limitations of this scheme include: (1) the local solar energy resources are required to be abundant; (2) the local financial support is required to be sufficient; (3) the local commodity-market-service chain is complete and sustainable. Table 6 records the scheme details using Gaodunying Village as the case [18].

**Table 6.** Scheme details in Gaodunying Village.

| Name | Unit | Value |
|---|---|---|
| The number of solar thermal and photovoltaic panels | / | 26 |
| Total effective area of the panels | $m^2$ | 20.8 |
| The volume of heatingstorage tank with thermal insulation | $m^3$ | 1.0 |
| The number of farmhouses involved in statistics | / | 50 |
| Power generation per year | kWh | 3041 |
| On-grid offer | RMB/kWh | 0.2987 |
| Electric power price with subsidies by the government | RMB/kWh | 0.42 |
| Annual earnings from power generation | RMB | 2185.6 |
| Average ambient temperature in tests | °C | −2.0 |
| Mean indoor air temperature in tests | °C | 13.0 |
| Span of heating season | Month | 5 |
| Total power consumption in heating season | kWh | 6000–8000 |
| Total cost for power consumption in heating season | RMB | 3000–4000 |
| Net cost for power consumption in heating season | RMB | 1000–2000 |

Note: the raw data in this table is part of field investigations in Gaodunying Village, Yuzhong County, provided by Reference [18]; Chinese Society for Urban Studies. 2020 Annual report on China building energy efficiency. China Architecture & Building Press, Beijing, China. 2020.

As to Plan B, passive solar house, the working temperature range of this system is wide, and there is no anti-freezing problem in winter. The air will not corrode the heat collector and pipeline, and the air system has low requirements for air-tightness, which has little impact on the overall operation efficiency. The thermal collection rate of air is high, and the heating efficiency is enhanced without secondary heat exchange. Beyond heating season, indoor ventilation is realised through changing pipeline and corresponding valves to avoid overheating of the system. This scheme also provides fresh air which improves indoor air quality at the same time. The limitations of this scheme include requirements of rich solar resources and longer payback period. Table 7 preliminary estimates the heating cost by passive solar use in North China [18].

**Table 7.** Preliminary estimates of heating costby passive solar use in North China.

| Location | Beijing | Lasa | Dunhuang | Lanzhou |
|---|---|---|---|---|
| Solar resources in heating season | 1267 MJ/$m^2$ | 2917 MJ/$m^2$ | 2221 MJ/$m^2$ | 1244 MJ/$m^2$ |
| Heating cost | 0.379 RMB/kWh | 0.176 RMB/kWh | 0.229 RMB/kWh | 0.383 RMB/kWh |
| Payback period | 14.64 a | 6.01 a | 7.99 a | 14.83 a |

Note: the solar resources involved in statistics require that radiation intensity is higher than 400 W/$m^2$, and these data cited in the table is based on the ideal constructions under optimum solar collection states.

## 5.6. Yellow River Area

### 5.6.1. General Principles

Due to the high terrain, water can only be diverted for irrigation. Thus, water conservancy facilities are the foundation of development. It is necessary to improve the multiple cropping indexes, concentrated on the development of high-quality and high-efficiency agriculture, and plant high-quality grain varieties in the rivers and valleys where there are more people and less land and good water and heat.

The rural development in this zone is highly dependent on the neighbouring cities and towns, and a variety of resources cannot be self-sufficient. In accordance with the regional economy, the key of clean heating is clean electricity. In economically underdeveloped areas, peasant farmers' income mainly comes from migrant workers. It is necessary to adhere to the ecological basis and improve the living environment of peasant farmers gradually.

### 5.6.2. Energy Saving Based on Thermal Insulation of Rural Buildings

The rural houses in this area mainly focus on energy-saving transformation, and adopt economic targeted thermal insulation system, which is consistent with the technical route adopted in Longxi area.

### 5.6.3. Clean Heating Project

There are two plans for RCHP including household photovoltaic power generation integrated with photo-thermal heating, and low-temperature ASHP air heater. As for Plan A, we have detailed in the case of Hexi Corridor area. As to Plan B, low-temperature ASHP air heater, we can install the unit in accordance with heating demand of peasant farmers. The mentioned installation is controlled independently when it is operated with the part-time and part-space patterns, which saves considerable costs. Since the indoor unit is installed on the ground, the hot air at the lower outlet flows close to the ground and rises slowly and naturally, so that the peasant farmers' ankles are warm and their feelings are comfortable. The upper air outlet of the installation is designed for the working area of peasant farmers, and the temperature in this zone rises quickly. Due to the indoor unit is terminal equipment, there is no need to add other devices for heating. In addition, the ASHP air heater has realised one key operation, and there is no pollutant emission. Hence this scheme is convenient and environmentally friendly.

The limitations of this scheme include: (1) local commodity-market-service chain is required to be complete and sustainable, including scheme design, equipment installation, adjustment, training for use, after-sales maintenance, and so forth; (2) local financial support is required to be sufficient and financing channels are unobstructed; (3) peasantfarmers' disposable property is required to be abundant. Table 8 reports the scheme details based on the Gaodunying Village Data [18].

**Table 8.** Scheme details about low-temperature ASHP air heater used in Gaodunying Village.

| Name | Unit | Value |
|---|---|---|
| The number of farmhouses involved in statistics | / | 50 |
| Electric power price with subsidies by the government | RMB/kWh | 0.42 |
| Average ambient temperature in tests | °C | −2.0 |
| Mean indoor air temperature in tests | °C | 13.0 |
| Span of heating season | Month | 5 |
| Total power consumption in heating season | kWh | 6000–8000 |
| Total cost for power consumption in heating season | RMB | 3000–4000 |

Note: the raw data in this table is part of field investigations in Gaodunying Village, Yuzhong County, provided by Reference [18]; Chinese Society for Urban Studies. 2020 Annual report on China building energy efficiency. China Architecture & Building Press, Beijing, China. 2020.

## 6. Conclusions and Policy Implications

### *6.1. Conclusions*

Rural Gansu has long been suffered from limited energy supply, unreasonable energy-use structure, unavailable or inefficient biomass energy use, and great pressure on ecological environment, due to its special natural geographical environment and economic factors.

According to the resource distribution and planning situation of the province, the execution of RCHP is zoned for six regions: Longdong zone with grass field ecological agriculture and animal husbandry industrial area; Longxi dry farming and rain collecting agricultural and livestock industrial area; Longnan Mountainous 3D ecological characteristic agricultural area; Gannan-Linxia artificial ecological grass and livestock industry area; Hexi Corridor oasis ecological agriculture area; and Yellow River area.

We can build a "gas-electricity-heat-fertiliser" cogeneration pattern with near zero carbon emission, which aims at ecological circular economic zone. Similarly, an "organic agriculture integrated with biomass briquette fuel forming and burning" mode can be set up towards characteristic agricultural and side line products processing circular economic zone. In the biomass-scarcity areas, by contrast, what we can do is multi-dimensionally comparing various technical schemes, and gradually boosting RCHP in accordance with local conditions.

More importantly, no matter in which zone, the thermal insulation of housing envelope needs improving urgently, which is the fundamental to RCHP in rural Gansu.

### *6.2. Policy Implications*

In general, policy formulation has to be well-directed to fully consider the uneven distribution of resources, population, and economic development in Gansu Province. Meanwhile, the practice of RCHP in other provinces and states should be thoroughly concerned.

Both economic construction and livelihood improvement have to be based on ecological sustainability. The narrow pollution-treatment approach must be prohibited owing to the originally fragile ecology in Gansu Province.

In the areas with abundant biomass energy, building an ecological circulation cogeneration mode will take priority over all others. First off, a complete regional industrial chain has been set up, which enhances the flexibility and toughness of the economic ecological environment. Then the reasonable consumption of biomass resources has been realised, and a sustainable material cycle has been established.

In the regions with general application of biomass, it is necessary to explore biomass potentials and enhance agricultural economy chain, which completes the organic agriculture plus biomass fuel industry. Firstly, it strengthens the risk resistance ability of the regional economy. Secondly it reasonably absorbs the surplus agricultural waste and set up a sustainable ecological cycle.

As for the areas where all kinds of resources are very scarce, the peasant farmers' life is hard, and the primary work is the repair and insulation of rural houses. Thereafter, we can gradually promote poverty alleviation and RCHP.

The coordination among different departments needs to be strengthened. First of all, policy conflicts must be eliminated.

When implementing financial subsidies one has to fully consider the differences (economic, environmental, sustainable, etc.) between various technical schemes, and the scheme review should be taken the form of anonymous experts (Delphi technique).

We should make clear economic incentive policies and ensure subsidy security. A group of sustainable policies are the basis that the RCHP can be effectively implemented, since peasant farmers might abandon RCHP when their expenditure on heating rises steeply.

The government needs to strengthen scientific guidance of planning and technical path. The follow-up operation and maintenance have to be concerned too. Based on advanced experience, the government should also guide professionals to build clean heating standards. In the meantime, there should be third party to monitor RCHPs to revise the technical route based on quantitative measurement and objective evaluation.

Refine the implementation plan and encourage peasant farmers to participate actively. A clear and feasible plan can help peasant farmers to understand that the RCHP will not only reduce their expenses of heating but also improve their living conditions. Hence, the RCHP is practicable.

We needto gradually improve the market-oriented operation mechanism, get rid of the excessive dependence on government subsidies, open up financing channels, and set up a benign cyclepattern of "enterprise oriented, government driven, and residents affordable".

The government has to guide the building of green development funds for poor and backward areas, support the issuance of green bonds by clean heating enterprises, finance the flow and exchange of talents, and actively encourage the introduction of private capital for the benefit of people.

Finally, tax levers should also be considered to support and guide peasant farmers' self-organised biomass cooperatives to conduct the "biomass energy poverty alleviation project".

**Author Contributions:** D.Q.: Conceptualization, Methodology, Investigation, Formal analysis, writing original draft, writing review and editing. X.D.: Investigation, Micro survey. J.W.: Writing review and editing, Micro survey. C.H.: Investigation, writing review and editing. G.W.: Project administration. F.Z.: Funding acquisition. S.L.: Project administration. All authors have read and agreed to the published version of the manuscript.

**Funding:** This work was financially supported by the National Natural Science Foundation of China (51978200& 11962016) and the Science Foundation of Gansu Province Building Energy Saving (JK2021-10).

**Institutional Review Board Statement:** Not applicable.

**Informed Consent Statement:** Not applicable.

**Data Availability Statement:** The paper have stated all owners' name and accepted their allowance for publication.

**Acknowledgments:** The authors are grateful to all the research participants for generously sharing their experiences.

**Conflicts of Interest:** The authors declare that they have no known competing financial interest or personal relationships that could have appeared to influence the work reported in this paper.

## Nomenclature

| | |
|---|---|
| BES | Building energy saving |
| CMS | Conservation-minded society |
| IT IS | Internal thermal insulation |
| NRC | New Rural Construction |
| RCHP | Rural clean heating project |

## References

1. Romero, C.A.T.; Castro, D.F.; Ortiz, J.H.; Khalaf, O.I.; Vargas, M.A. Synergy between Circular Economy and Industry 4.0: A Literature Review. *Sustainability* **2021**, *13*, 4331.
2. Rojas, C.N.; Peñafiel, G.A.A.; Buitrago, F.L.; Romero, C.A.T. Society 5.0: A Japanese Concept for a Superintelligent Society. *Sustainability* **2021**, *13*, 6567.
3. Dantas, T.E.T.; de-Souza, E.D.; Destro, I.R.; Hammes, G.; Rodriguez, C.M.T.; Soares, S.R. How the combination of Circular Economy and Industry 4.0 cancontribute towards achieving the Sustainable Development Goals. *Sustain. Prod. Consum.* **2021**, *26*, 213–227. [CrossRef]
4. Ding, Y.-J.; Li, C.-Y.; Wang, X.; Wang, Y.; Wang, S.-X.; Chang, Y.-P.; Qin, J.; Wang, S.-P.; Zhao, Q.-D.; Wang, Z.-R. An overview of climate change impacts on the society in China. *Adv. Clim. Chang. Res.* **2021**, *12*, 210–223. [CrossRef]
5. Mofijur, M.; Fattaha, I.M.R.; Alam, M.A.; Islam, A.B.M.; Ong, H.C.; Rahman, S.M.A.; Najafi, G.; Ahmed, S.F.; Uddin, M.A.; Mahlia, T.M.I. Impact of COVID-19 on the social, economic, environmental andenergy domains: Lessons learnt from a global pandemic. *Sustain. Prod. Consum.* **2021**, *26*, 343–359. [CrossRef] [PubMed]

6. Asante, D.; He, Z.; Adjei, N.O.; Asante, B. Exploring the barriers to renewable energy adoption utilising MULTIMOORA-EDAS method. *Energy Policy* **2020**, *142*, 111479. [CrossRef]
7. Matthäus, D. Designing effective auctions for renewable energy support. *Energy Policy* **2020**, *142*, 111462. [CrossRef]
8. Gozgor, G.; Mahalik, M.K.; Demir, E.; Padhan, H. The impact of economic globalization on renewable energy in the OECD countries. *Energy Policy* **2020**, *139*, 111365. [CrossRef]
9. Clausen, L.T.; Rudolph, D. Renewable energy for sustainable rural development: Synergies and mismatches. *Energy Policy* **2020**, *138*, 111289. [CrossRef]
10. Haar, L. An empirical analysis of the fiscal incidence of renewable energy support in the European Union. *Energy Policy* **2020**, *143*, 111483. [CrossRef]
11. National Development and Reform Commission of China. Action Plan of Energy Revolution and Innovation (2016–2030). Available online: http://www.nea.gov.cn/2016-06/01/c_135404377.htm (accessed on 1 June 2016).
12. National Energy Administration of China. 13th Five-Year Plan for Biomass Energy Development. 2016. Available online: http://www.nea.gov.cn/2016-12/14/c_135904504.htm (accessed on 14 December 2016).
13. Qi, Y.; Zhang, X.L. *Blueprint of Low Carbon Development*; Social Sciences Academic Press: Beijing, China, 2018.
14. National Development and Reform Commission of China. Strategy of Revolution of Energy Production and Consumption (2016–2030). Available online: http://www.gov.cn/xinwen/2017-04/25/5230568/files/286514af354e41578c57ca38d5c4935b.pdf (accessed on 25 April 2017).
15. National Development and Reform Commission of China. Notice on the Establishment and Improvement of Renewable Energy Electricity Consumption Guarantee Mechanism. 2019. Available online: http://zfxxgk.nea.gov.cn/auto87/201905/t20190515_3662.htm (accessed on 15 May 2019).
16. National Development and Reform Commission of China. Overall Plan of Green Life Creation Action. 2019. Available online: http://www.gov.cn/xinwen/2019-11/05/5448936/files/7a105ee3d9b24dec8a430dd9f64ef97f.pdf (accessed on 5 November 2019).
17. National Bureau of Statistics of China. China Statistical Yearbook. Available online: http://www.stats.gov.cn/tjsj/ndsj/ (accessed on 3 March 2021).
18. Chinese Society for Urban Studies. *2020 Annual Report on China Building Energy Efficiency*; China Architecture & Building Press: Beijing, China, 2020.
19. Ziemele, J.; Cilinskis, E.; Blumberga, D. Pathway and restriction in district heating systems development towards 4th generation district heating. *Energy* **2018**, *152*, 108–118. [CrossRef]
20. Sernhed, K.; Lygnerud, K.; Werner, S. Synthesis of recent Swedish district heating research. *Energy* **2018**, *151*, 126–132. [CrossRef]
21. Werner, S. International review of district heating and cooling. *Energy* **2017**, *137*, 617–631. [CrossRef]
22. Li, H.W.; Svendsen, S. Energy and exergy analysis of low temperature district heating network. *Energy* **2012**, *45*, 237–246. [CrossRef]
23. Soltero, V.M.; Chacartegui, R.; Ortiz, C.; Velazquez, R. Potential of biomass district heating systems in rural areas. *Energy* **2018**, *156*, 132–143. [CrossRef]
24. Hendricks, A.M.; Wagner, J.E.; Volk, T.A.; Newman, D.H.; Brown, T.R. A cost-effective evaluation of biomass district heating in rural communities. *Appl. Energy* **2016**, *162*, 561–569. [CrossRef]
25. Vallios, I.; Tsoutsos, T.; Papadakis, G. An applied methodology for assessment of the sustainability of biomass district heating systems. *Int. J. Sustain. Energy* **2016**, *35*, 267–294. [CrossRef]
26. Long, N.; Qu, D.H.; Li, W.Z.; Fu, X.H.; Yang, Y. Moisture proof calculation of envelope in severe cold zone. *Heat. Vent. Air Cond.* **2018**, *48*, 107–112.
27. Jin, S.L. Investigation on the model of dry farming circular ecological agriculture with farmer as unit in semi-arid rain fed agricultural area of Gansu Province. *Commun. Agric. Sci. Technol.* **2011**, *2*, 11–13.
28. He, W. A preliminary study on the mode of comprehensive agricultural development for poverty alleviation and ecological environment improvement in the poverty stricken mountainous areas of central and southern Gansu Province. *Commun. Agric. Sci. Technol.* **2008**, *16*, 16–17.

*sustainability*

*Article*

# Current Trends in Urban Heritage Conservation: Medieval Historic Arab City Centers

**Ahmed Mohamed Shehata**

Department of Islamic Architecture, Umm Al-Qura University, Makkah al-Mukarramah 24382, Saudi Arabia; amshehata@uqu.edu.sa; Tel.: +966-500-016-083

**Abstract:** Traditional conservation efforts did not improve the conditions in most historic urban centers of Arab cities. The internationally adopted shift in historic urban conservation grants better urban vitality and sustainability for these areas. This study investigates the existing trends and forthcoming changes in urban conservation and their implication on historical centers. Urban Heritage Conservation UHC trends were reviewed, conservation parameters were defined, and quality aspects of successful historic urban conservation were identified, and an assessment framework was developed to evaluate the resulting conserved urban heritage. Two case studies of Arab cities, Jeddah and Aman, were analyzed. The findings highlight the most common urban issues such as reusing historic buildings, traffic congestion, and lack of funds. The impact of urban management on historic areas' quality was revealed. Moreover, the paper ends with recommendations for conservation authorities. These include engaging residents in the conservation efforts, adopting more innovative traffic solutions to ease congestions, turning the historic area into a pedestrian-friendly space, attracting visitors through arranging cultural events, creating new job opportunities through heritage, and improving the image of the areas through urban regulations. The paper's findings would contribute to the knowledge related to Urban Heritage Conservation (UHC), and its recommendations would help practitioners and decision-makers.

**Keywords:** MENA Islamic cities; urban management; sustainable built environment

**Citation:** Shehata, A.M. Current Trends in Urban Heritage Conservation: Medieval Historic Arab City Centers. *Sustainability* **2022**, *14*, 607. https://doi.org/10.3390/su14020607

Academic Editor: Shervin Hashemi

Received: 30 November 2021
Accepted: 17 December 2021
Published: 6 January 2022



## 1. Introduction

No matter how old the urban heritage is, historical urban areas are considered an essential part of the local culture, a crucial base for cultural tourism, and local society prosperity [1]. It is an integral part of the collective memory of the local communities. Any development project of such urban areas should carefully consider the symbology of the urban fabric and its surrounding buildings [2]. Heritage definition, along with the conservation and restoration scope of monuments and sites, evolved since the 1964 Venice charter. There has been a shift among international organizations and conservation authorities to consider heritage as urban areas rather than a single monument. This shift is accompanied by increasing efforts toward engaging the local communities in preserving, managing, and controlling urban conservation plans [2].

From 2003 to 2019, UNESCO and ICOMOS made continuous efforts to revise the heritage definition and scope besides executing the conservation activities according to more sustainable approaches [3–5]. In parallel, The Seventeen Sustainable Development Goals (SDG), adopted anonymously by member states of the United Nations, repeatedly mention cultural aspects as the base for sustainable urban development. Moreover, the adopted Urban Agenda of the United Nations as of 2016 acknowledges tangible and intangible heritage as a critical parameter in developing sustainable urban economies to transition towards higher productivity [6].

Urban characteristics result from socio-cultural aspects such as values and traditions; environmental conditions such as climate and geographical location; and technological factors such as materials, economics, and construction technology, contribute to the physical form and architectural details [6]. Historic cities' traditional built environments reflect local community values in their urban composition and spatial arrangement.

Despite the differences in climate, geographical location, and topography, similarities in historic Arab cities resulted from sharing the same socio-cultural characteristics, governing religious roles, and local construction materials and systems [7]. It should be noted that the urban morphology of these historic cities is impacted by the social aspects of private life rather than their geometric forms [8]. The urban fabric of Arab historical centers is a solid-built mass on which courtyards and narrow passageways are engraved [9]. These compact urban fabric, narrow shaded street networks, and buildings' protected openings were environmental and social responses [10]. The satellite image in Figure 1 illustrates the shared urban characteristics of three different cities, Rabat, Cairo, and Damascus, in terms of inner-oriented buildings with courtyards and narrow irregular-shaded pedestrian networks.

(a) (b) (c)

**Figure 1.** Historic urban fabric of three Arab cities having the same inner-oriented openings and narrow irregular shaded pedestrian networks. (**a**) Historic urban fabric of Rabat; (**b**) historic urban fabric of Cairo; (**c**) historic urban fabric of Damascus.

Traditional conservation efforts for historical centers were neither successful nor sustainable in most of these centers for many reasons, such as a lack of funds, unsuitable urban management policies, and the narrow scope of development projects. Therefore, this paper investigates the existing trends in conservation and the forthcoming changes in urban heritage conservation approaches, and their implications on the historic urban city centers in Arab cities. This paper is structured in five main sections. Following the introduction, the literature review highlights the Arab historical centers in terms of their urban characteristics, threats, and problems. The third section presents the theoretical Background and Contextual Approach that discusses the international trends in historic urban landscape conservation. The fourth section is the methodology and assessment framework that presents the main aspects of saucerful conservation, the main success indicators, and the UHC assessment framework. In the fifth section, two historical centers of Arab case study cities were introduced and analyzed, and the resulting urban heritage was assessed. The sixth section discusses the research findings and presents the final recommendations. The seventh section concludes the research outcomes while highlighting the research limitation and its future extensions.

## 2. Literature Review: Arab Historic City Centers

### *2.1. Characteristics of Arab Historic City Centers*

The medieval urban fabric was built inward, where every level of its components is contained, from its gated city walls to its quarters and local residential clusters. Even central public spaces are tightly confined. Main streets lead to local, extremely narrow cul-de-sacs, which lead by tight, dog-leg corridors to private patios. Public open spaces are enclosed by shop fronts or solid walls of the houses and gardens [6]. Narrow dead-end alleys lead pedestrians from public spaces to semi-public and semi-private circulation spaces. Most urban traffic serves areas for commercial and religious purposes, and city gates use main walkways. Streets are sided by many blind facades, while outside openings are covered with wooden curtains on the upper floors [11]. Historic urban areas have two types of public spaces. The first is the open public space appearing in open plazas in front of key buildings such as the mosque, market, and school. The second type is the traffic streets, represented by the narrow-structured pattern of vehicular pedestrian-oriented alleys [4]. The narrow streets "were a traffic control system that effectively arranged different urban functions [12]. The "Friday prayer" courtyard acts as a gathering space [11].

In contrast, the plaza in front of the grand mosque and other public buildings serves as circulation distribution nodes to and from the lower category neighboring paths. A wider space often appeared along the central path spine of the fabric. The size of this space depended on its location within the urban fabric and its accommodated pedestrian traffic [8]. Physical borders are loosened in those spaces and extend the bazaar to accommodate the informal commercial and social activity. Urban heritage researchers indicated a typical unwritten relationship between the width of streets and their bounding buildings' heights [12]. The main feature of the city's land use is the clear separation of commercial from residential areas. All markets are within the central zone, and the city's central urban area is distinctive by its relatively large, straight streets leading directly to the old main gates [10].

In historic Arab cities, there has been a significant contrast between commercial crowded streets and the quiet calm of the private courtyards [6]. The network structure of the open spaces is based on controlling access and mobility, and doors were used to manage social contact, even in streets where no entries were allowed to open in front of another [13]. Moreover, the courtyard concept is crucial, utilizing open space surrounded by private usage. The effective use of courtyards allows narrow lanes and interrupts perspectival alignments [11].

### 2.2. *Endangered Arab Historic City Centers*

Urban settlements evolve and change to fulfill their inhabitants' requirements. Over the past century, historic city centers have struggled between their symbolic significance as heritage with a distinctive sense of place and many political and economic development programs [11]. The historical city centers' "Albalad" areas accommodated the city's central markets, wholesale shops, and traditional crafts [14]. The historic urban centers are not isolated individual areas. Areas surrounding historic centers constitute an interface between the historic center and other city districts and complement its character [15]. These districts adopt the same urban identity. Over the course of time, climate and religion have lost their influence over historical cities and their urban character. Inspired by aspirations of modernity, following the lead of western globalization and the impact of the powerful private corporations, cities enforced certain practices, spaces, and forms on their traditional urban fabric [16]. The requirements of everyday life coupled with economic growth had a destructive impact on the urban heritage of the historic centers of Middle Eastern cities [7]. The physical characteristics of open spaces have changed, but they still follow the same traditional socio-cultural structure.

The modernization shift in Arab cities' urban form, architecture, and housing started with the beginning of the railway transportation systems in the early nineteenth century linking the region to European cities. This system had a significant role in modernizing many Arab cities such as Mosul, Amman, Cairo, Alexandria, and Damascus. Consequently, it creates an interactive, diverse environment for people from diverse ethnic and religious groups, and opportunities for various commercial activities [17,18]. The modern grid pattern forms the most common type of urban fabric in Arab cities of the nineteenth century, where residence units vary between row-house, semi-detached, and detached villas. These units are distributed on a regular grid pattern with horizontal and vertical streets. The new type followed in the expansion of districts away from the historic center districts [17].

Analyzing tangible and intangible factors that affected the Arab cities' transition between the medieval and early twentieth centuries reveals that the transformation was a gradual shift from a harmonious pedestrian-oriented built environment into a fragmented vehicular-oriented environment [19,20].

### 2.3. *Urban Problems of Arab Historic Centers*

Historic centers within Arab cities have something in common; they are all about to lose their historical identity and urban character. Many researchers have identified more problems related to the urban changes of historical centers [8]. In addition to the previously mentioned impact of railways in cities' modernization during the ninth century, the car has had significantly impacted city planning through the last century [21]. In the case of Arab cities, what was once a pedestrian-friendly traffic network has had to accommodate a vehicular-based traffic system. During this transition, economic, technological changes substituted the city's traditional practices with western practices producing an international model known later as globalization. These changes happened in traditional centers of the Arab world, and new administrative capitals such as Cairo, Baghdad, Rabat, and Damascus occurred the same way [18]. One of three main challenges was accommodating emergency and service vehicles in these allays. In cities like Cairo, using smaller mobilization services units was the solution, while in Jeddah, using small carts to serve the shops was the solution. The second issue was accommodating residents and visitors' vehicle traffic and parking. As a result, historic city centers ended up with streets without sidewalks where pedestrians and vehicles share the same space. The third issue was that historic buildings were not built with modern infrastructures such as drainage, water supply piping systems, fire protection, and electricity. Municipalities had to put these outdoor services' equipment in the narrow alleys to fulfill residents' requirements, causing, as shown in Figure 2, visual pollution and a negative image for the historic areas. Moreover, residents of historic areas in hot climate zones started to add air-conditioning units, causing a further negative visual impact for the historic areas.

(**a**) Historic Jeddah (**b**) Historic Amman (**c**) Historic Jeddah

**Figure 2.** (**a**) Services in historic Jeddah allays, (**b**) services in the historic area of Amman, (**c**) outdoor electric systems in historic Jeddah.

Moreover, literature on the current conditions of historic Arab centers reveals certain physical phenomena. Large buildings are taking over the city scene bringing more traffic [8,16]. More traffic has led to a more congested traffic network, as shown in Figure 3, creating wider spaces increasing thermal loads on buildings, and extra thermal loads pushed for a higher air-conditioning dependency. More air conditioning left the outdoor spaces subject to larger thermal loads [1,22].

**Figure 3.** Two photos demonstrate the difference in heights between old and new buildings, causing congested traffic networks.

Socioeconomically, as part of the vertical immigration within the cities, original residents deserted their historic low service areas and were replaced with residents from different classes with different cultural values and habits [23]. The informal economy flourished and impacted local business owners poorly [24]. Environmentally, the internally oriented built environment used passive cooling strategies, today transformed into outward-looking massive buildings with unprotected open spaces heavily dependent on mechanical air-conditioning systems. Figure 4 illustrates the historic area where new massive buildings replaced the historical ones [10]. Un-serviced areas with deteriorated built environments began to be deserted by their residents.

**Figure 4.** A satellite image showing some of the newly added massive buildings to the historic area of Jeddah.

## 3. Theoretical Background and Contextual Approach

### *3.1. Evolution of Heritage Definition*

Bruno Favel [25] cited three different changes for the heritage definition: The first defined it as a monument embodying the nation's values. Later, the definition expanded to include private monuments, monuments in their environmental context, neighborhoods, and everyday immaterial heritage. Since then, the heritage definition has expanded to include all aspects of culture, such as language, food, and arts [26]. Since then, conservation authorities worldwide have begun to consider protecting the intangible components of urban heritage in their conservation and restoration projects [25].

### *3.2. Historic Urban Landscape Approach (HUL)*

In 2005, the fifth general assembly of state parties to the convention concerning the protection of the world's cultural and natural heritage expanded the heritage definition to include the historic urban landscape; the definition was widened to include a profound understanding of the living and inhabited landscape. The historic urban landscape is embedded in the current and past of the place-based social expressions and composed of character-defining elements including spatial organization, land uses, visual relationships, natural elements such as topography, soils, and vegetation, in addition to all technical infrastructure elements, including construction details [27]. As the concept of heritage evolved, intangible heritage and culture have been recognized as heritage [28].

The Historic Urban Landscape (HUL) is defined as the "*historical layering of cultural and natural values and attributes*" (Article 8), integrating the intangible aspects of heritage and its related processes. This approach distinguishes the need to protect, conserve, and valorize cultural and natural heritage in a rapid and uncontrolled urbanized world. Moreover, this approach integrates heritage conservation plans into the development strategies [27].

### *3.3. Urban Heritage*

In 1987, historic urban areas were defined in the ICOMOS Washington Charter as either large or small, including cities, towns, and historical centers. Their natural and built environments embody the values of traditional urban cultures [29].

Urban heritage reflects the local community's identity, memories, and authenticity. It has been the source of civic pride and should be the driving force behind any restoration efforts [30]. Prosperous urban areas attract visitors to participate in cultural activities within their open public spaces [31]. There is a growing shift in heritage research trends toward a dynamic process [24]. Through this process, urban heritage should first be defined for its cultural value to a group of people; then, effort should be made to conserve it and pass it on to future generations [25]. Hall proposed this process as a "cultural circuit" that explains how heritage is constructed through cultural production, representation, and consumption. Smith also emphasizes urban heritage as a cultural process of remembering the work, creating ways to understand it, and engaging it with the present. He also argues that the tangible elements are the cultural tools for facilitating this process. Moreover, he claimed that the potential social value of heritage was first recognized in 2005, the Australia ICOMOS Article 1.2, when the previously used terms "monument" and "site" were replaced by the broad term of "place" as a heritage subject for conservation. In the URBAN II urban regeneration funding program, deteriorated urban areas were defined as areas with deteriorated economy and high unemployment, crime, school dropout rates, and a deteriorated physical and social environment [32].

Later, as of 2011, in the Recommendation on the Historic Urban Landscape Report of the Second Consultation on its implementation by the Member States, urban heritage was categorized into three main types. Firstly, a monumental heritage of outstanding cultural value; secondly, non-exceptional heritage elements that are present in a coherent way with a relative plenitude; thirdly, the new urban elements such as the urban built form, streets, or regular/public open spaces; and lastly, urban infrastructures such as material networks and equipment [33].

In 2019, cultural heritage was defined in the first article of UNESCO's Operational Guidelines of the World Heritage Convention as separate or connected buildings. This heritage includes archaeological sites characterized as ethnological, archaeological, historical, or aesthetic [30]. The unique universal value of these monuments comes from their architecture, homogeneity, or place in the landscape, and this value is evaluated in light of history, art, or science [30]. Moreover, article 47, in the same document, defined the cultural landscape as manmade, either integrated within natural sites or not. Moreover, the third annex of the guidelines listed five categories as heritage: Cultural landscape, historic towns and downtowns, heritage canals, and heritage routes [34].

### 3.4. Urban Heritage Conservation (UHC)

The Burra Charter identified the need to involve people in different phases, particularly those with solid cultural significance. In the 1999 Australia ICOMOS charter, value-based management was divided into three phases: Significance assessment, policy development, and management [35]. Further revisions of these charters included a fourth phase for assessing vulnerability and identifying threats to cultural significance [36]. As cited by K. Ashour, Roberts' successful regeneration was defined as conserving the physical urban fabric, creating an economic base, supporting social structures, and improving environmental conditions, following sustainability principles. It must have a strategic vision, clear, measurable objectives, and tools to measure them, utilize resources efficiently, and involve all stakeholders [37].

As stated in the report of member states' national survey, conducted by the UNESCO World Heritage Centre and titled "the second consultation on implementing the historic urban landscape approach by member states" dated in 2019, none of the Arab members adopted the concept to make use of the Heritage Urban Landscape approach [33]. Heritage conservation efforts within Arab cities have been focused on physical and materialized forms.

## 4. Methodology and Assessment Framework

A few researchers discuss the assessment of conservation activities. Efficiency and effectiveness measures are the basis of any acceptable assessment framework, and an empirical assessment provides information on issues of the measurement of conservation outputs. Moreover, it highlights the usefulness of adopting specific indicators assessing the conservation decision-making process [28]. This assessment process is straightforward in cases of physical heritage conservation. Ilde Rizzo, as cited by Pietro, highlighted the following arguable assessment issues: Scope, priorities, intervention strategies, conservation strategies, and long-term outcomes [38]. The currently adopted definition of heritage comprises the physical urban environment and intangible cultural values. In such cases, conservation activities must be embedded within the urban development plans, and residents' participation in different phases of conservation is required [39].

L. Veldpaus et al. extracted the main phases of urban heritage conservation out of 28 international policy documents. She concluded that the conservation evaluation framework needs to be based on the four main pillars of conservation projects. First, the heritage definition must be broadened to be inclusive of the built environment and the geographical setting. Second, the main objective of the conservation project must depend on a comprehensive, integrated approach in identifying, assessing, conserving, and managing the urban heritage significance. Third, strategies must be utilized for engaging localized national authorities to develop tools and instruments to respond positively to local values and needs. Such values and needs can be associated with governing systems, the participatory approach, environmental impact assessment, and sustainable socio-economic urban heritage development. Fourth, various local, national, and international stakeholders must be positively engaged in policymaking, governance system, and management of the conservation efforts of the urban heritage area [40,41]. Successful UHL should be built on involving communities, decision-makers, planners, and managers [36]. Figure 5 summarizes the main sustainable urban heritage conservation aspects.

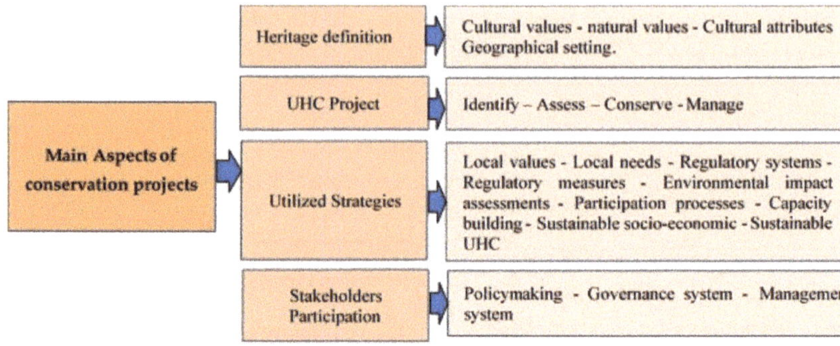


**Figure 5.** Successful sustainable UHC aspects.

Samrqandi, in her urban quality evaluation workflow, defined six attributes for the evaluation of heritage places: Access and linkage, comfort and image, uses and activities, sociability, context and conservation, and site interpretation [42]. Abou-Soliman suggested five physical features for vibrant urban heritage: Accessibility, having a suitable scale and form for located activities, accommodating appropriate usage, encouraging visitors' activities, and infrastructure and services to support designated activities. He also suggested seven intangible qualities for successful public space features for thriving urban heritage: Vibrant, readable, defined, meaningful, safe, symbolic, and encouraging [43]. The concept of intangible attributes ranges from asset-related features such as character, style, and uses to 'people' and processes. As international interest in community sustainability increased over two decades, practices, traditions, people, and communities were added as independent attributes [41].

Indicators are required to evaluate the success of any culture or heritage conservation-related programs. They must be conceptually based and simplified to be practical. Traditionally, experts and professionals make the indicators' selection [28]. Researchers in conservation, economics, sociology, and environmental fields have proposed different indicators to assess their areas of study activities, with the leading indicators used for each of these intersected fields of heritage conservation assessment [1]. Figure 6 presents the main attributes for urban quality assessment and indicators of successful sustainable UHC as extracted from the literature review [1,43].

A proposed framework that relates the UHC aspects and the resulting qualities of the targeted urban area is presented in Table 1 where:

- As concluded from the theoretical background, five aspects of the UHC project have to be accomplished through twenty practices. Each one of these development practices has an impact on the success of the resulting urban heritage. These practices comprise the table header.
- Urban quality attributes, representing the development success, fell into seven main areas and can be estimated through thirty-three indicators. These quality indicators constitute the table rows.
- The proposed relation between practices and quality indicators is categorized into five different shades; the darker color represents a strong impact and has a weight of five, while the lighter shade represents a lower impact and has the value of one.

Some of these indicators are objective and can be calculated or measured statistically, such as the unemployment rate and created job opportunities. Others will be represented with no or yes, such as public transportation and car parking areas; others are subjective, and their value will depend on the degree of practice, such as the sense of pride and sense of belonging and need to be collected through a public survey.

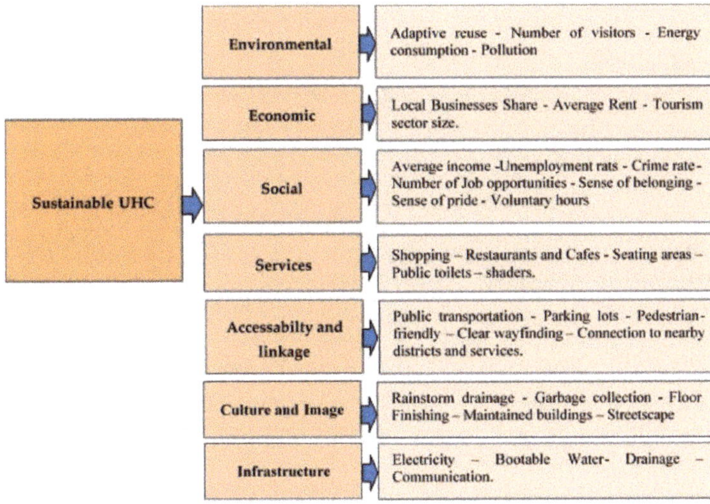


**Figure 6.** Successful sustainable UHC attributes.

**Table 1.** Areas and indicators of urban conservation impact assessment.

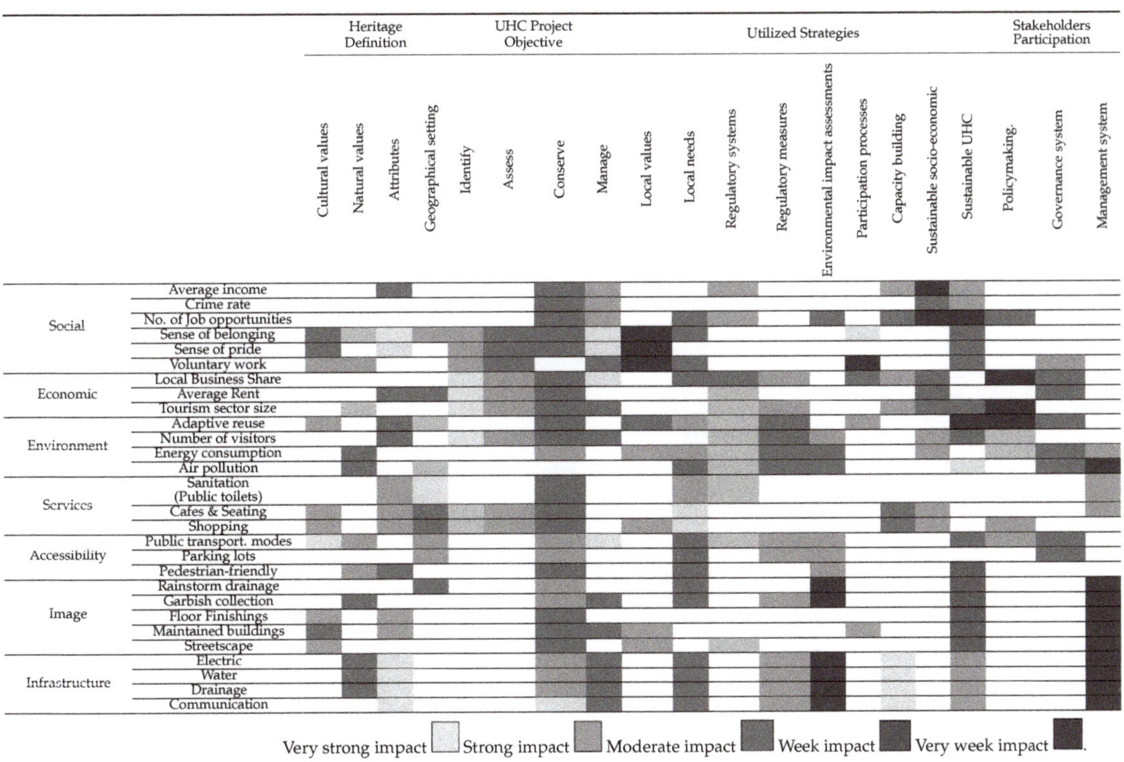


## 5. Case Studies

Two historic sample centers were selected to investigate the impact of the management policies on the urban condition of these fragile built environments. As mentioned before, most Arab countries did not take advantage of HUL. Despite the subsequent case studies not considering this approach in their conservation efforts, they have some similar aspects. For example, Amman conservation efforts integrated conservation into the city strategic plan, and there is a clear vision for touristic attraction and creating job opportunities. Meanwhile, in Jeddah, a dedicated management administration is taking care of the conservation efforts.

Jeddah's historic center was chosen to represent cities within the wealthy Arab countries, and historic Jeddah received much attention from both the national and international heritage authorities. Despite the availability of generous governmental funds, the historic center suffers as much social and environmental deterioration as the rest of the region's historical centers [1,45]. The second case study was the Amman Historic center, which lacked funds and depended mainly on international aid for conservation activities. As mentioned by Smith, international aid usually comes with predefined priorities and scope of conservation [46]. Conservation efforts and urban management plans aimed for socio-economic uplifting to mitigate the requirements of the booming number of refugees within the city [19,45,46].

### 5.1. Amman City

Despite Amman City's historical background, its first existence as the city we know today started in the early twenties of the last century, the Ottoman era, as an intermediate Junction on the Hijaz railway line. Amman's population at that time was about 5000 inhabitants, in an urban area of 0.321 km$^2$. Amman has been the fastest growing city globally in the past few decades due to several immigration waves from neighboring countries [19,21]. In 2020, Amman had an urban area of around 0.321 km$^2$ with less than 5 million inhabitants in the latest available statistics. Figure 7 illustrates a tempo map for the urban city growth [46,47].

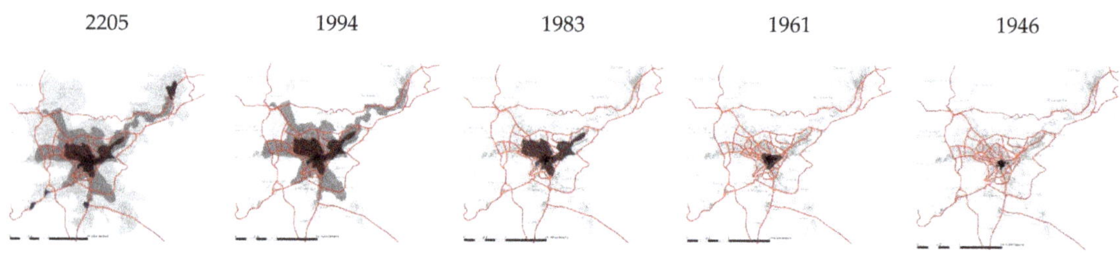


**Figure 7.** Amman City's urban growth (1946–2005) [19].

#### 5.1.1. Amman Historical Center "Albalad"

As shown in Figure 8, the historic Amman or "Albalad" used to be the central market, administrative center, and transportation hub, surrounded by residential areas on the neighboring hills. The hills' topography and valley shape influenced the pattern of the physical development of the 182 hectares of the historic city. The urban pattern adopted the natural physical form. The city expanded following the contours up to the surrounding hilltops. The neighborhoods were connected via stepped alleys as pedestrian links to the city center [19,36]. In addition to its traditional urban, wholesale market, and leisure facilities, the "Albalad" area is near archeological sites with monuments from different eras, such as the mountain castle, the citadel, the Hercules temple, the Roman theater, the Umayyad palace, and the Alhosainy Grand Mosque [19,30].

**Figure 8.** Historic urban features of Amman City. Dated 1931. Source: Adapted from Department of Lands and Survey, Transjordan cited in [19], reproduced by the author.

5.1.2. Conservation Activities

Amman City was subject to several development plans. In 1955, a development plan was approved by the Greater Amman Municipality (GAM) to mitigate the growing demand for housing, services, and the lack of job opportunities [19,47]. Later, in 1968, another master plan was suggested to manage the fast-growing informal settlements, lack of services, road congestion, centralization of business and commercial activities, and high unemployment rates. In 1978, a Japanese firm (IECA) proposed relieving the growing traffic congestion and increasing municipal revenues. GAM proposed and approved the city's final master plan from 2008 to 2025. This master plan was a comprehensive, multilevel sustainable plan [48].

Moreover, this plan came as a participatory effort of all stakeholders. The master plan framework defines urban policies, strategies, land use, legislation, and public participation [21]. The main objective for the historic central conservation was to improve its local economy and social cohesion, recover its centrality within the city, and revive its authenticity and social cohesion. The tools to achieve these objectives were creating spaces for everyone and preserving the cultural heritage with symbolic values. Creating a high-quality environmental and social urban fabric would create a tourist commodity. Increasing tourism activities would create a prosperous community [19]. Participatory community activities helped attract private investment and created an environment for businesses and community involvement projects [19]. The action plan's proposed projects shown in Figure 9 were adopted to achieve the objectives for the Albalad area.

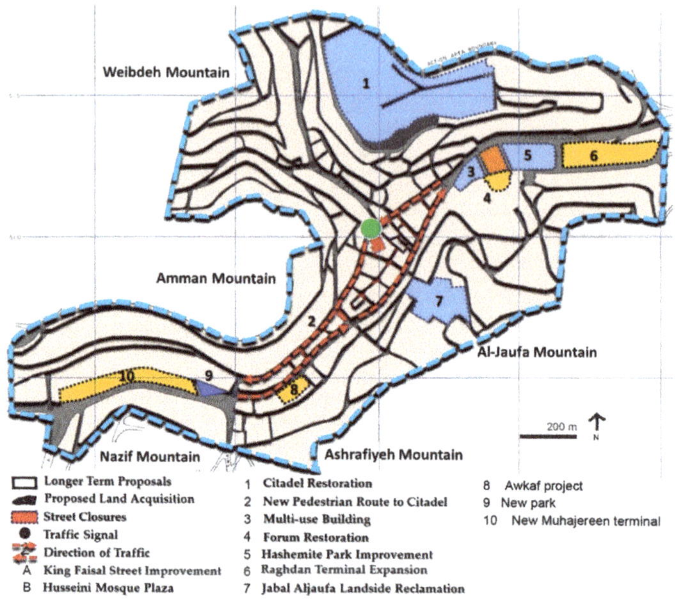


**Figure 9.** Spatial distribution of "Albala" Conservation Action Plan.

### 5.1.3. Current Urban Management Plans

The current master plan includes the following multidirectional urban management programs:

- Re-establish the Albalad as a commercial destination by conserving the existing traditional market's character. Regulate the informal street retailers to achieve a vibrant street mall with good urban design qualities and enhance the specialized and thematic commercial areas.
- Define cultural heritage buildings and districts. Conserve and adapt the reuse of heritage buildings. Implement guidelines for heritage protection and encourage artistic and cultural activities.
- Establish connections between the public realm and its surrounding areas through pedestrian and public transport routes and public and commercial spaces. Improve the image quality of public space and create gateways and thematic landmarks.
- Preserve the existing urban physical character in form and visual relationship with surrounding hilltops.
- Support tourism by enhancing the heritage district spirit, defining the tourist pedestrian routes, and providing tourist services and cultural activities.
- Create an entertaining district by establishing a mixed-use character of specialty retail, cafes, and restaurants. At the same time, make arrangements to hold entertainment venues, festivals, and cultural events.

### 5.1.4. Impact of Implemented Development Actions

Amman's downtown development plan is still in progress. Urban management in terms of enforcing regulations and legalizations successfully revitalized the area as a traditional commercial center. It has created cultural and public spaces facilitating cultural and artistic events. The adopted public uses for the first floor of buildings on selected historical paths for cafes and restaurants created unique and vibrant public spaces. Traffic arrangements and transportation plans helped in easing congestion within the area. Table 2 presents a subjective assessment of the urban quality of the conserved historical center.

**Table 2.** Quality assessment of the "Albalad" conserved urban area (Amman City).

| Area | Assessment |
| --- | --- |
| Environment | Adaptive reuse: Commercial ground floors spread evenly through the narrow allays as using first floors as cafes and restaurants attracted more visitors shown in Figure 10<br>Traditional food and specialized cafes and restaurants and gift shops and bazaars are the majority of the ground floor uses, as shown in the map in Figure 11. At the same time, some famous restaurants are situated on the first floor, as shown in Figure 10.<br>The number of visitors: As the COVID-19 pandemic impacts started to decrease, the area flourished with visitors from other cities and regions. The statistics revealed a 30% increase in tourists' numbers between 2016 and 2019 [49]. Photos in Figure 12 show the vibrant spaces of the city.<br>Energy consumption: Despite that the area climate is considered moderate, adding decorative lights throughout the streets is shown in Figures 12 and 13. Extending the working hours for the commercial area increases the electricity consumption.<br>Pollution: Adding a ring road and two central stations and arranging an internal taxi shuttle service helped ease traffic congestion within the area and reduce air pollution. Preserving the ever-green trees improved the visual and environmental qualities of the area. |
| Economic | Local Businesses Share: Statistics are not available at this stage.<br>Average Rent: Hotel rooms' rent varies at 150–250 JD. Exact statistics are not available at this stage.<br>Tourism sector size: Statistics are not available at this stage. |
| Social | Average income: Lower-middle-class income is 4850 JD average income while poverty line is above 3737 JD statistics within the area ranges is not available at this stage. [50]<br>Unemployment rates: Department of Jordanian statistics show that the unemployment rate in Amman in 2021 in the second quarter was 24.8% [49]<br>Crime rate: Not available at this stage.The number of job opportunities: One of the objectives of the area development was to create jobs. [48]. The ongoing development of the Albalad allowed Hawkers to use the sidewalks to create more jobs, as shown in Figure 14.<br>Sense of belonging: Many residents do not originate in the city and have no sense of belonging; they still enjoy the center and benefit economically, as shown in Figure 15.<br>Sense of pride: Native residents are not happy with the rising refugee numbers because they share their services and the limited job opportunities.<br>Volunteer hours: Statistics are not available at this stage. |
| Services | Shopping: Traditional food, touristic gift shops, cultural events, and cleanly maintained walkways give the public spaces a vibrant image, as illustrated in Figures 13–16.<br>Restaurants and Cafes: As per the land use map in Figure 11<br>Seating areas: Publish seating areas are limited, but outdoor cafes are distributed on the main streets and the stepped alleys.<br>Public toilets: No public toilets except the Mosques and restaurants.<br>Shaders: All the alleys have shaders with different designs and formations. Figure 17 illustrates some of these shading units. |
| Accessibility and Linkage | Public transportation: The area has a taxi shuttle service and two main public transportation stations to link the area to the city districts.<br>Parking lots: There is parking on the streets and parking lots near the two newly refurbished stations.<br>Pedestrian-friendly: The walkways are separated from the cars<br>Clear wayfinding: Streets' slopes and surrounding hills make clear landmarks for visitors. These slops make it easy to move around.<br>Connection to nearby districts and services: The neighboring areas are mainly on the surrounding slopes, meaning the linkage is not direct or easy. |

**Table 2.** *Cont.*

| Area | Assessment |
| --- | --- |
| Culture and image | Rainstorm drainage: The area is a valley surrounded by hills. Several projects to make the best use of the floodwater.<br>Garbage collection: The area is serviced by an effective garbage collection system.<br>Floor Finishing: Walkways using finishing are not expensive, but their details are well designed and constructed.<br>Maintained buildings: Archeological sites lack services, and their surroundings are not well-maintained, as shown in Figure 10, which exhibits the physical conditions of the nearby Castel mountain archeological site. Preserving the contours helped to maintain the original character of the Albalad area Figure 18. Moreover, limiting the building heights to three or four floors maintained the visual contact with the surrounding hilltops and the ratios of the width of the spaces to their surrounding buildings, as presented in Figure 13. Preserving the ever-green trees improved the visual and environmental qualities of the spaces.<br>Streetscape: Walking around gives a rich visual experience either day or night, despite some visual pollution resulting from the suspended electric cables and shops signs with different colors and sizes. |
| Infrastructure | Electricity: The area is covered by electricity. Electric cables are suspended along the many streets within the area.<br>Bootable Water: The area is served by a water network despite the issue of hell areas.<br>Drainage: The area is covered by a sewerage network despite the issue of hell areas.<br>Communication: There is well covered by phone and internet service. |

**Figure 10.** First-floor entertaining activities doubled the commercial use of the area. Source: The Author.

**Figure 11.** Specialized shopping and markets. Source: Amman institute as cited by Ashour [36].

**Figure 12.** Commercial activities created vibrant streets. Source: The Author.

**Figure 13.** Food, touristic gift shops, cultural events.

**Figure 14.** Allowing Hawkers to use the sidewalks creates more jobs: The Author.

**Figure 15.** Many migrants enjoy the benefits of the commercial center. Source: The Author.

**Figure 16.** Preserving the contours maintained the city's original character. Source: The Author.

**Figure 17.** Alleys have shaders with different designs. Source: The Author.

**Figure 18.** The physical conditions of the nearby Castel mountain archeological site.

### 5.2. Jeddah City

With more or less 5 million inhabitants, Jeddah ranks second in size as a Saudi city after Riyadh. It is the economic capital and the main port of the Kingdom of Saudi Arabia. Additionally, it is part of the Red Sea civilization. Jeddah is one of the oldest cities in the region, and it was inhabited about one thousand years before the Crist and has been a trading hub for the region since then [51]. Most of the urban fabric, city walls, the six gates, and residential and public buildings belong to the Ottoman era of the 15th century. Jeddah's modernization started with the Saudi Kingdom. In this area, the city has witnessed several changes.

Historic city walls and gates were demolished in 1948: The city expanded to the north, a new port was established to the south, and an airport was constructed. Later in the 1970s, shopping malls and office buildings were built in the demolished western part of the historic city. A new road network and parking facilities were built to meet the newly growing demand of its inhabitants [51–53]. Figure 19 illustrates the city's growth over the six decades. Due to these modernization activities, the historic district of Jeddah that was once an entire walled city had become a tiny area within the city itself. The historical Jeddah lost whole sections of its original urban fabric. The new modern buildings were concentrated next to the seashore [1,51,53]. During the 1980s and 1990s, the area was no longer able to provide the requirements of contemporary life for its original residents, and they abandoned the area and moved gradually to the wealthy, newly developed areas to the north. Over time, the area started to decline in its economic and social characteristics and converted into an entirely rented area for poor foreign immigrants [52,53].

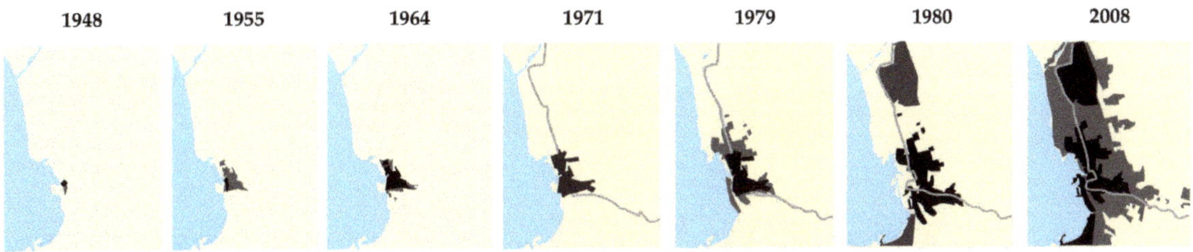


**Figure 19.** Historic Jeddah's urban growth through the last five decades [33].

#### 5.2.1. Jeddah Historical Center

Historic Jeddah, known as Albalad, extends over seventeen hectares, and it symbolizes the heritage and culture of the city. The cityscape of historic Jeddah resulted from an exchange of local culture, environmental conditions, technical construction systems, and materials [54]. Its architectural style is characterized by three- to four-story coral stone residences, decorated with openings covered with wooden "Roshan" [55]. Figure 20 shows the location and spatial organization of these houses. The area's outstanding universal value is based on its unique architectural style and urban character [54,56].

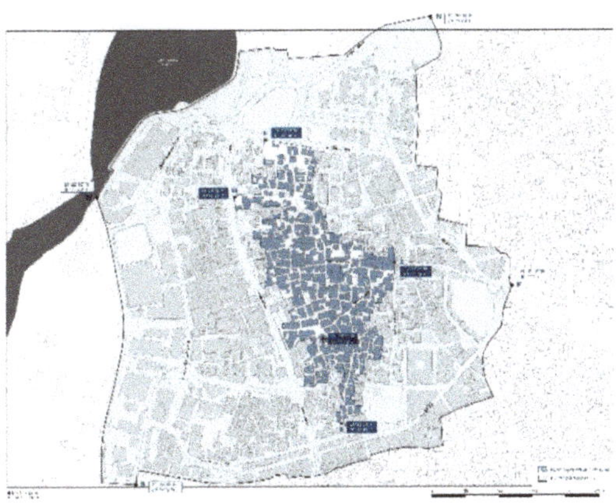

**Figure 20.** The listed buildings within the Albalad area [56].

Furthermore, Historic Jeddah "Albalad" is considered the region's last preserved original urban context. Figure 21 illustrates historic subdistricts and the unique premodern urban landscape, with narrow irregular alleys and small public spaces. The pictures in Figure 22 illustrate the urban character, diverse, multicultural population, and active, well-maintained Alalawy Street—vibrant public spaces that still play a significant cultural symbology role in the local community.

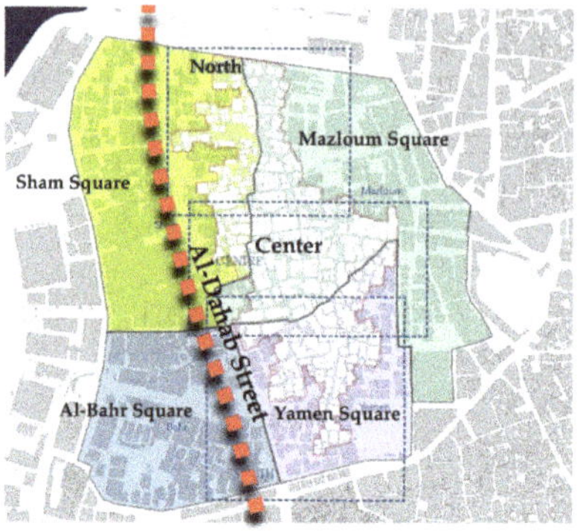


**Figure 21.** Districts of Albalad area.

(**a**) Qabil Street. (**b**) Alalawy Street. (**c**) Algamea Street.

**Figure 22.** (**a**) Area's urban & architectural unique character; (**b**) maintained urban spaces as in Alalawy Street, (**c**) commercial activities at the whole sale market in the Algamea street.

5.2.2. Conservation Activities

Efforts to maintain, conserve, and protect the Albalad area has continued through the last decades. In the 1970s, the area was subject to urban upgraded plans where new malls and office buildings replaced the coastal sector of the site. The city walls and gates were removed to ease the accessibility, and a new main road, "Al-Dahab Street", created parallel to the sea, was constructed to meet the growing traffic demand requirements. Some buildings were subject to significant restoration, and three of the city gates and the remaining city walls were restored [51,57]. From 2003 to 2011, the King Abdul Aziz Project for the Conservation of the Historic District of Jeddah was launched at the organizational and administrative levels. Policies related to heritage buildings were developed. The Department of Architectural Development was established in the region, which formed the nucleus of the current Jeddah Historic District Municipality. A reference guide was adopted to restore heritage buildings and storefronts. The heritage buildings in the Jeddah Historic District Municipality complex were also restored [54,57].

Moreover, through collaborative efforts between the municipality and the wealthy merchants within the city, some residential buildings, mosques, rabats, and markets were restored. Parallel to these efforts, a major urban landscape project was carried out to update spaces within the area. This project includes unifying the colors and size of the shop signs, adding designed lighting units and infrastructure, implementing the first phase of the firefighting network, maintaining the alleys' paving work, and reducing the vehicle traffic within the area [22,58]. In 2012, a new urban legislative framework was enforced, providing legal tools to control the speculative movements that once triggered the loss of some historic buildings. Figure 23 shows the protected urban zone and its neighboring four districts. Each one of these districts follows different building regulations. In 2014, the site was granted UNESCO World Heritage status; several traditional buildings have been restored and opened to the public [51]. In 2019, the Saudi crown prince issued a royal decree to restore 50 historic buildings within the historical site [57].

**Figure 23.** The buffer zone around the listed heritage where areas 1 to 5 represent different applied buildings regulations. [36].

5.2.3. Current Urban Management Plans

UNESCO did not list the whole area but selected a group of buildings. As a result, there are two different lists of heritage buildings, one listed by the municipality and one listed by UNESCO. Based on this, the protection system divides the urban area into four categories [25,59]. Figure 24 illustrates these categories and what Jeddah Municipality defined a larger urban ring as a transitional buffer area around the perimeter of the area listed by UNESCO. This area comprises the 1950s and1960s' expansion of Jeddah beyond the city walls. Building regulations for this outer "ring" aimed at creating a smooth transition between the historic area and the city's modern designed expansion. This urban ring development program creates large-scale structured urban blocks [25,51]. The plan aims to increase accessibility to the inner neighborhoods while preserving the original organic urban fabric, land use, building heights, and densities. Moreover, it targeted updating infrastructure and providing all the necessary social and technical infrastructure to the residents [54].

**Figure 24.** The buffer zone around the listed heritage [37].

#### 5.2.4. Impact of Implemented Development Actions

The enforced land-use legalization for ground floors as commercial and the first floor as offices helped create vital active spaces during shop working hours [29–33]. The typical floor spaces did not attract companies to utilize the traditional residential spaces with tiny rooms and poor utilities. Shop renters used the spaces informally as a store and housing for their workers. This use endangered the valuable, fragile buildings and created risky alleys to walk through after the shops' opening hours [1]. It is worth mentioning that several fires in the last five years resulted in the demolition of several of the listed buildings [57]. At the same time, buildings that were subject to maintenance were locked without any usage, resulting in the fast decay of these buildings. The building regulations for areas one and two, shown on the map in Figure 23, permit owners to have up to twelve floors to combine plots to achieve a larger plot area. This building permit system negatively impacted the narrow alleys' homogenous skyline and width-to-height ratios. Otherwise, the urban spaces are well maintained and vibrant, especially during the Ramadan night season, when daily activities extend for a whole month in the northern part of the historic area. The area is still considered the city's wholesale market for honey, perfume, cloth, toys, etc. [44,51]. Table 3 summarizes the subjective assessment of the implemented conservation strategies and the ongoing activities in the Albalad area of Jeddah city.

**Table 3.** Quality assessment of the "Albalad" conserved urban area (Jeddah city).

| Area of Assessment | Assessment |
|---|---|
| Environmental | Adaptive Reuse: Selecting commercial ground floor and offices as adaptive reuse for the historical buildings did not work well because they did not meet the requirements for a well-equipped environment for modern office buildings. The whole floor above the ground turned into a store and housing for shops' workers. They put the valuable buildings at risk of repeated fires and did not bring the expected users. Figure 25 illustrates commercial activities in the Aljamea streets<br>Number of Visitors: The area has been the whole city market, and increasing the commercial area attracted more shoppers. Annual events such as the Ramadan month cultural festival brought different categories of visitors (looking for entertainment visitors rather than shoppers)<br>Energy consumption: Although these historic buildings were designed to utilize passive systems, heavy air-conditioning systems can be noticed, as shown in Figure 2a.<br>Pollution: Creating paid parking along the surrounding streets pushed shop owners and shoppers to use the narrow alleys to make a full-time traffic-congested zone. This traffic congestion increased air temperature, heat, and air pollution. |
| Economic | Local Businesses Share: The area receives many shoppers, the wholesale market of many products such as honey, makeup, clothes, spices, stationeries, and gifts. Exact statistics are not available at this stage<br>Average Rent: Even though the area is viewed as the old cheap market, the area's rent is considered the highest within the surrounding districts. It ranges between 2500 and 3500 SR per meter a month.<br>Tourism sector size: The area does not attract tourists but shoppers; conservation does not impact visitors' numbers. The exact visitors' statistics are not available at this stage. |
| Social | Average income: The native residents deserted the area and rented their properties to froggier merchants. There are no available statistics for income.<br>Unemployment rats: There are no available statistics for income.<br>Crime rate: There are no available statistics for income.<br>The number of Job opportunities: Conservation did not impact the area's job opportunity. Sense of belonging: Current area residents do not originate in the city and have no sense of belonging; they still benefit as shop vendors. Nevertheless, survey visitors of the Ramadan festival visit the area several times during the month, implying that their sense of belonging increased [60].<br>Sense of pride: Ramadan festival brought old residents back to the area, and the new generation started to be aware of their inherited culture in food, art, and traditions. Figure 26 illustrates Ramadan festival celebrations.<br>Volunteer hours: There are no statistics for volunteers, but the government started a national program for volunteer work with good incentives. It is expected to boost the volunteer work in the area's conservation efforts. |

**Table 3.** *Cont.*

| Area of Assessment | Assessment |
| --- | --- |
| Accessibility and linkage | Public transportation: The area is not covered by any public transportation. The city is studying different modes of public transportation that cover the whole city, including the Albald historic district.<br>Parking lots: There is multi-story car parking, and visitors use some empty lots as parking. It is worth mentioning that the municipality decided to put parking meters on the streets surrounding the area. This creates a continuous traffic jam within the area's narrow streets, as shown in Figure 27.<br>Pedestrian-friendly: The street network is mostly pedestrian-oriented. The municipality created vehicle access through a small number of the alleys to produce the needed services for the residents.<br>Clear wayfinding: Wayfinding is not an easy task since the visual character has not been studied before.<br>Connection to nearby districts and services: The area is well connected to the surrounding districts, especially the neighboring old shopping malls. |
| Culture and Image | Shopping: The area is considered the wholesale market. Conservation did not attract any other new commercial activities.<br>Restaurants and Cafes: Conservation efforts did not create any food or cafes commercial activities.<br>Seating areas: New seating areas next to some of the restored buildings were added.<br>Public toilets: The only available toilets are within the mosques or the neighboring shopping malls.<br>Shaders: The narrow alleys are already shaded, and there is no need for extra shaders. |
| Cultural | Rainstorm drainage: The area has no rain drainage system; this caused the shops' owners much trouble during the last years when the city received more heavy rains than usual.<br>Garbage collection: There is a garbage collection system, but it is not well-managed. Sometimes, in specific locations, garbage accumulates.<br>Floor Finishing: The whole area was covered with granite tiles, which gives the area richness and a good image, as illustrated in Figure 28.<br>Maintained buildings: The buildings have a unique architectural style, and the in-between urban spaces and their composition is one of the oldest and most unique in the region. Figure 29 illustrates the sample of the neglected deteriorated buildings, while Figure 30 shows a sample of the authority's efforts to maintain it.<br>Streetscape: Cleanly tiled alleys, stylish lighting units, and unified shop signs give the visitors a good impression, as in Figure 28. |
| Infrastructure | Electricity: The area is covered with electricity, but the buildings could not accommodate the transformers. Suspended cables and outdoor transformers could be seen along the alleys.<br>Bootable Water: The whole area is connected to the municipality network.<br>Drainage: The area is connected to the sewerage network, but there is no rain drainage, and the residents complained about the repeated sewerage problems.<br>Communication: The area is covered with a good network for both internet and phones. |

**Figure 25.** Commercial activities spread along all the narrow alleys.

**Figure 26.** Ramadan nights, unified shop signage, and stylish lighting units enhance the image quality of the maintained spaces.

**Figure 27.** Narrow alleys turned into car parking.

**Figure 28.** Tiled flooring, unified shop signage, and stylish lighting units.

**Figure 29.** Buildings' owners are not motivated to conserve their properties.

**Figure 30.** Sample of the continuous efforts of the Conservation Authorities to maintain the listed buildings.

## 6. Discussion

The paper proposed an assessment framework for urban heritage conservation that links four different conservation parameters to seven aspects of urban qualities. The first conservation parameter defines the urban heritage and its attributes; the second parameter is the assessment of planning, conserving, and managing the urban heritage;

the third parameter is the institutional, communal, legal, and sustainable strategies for implementation; the fourth parameter is the stakeholders' management and the governance framework. The paper proposed twenty practices that may impact the resulting urban quality [39–41,61]. Successful urban heritage conservation was suggested to be assessed through its urban qualities. Seven aspect areas were identified through the study. The first is the environmental aspects; the second is the economic aspects; the third is the social aspects; the fourth is the services to support the expected activities; the fifth is the accessibility and linkage of the developed area; the sixth is culture and image of the resulted urban heritage; the seventh is the infrastructure needed to support the predicted number of visitors. The paper suggested thirty-three indicators for measuring urban conservation success. The impact of each development parameter on the different quality indicators was presented where five different shades were used to represent the strength of the relation between them [1,29,41,42]. The literature review and case studies' investigation showed that, except in Qatar, none of the conservation authorities of the Arab countries had adopted the Historic Urban Landscape approach in their heritage conservation scope or policies yet [33]. Urban heritage areas and their neighboring districts within certain historic cities in many developing countries suffer from deteriorated social fabric, low-standard services, and utility problems [62]. Urban heritage areas and their neighboring districts within the case study urban areas suffered from deterioration, low-standard services, and utility problems [62]. The planning departments of local municipalities were responsible for these historic urban areas [63]. Urban planning authorities spent continuous efforts upgrading them and solving their urban problems [23].

In Amman's case, the final strategic plan of the city recognized the historic area as a cultural asset and touristic commodity [48]. The investigation of Amman city shows that rational urban management policies and suitable adaptive reuse are key to prosperous sustainable historic urban heritage. The historic area has been subject to subsequent management plans since 1995 [30]. These plans were oriented to improve living conditions, create more job opportunities, ease traffic congestion, and improve environmental quality. Increasing tourism activities were embedded within the current management plan [64]. The development in Amman city in general and the Albalad area, in particular, was based directly on foreign aid, and actions were selected based on the significance of their impacts on residents [37]. Facilities to support cultural and touristic activities were developed, and adaptive reuse for buildings was implemented. The result was a vibrant urban and commercial area. However, Islamic and Roman monuments in the surrounding areas are still undeveloped and require dedicated funds for conservation [13]. Gad argues that ring roads are ineffective in defining urban area boundaries and their role as catalysts for urban sprawls [23]. Analyzing the Amman ring road impacts the Albald area positively by decreasing traffic congestion since the surrounding topology already defines the area.

The significance of the historic area of Jeddah allows it to play a significant role in boosting the city's economy [60]. Jeddah's historic area had a dedicated administration for the area management, but up to 2020, the effort still focused on restoring and preserving the listed buildings. Like many other Arab cities, there is no clear multilevel sustainable strategic vision for urban heritage [58]. Mandeli argues that the Jeddah planning authority's adoption of Western approaches to city planning was at the expense of culture and traditions, failing to preserve the historic center or render the nearby Corniche to the local context [63]. Throsby suggested an independent administrative structure, as in the case of the Petra site in Jordan, rather than leaving it to be managed by local management with limited resources. The Kingdom's newly adopted 2030 vision considers the historic area as a tool for development and heritage as a national asset [57]. Such areas usually require expensive solutions beyond the local governments' financial resources, resulting in partial solutions or long-term phased development plans [65]. The case of Jeddah proved that fund availability is not enough. As fire accidents hit the historic area several times over in recent years resulting in the loss of several historical buildings, the Crown Prince, earlier this year, provided funds to conserve fifty buildings of the listed heritage buildings [62].

There is a need for a legal and institutional framework to maintain the sustainability of the conserved heritage urban areas [60,62]. Neither the selected adaptive reuse for the listed heritage buildings in Jeddah, as shopping ground floors and offices for typical floors, nor the financial incentives for owners of heritage buildings provide the desired sustainability. On the contrary, implemented planning regulations and building codes gave owners of heritage buildings motives to neglect and even demolish their buildings to receive a higher floor area ratio and different floor usability [9]. These regulations changed the urban character of the historic area and its urban buffer. [63] Despite some recent and ongoing improvements, area residents' survey results showed the need for a better infrastructure regarding sewerage, lighting, parking, and waste disposal. Throsby et al. concluded that there is no strategic planning for the area based on giving permission for many large shopping centers without facilitating the required parking facilities. This worsens the traffic congestion within the historic area [60].

Upon comparing the urban management plans in Amman and Jeddah it can be concluded that good design and details can produce a positive image, irrespective of the richness of the used materials. Moreover, Amman benefited from being part of GAM, allowing the historic area to be part of the strategic vision of the city master plan [19,48]. In Jaddah, where a dedicated district administration was initiated, limited resources and planning expertise did not help as much as the administration structure of Amman did [60]. Mateu et al. argue that the management decentralization of historic urban areas is inefficient and unreasonable. He also suggested a participatory governance approach, where the relevant national institutions' representatives and local bodies can work together with the key local stakeholders to develop and manage urban heritage [66].

The study proved that adaptive reuse should be selected carefully. Itohan et al. proposed a performance-based adaptive reuse selection framework to achieve sustainable and resilient urban areas. This suggested framework prioritizes these aspects as follows: Economic sustainability, heritage conservation, socio-cultural, historic building usability, and buildings regulations [67].

Miriam concluded from her survey on a case study in Morocco that development projects that impose strict regulations always face dissatisfaction and resistance from residents because of the lack of freedom in living within traditional socio-cultural norms [20]. The investigated case studies showed that adopting a legal and financial multi-dimensional framework with an area of freedom and participation is essential for any sustainable urban heritage conservation plans [60]. The study showed that conservation activities improve the overall market value of the proprieties within the area and attract a new category of residents [68]. The motive for the local community is the sense of belonging that can be achieved through programs of preserving the social structure and conserving local heritage. Such programs require smaller budgets than conserving the physical environment and can be economically sustainable [30,44]. The study investigation's most important lesson was that sustaining and preserving the cultural and social aspects does not require capital investment. As much as conserving the physically built environment, social programs raise the sense of belonging and cultural identity within the local community [14], such as Ramadan nights organized by the Saudi Antiquities Authorities in the Albalad area in Jeddah and the cultural events and activities held in the Albalad area in Amman. These programs proved to be very helpful in raising the local community's awareness of their heritage and gaining their willingness to participate in the conservation process [1,13].

Urban spaces within historic areas are not built to accommodate contemporary traffic, and reducing traffic congestion must be one of the primary targets of any development plan, which will have a direct environmental impact through reducing air pollution [7,21]. Moreover, the literature review indicates that improving the quality of the outdoor spaces and enhancing the public space image increase the overall market value of the properties within the area and attract a new category of residents and activities [32,36].

Literature on the subject implies that urban buffer areas should be treated so that smoother transactions can be facilitated between the historic urban and the rest of the city urban areas [25]. Gad concluded that planning mega projects in urban areas could significantly change their surrounding areas' land use [23]. This conclusion pushed heritage decision-makers toward having such projects within urban areas surrounding the historical centers. He also argues that urban growth would occur at the expense of valuable resources, either green space or cultural assets within or surrounding the city. This argument alerts the heritage authorities to initiate protective legalization for heritage areas as soon as they are defined. Dailia suggested planners apply contemporary trends while engaging the urban heritage through analysis to end up with a balanced urban management plan that satisfies the technical, environmental, and cultural needs [18]. Moreover, she suggests bridging the gap between the need for shaded winding streets and wide, straight streets to facilitate vehicles' movement through utilizing the traditional environmental techniques and local features creating "place-related social identity." [18].

"What can be measured can be accomplished": This quote was one of the fundamental outputs of the Millennium Development Goals (MDGs) final report, adopted by the United Nations' Millennium Declaration in 2000 [69]. The study showed that the best way to perform conservation of historical urban areas is to embed it within the urban planning strategies of the city [67,68]. Therefore, the suggested objective and subjective indicators need to be used as soon as the conservation objectives are defined. Data collection of all the existing conditions is required before any implementation efforts. The data are the base for accurately measuring the outcomes [34].

Miriam argues that development policies that promote residents' participation in Morocco were not entirely successful. Moreover, she concluded that achieving successful development initiatives with residents' participation rarely reaches its planned objectives and constitutes a repeated dilemma for urban management authorities [20]. She suggested that engaging residents at the early stages of the development project would allow them to voice their requirements, bridge the gap between their needs and the planners' vision, build trust among the stakeholders, and ensure the delivery of what was promised. Moreover, the case study analysis proved that local awareness and public participation contributed positively to improving and revitalizing the place's character. Indeed, urban heritage is an essential part of any local culture. Residents' sense of belonging and pride can be the driving force for any conservation efforts. Residents' participation is necessary for any urban management plans. Locals should be part of the team to identify the heritage objects, define conservation priorities, management plans, and activities. Policymakers must induce regulations that ensure the effective conservation of cultural values in such historic city centers.

Consequently, these cultural values can generate more attraction to those centers and, hence, sustain these places. Some cultural activities, such as social gathering at night, requires special urban qualities to support these social activities. Table 4 summarizes the study recommendations for conservation authorities and city planning departments.

**Table 4.** Recommendations to city managers and planners.

| | | Recommendation | Anticipated Result |
|---|---|---|---|
| Management recommendations | 1. | Adopt a participatory approach for conservation, engaging related national and local institutions and all stakeholders. | Ensure achieving support from related stakeholders. |
| | 2. | Develop a strategic vision and phased action plan for the city, including the historic center. | Ensure clear objectives and sustainable phased development. |
| | 3. | Define heritage conservation and development measurable objectives. | Ensure good assessment and feedback for corrections. |
| | 4. | Define assessment indicators for assessment in the early phases of the project | Provide good reliable feedback for the project management. |
| | 5. | Data collection of existing conditions must be conducted before development implementation. | Developed KPIs can be measured, and conservation efforts can be assessed and adjusted. |
| Execution recommendations | 6. | Arrange social activities and cultural events programs. Engage residents in the urban management plans process. | Increase sense of belonging and pride; create a local driving force for the development plans. |
| | 7. | Engage residents in identifying the heritage and categorizing its value. | Increased sense of belonging and creating a driving force for preserving the heritage asset. |
| | 8. | Select the adaptive reuse of historical buildings carefully. | New usage can attract visitors and boost the local economy. |
| | 9 | Follow a sustainable multidirectional UHC. | Ensure the sustainability of the development. |
| | 10 | Create a car-free historic area and make it pedestrian-friendly. | Create an excellent urban image of a safe, less polluted urban environment. |

## 7. Conclusions

The main research findings and interpretations are:

1. The proposed assessment framework showed that conservation parameters could be assessed through the resulting urban qualities.
2. The presented parameters of Urban Heritage Conservation need to be defined and followed from the early stages of the project.
3. A set of assessment indicators needs to be defined as soon as the strategic vision is identified. The proposed framework provides a good base for such an assessment process.
4. It has been proven that the conservation efforts of most Arab cities' historical centers are outdated in benefiting from the historic urban landscape approach and intangible cultural assets.
5. Case studies' analysis shows that social activities and heritage-based events can help build a sense of belonging and a sense of place. Seasonal programs with weekly events of various types can bring visitors from different backgrounds to the targeted areas and help promote them. Moreover, it would build the required image of the place.
6. Creating outdoor quality spaces is a good step in developing the historic urban area, but it will not bring visitors. Creating activities, adopting new uses for buildings, adding commercial activities, restaurants, cafes, and seating areas can create a vital active place.
7. Adaptive reuse for historic buildings should be selected carefully. Such reuse should offer the safety and security of the historic buildings simultaneously; it should support their sustainability. Selected functions could significantly influence the livability and vitality of the open spaces between these buildings.
8. Case studies suggested that repeated efforts to accommodate the car traffic systems within the historic urban spaces proved to fail. Creating pedestrian-friendly urban areas contributes positively to safe, less congested, and less polluted urban areas.
9. The Amman case study analysis proved that creating a ring road around the area eases traffic congestion and environmental pollution. Providing a mass-transit system

linked the area to the neighboring districts. Having parking areas at the area borders contributes to easing traffic.

10. The case of Amman proved that having a public transit system for the historic area can encourage visitors to leave their cars on the area's borders.
11. The Jeddah case study showed that the unavailability of mass transit and the enforcement of a paid parking policy along the streets around the area pushed visitors to park their cars inside the historic area narrow alleys causing a congested traffic jam, where pedestrians and vehicles share the same spaces without any separation.
12. A legal and economic framework must be defined and put to action before any conservation efforts start. This framework should deal with ownership issues and the buildings' regulations to guarantee the security of these buildings.

The limitations of this research are that the case study analysis was based on assessing the resulting urban quality and not examining the impact of the process aspects on the quality indicators since both did not fully use the UHC approach. Some of the indicator data were not available at this research stage. The next stage is expected to include residents' surveys and secondary data. For more validation, full implementation of the framework in assessing UHC development should be used to examine a case study that implements the HUL approach. Further research is needed in two main areas. The first is investigating the economic bases of urban heritage conservation projects in terms of their validity, alternatives, impacts on the local community prosperity, and the generated wealth distribution. The second area is to examine more specialized indicators in the social and economic fields.

**Funding:** This research received no funds.

**Institutional Review Board Statement:** Not Applicable.

**Informed Consent Statement:** Not applicable.

**Conflicts of Interest:** The author declares no conflict of interest.

## References


1. Badawy, S.; Shehata, A.M. Sustainable Urban Heritage Conservation Strategies—Case Study of Historic Jeddah Districts. In *Cities' Identity through Architecture and Arts*; Anna Catalani, Z.N., Versaci, A., Hawkes, D., Bougdah, H., Sotoca, A., Ghoneem, M., Trapani, F., Eds.; Routledge: Cairo, Egypt, 2020; Volume 1, p. 17.
2. Al-Saffar, M. Urban heritage and conservation in the historic center of Baghdad. *Int. J. Herit. Arch. Stud. Repairs Maint.* **2017**, *2*, 23–36. [CrossRef]
3. Yildirim, E.; Culture in the Implementation of the 2030 Agenda: A Report by Culture 2030 Global Goals Movement. ICOMOS. 2018, p. 2. Available online: https://www.icomos.org/images/DOCUMENTS/Secretariat/2021/SDG/ICOMOS_SDGs_Policy_Guidance_2021.pdf (accessed on 12 October 2021).
4. UNESCO. Item 12 of the Provisional Agenda: Revision of the Operational Guidelines. In *Convention Concerning The Protection of the World Cultural and Natural Heritage—Extended Forty-Fourth Session*; 2021; Available online: https://whc.unesco.org/document/173608 (accessed on 8 October 2021).
5. UNESCO. Adopted decisions during the 42nd session of the World Heritage Committee. In *Convention Concerning The Protection of the World Cultural and Natural Heritage-Forty-Second Session*; WHC/18/42.COM/18; 2018; Available online: https://whc.unesco.org/document/168796 (accessed on 8 October 2021).
6. Mandeli, K. Interpretation of Urban Design Principles in a Traditional Urban Environment: The Role of Social Values in Shaping Cities. Ph.D. Thesis, University of Southern California, Los Angeles, CA, USA, 1993.
7. Jasim, S. The Issue of Standardization in Tradition Urban Fabric of Islamic City. In Proceedings of the 13th International Conference "Standardization, Prototypes, and Quality: A Means of Balkan Countries' Collaboration", Brasov, Romania, 3–4 November 2016.
8. Khalaf, R.W. Traditional vs. modern Arabian morphologies. *J. Cult. Herit. Manag. Sustain. Dev.* **2012**, *2*, 27–43. [CrossRef]
9. Nasralden, M.K. The Transformation of Public Spaces in Saudi Cities: A Case Study of Jeddah. In Proceedings of the Saudi International Conference, Coventry, UK, 23–26 June 2011.
10. Mahmoud, M.L. The Egyptian City Centers in the Islamic Era: Image Analysis, Evaluation, and Contemporary Reflections. *J. Eng. Sci.* **2020**, *48*, 538–553.
11. Correia Jorge, M.T. Traditional Islamic Cities Unveiled: The Quest for Urban Design Regularity. *Gremium. Mag.* **2015**, 2. Available online: https://editorialrestauro.com.mx/traditional-islamic-cities-unveiled-the-quest-for-urban-design-regularity/ (accessed on 20 October 2021).

12. AlSayyad, N. Space in an Islamic City: Some Urban Design. *J. Archit. Plan. Res.* **1987**, *4*, 11.
13. Hannah, D. The Rights to Amman: An Exploration of the Relationship between a City and Its Inhabitants. Master's Thesis, Leiden University, Leiden, The Netherlands, 2021.
14. Yung, E.H.K.; Zhang, Q.; Chan, E.H.W. Underlying social factors for evaluating heritage conservation in urban renewal districts. *Habitat Int.* **2017**, *66*, 135–148. [CrossRef]
15. Ertan, T.; Eğercioğlu, Y. Historic City Center Urban Regeneration: Case of Malaga and Kemeraltı, Izmir. *Procedia Soc. Behav. Sci.* **2016**, *223*, 601–607. [CrossRef]
16. Abu-Hamdi, E. Neoliberalism as a site-specific process: The aesthetics and politics of architecture in Amman, Jordan. *Cities* **2017**, *60*, 102–112. [CrossRef]
17. Saeed, Z.O.; Almukhtar, A.; Abanda, H.; Tah, J. Mosul City: Housing Reconstruction after the ISIS War. *Cities* **2022**, *120*, 103460. [CrossRef]
18. Mohamed, D.H.I.L. Cairo: An Arab city transforming from Islamic urban planning to globalization. *Cities* **2021**, *117*, 103310. [CrossRef]
19. Alnsour, J. Managing urban growth in the city of Amman, Jordan. *Cities* **2016**, *50*, 93–99. [CrossRef]
20. Keep, M.; Montanari, B.; Greenlee, A.J. Contesting "inclusive" development: Reactions to slum resettlement as social inclusion in Tamesna, Morocco. *Cities* **2021**, *118*, 103328. [CrossRef]
21. Abdeljawad, N.; Nagy, I. Urban Environmental Challenges and Management Facing Amman Growing City. *Rev. Int. Geogr. Educ.* **2021**, *11*, 2021. [CrossRef]
22. Abdulaal, W.A. Large urban developments as the new driver for land development in Jeddah. *Habitat Int.* **2012**, *36*, 36–46. [CrossRef]
23. Jaad, A.; Abdelghany, K. The story of five MENA cities: Urban growth prediction modeling using remote sensing and video analytics. *Cities* **2021**, *118*, 103393. [CrossRef]
24. Koorosh, S.S.; Sza, I.; Ahad, F. Evaluating Citizens' Participation in the Urban Heritage Conservation of Historic Area of Shiraz. *Procedia Soc. Behav. Sci.* **2015**, *170*, 390–400. [CrossRef]
25. UNESCO World Heritage Centre. *Developing Historic Cities Key Understanding and Taking Actions*; UNESCO: Paris, France, 2014.
26. Hoff, A.G. Sacred Placemaking and Urban Policy the Case of Tepoztlán, Mexico. Ph.D. Thesis, The University of California, Irvine, CA, USA, 2020.
27. UNESCO. *Fifteenth General Assembly of States Parties to the Convention Concerning the Protection of the World Cultural and Natural Heritage*; UNESCO: Paris, France, 2005.
28. Zancheti Silvio Mendes, K.S. Measuring Heritage Performance. In Proceedings of the In 6th International Seminar on Urban Conservation, Recife, Brazil, 29–31 March 2011.
29. Janset, S. Al-Balad as a Place of Heritage: Problematizing the Conceptualization of Heritage in the Context of Arab Muslim Middle East. Ph.D. Thesis, University College London, London, UK, 2011.
30. Richards, G.; King, B.; Yeung, E.Y.M. Experiencing culture in attractions, events and tour settings. *Tour. Manag.* **2020**, *79*, 104104. [CrossRef]
31. Francesca, N. The Role of Cultural Heritage in Sustainable Development Multidimensional Indicators as Decision Making Tool Enhanced Reader. *Sustainability* **2017**, *9*, 28.
32. UNESCO World Heritage Centre. *Recommendation on the Historic Urban Landscape-Report of the Second Consultation on Its Implementation by the Member States*; UNESCO: Paris, France, 2019.
33. UNESCO World Heritage Centre. *Operational Guidelines for the Implementation of the World Heritage Centre*; UNESCO: Paris, France, 2019; pp. 1–177.
34. Australia ICOMOS. *The Burra Charter Process, Flow Chart from the Australia ICOMOS Burra Charter*. 2013. Available online: https://australia.icomos.org (accessed on 18 November 2021).
35. Ashour, K.N. Urban Regeneration Strategies in Amman's Core: Urban Development and Real Estate Market. Ph.D. Thesis, Dortmund Technical University, Dortmund, Germany, 2016.
36. Pietrostefani, E.; Holman, N. The politics of conservation planning: A comparative study of urban heritage making in the Global North and the Global South. *Prog. Plan.* **2020**, *152*, 100505. [CrossRef]
37. Katapidi, I. Heritage policy meets community praxis: Widening conservation approaches in the traditional villages of central Greece. *J. Rural. Stud.* **2020**, *81*, 47–58. [CrossRef]
38. Veldpaus, L.; Roders, A.P. Historic Urban Landscapes—An Assessment Framework. In Proceedings of the 33rd Annual Meeting of the International Association for Impact Assessment, Calgary Stampede B.M.O Centre, Calgary, AB, Canada, 13–16 May 2013.
39. Silva, A.T.; Roders, A.P. Cultural Heritage Management, and Heritage (Impact) Assessments. In Proceedings of the Joint CIB W070, W092 & TG72 International Conference on Facilities Management, Procurement Systems, and Public-Private Partnership, Cape Town, South Africa, 23–25 January 2012.
40. Chau, K.W.; Choy, L.H.T.; Lee, H.Y. Institutional arrangements for urban conservation. *Neth. J. Hous. Environ. Res.* **2018**, *33*, 455–463. [CrossRef]
41. Samargandi, S. Integral Placemaking in Sensitive Heritage Sites for Successful Cultural Tourism. Master's Thesis, Effat University, Jeddah, Saudi Arabia, 2018.

42. Yousof, A.S. Plaza Design Criteria—Applied Study of South East Plaza of the Grand Mosque at Makkah. Master's Thesis, Umm Al-Qura University, Makkah, Saudi Arabia, 2018.
43. Shehata Ahmed, A.E. Impact of Urban Development Policies on the Produced Urban Characteristics. In Proceedings of the 3rd International Conference on Conservation of Architectural Heritage (C.A.H.), Aswan, Egypt, 19–22 February 2019.
44. Smith, L. *Uses of Heritage*; Routledge: New York, NY, USA, 2006.
45. Beauregard, R.A.; Marpillero-Colomina, A. More than a master plan: Amman 2025. *Cities* **2011**, *28*, 62–69. [CrossRef]
46. Shwartz, A.; Turbé, A.; Julliard, R.; Simon, L.; Prévot, A.-C. Outstanding challenges for urban conservation research and action. *Glob. Environ. Chang.* **2014**, *28*, 39–49. [CrossRef]
47. Jordanian Department of Statistics. Amman Vital Statistics: Jordan Statistical Yearbook. 2020. Available online: http://dosweb.dos.gov.jo/databank/yearbook/YearBook_2020.pdf (accessed on 20 November 2021).
48. Abu-Ghazalah, S. Le Royal in Amman: A new architectural symbol for the 21st century. *Cities* **2005**, *23*, 149–159. [CrossRef]
49. Jordanian Department of Statistics. Tourism Statistics: Jordan Statistical Yearbook. 2020. Available online: http://dosweb.dos.gov.jo/products/jordan-statistical-yearbook-2020/ (accessed on 20 November 2021).
50. Jordanian Department of Statistics. Population Jordan Statistics. 2018. Available online: http://dosweb.dos.gov.jo/databank/yearbook/YearBook_2018.pdf (accessed on 20 November 2021).
51. Ministry of Municipal Affairs. *Future Saudi Cities Programme City Profiles Series: Jeddah City Profile*; Ministry of Municipal Affairs: Jeddah, Saudi Arabia, 2019.
52. Alshoaibi, A. Albeeah Architects and Engineers. In *Proposed Urban Strategy Report*; Jeddah Governorate: Jeddah, Saudi Arabia, 2004.
53. Bokhari, A.Y. Conservation in the Historic District of Jeddah. In Proceedings of the Redeveloping and Rebuilding Traditional Areas, Boston, MA, USA, 16–20 August 1982; pp. 60–68.
54. Saudi Heritage Commission. *Executive Summary of State of Conservation Report—Kingdom of Saudi Arabia World Heritage Sites*; UNESCO: Jeddah, Saudi Arabia, 2020.
55. Mady Mohamed, E.M. Investigating the Environmental Performance of the Wind Catchers in Jeddah. *WIT Trans. Built Environ.* **2018**, *177*, 11.
56. Geographic Information Center at Jeddah Minucipality. *Map of the Nominated Property and the Buffer Zone of Historic Jeddah*; Saudi Heritage Commission: Al Bahah, Saudi Arabia, 2012.
57. UNESCO World Heritage Centre. *State of Conservation: Historic Jeddah, the Gate to Makkah (Saudi Arabia)*; UNESCO: Paris, France, 2021.
58. Ahmed, M.K.N. Public Spaces in a Contemporary Urban Environment: Multi-Dimensional Urban Design Approach for Saudi Cities. Ph.D. Thesis, University of Newcastle, Newcastle, UK, 2011.
59. The Supreme Commission for Tourism, Kingdom of Saudi Arabia. *National Tourism: Development Project Phase 1: General Strategy*; The Supreme Commission for Tourism: Riyadh, Saudi Arabia, 2002.
60. Throsby, D.; Petetskaya, K. Heritage-led urban rehabilitation: Evaluation methods and an application in Jeddah, Saudi Arabia. *City, Cult. Soc.* **2021**, *26*, 100397. [CrossRef]
61. Landorf, C. A Framework for Sustainable Heritage Management: A Study of U.K, Industrial Heritage Sites. *Int. J. Heritage Stud.* **2009**, *15*, 494–510. [CrossRef]
62. Lai, L.W.C.; Lorne, F.T. Sustainable Urban Renewal and Built Heritage Conservation in a Global Real Estate Revolution. *Sustainability* **2019**, *11*, 850. [CrossRef]
63. Mandeli, K. Public space and the challenge of urban transformation in cities of emerging economies: Jeddah case study. *Cities* **2019**, *95*, 102409. [CrossRef]
64. Ali Al-Shomali, M.A. Establishing Evaluation Criteria of Modern Heritage Conservation in Historic City Centers in Jordan. *Int. J. Environ. Sci. Dev.* **2020**, *11*, 561–571. [CrossRef]
65. Udeaja, C.; Trillo, C.; Awuah, K.G.; Makore, B.C.; Patel, D.A.; Mansuri, L.E.; Jha, K.N. Urban Heritage Conservation and Rapid Urbanization: Insights from Surat, India. *Sustainability* **2020**, *12*, 2172. [CrossRef]
66. Mirzakhani, A.; Turró, M.; Jalilisadrabad, S. Key stakeholders and operation processes in the regeneration of historical urban fabrics in Iran. *Cities* **2021**, *118*, 103362. [CrossRef]
67. Aigwi, I.E.; Ingham, J.; Phipps, R.; Filippova, O. Identifying parameters for a performance-based framework: Towards prioritising underutilised historic buildings for adaptive reuse in New Zealand. *Cities* **2020**, *102*, 102756. [CrossRef]
68. Jayantha, W.M.; Yung, E.H.K. Effect of Revitalisation of Historic Buildings on Retail Shop Values in Urban Renewal: An Empirical Analysis. *Sustainability* **2018**, *10*, 1418. [CrossRef]
69. United Nations. *The Millennium Development Goals Report*; United Nations: New York, NY, USA, 2015.

*sustainability*

*Article*

# Role of Ethical Marketing in Driving Consumer Brand Relationships and Brand Loyalty: A Sustainable Marketing Approach

**Muhammad Tanveer [1,*], Abdul-Rahim Ahmad [2], Haider Mahmood [3] and Ikram Ul Haq [4]**

[1] Prince Sultan University, Riyadh 11586, Saudi Arabia
[2] KFUPM Business School, Dhahran 31261, Saudi Arabia; abdulrahim.ahmad@kfupm.edu.sa
[3] Department of Finance, College of Business Administration, Prince Sattam bin Abdulaziz University, Alkharj 11942, Saudi Arabia; h.farooqi@psau.edu.sa
[4] King Abdullah International Medical Research Center, College of Dentistry, King Saud bin Abdulaziz University for Health Sciences, Riyadh 11481, Saudi Arabia; ikram34439@yahoo.com
* Correspondence: mtanveer@psu.edu.sa

**Citation:** Tanveer, M.; Ahmad, A.-R.; Mahmood, H.; Haq, I.U. Role of Ethical Marketing in Driving Consumer Brand Relationships and Brand Loyalty: A Sustainable Marketing Approach. *Sustainability* **2021**, *13*, 6839. https://doi.org/10.3390/su13126839

Academic Editor: Shervin Hashemi

Received: 6 June 2021
Accepted: 12 June 2021
Published: 17 June 2021

**Publisher's Note:** MDPI stays neutral with regard to jurisdictional claims in published maps and institutional affiliations.



**Abstract:** This research is focused on studying the impact of ethical marketing practices on value-adding product sustainability and customer brand relationship sustainability. It further investigates the consequent effects of value-adding product sustainability and customer brand relationship sustainability on brand loyalty. Data for this study were collected from a sample of 1500 customers having multiple interactions with goods and brands of retail organizations in Pakistan. We employed structural equation modeling (SEM) using SPSS 24.0 to analyze our data. The findings of this paper provide empirical support to the proposed relationships. More specifically, ethical marketing practices were found to have a significant impact on value-adding product sustainability and customer-value brand relationship sustainability. The findings also support a positive impacts of value-adding product sustainability and customer-value brand relationship sustainability on brand loyalty. This study provides some valuable implications for the theory and practice in that it identifies and empirically validates key ethical marketing factors affecting loyalty in business-to-consumer interactions. Besides, this study advocates implications for firms regarding some key aspects of ethical marketing practices that should be strengthened to achieve sustained brand loyalty.

**Keywords:** ethical marketing; extended marketing mix; consumer brand relationships; brand loyalty; sustainability

## 1. Introduction

Building strong customer relationships and loyalty is increasingly important for companies in today's rapidly changing marketing environment [1,2]. Developing sustained brand loyalty has attained such a staggering attention because it helps firms in developing advantages that are viable in the markets [3]. It has been observed that the attainment of brand loyalty is based on programs related to the corporate marketing [4–7]. Typically, brand loyalty is reflected in how customers evaluate the company's outlook towards the product evaluation and consumer-brand relations [8–10]. Certain studies have argued that companies should give their brand message while considering social and environmental problems, and sell their products to consumers [11,12]. Strategies focused on common issues increase the interest of consumers in buying the products or services from such companies. Given that modern societies require companies to be responsible and ethical to their stakeholders. Therefore, the maintenance of consumer relationships and brand loyalty are not easy to attain and pose major challenges to firms in today's marketing environment.

Ethical marketing strategies have been studied for nearly all business areas which are formulated to gain a competitive advantage [13–17]. A company's ethical marketing practices affect the daily routine of consumer activity. Every company's ethical marketing practices are closely related to the purchase of products or services, regardless of whether it is conscious of consumer purchasing power strengths and weaknesses. The importance of ethics in the advancement of business sustainability, and general marketing issues (including product safety, price tags and advertising) has duly been recognized by corporate managers and vendors [18]. As an outcome, an economic behavior, whether ethical or non-ethical, is inherently linked with a company's overall reputation and assessment, and stresses the fundamental factors in keeping the company competitive on the market [19–21].

Despite the apparent significant role of ethical marketing practices on relationship building, product evaluation, and high-brand loyalty, only a few studies have examined the effects of extended marketing mix elements (viz. product, price, place, promotion, person, physical evidence, and packaging) on consumption and brand loyalty [22–26]. Moreover, ethical marketing practice plays a significant role in improving consumer-firm ties, product assessment, and brand loyalty [27–29]. Only a few studies have examined the role of ethical marketing mix elements on such attitudinal and behavioral outcomes [18,30,31]. Moreover, the literature has explored the effects of extended marketing mix elements on value-adding product sustainability and customer-brand relationship sustainability in B2C transactions from an ethical marketing perspective [32,33]. Nevertheless, the focus of the literature was on the 4Ps of the marketing mix in this regard. Hence, the present study highlighted this literature gap and contributed by exploring the role of extended marketing mix with the 7Ps approach from an ethical marketing perspective. The findings of this study would be attractive for the companies to build an ethically rich culture, adopting the proposed extended marketing mix with the 7Ps approach that offers fair and trust-oriented transactions. Such an ethically sound approach can help companies facilitate strong relationship building and brand loyalty, which eventually offers them a competitive advantage to survive and sustain for the long term.

## 2. Theoretical Background

### 2.1. Ethical Marketing

In the literature, "ethics" describes concepts like good and evil, correct and false, virtue and sin, justice and crime [33–36]. Ethics seeks to address the human moral issues. The fields of moral psychology, descriptive ethics, as well as value theory are all connected to ethical philosophy [37]. The words "ethics" or "morals" in the literature refer to a collection of moral standards [38], values and beliefs and moral principles, such as moral decisions, standards, norms and laws, and the essence and motivations that direct people to behave a certain way [39–44]. Gaski [45] defined ethical marketing as "a code of morals and conduct used in marketing practices". Ethical marketing drives companies make marketing decisions that are morally acceptable in terms of ties between marketing managers and other stakeholders including customers, employees, competitors, and general public [39]. Ethical marketing is an important subject for both academia and practitioners because ethical principles require companies abide by the minimum standards of liability and conduct their marketing activities in ways that make business transparent and acceptable to all [46–51].

The significance of ethical marketing is further highlighted by the fact that consumers in modern societies continue to require high-quality products and prefer socially renowned brands even though they may aquire those products at higher prices [52]. Ethical marketing typically contributes to a more culturally sensitive and socially conscious business culture. Given that marketing is a big part of any business model, ethical marketing forms an integral part of corporate ethics [53–56]. Built on the theoretical underpinnings of ethical marketing, ethical practices should be applied while analyzing whether the product or the service is portrayed accurately and factually in terms of cultural and social values. Ethical marketing practices provide directions to managers and marketers on how to deal with an ethical issue [57]. Certain researchers consider the use of moral principles in marketing decisions a systematic approach.

This study has considered practices encompassing ethics related to the extended marketing mix elements, viz. product, price, place, promotion, person, physical evidence, and packaging. Several scholars have highlighted the ethics approach of a marketing combination. Ethics related to product decisions offer a competitive advantage and determine if such products are detrimental to customers. The concept of proportion and justice is included in price-related ethics. Ethical pricing should be equivalent to the advantage gained from the product by the consumer [58]. Unreasonable pricing may influence the structure of competition. Ethics related to place-specific decisions emerge largely in relationships along distribution channels as some organizations sometimes follow ethical actions, including abuse of power [59,60]. Promotional ethics issues can be analyzed in marketing and personal sales. Promotional ethics covers questions of marketing, sales and public relations [61]. Other examples of person-specific ethical issues include difficulties with sales advertisements for customers by brokers and suppliers as well as conflicts with advertising media organizations. Physical evidence and packaging specific ethical approach are directed towards a more socially conscious way of thinking; wherein adjustments are made because of its concerns about the ethical issues such as child labor, working climate, ties with third-world countries and environmental problems [62–64].

Considering these arguments, this study intends to build a theoretical framework that explains the relationship of the above discussed extended marketing mix elements from an ethical perspective with some specific consumer behavioral outcomes, which we discuss in the following sections.

### *2.2. Consumer-Value Brand Relationship Sustainability*

To illustrate the common elements of relationship sustainability (RS) which have evolved over the past few decades are relationship marketing and branding [65–67]. Therefore, it is important to analyze the recent branding context description for consumer product identification (CPI). The questions arising with respect to its nature and its connection to RS are equally important. The significance of the relationship between brand and customer are reflected by various terms viz. brand influence, brand identity, personal relationships, brand trust, brand interdependence, and brand loyalty. From a sustainable relationship perspective, if customers are happy, it leads to an increase in the potential of the brand in particular as well as the company in general [68,69]. Previous models of this relationship provide hierarchical aspects of the cognitive, affective and conative elements. This means that connections between customers and brands are formed and maintained over a period of time. The main factors influencing the quality of brand relations between customers include acquisition experience, emotional experience, experience in action, cognitive convictions, and brand commitment [70]. Prior research suggests that in order to enhance the consumer-brand relationship and strengthen the moral aspect and ethical relationships of a business with its consumers, factors specific to the quality of a consumer-brand connection need to be considered [71–74]. Consumer brand partnership sustainability is an alliance created through a mechanism in which customers and brands play their part and engage on a marketplace, as two equal parties. Relationship marketing theory further propounds that a product can longer win a customer's favor through just quality, rather customers often look for brands that focus on building an engaging and sustainable relationship with them.

### *2.3. Brand Loyalty*

Scholars have investigated the concept of brand loyalty across contexts extensively over the past few decades [75,76]. The concept of loyalty is used in different situations and, therefore, varies across contextual dimensions. Different conceptualizations of brand loyalty have been offered by scholars over time. For instance, brand loyalty is "the probability that a consumer will purchase or recommend a particular product or service" [77,78]. Similarly, the 1999 Oliver in defined loyalty as "a deeply held commitment to re-buy or repatronize a preferred product or service consistently in the future, despite situational influences and marketing efforts having the potential to cause switching behavior" [79].

Furthermore, some scientists consider brand loyalty to be a strong commitment to disposing, contributing to customers avoiding situational pressures and marketing strategies that may help to brand change. The commitment to action is demonstrated by the customer's true buying conduct towards a certain brand. Products may have the potential to engage customers and to make them feel emotionally attached [80,81]. Customer beliefs and perceptions build brand images, and this has an effect on the way they view brands they come in contact with. Therefore, companies cannot just rely on informing customers about their products/services, rather they also have to provide exceptional customer experiences to make them sustainably loyal over a period of time [2]. This loyal base of customers will serve as a sustainable competitive advantage for companies across industries.

### 2.4. Contribution of the Present Study

The literature has signified the role of ethics in basic marketing principles [33–36]. Hence, marketing ethics should satisfy all business stakeholders [39]. Therefore, a majority of the literature has investigated the role of ethical marketing in satisfying the business stakeholders [46–51]. The literature has also integrated ethical marketing with overall corporate ethics [53–56]. Ethical marketing may win the customer's confidence, which would help develop a sustainable relationship between brand and customer [68–74]. A vast literature has investigated the role of ethical standards in extended marketing mix elements on brand loyalty [22–29], on the attitudinal and behavioral outcomes [30,31], and on the customer-brand relationship sustainability in B2C transactions [32,33]. However, the literature has investigated the ethical marketing in limited marketing mix elements. This present study extended the scope of empirical literature, focusing on the 7 Ps, i.e., product, price, place, promotion, person, physical evidence, and packaging, in the extended marketing mix to improve consumer brand relationship sustainability. The literature has floated the ethical marketing implications for a limited extended marketing mix. However, the present study provided more holistic view of the extended marketing mix in the theoretical debate of ethical marketing. Moreover, it also opens the doors for practitioners to adopt the greater dimensions of managerial and marketing tools to win the customer loyalty in the campaigns of their brands to accelerate the consumer brand relationships sustainability.

## 3. Conceptual Model and Hypothesis Development

Marketing mix strategy explains how a company offers services and products to satisfy the requirements of divisions of the target market [82]. Marketing strategy with ethical perspectives concern with the righteous behavior and high minded beliefs that enlighten the venture which is marketable so that marketing strategies are applicable for ethical deliberations. Problems in ethical marketing take place further continually in such zones where legitimate acts can be sometimes immoral or in which the lawfulness or morality of actions is undetermined [1]. In order to create positive effects on consumers and brands relationship sustainability, the role of marketing ethics has become progressively important because of this uncertainty in those regions [82,83]. For increasing the trust between multiple stakeholders of the company (including customers), the superior level of ethical authority leads [80,81].

Consumers and brands interact with each other continually [84]. The relational value that a company offers through ethical marketing practices determinses the intellectual and psychological procedure that enhances the interaction between customers and the brand on a sustainable basis. Correspondingly, in this study we deal with a proximity point between the ethical marketing practice and customer value-brand relationship sustainability because consumers estimate the brand and builds the relationship sustainability through some processes in which experience and interaction with the brands occur. Extended marketing mix often serves as the first contact impression through which customers know a brand and come in terms with it [11,16]. Consumers observe the ethical marketing practices because it is the most visible in a brand's manifold schemes and play important role in shaping consumer perceptions towards a brand [85–88].

Prior studies reveal that ethical problems in marketing have involved issues pertaining to product safety and quality, price fixing, deceptive promotions, illegal product placement, child labor and unfair person treatment, misleading information provided through fake physical evidence and packaging [7,11,12,63]. Such ethical issues revolving around fairness and human resource management affect subjective evaluations of product, and relationship value of customers [11]. Fraudulent prices are main cause of misleading the consumer's basic features to purchase, so that is why to improve the relationship between brands and consumers, right prices are needed to be set in the marketplace. Information about product should be truthful to maintain the relationship with customer because negative effects of displays could lead to unreasonable decisions by consumers [11,37,40]. It is suggested that ethical marketing practices related to the marketing mix elements could lead to favorable decisions by consumers which may further derive positive outcomes for the company/brand. Ethical marketing practices results in a more socially responsible and sensitive business community which eventually has the potential (in both the short and long run) to benefit the society as a whole [55,71]. Building upon these arguments, it is suggested that positive relationship occurs through a company's/brand's ethical marketing practice and their customer value-brand relationship sustainability. Therefore, we hypothesize the following:

**Hypothesis 1a (H1a).** *Product-associated ethics shows a positive effect on customer value-brand relationship sustainability.*

**Hypothesis 1b (H1b).** *Price-associated ethics shows a positive effect on customer value-brand relationship sustainability.*

**Hypothesis 1c (H1c).** *Place-associated ethics shows a positive effect on customer value-brand relationship sustainability.*

**Hypothesis 1d (H1d).** *Promotion-associated ethics shows a positive effect on customer value-brand relationship sustainability.*

**Hypothesis 1e (H1e).** *Person-associated ethics shows a positive effect on customer value-brand relationship sustainability.*

**Hypothesis 1f (H1f).** *Physical evidence-associated ethics shows a positive effect on customer value-brand relationship sustainability.*

**Hypothesis 1g (H1g).** *Packaging-associated ethics shows a positive effect on customer value-brand relationship sustainability.*

Value-adding product sustainability should persuade customer's wishes and demands in which safety of the product, well-being levels and habitats take place. The brands that convey ethical values with responsible behavior are the brands which are surrounded by high quality products and are chosen by the customer. Regarding the idea of product sustainability, this is referred to as the personalized sustainability that customers expect in the product. The brand that organizes the responsible actions for marketing in their products and services, receive warm respect and trust from their customers as well as from their shareholders [89–91]. Companies can achieve sustained benefits and trust from their customers only when their deals and agreements are based on the ethical marketing practices with their interested parties. As an outcome, customers prefer brands that are perceived to deliver ethical values and behave responsibly [89]. Based on these arguments, we suggest a positive association between a brand's ethical marketing practices with their value adding product sustainability. So, we propose the following hypotheses:

**Hypothesis 2a (H2a).** *Product-associated ethics shows a apositive effect on value adding product sustainability.*

**Hypothesis 2b (H2b).** *Price-associated ethics shows a positive effect on value adding product sustainability.*

**Hypothesis 2c (H2c).** *Place-associated ethics shows a positive effect on value adding product sustainability.*

**Hypothesis 2d (H2d).** *Promotion-associated ethics shows a positive effect on value adding product sustainability.*

**Hypothesis 2e (H2e).** *Person-associated ethics shows a positive effect on value adding product sustainability.*

**Hypothesis 2f (H2f).** *Physical-associated ethics shows a positive effect on value adding product sustainability.*

**Hypothesis 2g (H2g).** *Packaging-associated ethics shows a positive effect on value adding product sustainability.*

Customers always observe the brand's marketing strategies because it is one of the most perceptible fields [92–94]. Moreover, customers nudge the brand by applying mutual and emotional standards to activities, companies and stories when they chose a brand. Later on, a customer value-brand relationship builds up. Customers mostly evaluate the brand and prefers based on their experiences rather than the particular characteristics of the product. Brand loyalty also has a major role in the establishment of the brand's asset. Previous studies show that the brand loyalty and sustainability improve focusing with the B2C connection perfectly [67,79]. Maintaining the high quality of customer-brand relationship is a major aim of branding because relationship between customer and brand provides a sustainable competitive advantage to the companies.

Previous studies reveal that customers are always satisfied with the positive effects of product quality [11,38]. Also, the value of the product is intimately significant to the customer's contentment and trust because product's value can be seen as the dissimilarity between general and unexplored standards. When customers build a relationship with the brand due to its favorable brand experience and the sustained product value, it drives customers to develop loyalty intentions and stay with the brand for a longer period of time [2,54,60]. Therefore, value-brand relationship sustainability and value adding product sustainability play a dominant role on brand loyalty [89]. Hence, we propose the following hypotheses:

**Hypothesis 3 (H3).** *Customer value-brand relationship sustainability have a positive effect on brand loyalty.*

**Hypothesis 4 (H4).** *Value adding product sustainability has a positive effect on brand loyalty.*

Based on the above hypotheses, we developed our research model as shown in Figure 1 below.

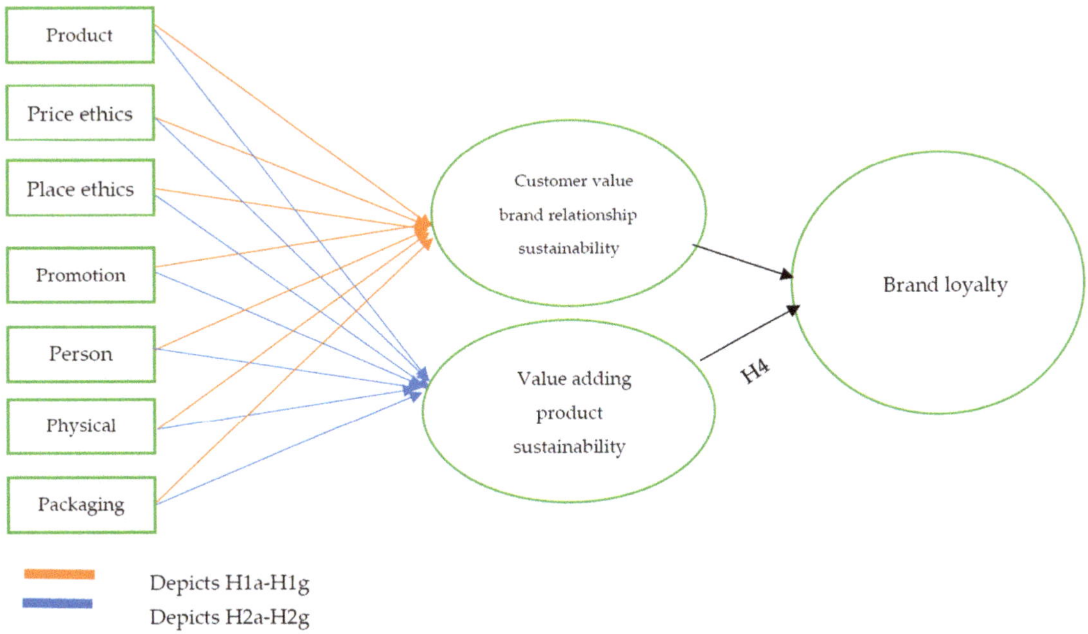


**Figure 1.** Proposed conceptual model.

## 4. Methodology

### *4.1. Questionnaire Development and Data Collection*

The questionnaire in this study was based on independent and dependent variables. We use a non-probability snowball sampling technique. We use this technique for two reasons. First, we did not know the exact size of the population at the start of sampling. Secondly, we expected to have more respondents once we contacted any respondent. We collected 1500 participants from Islamabad—the capital city of Pakistan—who answered the questions. These participants were contacted and asked to participate in the survey by telephone calls. The participants in this survey were any brand's customers who have an understanding of the brand's ethical practices and had involvement with their outcomes more than once. After researchers and professors checked whether there is any conceptual errors or limitations in our questionnaire, we distributed our survey. The questionnaire consisted of two parts; the first part comprised the demographic questions and the second part was about the ethical marketing practices and other constructs under consideration.

The demographic profile of the participants in which their gender, age, education and household income is shown is presented in Table 1.

**Table 1.** The demographic profile.

| | | Frequency | Percent | Valid Percent | Cumulative Percent |
|---|---|---|---|---|---|
| Gender | | | | | |
| Valid | Male | 762 | 50.8 | 50.8 | 50.8 |
| | Female | 738 | 49.2 | 49.2 | 100.0 |
| | Total | 1500 | 100.0 | 100.0 | |
| Age | | | | | |
| Valid | 20–29 years | 332 | 22.1 | 22.1 | 22.1 |
| | 30–39 years | 461 | 30.7 | 30.7 | 52.9 |
| | 40–49 years | 442 | 29.5 | 29.5 | 82.3 |
| | 50 or above | 265 | 17.7 | 17.7 | 100.0 |
| | Total | 1500 | 100.0 | 100.0 | |
| Education | | | | | |
| Valid | school | 418 | 27.9 | 27.9 | 27.9 |
| | college | 673 | 44.9 | 44.9 | 72.7 |
| | university | 409 | 27.3 | 27.3 | 100.0 |
| | Total | 1500 | 100.0 | 100.0 | |
| Household income | | | | | |
| Valid | up to 30,000 | 273 | 18.2 | 18.2 | 18.2 |
| | 30,000–39,000 | 444 | 29.6 | 29.6 | 47.8 |
| | 40,000–49,000 | 485 | 32.3 | 32.3 | 80.1 |
| | above 50,000 | 298 | 19.9 | 19.9 | 100.0 |
| | Total | 1500 | 100.0 | 100.0 | |

### 4.2. Research Instrument

This study borrowed items from previously developed scales (for example, [11,31]) and made minor modifications to fit to the context. Ethical marketing practices are a base in this study which is set up on the righteous and behavioral standards in expressed marketing practices in the marketplaces [94,95]. In this study, the marketing mix strategy from ethical perspectives, is categorized as product-ethics, price-ethics, place-ethics, promotion-ethics, person-ethics, physical-ethics, and packaging-ethics. Questions pertaining to marketing mix elements were based around product-related ethics such as product's package, branding, and warranty of the product (e.g., the sample item is whether the company/brand offers a warranty to cover any faults). Price-related ethics comprises the pricing policy which shows if illegal prices or false prices are being used (the sample item includes if the company/brand refrains from using predatory pricing). Place-related ethics comprises the sales strategy, extent of collaboration and the transactions degree with sample item reading as "the company/brand refrains from controlling transactions by abusing its status". Promotion-related ethics based on the advertising companies with legal notices or with the leading and non-deceptive aspects (e.g., the company/brand refrains from providing false and exaggerated advertisements). Person-related ethics promotes the righteous principles that control the person's behavior and their activities (e.g., the employees of the company/brand treat its customers well). Physical-related ethics refers to the physical ambiance and workplace-related decor that promotes the service experience as well as human rights (e.g., the company/brand offers a great physical presence and environment). Packaging-related ethics is a role of company's branding and marketing policies in the message sent by company through packaging (e.g., the company/brand uses environmentally friendly materials for packaging its products and refrains from making fake promises through its packaging). There were a total of 25 items related to ethical marketing mix elements with five questions on product ethics, four questions on price ethics, four questions on place ethics, three questions on promotion ethics, three questions on person ethics, three questions on physics ethics, and three questions on packaging ethics [92,93,96,97].

We measured customer value-brand relationship sustainability with 12 questions with sample items like: "I really like to talk about the company/brand with others" [58,63,64]. Value-adding product sustainability defines the perception of the customer regarding products and its overall supremacy and leadership, which was measured using four questions, with a sample item reading as "this company's/brand's product has a unique design" [93]. Brand loyalty was measured with five questions (e.g., I am loyal to this company/brand). We used 5-point Likert-scales range from 1 = strongly disagree to 5 = strongly agree to measure the various constructs of this study.

## 5. Data Analysis

### *5.1. Measurement Model*

In our study we used the SPSS software version 22nd IBM for all the analysis. We tested the hypotheses, the goodness of fit test and also verified the exploratory factor analysis of the proposed measurement model. We also examined the relationship between each construct employing structure equation modeling. First, we performed an exploratory factor analysis (EFA) which showed the validity and reliability of the ethical marketing constructs. Cronbach's alpha was used for this purpose because it is the most commonly used method when researchers have multiple Likert questions in a questionnaire. Cronbach's alpha determines the covariance between constructs [98], the relationship should be greater the 0.7 because it shows that the relationship is sufficiently reliable. Tables 2 and 3 below show the results.

**Table 2.** Results of consistency on the construct.

| Constructs | Items | Cronbach's Alpha |
|---|---|---|
| Product ethics | 5 | 0.751 |
| Price ethics | 4 | 0.769 |
| Place ethics | 4 | 0.810 |
| Promotion ethics | 3 | 0.799 |
| Person ethics | 3 | 0.902 |
| Physical ethics | 3 | 0.658 |
| Packaging ethics | 3 | 0.973 |
| CVBRS | 12 | 0.836 |
| VAPS | 4 | 0.817 |
| BL | 5 | 0.796 |

Note: CVBRS = Customer value brand relationship sustainability, VAPS = Value adding product sustainability, BL = Brand loyalty and sustainability.

**Table 3.** Discriminant Validity.

| | PRODUCT | PLACE | PROMOTION | PERSON | PRICE | PHYSICAL | PACKAGING | CVBRS | VAPS | BL |
|---|---|---|---|---|---|---|---|---|---|---|
| PRODUCT | 1.000 | 0.026 | −0.027 | 0.018 | 0.001 | −0.013 | −0.071 | 0.051 | −0.018 | −0.016 |
| PLACE | 0.026 | 1.000 | 0.010 | 0.018 | 0.003 | −0.033 | −0.027 | −0.003 | −0.040 | 0.004 |
| PROMOTION | −0.027 | 0.010 | 1.000 | 0.024 | −0.038 | −0.025 | −0.018 | 0.030 | 0.019 | −0.018 |
| PERSON | 0.018 | 0.018 | 0.024 | 1.000 | 0.012 | −0.049 | 0.013 | −0.003 | −0.024 | −0.041 |
| PRICE | 0.001 | 0.003 | −0.038 | 0.012 | 1.000 | 0.024 | −0.056 | −0.015 | 0.048 | 0.010 |
| PHYSICAL | −0.013 | −0.033 | −0.025 | −0.049 | 0.024 | 1.000 | −0.014 | 0.011 | 0.045 | 0.567 |
| PACKAGING | −0.071 | −0.027 | −0.018 | 0.013 | −0.056 | −0.014 | 1.000 | −0.017 | −0.005 | 0.019 |
| CVBRS | 0.051 | −0.003 | 0.030 | −0.003 | −0.015 | 0.011 | −0.017 | 1.000 | 0.031 | 0.017 |
| VAPS | −0.018 | −0.040 | 0.019 | −0.024 | 0.048 | 0.045 | −0.005 | 0.031 | 1.000 | 0.069 |
| BL | −0.016 | 0.004 | −0.018 | −0.041 | 0.010 | 0.567 | 0.019 | 0.017 | 0.069 | 1.000 |

The Cronbach's alpha results show that all the constructs are reliable, which means there is a good probability that the constructs will perform in a defined environment without failure. The correlation coefficients are between 0.567 and −0.071 in the table of discriminant validity. The systematic assessment of discriminant validity shows that these constructs are not highly correlated to each other. As the correlation of constructs are between −0.071 and 0.567 it means there is a negative relationship between each construct with other construct, although the correlation of any construct with itself is 1.000, which shows that the data used in this study is valid and can be used for further analysis.

In Table 2, the Cronbach's alpha values of all constructs are greater than 0.7 which clearly shows that the reliability is sufficient. We also performed principal component factor analysis which is used to diminish the dimensionality of the constructs, it is used to simplify the data like reducing the number of variables. We performed principal factor analysis with varimax rotation on the constructs, where 0.50 was taken as the minimal cutoff value. The items could be deleted from the further analysis because of the cross loading. Results of factor loadings of dependent and independent constructs are shown in the Tables 4–7.

**Table 4.** Independent variables factor analysis.

| Constructs | Items | Factor Loadings |
|---|---|---|
| | PROD1 | 0.695 |
| | PROD2 | 0.701 |
| Product ethics | PROD3 | 0.683 |
| | PROD4 | 0.662 |
| | PROD5 | 0.709 |
| | PRICE1 | 0.882 |
| Price ethics | PRICE2 | 0.796 |
| | PRICE3 | 0.763 |
| | PRICE4 | 0.851 |
| | PLACE1 | 0.765 |
| Place ethics | PLACE2 | 0.721 |
| | PLACE3 | 0.893 |
| | PLACE4 | 0.812 |
| | PROM1 | 0.699 |
| Promotion ethics | PROM2 | 0.701 |
| | PROM3 | 0.687 |
| | PER1 | 0.851 |
| Person ethics | PER2 | 0.799 |
| | PER3 | 0.837 |
| | PHYS1 | 0.712 |
| Physical ethics | PHYS2 | 0.735 |
| | PHYS3 | 0.802 |
| | PCKG1 | 0.871 |
| Packaging ethics | PCKG2 | 0.865 |
| | PCKG3 | 0.797 |

Tables 6 and 7 show that the eigenvalues of each construct are higher than 1.00, which shows that the data which is selected in this study has no common method bias.

### 5.2. Structural Model

We test the hypotheses between constructs by using SEM [99], to test the casual relationship between each construct we made the hypothesis path casually to estimate what kind of relationship is present in these constructs, the results are presented in Table 8 below:

**Table 5.** Dependent variables factor analysis.

| Constructs | Items | Factor Loadings |
|---|---|---|
| CVBRS | CVBRS1 | 0.758 |
| | CVBRS2 | 0.893 |
| | CVBRS3 | 0.776 |
| | CVBRS4 | 0.852 |
| | CVBRS5 | 0.763 |
| | CVBRS6 | 0.892 |
| | CVBRS7 | 0.781 |
| | CVBRS8 | 0.741 |
| | CVBRS9 | 0.861 |
| | CVBRS10 | 0.771 |
| | CVBRS11 | 0.773 |
| | CVBRS12 | 0.796 |
| VAPS | VAPS1 | 0.821 |
| | VAPS2 | 0.854 |
| | VAPS3 | 0.743 |
| | VAPS4 | 0.861 |
| BL | BL1 | 0.736 |
| | BL2 | 0.751 |
| | BL3 | 0.763 |
| | BL4 | 0.811 |
| | BL5 | 0.835 |

**Table 6.** Eigenvalues in factor loadings (independent variables): total variance explained.

| Component | Initial Eigenvalues | | |
|---|---|---|---|
| | Total | % of Variance | Cumulative % |
| Product | 7.105 | 15.785 | 15.785 |
| Price | 6.092 | 15.597 | 31.382 |
| Place | 6.007 | 14.384 | 45.765 |
| Promotion | 4.997 | 14.246 | 60.011 |
| Person | 4.975 | 13.930 | 73.941 |
| Physical | 3.942 | 13.451 | 87.391 |
| Packaging | 2.883 | 12.609 | 100.000 |

Note: Extraction Method: Principal Component Analysis.

**Table 7.** Eigenvalues in factor loadings (dependent variables): total variance explained.

| Component | Initial Eigenvalues | | |
|---|---|---|---|
| | Total | % of Variance | Cumulative % |
| CVBRS | 6.084 | 36.126 | 36.126 |
| VAPS | 2.987 | 32.908 | 69.034 |
| BL | 2.929 | 30.966 | 100.000 |

Note: Extraction Method: Principal Component Analysis.

**Table 8.** Hypothesis test results.

| Hypothesis | Sig. Values | Results |
|---|---|---|
| H1a (PROD-CVBRS) | 0.251 | Significant |
| H1b (PRICE-CVBRS) | 0.135 | Significant |
| H1c (PLACE-CVBRS) | 0.119 | Significant |
| H1d (PROM-CVBRS) | 0.047 | Not significant |
| H1e (PER-CVBRS) | 0.310 | Significant |
| H1f (PHYS-CVBRS) | 0.227 | Significant |
| H1g (PCKG-CVBRS) | 0.094 | Not significant |
| H2a (PROD-VAPS) | 0.310 | Significant |
| H2b (PRICE-VAPS) | 0.026 | Not significant |
| H2c (PLACE-VAPS) | 0.222 | Significant |
| H2d (PROM-VAPS) | 0.300 | Significant |
| H2e (PER-VAPS) | 0.119 | Significant |
| H2f (PHYS-VAPS) | 0.273 | Significant |
| H2g (PCKG-VAPS) | 0.303 | Significant |
| H3 (CVBRS-BL) | 0.481 | Significant |
| H4 (VAPS-BL) | 0.396 | Significant |

## 6. Discussion

This study aimed at investigating the elements of marketing mix and brand relationship from an ethical marketing perspective. To empirically validate our proposed conceptual model, sixteen hypotheses were tested in the current study, of which thirteen revealed a positively significant relationship and were supported. The remaining three hypotheses were not supported by our results. Hypotheses H1a, H1b, H1c, H1e, and H1f specifying the areas of ethical marketing practices i.e., product, price, place, person and physical ethics have a positively significant effect on the customer value-brand relationship sustainability. However, H1d was not supported, revealing that promotion ethics has no significant effect on customer value-brand relationship sustainability. ($\gamma = 0.047 < p = 0.05$). Similarly, H1g also revealed a non-significant effect of packaging ethics on customer value-brand relationship sustainability ($\gamma = 0.094 > p = 0.05$). In a similar vein, H2a, H2c, H2d, H2e, H2f and H2g identify that the ethical marketing practices in terms of product, place, promotion, person, physical evidence, and packaging are significantly affecting the value adding product sustainability, whereas H2b shows that price ethics has no significant effect on value adding product sustainability ($\gamma = 0.026 < p = 0.05$). Lastly, both H3 identifying the positive relationship between customer value-brand relationship sustainability and brand loyalty and H4 examining the effect of value adding product sustainability on brand loyalty were supported by the results.

The current study studies the role of ethical marketing practices to build positive relationships between customers and brands. Hypotheses testing shows the relationship between these variables and the findings show that, except for promotion ethics and packaging ethics, all other ethical marketing practices have a positive effect on customer value-brand relationship sustainability. Similarly, except for price ethics, all other ethical practices have a significantly positive effect on value-adding product sustainability, which is in sync with some prior studies in different geographic and industrial contexts (e.g., [11,100–102]). Our results suggested that product, price, place, person, and physical ethics are very important for customer value-brand relationship sustainability. This shows that products should be free of harm for users and environment. This concept would increase the value to the customer. In the same manner, the ethical place is also equally crucial as a place should not harm any inhabitant in the environment. Moreover, physical and person ethics are essential to customer value-brand relationship sustainability. Nevertheless, ethical price is also essential for the customer, which can be achieved by avoiding any fraudulent or illegal practices in pricing. Furthermore, our findings suggest that the ethical marketing practices regarding product, place, promotion, person, physical evidence, and packaging significantly affect the value-adding product sustainability. Our results reveal that product ethics significantly affect customer value-brand relationship sustainability and value adding product sustainability, suggesting that customers are inclined to purchase the product with the highest level of product safety and quality. Also, the product's packaging and branding affect customer's trust and satisfaction, which is why it is important to evaluate the product ethics regarding customer value-brand relationship and sustainability and value-adding product sustainability. Besides, value-adding sustainability is established when the company hones authentic communication under legal legislation and leading abilities, i.e., truthfulness and accurate information about the product, which exhibits trustworthiness in the brand. Packaging ethics has an insignificant effect on customer value-brand relationship sustainability as customers want to see beyond the message the company is sending through their packages. However, packaging ethics have a positive effect on value-adding product sustainability, so it is worthwhile for companies to build up a strong relationship with customers. If a company enhances its relationship quality through product, price, place, promotion, person, physical and packaging ethics, we expect more effectiveness on brand loyalty through exciting relationships with customers.

### 6.1. Implications

The academic literature on B2C transactions concludes that there is some evidence of positive effects of traditional marketing practices (in the form of product, price, place and promotion) on brand loyalty. However, our contribution to the literature is first of its kind because an improved model for analysis has been used. The focus is on studying the relationship between ethical marketing practices of an extended marketing mix elements and consumer-brand relationship and brand loyalty, which is common in B2C transactions. Our additional theoretical contribution lies in the fact that most of the studies about ethical marketing practices have been conducted in the Western countries, leaving a huge gap in the literature, which this study fills by conducting in a developing economy context of Pakistan. Besides, this study proposes and empirically validates a novel model from a sustainable marketing approach incorporating the role of relationship marketing, which further reflects its theoretical contributions. Lastly, this study further contributes by employing different theories (e.g., ethical marketing theory, relationship marketing theory) to the context of marketing mix elements and brand loyalty, thereby further generalizing the use of these theories.

As for the practical implications are concerned, managers should consider the ethical marketing practices to ensure sustainable success through relationship building and by facilitating long-term brand loyalty. Companies need to focus on improving product parameters like quality, safety, warranties, and eco-friendliness. Companies also need to take price regulations seriously in building sustained loyalty towards the brand. If companies offer a suitable price, they can attract more customers to tighten the relation between them. Customers most probably need the products with more viable prices. If a supplier does not pay heed to the customers with feasible price-concerned morals, the cons of brand truthfulness can be seen at the end. Furthermore, given that promotion-related ethics signifies the vital role in bringing value adding product sustainability. Therefore, firms need to evaluate the promotion related ethics to maximize the trust-based activities to cement the associations with firms. The summarized implication here is that companies/brands can sustain even during the fierce market competition if they establish an ethical culture which gives due preference to trust-based transactions, thereby strengthening customer-firm relationships. Our findings are expected to provide valuable implications to companies so that they support an effective marketing strategy from an ethical perspective.

### 6.2. Limitations and Suggestions

Regardless of its contribution and implications, this study has several limitations, Firstly, while examining the interplay of marketing ethics and brand loyalty, we didn't consider B2B transactions. Moreover, our proposed relationships may vary across industrial contexts types, therefore, future studies may further validate our model across study settings. Moreover, future analysis should consider the more ethical practices from the point of government ethics, culture ethics, individual ethics, and religiosity to present further insights of the domain. Some circumstantial and individual factors (e.g., personality type) may act as potential moderators, which future studies can also consider. A longitudinal study may also be conducted to gauge loyalty formation over a period of time.

**Author Contributions:** Conceptualization, M.T. and A.-R.A.; methodology, M.T.; software, A.-R.A.; validation, M.T., A.-R.A. and H.M.; formal analysis, A.-R.A.; investigation, M.T.; resources, M.T.; data curation, M.T.; writing—original draft preparation, M.T. and A.-R.A.; writing—review and editing, H.M. and I.U.H.; visualization, M.T. and I.U.H.; supervision, M.T. All authors have read and agreed to the published version of the manuscript.

**Funding:** This research received no external funding.

**Institutional Review Board Statement:** Not applicable.

**Informed Consent Statement:** Not applicable.

**Data Availability Statement:** Not applicable.

**Acknowledgments:** All authors of this article would like to thank the Prince Sultan University for their financial and academic support and publish in "Sustainability".

**Conflicts of Interest:** The authors declare no conflict of interest.

## References

1. Islam, J.U.; Rahman, Z. The impact of online brand community characteristics on customer engagement: An application of Stimulus-Organism-Response paradigm. *Telemat. Inform.* **2017**, *34*, 96–109. [CrossRef]
2. Islam, J.U.; Shahid, S.; Rasool, A.; Rahman, Z.; Khan, I.; Rather, R.A. Impact of website attributes on customer engagement in banking: A solicitation of stimulus-organism-response theory. *Int. J. Bank Mark.* **2020**, *38*, 1279–1303. [CrossRef]
3. Kaur, H.; Paruthi, M.; Islam, J.; Hollebeek, L.D. The role of brand community identification and reward on consumer brand engagement and brand loyalty in virtual brand communities. *Telemat. Inform.* **2020**, *46*, 101321. [CrossRef]
4. Andreassen, T.W.; Lindestad, B. Customer loyalty and complex services: The impact of corporate image on quality, customer satisfaction and loyalty for customers with varying degrees of service expertise. *Int. J. Serv. Ind. Manag.* **1998**, *9*, 7–23. [CrossRef]
5. Dowling, G.R.; Uncles, M. Do customer loyalty programs really work? *Sloan Manag. Rev.* **1997**, *38*, 71–82.
6. Utami, S. The Influence of Customers' Trust on Customer Loyalty. *Int. J. Econ. Commer. Manag.* **2015**, *3*, 638–653.
7. Alrubaiee, L.; Al-Nazer, N. Investigate the impact of relationship marketing orientation on customer loyalty: The customer's perspective. *Int. J. Mark. Stud.* **2010**, *2*, 155. [CrossRef]
8. Kumar, V.; Sharma, A.; Shah, R.; Rajan, B. Establishing profitable customer loyalty for multinational companies in the emerging economies: A conceptual framework. *J. Int. Mark.* **2013**, *21*, 57–80. [CrossRef]
9. Jiménez-Zarco, A.I.; Rospigliosi, A.; Martínez-Ruiz, M.P.; Izquierdo-Yusta, A. Marketing 4.0: Enhancing consumer-brand engagement through big data analysis. In *Web Services: Concepts, Methodologies, Tools, and Applications*; IGI Global: Hershey, PA, USA, 2019; pp. 2172–2195.
10. Tanveer, M. Analytical approach on small and medium Pakistani Business based on E-Commerce Ethics with effect on customer repurchase objectives and loyalty. *J. Leg. Ethical Regul. Issues* **2021**, *24*, 1–10.
11. Kotler, P.; Lee, N. Best of breed: When it comes to gaining a market edge while supporting a social cause, "corporate social marketing" leads the pack. *Soc. Mark. Q.* **2005**, *11*, 91–103. [CrossRef]
12. Brunk, K.H. Exploring origins of ethical company/brand perceptions—A consumer perspective of corporate ethics. *J. Bus. Res.* **2010**, *63*, 255–262. [CrossRef]
13. David, F.R.; David, F.R.; David, M.E. *Strategic Management: Concepts and Cases: A Competitive Advantage Approach*; Pearson: Upper Saddle River, NJ, USA, 2013.
14. Kaplan, R.S.; Norton, D.P. *The Execution Premium: Linking Strategy to Operations for Competitive Advantage*; Harvard Business Press: Boston, MA, USA, 2008.
15. Weeden, C. Ethical tourism: An opportunity for competitive advantage? *J. Vacat. Mark.* **2002**, *8*, 141–153. [CrossRef]
16. Berger, L.A.; Berger, D.R. *The Talent Management Handbook: Creating a Sustainable Competitive Advantage by Selecting, Developing, and Promoting the Best People*; McGraw-Hill: New York, NY, USA, 2011.
17. Papadimitri, P.; Pasiouras, F.; Tasiou, M. Do national differences in social capital and corporate ethical behaviour perceptions influence the use of collateral? Cross-country evidence. *J. Bus. Ethics* **2020**. [CrossRef]
18. Lee, J.Y.; Jin, C.H. The role of ethical marketing issues in consumer-brand relationship. *Sustainability* **2019**, *11*, 6536. [CrossRef]

19. Vallaster, C.; Kraus, S.; Lindahl, J.M.M.; Nielsen, A. Ethics and entrepreneurship: A bibliometric study and literature review. *J. Bus. Res.* **2019**, *99*, 226–237. [CrossRef]
20. Rekker, S.A.; Benson, K.L.; Faff, R.W. Corporate social responsibility and CEO compensation revisited: Do disaggregation, market stress, gender matter? *J. Econ. Bus.* **2014**, *72*, 84–103. [CrossRef]
21. Sahin, A.; Zehir, C.; Kitapçı, H. The effects of brand experiences, trust and satisfaction on building brand loyalty; an empirical research on global brands. *Proc. Soc. Behav. Sci.* **2011**, *24*, 1288–1301. [CrossRef]
22. Tanveer, M.; Hassan, S.; Bhaumik, A. Academic Policy Regarding Sustainability and Artificial Intelligence (AI). *Sustainability* **2020**, *12*, 9435. [CrossRef]
23. Kuo, Y.F.; Feng, L.H. Relationships among community interaction characteristics, perceived benefits, community commitment, and oppositional brand loyalty in online brand communities. *Int. J. Inf. Manag.* **2013**, *33*, 948–962. [CrossRef]
24. Nandan, S. An exploration of the brand identity–brand image linkage: A communications perspective. *J. Brand Manag.* **2005**, *12*, 264–278. [CrossRef]
25. Cuomo, M.T.; Foroudi, P.; Tortora, D.; Hussain, S.; Melewar, T.C. Celebrity Endorsement and the Attitude Towards Luxury Brands for Sustainable Consumption. *Sustainability* **2019**, *11*, 6791. [CrossRef]
26. Alwi, S.F.; Ali, S.M.; Nguyen, B. The importance of ethics in branding: Mediating effects of ethical branding on company reputation and brand loyalty. *Bus. Ethics Q.* **2017**, *27*, 393–422. [CrossRef]
27. Hur, W.M.; Ahn, K.H.; Kim, M. Building brand loyalty through managing brand community commitment. *Manag. Decis.* **2011**, *49*, 1194–1213. [CrossRef]
28. So, K.K.F.; King, C.; Sparks, B.A.; Wang, Y. The influence of customer brand identification on hotel brand evaluation and loyalty development. *Int. J. Hosp. Manag.* **2013**, *34*, 31–41. [CrossRef]
29. Haryanto, J.O.; Moutinho, L.; Coelho, A. Is brand loyalty really present in the children's market? A comparative study from Indonesia, Portugal, and Brazil. *J. Bus. Res.* **2016**, *69*, 4020–4032. [CrossRef]
30. Hem, L.E.; Iversen, N.M. Transfer of brand equity in brand extensions: The importance of brand loyalty. ACR North American Advances. In *NA—Advances in Consumer Research*; Keller, P.A., Rook, D.W., Valdosta, G.A., Eds.; Association for Consumer Research: Columbus, OH, USA, 2003; Volume 30, pp. 72–79.
31. Sebastianelli, R.; Tamimi, N.; Rajan, M. Perceived quality of online shopping: Does gender make a difference? *J. Int. Commer.* **2008**, *7*, 445–469. [CrossRef]
32. Thirumalai, S.; Sinha, K.K. Customer satisfaction with order fulfillment in retail supply chains: Implications of product type in electronic B2C transactions. *J. Oper. Manag.* **2005**, *23*, 291–303. [CrossRef]
33. Bostrom, N.; Roache, R. Ethical issues in human enhancement. In *New Waves in Applied Ethics*; Ryberg, J., Petersen, T., Wolf, C., Eds.; Palgrave Macmillan: Basingstoke, UK, 2008; pp. 120–152.
34. Tanveer, M.; Hassan, S. The role of new and creative ideas in developing industries of education, software and manufacturing in Pakistan. *J. Entrep. Educ.* **2020**, *23*, 1–10.
35. Floridi, L.; Sanders, J.W. Artificial evil and the foundation of computer ethics. *Ethics Inf. Technol.* **2001**, *3*, 55–66. [CrossRef]
36. Connolly, W.E. Beyond good and evil: The ethical sensibility of Michel Foucault. *Polit. Theory* **1993**, *21*, 365–389. [CrossRef]
37. Cooper, J.M. *Reason and Emotion: Essays on Ancient Moral Psychology and Ethical Theory*; Princeton University Press: Princeton, NJ, USA, 1999.
38. Resnik, D.B. *What Is Ethics in Research and Why Is It Important*; National Institute of Health: Stapleon, NY, USA, 2011. Available online: https://www.niehs.nih.gov/research/resources/bioethics/whatis/index.cfm#:~{}:text=There%20are%20several%20reasons%20why,the%20truth%20and%20minimize%20error (accessed on 12 February 2021).
39. Goulet, D. Development ethics: A new discipline. *Int. J. Soc. Econ.* **1997**, *24*, 1160–1171. [CrossRef]
40. Lund, D.B. An empirical examination of marketing professional's ethical behavior in differing situations. *J. Bus. Ethics* **2000**, *24*, 331–342. [CrossRef]
41. Kock, C. Choice is not true or false: The domain of rhetorical argumentation. *Argumentation* **2009**, *23*, 61–80. [CrossRef]
42. Bowie, N.E. Business ethics as a discipline: The search for legitimacy. In *The Ruffin Series in Business Ethics*; Oxford University Press: Oxford, UK, 1991.
43. Wright, C. Truth in Ethics. *Ratio* **1995**, *8*, 209–226. [CrossRef]
44. Zwingli, U. *Commentary on True and False Religion*; Wipf and Stock Publishers: Eugene, OR, USA, 2015.
45. Gaski, J.F. Does marketing ethics really have anything to say? A critical inventory of the literature. *J. Bus. Ethics* **1999**, *18*, 315–334.
46. Tanveer, M.; Khan, N.; Ahmad, A.R. AI Support Marketing: Understanding the Customer Journey towards the Business Development. In Proceedings of the 2021 1st International Conference on Artificial Intelligence and Data Analytics (CAIDA), Riyadh, Saudi Arabia, 6–7 April 2021; pp. 144–150. [CrossRef]
47. Pearson, R. Business ethics as communication ethics: Public relations practice and the idea of dialogue. *Public Relat. Theory* **1989**, *27*, 111–131.
48. Ferrell, O.C.; Harrison, D.E.; Ferrell, L.; Hair, J.F. Business ethics, corporate social responsibility, and brand attitudes: An exploratory study. *J. Bus. Res.* **2019**, *95*, 491–501. [CrossRef]
49. Aaleya, R.; Farooq, A.S.; Muhammad, T. Relational Dynamics between Customer Engagement, Brand Experience, and Customer Loyalty: An Empirical Investigation. *J. Internet Commer.* **2021**. [CrossRef]

50. Domino, M.A.; Wingreen, S.C.; Blanton, J.E. Social cognitive theory: The antecedents and effects of ethical climate fit on organizational attitudes of corporate accounting professionals—A reflection of client narcissism and fraud attitude risk. *J. Bus. Ethics* **2015**, *131*, 453–467. [CrossRef]
51. Ferrero, I.; Sison, A.J.G. A quantitative analysis of authors, schools and themes in virtue ethics articles in business ethics and management journals (1980–2011). *Bus. Ethics A Eur. Rev.* **2014**, *23*, 375–400. [CrossRef]
52. McGuire, J.B.; Sundgren, A.; Schneeweis, T. Corporate social responsibility and firm financial performance. *Acad. Manag. J.* **1988**, *31*, 854–872.
53. Dunfee, T.W.; Smith, N.C.; Ross, W.T., Jr. Social contracts and marketing ethics. *J. Mark.* **1999**, *63*, 14–32. [CrossRef]
54. Singhapakdi, A.; Vitell, S.J. Marketing ethics: Factors influencing perceptions of ethical problems and alternatives. *J. Macromark.* **1990**, *10*, 4–18. [CrossRef]
55. Saeed, M.; Ahmed, Z.U.; Mukhtar, S.M. International marketing ethics from an Islamic perspective: A value-maximization approach. *J. Bus. Ethics* **2001**, *32*, 127–142. [CrossRef]
56. Nantel, J.; Weeks, W.A. Marketing ethics: Is there more to it than the utilitarian approach? *Eur. J. Mark.* **1996**, *30*, 9–19. [CrossRef]
57. Reidenbach, R.E.; Robin, D.P.; Dawson, L. An application and extension of a multidimensional ethics scale to selected marketing practices and marketing groups. *J. Acad. Mark. Sci.* **1991**, *19*, 83–92. [CrossRef]
58. Perrea, T.; Grunert, K.G.; Krystallis, A. Consumer value perceptions of food products from emerging processing technologies: A cross-cultural exploration. *Food Qual. Prefer.* **2015**, *39*, 95–108. [CrossRef]
59. Fournier, S. Consumers and their brands: Developing relationship theory in consumer research. *J. Consum. Res.* **1998**, *24*, 343–373. [CrossRef]
60. Tanveer, M.; Karim, M.A. Higher Education Institutions and the Performance Management. Library Philosophy and Practice. 2018, p. 2183. Available online: https://digitalcommons.unl.edu/libphilprac/3010/ (accessed on 20 March 2021).
61. Dudiak, J. *The Intrigue of Ethics: A Reading of the Idea of Discourse in the Thought of Emmanuel Levinas*; Fordham University Press: New York, NY, USA, 2001.
62. Slade, S.; Prinsloo, P. Learning analytics: Ethical issues and dilemmas. *Am. Behav. Sci.* **2013**, *57*, 1510–1529. [CrossRef]
63. Steinbock, B.; London, A.J.; Arras, J. *Ethical Issues in Modern Medicine: Contemporary Readings in Bioethics*; McGraw-Hill: New York, NY, USA, 2013.
64. Zimmermann, M. Ethical guidelines for investigations of experimental pain in conscious animals. *Pain* **1983**, *16*, 109–110. [CrossRef]
65. Roberts, K.; Varki, S.; Brodie, R. Measuring the quality of relationships in consumer services: An empirical study. *Eur. J. Mark.* **2003**, *37*, 169–196. [CrossRef]
66. Rogge, R.D.; Fincham, F.D.; Crasta, D.; Maniaci, M.R. Positive and negative evaluation of relationships: Development and validation of the Positive–Negative Relationship Quality (PN-RQ) scale. *Psychol. Assess.* **2017**, *29*, 1028. [CrossRef] [PubMed]
67. Athanasopoulou, P. Relationship quality: A critical literature review and research agenda. *Eur. J. Mark.* **2009**, *43*, 583–610. [CrossRef]
68. Crosby, L.A.; Evans, K.R.; Cowles, D. Relationship quality in service selling: An interpersonal influence perspective. *J. Mark.* **1990**, *54*, 68–81. [CrossRef]
69. Krishnamurthi, L.; Raj, S.P. An empirical analysis of the relationship between brand loyalty and consumer price elasticity. *Mark. Sci.* **1991**, *10*, 172–183. [CrossRef]
70. Thaichon, P.; Quach, T.N. From marketing communications to brand management: Factors influencing relationship quality and customer retention. *J. Relatsh. Mark.* **2015**, *14*, 197–219. [CrossRef]
71. Hess, J.; Story, J. Trust-based commitment: Multidimensional consumer-brand relationships. *J. Consum. Mark.* **2005**, *22*, 313–322. [CrossRef]
72. Dwivedi, A.; Johnson, L.W.; McDonald, R. Celebrity endorsements, self-brand connection and relationship quality. *Int. J. Advert.* **2016**, *35*, 486–503. [CrossRef]
73. Van der Westhuizen, L.M. Brand loyalty: Exploring self-brand connection and brand experience. *J. Prod. Brand Manag.* **2018**, *27*, 172–184. [CrossRef]
74. Thakor, M.V.; Lavack, A.M. Effect of perceived brand origin associations on consumer perceptions of quality. *J. Prod. Brand Manag.* **2003**, *12*, 393–407. [CrossRef]
75. Khan, I.; Hollebeek, L.D.; Fatma, M.; Islam, J.U.; Riivits-Arkonsuo, I. Customer experience and commitment in retailing: Does customer age matter? *J. Retail. Consum. Serv.* **2020**, *57*, 102219. [CrossRef]
76. Islam, J.U.; Rahman, Z.; Hollebeek, L.D. The role of consumer engagement in recovering online service failures: An application of service-dominant logic. In *Handbook of Research on Customer Engagement*; Edward Elgar Publishing: Cheltenham, UK, 2019.
77. Srinivasan, S.S.; Anderson, R.; Ponnavolu, K. Customer loyalty in e-commerce: An exploration of its antecedents and consequences. *J. Retail.* **2002**, *78*, 41–50. [CrossRef]
78. Yi, Y.; Jeon, H. Effects of loyalty programs on value perception, program loyalty, and brand loyalty. *J. Acad. Mark. Sci.* **2003**, *31*, 229–240. [CrossRef]
79. Oliver, R.L. Whence consumer loyalty? *J. Mark.* **1999**, *63*, 33–44. [CrossRef]
80. Islam, J.U.; Zaheer, A. Using Facebook brand communities to engage customers: A new perspective of relationship marketing. *PEOPLE Int. J. Soc. Sci.* **2016**, *2*, 1540–1551. [CrossRef]

81. Islam, J.U.; Rahman, Z. Examining the effects of brand love and brand image on customer engagement: An empirical study of fashion apparel brands. *J. Glob. Fash. Mark.* **2016**, *7*, 45–59. [CrossRef]
82. Nill, A.; Schibrowsky, J.A. Research on marketing ethics: A systematic review of the literature. *J. Macromark.* **2007**, *27*, 256–273. [CrossRef]
83. Laczniak, G.R.; Murphy, P.E. The role of normative marketing ethics. *J. Bus. Res.* **2019**, *95*, 401–407. [CrossRef]
84. Hennig-Thurau, T.; Gwinner, K.P.; Walsh, G.; Gremler, D.D. Electronic word-of-mouth via consumer-opinion platforms: What motivates consumers to articulate themselves on the internet? *J. Interact. Mark.* **2004**, *18*, 38–52. [CrossRef]
85. Vitell, S.J.; Lumpkin, J.R.; Rawwas, M.Y. Consumer ethics: An investigation of the ethical beliefs of elderly consumers. *J. Bus. Ethics* **1991**, *10*, 365–375. [CrossRef]
86. Kumar, P. Ethical marketing practices viewed through consumer spectacles. *Mark. Trž.* **2016**, *28*, 29–45.
87. Johnston, J. The citizen-consumer hybrid: Ideological tensions and the case of Whole Foods Market. *Theory Soc.* **2008**, *37*, 229–270. [CrossRef]
88. Smith, N.C.; Cooper-Martin, E. Ethics and target marketing: The role of product harm and consumer vulnerability. *J. Mark.* **1997**, *61*, 1–20. [CrossRef]
89. Keller, K.L. Branding and brand equity. In *Handbook of Marketing*; Sage Publishers: New York, NY, USA, 2002. [CrossRef]
90. Colquitt, L.; Wesson, O.B. *Improving Performance and Commitment in the Workplace*; McGraw-Hill: Irwin, FL, USA, 2009.
91. Isaksson, L. Corporate Social Responsibility: A Study of Strategic Management and Performance in Swedish Firms. Ph.D. Thesis, Bond University, Robina, Australia, 2012.
92. Ching, A.T. Consumer learning and heterogeneity: Dynamics of demand for prescription drugs after patent expiration. *Int. J. Ind. Organ.* **2010**, *28*, 619–638. [CrossRef]
93. Nash, J. Exploring how social media platforms influence fashion consumer decisions in the UK retail sector. *J. Fash. Mark. Manag.* **2019**, *23*, 82–103. [CrossRef]
94. Bodet, G.; Geng, H.E.; Chanavat, N.; Wang, C. Sport brands' attraction factors and international fans. *Sport Bus. Manag.* **2020**, *10*, 47–67. [CrossRef]
95. Cambra-Fierro, J.; Hart, S.; Polo-Redondo, Y. Environmental respect: Ethics or simply business? A study in the small and medium enterprise (SME) context. *J. Bus. Ethics* **2008**, *82*, 645–656. [CrossRef]
96. Ross, W.T.; Robertson, D.C. A typology of situational factors: Impact on salesperson decision-making about ethical issues. *J. Bus. Ethics* **2003**, *46*, 213–234. [CrossRef]
97. Lewenstein, B.V. What counts as a social and ethical issue in nanotechnology? In *Nanotechnology Challenges: Implications for Philosophy, Ethics and Society*; World Scientific Publishing: Singapore, 2006.
98. Gliem, J.A.; Gliem, R.R. Calculating, interpreting, and reporting Cronbach's alpha reliability coefficient for Likert-type scales. In Proceedings of the Midwest Research-to-Practice Conference in Adult, Continuing, and Community Education, Columbus, OH, USA, 8–10 October 2003. Available online: https://scholarworks.iupui.edu/handle/1805/344 (accessed on 26 April 2021).
99. Koufteros, X.A. Testing a model of pull production: A paradigm for manufacturing research using structural equation modeling. *J. Oper. Manag.* **1999**, *17*, 467–488. [CrossRef]
100. Luchs, M.G.; Naylor, R.W.; Irwin, J.R.; Raghunathan, R. The sustainability liability: Potential negative effects of ethicality on product preference. *J. Mark.* **2010**, *74*, 18–31. [CrossRef]
101. Yin, J.; Qian, L.; Singhapakdi, A. Sharing sustainability: How values and ethics matter in consumers' adoption of public bicycle-sharing scheme. *J. Bus. Ethics* **2018**, *149*, 313–332. [CrossRef]
102. Carmeli, A.; Brammer, S.; Gomes, E.; Tarba, S.Y. An organizational ethic of care and employee involvement in sustainability-related behaviors: A social identity perspective. *J. Organ. Behav.* **2017**, *38*, 1380–1395. [CrossRef]

*sustainability*

MDPI

*Article*

# Mechanical Properties of Silica Fume Modified High-Volume Fly Ash Rubberized Self-Compacting Concrete

**Wesam Salah Alaloul [1], Muhammad Ali Musarat [1,*], Sani Haruna [1], Kevin Law [1], Bassam A. Tayeh [2], Waqas Rafiq [1] and Saba Ayub [3]**

1 Department of Civil and Environmental Engineering, Universiti Teknologi PETRONAS, Bandar Seri Iskandar 32610, Malaysia; wesam.alaloul@utp.edu.my (W.S.A.); sani_17000823@utp.edu.my (S.H.); kevinlaw.97.kl@gmail.com (K.L.); waqas_18000277@utp.edu.my (W.R.)

2 Civil Engineering Department, Faculty of Engineering, Islamic University of Gaza, Gaza P.O. Box 108, Palestine; btayeh@iugaza.edu.ps

3 Department of Fundamental and Applied Sciences, Universiti Teknologi PETRONAS, Bandar Seri Iskandar 32610, Malaysia; saba_20000009@utp.edu.my

* Correspondence: muhammad_19000316@utp.edu.my

**Abstract:** The existing form of self-compacting concrete (SCC) comprises of a large amount of powdered and fine materials. In this study, a part of the cementitious material was replaced with constant high-volume fly ash, and a portion of fine aggregates was substituted by crumb rubber (CR). Besides that, silica fume (SF) was added, with the hope that by implementing a new type of nanomaterial, the loss in mechanical strength due to previous modifications such as rubberization and replacement will be prevented. Two variables were found to influence the constituent/component in the mix design: SF and CR. The proportion of SF varies from 0% to 10%, while that of CR from 0% to 30% by volume of the total river sand, where 55% of cement was replaced by the fly ash. A total of 13 rubberized SCC samples with CR and SF as controlling variables were made, and their design mix was produced by a Design of Experiment (DOE) under the Response Surface Methodology (RSM). The results reveal a slight increase in the mechanical properties with the addition of SF. The theoretical mathematical models and equation for each different mechanical strength were also developed after incorporating the experimental results into the software.

**Keywords:** self-compacting concrete; crumb rubber; strength; silica fume; response surface methodology

**Citation:** Alaloul, W.S.; Musarat, M.A.; Haruna, S.; Law, K.; Tayeh, B.A.; Rafiq, W.; Ayub, S. Mechanical Properties of Silica Fume Modified High-Volume Fly Ash Rubberized Self-Compacting Concrete. *Sustainability* **2021**, *13*, 5571. https://doi.org/10.3390/su13105571

Academic Editor: Shervin Hashemi

Received: 8 April 2021
Accepted: 13 May 2021
Published: 17 May 2021



## 1. Introduction

The origin of self-compacting concrete (SCC) can be traced back to the year 1986, when Hajime Okamura proposed his idea of having a concrete which is 'alive'. Then, his idea stimulated the development of the prototype of SCC, 2 years later, by Ozawa, to improve the durability of concrete structures [1]. SCC is special in the sense that by just relying on its weight, all process of vibration and compaction during pouring or concreting can be eliminated, by adding superplasticizers into the fresh concrete mixtures [2]. The adding of the superplasticizer is to ensure that the fresh concrete achieves high workability and a decent cohesiveness, so that the fresh concrete can flow and fill even to the narrowest gap between reinforcement bars. A similar level of quality with normal concrete will be produced even without the same degree of compaction. Besides reducing compaction noise, cost, and time in the construction site, the utilization of SCC induces advantages such as constructability, reliability, and low permeability [3–5]. By adding supplementary cementitious materials, the workability of SCC can be stabilized. Moreover, it can aid in reducing the quantity of the cement to be used, and thus help in reducing the environmental carbon footprint related to Portland cement [6–8].

In the past few decades, the waste rubber originated from scrap tyres and rubber products had accumulated to an explicitly large amount, until this issue started threatening the environment [9–11]. Approximately 1000 million (M) tyres reach their useful life each year. A vast number of tyres are previously stockpiled or landfilled, either whole or shredded, where the numbers are 3000 M in the EU region and 1000 M in the US alone. Tyres from the vehicle industry are expected to reach 5000 M until 2030, at which point they will be discarded on regular basis [12]. Burning the tyres might seem to be the most convenient and cheapest way. However, the air pollution caused by the emission of hazardous smoke and carbon dioxide ($CO_2$) during the open burning process slammed the door shut to this illegal method. Hence, one of the ideologies of recycling scientists and experts is to utilize the waste material as additive in cement-based materials. Crumb rubber (CR) as an additive in concrete is very common nowadays in the field of building materials, as it decreases the stiffness of the concrete [13], improves energy absorption and deformation capacity, and decreases chloride permeability [14,15]. The overall results of these findings show a considerable decline in the stiffness and strength of the concrete after adding CR, mainly due to physical properties and the interaction with aggregates [16–20].

The usage of fly ash to partially replace cement in the total mix is capable of improving the segregation and the strength of concrete [21,22]. Substituting cement with fly ash delivers impactful economic and environmental benefits, as less manufacturing of cement reduces the cost and the carbon footprint to the environment [23–25]. Fly ash also possesses low embodied energy compared to Portland cement. The properties of the cement pastes have a strong influence on the behaviour of the cement-based composites. Thus, controlling the cement hydration products can lead to a greater enhancement, especially in the strength properties of cement pastes.

Studies have shown that the optimal silica fume (SF) addition improves concrete performance [26,27]. SF, a by-product of the smelting process in the silicon and ferrosilicon industry, emerged as an attractive alternative that provides a high strength to concrete. The inclusion of SF in concrete gives an early high compressive, tensile, and flexural strength and impacts significantly on the modulus of elasticity, along with increasing the concrete's toughness. Besides that, SF makes the concrete resistive towards chemical attacks. That is why high-performance concrete containing SF is preferable for parking decks, highway bridges, and marine structures [28].

The incorporation of both SF and rubber particles potentially induces a coupling effect on the compressive properties of rubberized concrete. Typical means of enhancing the mechanical properties of concrete or rubberized concrete include mineral additives such as SF, fly ash, and ground granulated blast furnace slag. SF is regarded as a material belonging to the highly pozzolanic category, as the silica content inside exists in a non-crystalline form with a very highly specific surface. The concrete structure bears the interaction of several components, such as aggregates and cement paste. Several researchers [29–31] have stated that the addition of SF to rubberized concrete reduces the adverse impact of the rubber particles and enhances the mechanical properties of the rubberized concrete.

The ideology of incorporating crumb rubber into the cementitious composite comes from the overloading of unhandled rubber waste, which poses a threat to the environment. By introducing the reused crumb rubber, the rubberized concrete mixture is expected to have increased ductility and toughness, higher impact resistance, and improved cracking resistance, as rubber contributes to the enhanced elasticity. Thus, the effect of incorporating silica fume as an additive in the high-volume fly ash rubberized self-compacting concrete was examined to mitigate the loss of mechanical strength due to crumb rubber. This investigation aims to determine the influence of silica fume and crumb rubber on the compressive, tensile, and flexural strength of the rubberized self-compacting concrete with a high volume of fly ash. The addition of silica fume in self-compacting concrete will aid in regaining/mitigating the reduced strength of the rubberized SCC. Due to its fineness and highly reactive properties, it will bridge the inter transition zone between the crumb rubber aggregates and the cement paste, thus improving the strength properties of the rubberized SCC.

Response surface methodology (RSM) is a way of applying statistical techniques on the interaction between variables and their responses. RSM also entails the design of experiments such that optimized results of the interaction between the variables and between the variables and their responses are developed [32]. Response surface numerical models are also developed to validate the accuracy of the relationship between variables in achieving the desired response. Response surface methodology (RSM) was employed to study the relationship between silica fume and crumb rubber. An analysis of variance (ANOVA) was employed to evaluate the significance of the models developed by the RSM. The rubberized self-compacting concrete was designed in different mix proportions which consist of silica fume, crumb rubber, superplasticizer, cement, coarse aggregates, and river sand. Two variables were found to influence the constituent/component in the mix design: silica fume and crumb rubber. SF varies from 0% to 10% by volume, while the proportion of CR varies from 0% to 30% by volume of the total fine aggregates, which is the river sand. The idea of incorporating silica fume is to bridge the ITZ gap between the crumb rubber and the cement paste, which may result in the enhancement of the strength properties.

## 2. Materials and Methods

### 2.1. Materials

The materials utilized in the production of rubberized self-compacting concrete consist of Ordinary Portland Cement (OPC), silica fume (SF), fly ash, fine aggregate (F.A), crumb rubber (CR), coarse aggregate (C.A), superplasticizer (SP), and water. The OPC is categorized in Type 1 and met ASTM C1050 specifications. Class C fly ash conforming to the specifications of ASTM C618 was used. River sand with a specific gravity of 2.65 was used for the fine aggregates, and passed through sieve no. 4, whereas the majority of aggregates were retained by sieve no. 200. The crumb rubber (CR) used was obtained from waste tyres, and their size ranged from 600 µm to 2.36 mm. The oxide composition of the OPC and fly ash are presented in Table 1. Natural gravel was used as coarse aggregate with a 20 mm maximum size and a 2.68 specific gravity. A superplasticizer was added to the mixture in accordance with ASTM C230/C230M-03 to achieve the desired flowability.

**Table 1.** Oxide composition of OPC and Fly Ash.

| Oxide (%) | OPC | Fly Ash |
|---|---|---|
| $SiO_2$ | 20.76 | 37.8 |
| $Al_2O_3$ | 5.54 | 15.4 |
| $Fe_2O_3$ | 3.35 | 16.5 |
| MnO | - | 0.33 |
| CaO | 61.4 | 17.9 |
| MgO | 2.48 | 2.75 |
| $Na_2O$ | 0.19 | 0.26 |
| $K_2O$ | 0.78 | 2.85 |
| $TiO_2$ | - | 1.07 |
| $SO_3$ | - | 0.7 |
| Loss on ignition | 2.2 | 1.25 |

### 2.2. Mixing, Casting, and Test Procedure

The percentage of CR and SF in the modified high volume fly ash rubberized self-compacting concrete (HVFRSCC) was designed with the aid of a design of experiment (DOE) under the Design Expert software. The percentage of CR ranged from 0% to 30%, while that of silica fume (SF) ranged from 0% to 10% based on the RSM analysis. A total of 13 runs of combinations with different percentages of CR and SF were developed, as shown in Table 2. CR15SF5, the five duplications, are the central points which the software used to improve the precision of the experiment against any likely errors. Mixing took place in three stages; dry mixing for 2 min, wet mixing for 2 min, adding SP and final mixing for no less than 3 min. Afterwards, the fresh properties of HVFRSCC were determined.

The fresh mixture of HVFRSCC was then cast into standard moulds of 100 mm cubes, a 200 mm × 100 mm cylinder, and a 500 mm × 100 mm × 100 mm prism. The compressive strength test was performed on hardened cubes of 100 mm. According to BS 12390-3:2009, the loading rate was set at 3.0 kN/s, where the hardened cubes were checked at 7, 14, and 28 days for each combination, in which three specimens were checked for each curing time and procedure, and the average values were reported. The splitting tensile strength of HVFRSCC was measured according to BS EN 12390-6:2009, for every mixture, and assessed at 7, 14, and 28 days of curing. Prisms of 100 mm × 100 mm × 500 mm were prepared for three points loading bending test as per ASTM C293M-10 [33].

**Table 2.** Mix Proportions of high-volume fly ash self-compacting concrete (HVFSCC).

| Mix | CR (%) | SF (%) | CR, (kg) | F.A, (kg) | C.A (kg) | Fly Ash, (kg) | Water, (kg) | OPC, (kg) | SP (kg) |
|---|---|---|---|---|---|---|---|---|---|
| CR15SF0 | 15 | 0 | 2.42 | 13 | 15.582 | 5.929 | 3.234 | 4.851 | 0 |
| CR15SF5 | 15 | 5 | 2.42 | 13.70 | 15.582 | 5.929 | 3.234 | 4.609 | 0.24 |
| CR15SF5 | 15 | 5 | 2.42 | 13.70 | 15.582 | 5.929 | 3.234 | 4.609 | 0.24 |
| CR0SF5 | 0 | 5 | 0 | 16.11 | 15.582 | 5.929 | 3.234 | 583 | 0.24 |
| CR15SF10 | 15 | 10 | 2.42 | 13.70 | 15.582 | 5.929 | 3.234 | 4.366 | 0.49 |
| CR15SF5 | 15 | 5 | 2.42 | 13.70 | 15.582 | 5.929 | 3.234 | 4.609 | 0.24 |
| CR15SF5 | 15 | 5 | 2.42 | 13.70 | 15.582 | 5.929 | 3.234 | 4.609 | 0.24 |
| CR30SF10 | 30 | 10 | 4.83 | 11.28 | 15.582 | 5.929 | 3.234 | 4.366 | 0.49 |
| CR15SF5 | 15 | 5 | 2.42 | 13.70 | 15.582 | 5.929 | 3.234 | 4.609 | 0.24 |
| CR30SF5 | 30 | 5 | 4.83 | 11.28 | 15.582 | 5.929 | 3.234 | 4.609 | 0.24 |
| CR0SF0 | 0 | 0 | 0 | 16.11 | 15.582 | 5.929 | 3.234 | 4.851 | 0 |
| CR30SF0 | 30 | 0 | 4.83 | 11.28 | 15.582 | 5.929 | 3.234 | 4.851 | 0 |
| CR0SF10 | 0 | 10 | 0 | 16.11 | 15.582 | 5.929 | 3.234 | 4.366 | 0.49 |

CR: Crumb rubber, SF: silica fume, C.A: coarse aggregate, F.A: fine aggregate, SP: Superplasticizer.

## 3. Results and Discussion

### 3.1. Fresh Properties

The standard slump test (Flowability) was conducted according to the Specification and Guidelines for SCC, EFNARC 2002. This approach is the most widely used for calculating concrete's self-flow properties. This test chooses the consistency of SCC filling. The concrete is filled in without any concrete spitting, and then the slump funnel is pulled up where the concrete is expected to flow without any of its weight impact. The mixtures with a flow diameter between 600 mm and 800 mm were considered in the slump flow test based on the European SCC specifications (EFNARC 2002) [34]. During the mixing phase, slump flow test and $T_{50}$ time were obtained using the required quantity of fresh HVFRSCC before casting into the moulds as shown in Figure 1. The slump values of HVFRSCC were presented in Figure 2. The slump flow of the HVFRSCC decreases with the percentage increase of CR and SF. It is worth mentioning that all the slump values satisfy the EFNARC 2002 criteria. Besides that, The L-box test was also conducted to identify the passing ability of SCC. The apparatus needed for this test includes a rectangular-section box in the shape of an alphabet 'L', with a horizontal and a vertical component, trowel, scoop, and stopwatch which are separated by a movable door. Firstly, the vertical compartment was filled with concrete. Right after filling up, it was set for 1 min, and the door was lifted to let the concrete from the vertical compartment flow into the horizontal compartment. After the flowing is self-halted naturally, the height of the concrete at the end of the horizontal compartment was measured as H2, a proportion of that remaining in the vertical compartment and in the vertical section, H1. The blocking ratio, H2/H1, for this passing test should be 0.80 to 1.00. If the concrete being tested is truly self-levelling, like water, the value of the blocking ratio will be the same as that of water, which is 1. It will somehow form a slope of concrete when at rest. The result indicates the SCC's passing ability where the ratio of 0.76 was achieved, where the passage of concrete is restricted by the bars.

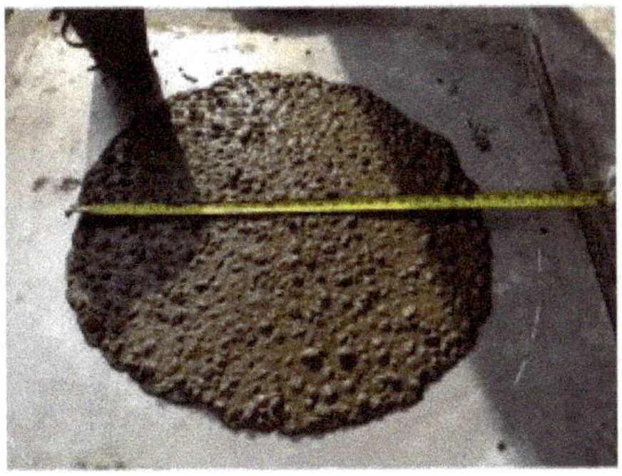

Figure 1. Slump flow measurement.

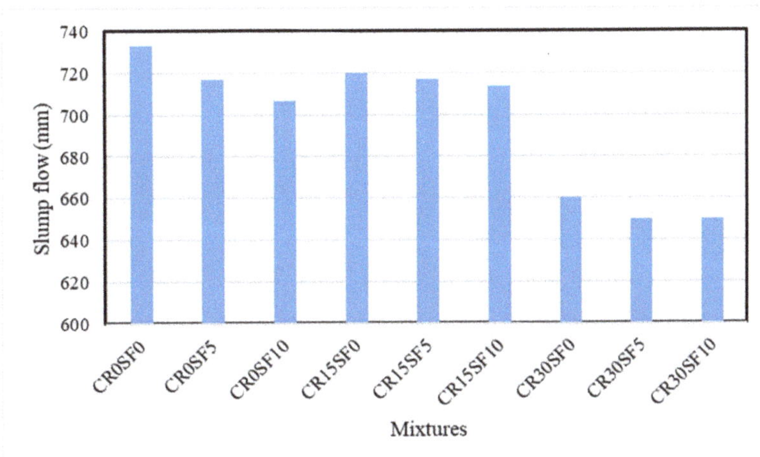


Figure 2. Slump flow of HVFRSCC.

### 3.2. Compressive Strength

The strength development of HVFRSCC from 7 to 28 days curing is shown in Figure 3. The highest compressive strength of the 28-day period, of 66.79 MPa, was achieved by CR0SF5, having 0% CR and 5% SF content. The lowest compressive strength of the 28-day period was achieved by CR30SF5, with the strength of 5.29 MPa, having 30% CR and 5% SF content. The control mix of HVFA-SCC (CR0SF0) possessed a compressive strength of 64.73 MPa during the 28-day period, which is in line with the work of Topçu [35]. It is also notable that by replacing a portion of fine aggregates with CR will lead to a fall-off in the compressive strength. From Figure 3, the mix with 30% CR achieved a 5.29 MPa strength only, which shows that the higher amount of CR as a replacer of fine aggregate induces a higher reduction of the compressive strength. This result corresponded well with the experiment by Guneyisi et al. [31], where the results showed a decrease in compressive strength with the increase in rubber content, with and without silica fume. Incorporating CR into SCC increases its porosity. Much void volume is induced during the mixing between CR and cement paste, as CR keeps on repelling water in the process. Besides, the reason causing a reduction in compressive strength regardless of the curing age is the increment of air entrapment volumes, which is caused by the inefficient binding of aggregates and CR. Apart from the air entrapped, the reduction in compressive strength is caused by the interfacial transition zone debility, the poor adhesion between crumb rubber and paste matrix, and the low stiffness. Because of the presence of the fine aggregate, and high-surface components such as the SF, HVFRSCC formed pores that allow the water to infiltrate the sample at a higher rate, which reduces the bonding automatically and thus decreases the compressive strength of the specimens [36]. However, the addition of SF can fill in the void present, which then produces a better bonding as well as a decent strength, or at least recovers some of the mechanical strength lost. For instance, CR30SF5 achieved a slight increase in compressive strength from 14 days to 28 days compared to CR30SF10 after doubling the amount of SF. It is deduced that SF has little influence on the compressive strength of HVFRSCC. The beneficial effect of SF is more evident at a lower CR percentage.

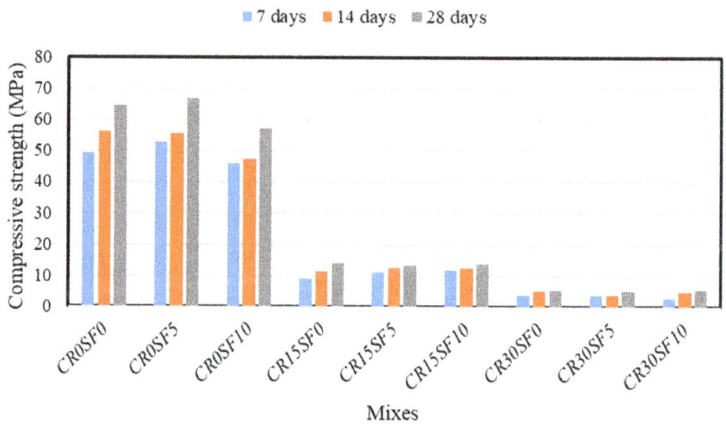


**Figure 3.** Compressive strength development of HVFRSCC.

As demonstrated in the 2D contour plot in Figure 4, the contour indicates where the red region symbolizes a higher strength region, whereas the blue region indicates the lower strength region of HVFRSCC strength. The higher value for compressive strength is within the red region, followed by the green region, while the lower value for compressive strength is within the blue region. The most optimum compressive strength achieved was 66.79 MPa at 0% and 5% for CR and SF, respectively. The inclined shape taken by the contours indicated that there is a weak interaction between the variables (percentages of CR and SF) [37,38]. Based on the 3D response surface plot as shown in Figure 4, the compressive strength significantly decreases with an increase in the content of the CR. Using ANOVA, the relationship between the crumb rubber and SF on the compressive strength of HVFRSCC was represented by the quadratic model shown in Equation (1).

$$Compressive\ strength - 28\ days\ (\text{MPa}) = 65.704 - 4.8205 \times CR - 0.1730 \times SF + 0.0242 \times CR \times SF + 0.0928 \times CR^2 - 0.0450 \times SF^2 \quad (1)$$

### 3.3. *Flexural Strength*

The flexural strength of the HVFRSCC is within the range of 0.42 MPa to 1.9 MPa, based on the percentage of CR and SF. From Figure 5, the CR0SF10 mix has the highest flexural strength, of 1.9 MPa. Similarly to the compressive strength, the flexural strength decreases significantly with the increase in the percentage of CR. The flexural strength decreases with the incorporation of silica fume and then increases. The incorporation of crumb rubber in the HVFRSCC reduces its flexural strength to almost half of its control mixture [39]. It is worth mentioning that the addition of SF in the HVFRSCC enhances the flexural strength by 8%. This is because of the nature of SF, which slightly improves the concrete strength loss due to crumb rubber addition. The presence of SF particles fills the pores that exist in the C-S-H gel structures whilst behaving as a nucleus to firmly bond with C-S-H gel particles. SF has an important pozzolanic reaction of its own. SF particles that act as both filler, which revamps the microstructure of mortar cement, and a medium that promotes the pozzolanic reaction could help enhance the strength development.

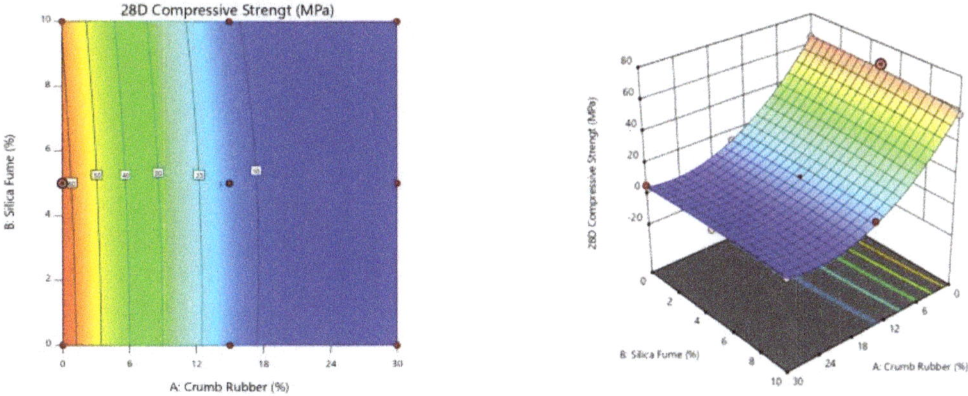


**Figure 4.** 2D Contour and 3D surface diagram for compressive strength.

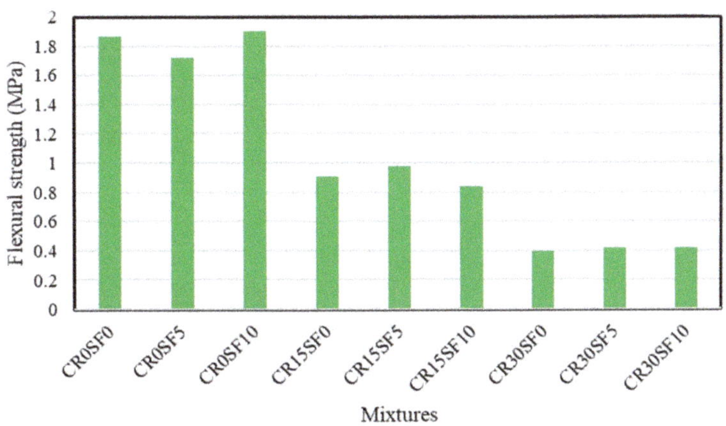


**Figure 5.** Flexural strength of HVFRSCC at 28 days.

The 2D contour and 3D response surface diagrams generated by RSM for flexural strength are demonstrated in Figure 6. The highest flexural tensile strength, of 1.9 MPa, was achieved at 0% CR and 10% SF content. However, from this model, the slope of the CR variable is quite steep, indicating that any minor changes in the amount of CR will result in a significant change in flexural tensile strength. By fixing the percentage of SF, an increasing percentage of CR in the design mix will decrease the flexural tensile strength. This justification can be supported by the straight-line contour plot, where each proportion of the red, green, and blue region is fairly distributed across the map. The model developed from the variable is showed to be satisfactory and significant. Equation (2) indicates HVFRSCC's flexural strength at 28 days.

$$\textit{Flexural Strength}\ (\text{MPa}) = 1.8308 + 0.0732 \times CR - 0.0005 \times SF + 0.00002 \times CR \times SF - 0.0009 \times CR^2 + 0.000006 \times SF^2 \quad (2)$$

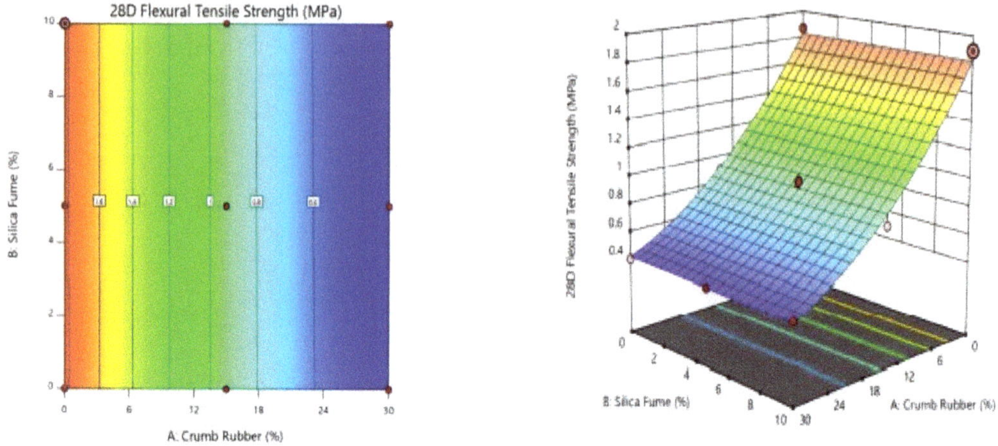


**Figure 6.** 2D Contour and 3D surface diagram for flexural strength.

### 3.4. *Splitting Tensile Strength*

The splitting tensile strength of high-volume fly ash rubberized self-compacting concrete (HVFRSCC) at 28 days of curing is shown in Figure 7. The splitting tensile strength of the HVFRSCC is between 0.77 MPa and 5.3 MPa, depending on the percentage of CR and SF. The highest split tensile strength of rubberized HVFA SCC was 5.3 MPa. It was achieved by taking 0% CR and 5% SF content. When the percentage of CR increases up to a certain range, the splitting tensile strength decreases significantly. The control mix sample (CR0SF0), which has a splitting tensile strength of 4.19 MPa, is quite similar to that of Guneyisi et al. [31]. The addition of 5% SF in the control mix enhances the splitting tensile strength by 27%, and the splitting tensile strength subsequently decreases by 14% when the amount of SF reaches 10%. The presence of crumb rubber significantly reduces the splitting tensile of the HVFRSCC to almost half of its control mixture. This significant reduction in the strength of the concrete after the addition of crumb rubber could be attributed to the poor adhesion of scrap tyre particles to the cement paste and a weak interaction with aggregates. Entrapped air on the crumb rubber surface makes the interfacial transition zone (ITZ) thicker than before, and this causes a weak bonding with the cement matrix, as the ITZ is a weak porous zone between cement paste and aggregate [18]. It was observed from Figure 7 that SF can mitigate the mechanical strength loss heavily induced by CR. However, after the SF exceeded the percentage of 5%, with a constant CR percentage, the splitting tensile strength became lower. Irrespective of the comparatively low splitting tensile strength of the HVFRSCC mixtures, rubberized specimens do remain intact after failure, unlike those which do not possess any CR content. This is due to the elongation of the ductile crumb rubber particles and their capability to bridge cracks in specimens, hence inhibiting a complete breakdown.

Based on the 2D contour plot and 3D response surface plot as demonstrated in Figure 8, the highest splitting tensile strength, of 5.3 MPa, can be achieved at the percentage of 0% and 5% for CR and SF, respectively. The trend of changes induced by both CR and SF is like the compressive strength test. It is deduced from the observation that the model built from the parameters is satisfactory and appropriate. The semi-elliptical lines in the contour diagram shown in Figure 8 indicate that there is a fair relationship between the crumb rubber and silica fume on the splitting tensile strength model. As demonstrated in the 3D surface diagram, the splitting tensile strength decreased with the increase of crumb rubber. Equation (3) represents the final equation in terms of actual factors for splitting tensile strength.

$$Splitting\ Tensile\ Strength\ (\text{MPa}) = 4.5767 - 0.2681 \times CR + 0.0701 \times SF + 0.0021 \times CR \times SF + 0.0046 \times CR^2 - 0.0135 \times SF^2 \quad (3)$$

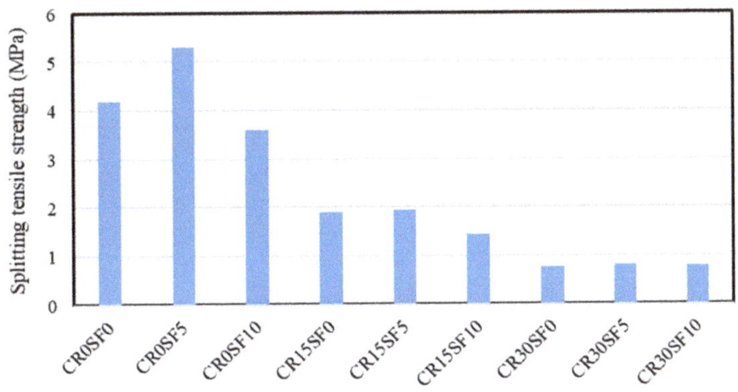


**Figure 7.** Twenty-eight-day Splitting tensile strength of HVFRSCC.

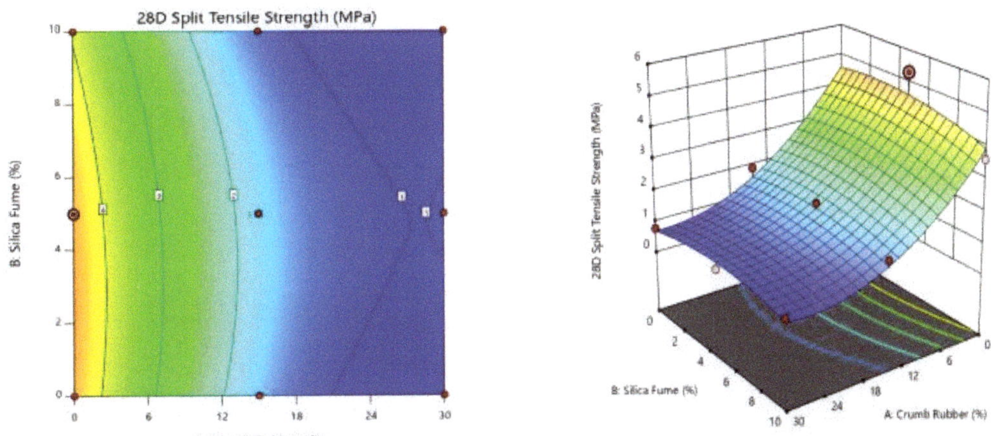


**Figure 8.** 2D Contour and 3D surface diagram of the splitting tensile strength of HVFRSCC.

## 4. Anova Analysis

Compressive strength, flexural strength, and splitting tensile strength were examined with the effect of the combination of two independent variables, namely, percentage of CR (A) and percentage of SF (B). The significance of the model was determined by an analysis of variance (ANOVA). As shown in Table 3, the F-value for the model of compressive strength is 385.92, for flexural strength is 139.54, and for splitting tensile strength is 5.45, where a high F-value indicates the significance and adequacy of the model. These values show that all the models are significant. The probability that noise would affect "Model F-value" is 0.01%. The $p$-value is less than 0.05, which indicates that the model is significant. The terms in the compressive strength, flexural, and splitting tensile strength models including A-CR and $A^2$ are significant to the models, while model terms AB, B, and $B^2$ were insignificant for all the output responses. The lack of fit F-values of 0.0007, 0.0267, and 0.0011 shows that all the strength models are significant with respect to lack of fit. Table 4 shows the validation of the models of compressive strength, flexural strength, and direct tensile strength. The coefficient of determination, which is also known as R-Squared ($R^2$) for the model should exceed 0.90 to show a high reliability for the equation. The variance among the adjusted $R^2$ and predicted $R^2$ is less than 0.2, which means the predicted $R^2$ is reasonably in agreement with the adjusted $R^2$. Besides, the minimum requirement for the Adequate (Adeq) precision, which is the measurement of the signal to noise ratio, is 4 [38]. Thus, Adeq Precision for compressive strength, flexural strength, and splitting tensile strength models were 49.643, 31.308, and 19.291, respectively. This shows that all the models can be used to navigate the design space.

**Table 3.** ANOVA analysis of the response models.

| Models | Source | Sum of Squares | df | Mean Square | F-Value | *p*-Value | Significance |
|---|---|---|---|---|---|---|---|
| Compressive Strength | Model | 6332.81 | 5 | 1266.56 | 385.92 | <0.0001 | Significant |
| | A-CR | 4954.78 | 1 | 4954.78 | 1509.73 | <0.0001 | Significant |
| | B-SF | 10.19 | 1 | 10.19 | 3.11 | 0.1214 | insignificant |
| | AB | 13.14 | 1 | 13.14 | 4.00 | 0.0855 | insignificant |
| | $A^2$ | 1204.05 | 1 | 1204.05 | 366.87 | <0.0001 | Significant |
| | $B^2$ | 3.50 | 1 | 3.50 | 1.07 | 0.3362 | insignificant |
| | Lack of Fit | 22.54 | 3 | 7.51 | 68.60 | 0.0007 | Significant |
| Flexural Strength | Model | 3.13 | 5 | 0.63 | 139.54 | <0.0001 | Significant |
| | A-CR | 3.00 | 1 | 3.00 | 670.05 | <0.0001 | Significant |
| | B-SF | $1.215 \times 10^{-4}$ | 1 | $1.215 \times 10^{-4}$ | 0.027 | 0.8739 | insignificant |
| | AB | $2.500 \times 10^{-5}$ | 1 | $2.500 \times 10^{-5}$ | $5.578 \times 10^{-3}$ | 0.9426 | insignificant |
| | $A^2$ | 0.11 | 1 | 0.11 | 23.75 | 0.0018 | Significant |
| | $B^2$ | $7.725 \times 10^{-6}$ | 1 | $7.725 \times 10^{-6}$ | $1.723 \times 10^{-3}$ | 0.9680 | insignificant |
| | Lack of Fit | 0.028 | 3 | $9.184 \times 10^{-3}$ | 9.61 | 0.0267 | significant |
| Splitting Tensile Strength | Model | 21.78 | 4 | 5.45 | 44.70 | <0.0001 | Significant |
| | A-CR | 19.13 | 1 | 19.13 | 157.03 | <0.0001 | Significant |
| | B-SF | 0.17 | 1 | 0.17 | 1.43 | 0.2664 | insignificant |
| | $A^2$ | 2.46 | 1 | 2.46 | 20.18 | 0.0020 | significant |
| | $B^2$ | 0.52 | 1 | 0.52 | 4.28 | 0.0724 | insignificant |
| | Lack of Fit | 0.96 | 4 | 0.24 | 51.25 | 0.0011 | significant |

**Table 4.** Models' validation.

| Model Terms | Compressive Strength | Flexural Strength | Splitting Tensile Strength |
|---|---|---|---|
| Std. Dev. | 1.81 | 0.067 | 0.35 |
| Mean | 23.21 | 1.02 | 2.16 |
| C.V. % | 7.80 | 6.59 | 16.17 |
| $R^2$ | 0.9964 | 0.9901 | 0.9572 |
| Adj. $R^2$ | 0.9938 | 0.9830 | 0.9358 |
| Pred. $R^2$ | 0.9673 | 0.9251 | 0.8250 |
| Adeq. precision | 49.643 | 31.308 | 19.291 |

The graph of Normal Plot of Residuals is one of the methods used to validate the models, while the graph of Normal Plot of Residuals is a straight line. Hence, the forecasted values will provide more precise results than expected when all the points roughly fall on the normality line [40,41]. Normal Plot of Residuals is used to identify the appropriateness of the model, and the ability of the model to determine the optimal extraction parameters based on the response results. The graph of Normal Plot of Residuals for all the responses is shown in Figure 9.

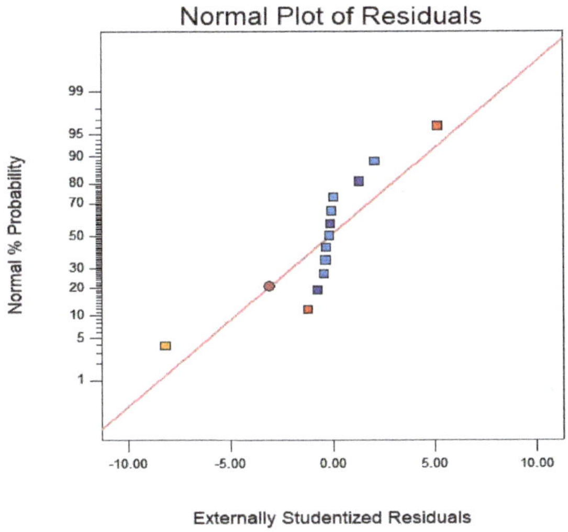


**Figure 9.** Normal plot of residuals.

*Optimization and Experimental Validation*

The response surface method (RSM) was used to model the effect of two variables in this study, which are the silica fume and crumb rubber, and their interactions with the mechanical strength properties. As mentioned above, once all the findings have been effectively obtained from the initial mixing, based on the 13 various mix designs conjured by the initial RSM mix design, they were inputted into the RSM once more to obtain a finalized model which is to be tested for its validity. The verification mix designs were acquired via a multi response optimization technique. In this response, the target was to maximize the response, which is the mechanical strength, and four solutions came up. The solution of the highest desirability is selected. The equation obtained from the RSM was validated experimentally. A total of four solutions developed from DOE to validate the equation are shown in Table 5, below, which shows the summary of the optimization mix. Figure 10 shows the optimization ramp with the desirability function. The desirability equals 0.609, which means that with 1.85% of CR and 7.4% of SF, the percentage for achieving 53.688 MPa for compressive strength, 1.69 MPa for flexural strength, and 3.903 MPa for split tensile strength is 60.9%.

**Table 5.** Summary of Optimization Mix.

| No. of Run | CR (%) | SF (%) | Compressive Strength (MPa) | Flexural Strength (MPa) | Splitting Tensile Strength (MPa) | Desirability |
|---|---|---|---|---|---|---|
| 1 | 1.850 | 7.401 | 53.688 | 1.695 | 3.903 | 0.609 |
| 2 | 1.847 | 7.419 | 53.686 | 1.695 | 3.901 | 0.609 |
| 3 | 1.854 | 7.362 | 53.701 | 1.694 | 3.907 | 0.609 |
| 4 | 29.895 | 5.223 | 6.170 | 0.415 | 1.015 | 0.046 |

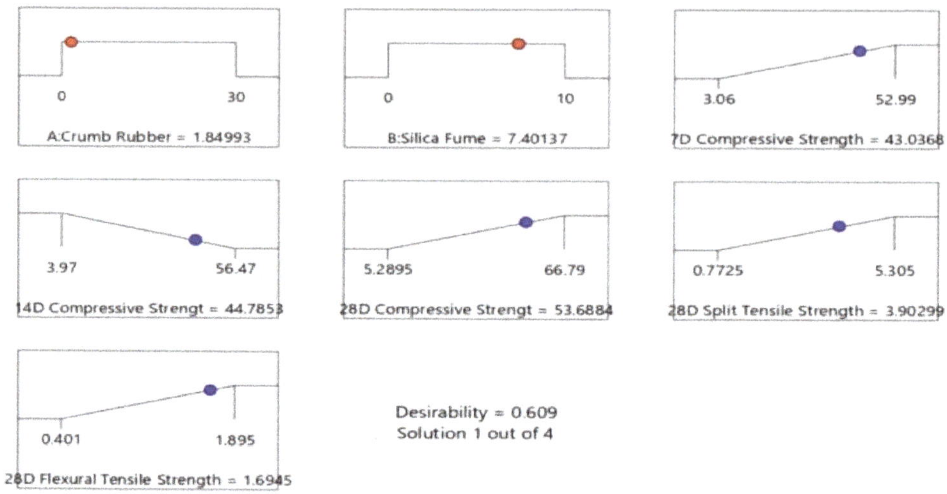


**Figure 10.** Optimization ramp showing the predicted results.

## 5. Conclusions

It was observed that crumb rubber in high volumes might not enable SCC composites to meet their strength requirement for any structural use. The incorporation of crumb rubber into concrete mixes is undeniably an eco-friendly move, but any reduction in mechanical strength can be disastrous for structural usage, and hence should not be used in structural elements that require a high strength. However, rubberized self-compacting concrete can be used in other areas, such as slab work, flooring, parking and driveways, compound construction, etc. The findings show that the addition of SF slightly enhances the mechanical strength of high-volume fly ash rubberized self-compacting concrete. SF improves the SCC strength loss due to CR by acting as both a filler that revamps the microstructure of mortar cement and a medium that promotes pozzolanic reaction. Mixtures with a higher quantity of CR experienced a drop in all the mechanical strength properties, especially design mixes which utilize 30% CR replacement. The lowest 28-day compressive strength of 5.29 MPa was achieved when 30% of CR and 5% of SF were utilized. Moreover, model equations were successfully established to predict the compressive, flexural strength, and splitting tensile strength of HVFRSCC, and the optimization was successfully carried out.

**Author Contributions:** Conceptualization, B.A.T., W.R. and S.A.; methodology, M.A.M., S.H. and K.L.; software, S.H. and K.L.; validation, S.A.; formal analysis, B.A.T. and W.R.; investigation, M.A.M., S.H. and K.L.; resources, W.S.A.; data curation, K.L., B.A.T. and S.A.; writing—original draft preparation, M.A.M., S.H., K.L., W.R. and S.A.; writing—review and editing, W.S.A., M.A.M. and B.A.T.; supervision, W.S.A.; project administration, W.S.A. All authors have read and agreed to the published version of the manuscript.

**Funding:** Not applicable.

**Institutional Review Board Statement:** Not applicable.

**Informed Consent Statement:** Not applicable.

**Data Availability Statement:** All the data are available within this manuscript.

**Acknowledgments:** The authors would like to appreciate the YUTP-PRF project (cost center # 015LC0-088) awarded to Wesam Alaloul for the support.

**Conflicts of Interest:** The authors declare no conflict of interest.

## References


1. Kamal, M.M.; Safan Mohamed, A.; Etman Zeinab, A.; Kasem Bsma, M. Mechanical properties of self-compacted fiber concrete mixes. *HBRC J.* **2014**, *10*, 25–34. [CrossRef]
2. Ramkumar, K.B.; Rajkumar, P.K.; Ahmmad, S.N.; Jegan, M. A Review on Performance of Self-Compacting Concrete–Use of Mineral Admixtures and Steel Fibres with Artificial Neural Network Application. *Constr. Build. Mater.* **2020**, *261*, 120215. [CrossRef]
3. Petit, J.Y.; Wirquin, E.; Duthoit, B. Influence of temperature on yield value of highly flowable micromortars made with sulfonate-based superplasticizers. *Cem. Concr. Res.* **2005**, *35*, 256–266. [CrossRef]
4. Musarrat, M.A.; Ullah, S.; Khan, S.H.; Ullah, K. Effect of bentonite on fresh and hardened property of self compacting concrete. *Sarhad Univ. Int. J. Basic Appl. Sci.* **2017**, *4*, 54–61.
5. Lussy, C.L.; Sugiarto, H. Pengaruh Penggunaan Polypropylene Fibre Terhadap Karakteristik Self Compacting Concrete. *J. Dimens. Pratama Tek. Sipil* **2020**, *9*, 49–55.
6. Mansour, S.M. Behavior of self-compacting concrete incorporating calcined pyrophyllite as supplementary cementitious material. *J. Build. Mater. Struct.* **2020**, *7*, 119–129.
7. Kumar, K.R.; Shyamala, G.; Awoyera, P.O.; Vedhasakthi, K.; Olalusi, O.B. Cleaner production of self-compacting concrete with selected industrial rejects-an overview. *Silicon* **2020**, 1–12. [CrossRef]
8. de Matos, P.R.; Oliveira, J.C.; Medina, T.M.; Magalhães, D.C.; Gleize, P.J.; Schankoski, R.A.; Pilar, R. Use of air-cooled blast furnace slag as supplementary cementitious material for self-compacting concrete production. *Constr. Build. Mater.* **2020**, *262*, 120102. [CrossRef]
9. Mohammed, B.S.; Yen, L.Y.; Haruna, S.; Huat, M.L.S.; Abdulkadir, I.; Al-Fakih, A.; Liew, M.S.; Zawawi, N.A.W.A. Effect of Elevated Temperature on the Compressive Strength and Durability Properties of Crumb Rubber Engineered Cementitious Composite. *Materials* **2020**, *13*, 3516. [CrossRef] [PubMed]
10. Valizadeh, A.; Hamidi, F.; Aslani, F.; Shaikh, F.U.A. The effect of specimen geometry on the compressive and tensile strengths of self-compacting rubberised concrete containing waste rubber granules. *Structures* **2020**, *27*, 1646–1659. [CrossRef]
11. Shaaban, I.G.; Rizzuto, J.P.; El-Nemr, A.; Bohan, L.; Ahmed, H.; Tindyebwa, H. Mechanical properties and air permeability of concrete containing waste tyres extracts. *J. Mater. Civ. Eng.* **2021**, *33*, 04020472. [CrossRef]
12. Sofi, A. Effect of waste tyre rubber on mechanical and durability properties of concrete—A review. *Ain Shams Eng. J.* **2018**, *9*, 2691–2700. [CrossRef]
13. Ismail, M.K.; Hassan, A.A.A. Use of Steel Fibers to Optimize Self-Consolidating Concrete Mixtures Containing Crumb Rubber. *ACI Mater. J.* **2017**, *114*, 581–594. [CrossRef]
14. Zhu, H.; Duan, F.; Shao, J.-W.; Shi, W.; Lin, Z. Material and durability study of a 10-year-old crumb rubber concrete bridge deck in Tianjin, China. *Mag. Concr. Res.* **2021**, *73*, 499–511. [CrossRef]
15. Strukar, K.; Šipoš, T.K.; Miličević, I.; Bušić, R. Potential use of rubber as aggregate in structural reinforced concrete element–A review. *Eng. Struct.* **2019**, *188*, 452–468. [CrossRef]
16. Alaloul, W.S.; Musarat, M.A.; A Tayeh, B.; Sivalingam, S.; Bin Rosli, M.F.; Haruna, S.; Khan, M.I. Mechanical and deformation properties of rubberized engineered cementitious composite (ECC). *Case Stud. Constr. Mater.* **2020**, *13*, e00385. [CrossRef]
17. Mohammed, B.S.; Hossain, K.M.A.; Swee, J.T.E.; Wong, G.; Abdullahi, M. Properties of crumb rubber hollow concrete block. *J. Clean. Prod.* **2012**, *23*, 57–67. [CrossRef]
18. Mohammed, B.S.; Adamu, M. Mechanical performance of roller compacted concrete pavement containing crumb rubber and nano silica. *Constr. Build. Mater.* **2018**, *159*, 234–251. [CrossRef]
19. Tjaronge, M.W.; Musarat, M.A.; Law, K.; Alaloul, W.S.; Ayub, S. Effect of Graphene Oxide on Mechanical Properties of Rubberized Concrete: A Review. In Proceedings of the Lecture Notes in Civil Engineering, Kuching, Malaysia, 13–15 June 2021; Springer Science and Business Media LLC: Singapore, 2021; pp. 484–492.
20. Alaloul, W.S.; Musarat, M.A.; Hui, C.J. Impact of Elevated Temperature on Rubberized Concrete: A Review. In Proceedings of the Lecture Notes in Civil Engineering, Kuching, Malaysia, 13–15 June 2021; Springer Science and Business Media LLC: Singapore, 2021; pp. 421–427.
21. Amat, R.C.; Ismail, K.N.; Ahmad, K.R.; Ibrahim, N.M. Effects of Metakoalin on Municipal Solid Waste Incineration (MSWI) Bottom Ash-Cement Composite. *Mater. Sci. Forum* **2020**, *1010*, 653–658. [CrossRef]
22. Kumar, M.; Sinha, A.K.; Kujur, J. Mechanical and durability studies on high-volume fly-ash concrete. *Struct. Concr.* **2021**, *22*, E1036–E1049. [CrossRef]
23. Niwas, R.; Mathur, D.; Agarwal, S. *A Study to Use of Waste Material in Bituminous Concrete (BC)*; IJTIMES: Mumbai, India, 2019.
24. Sandanayake, M.; Gunasekara, C.; Law, D.; Zhang, G.; Setunge, S.; Wanijuru, D. Sustainable criterion selection framework for green building materials–An optimisation based study of fly-ash Geopolymer concrete. *Sustain. Mater. Technol.* **2020**, *25*, e00178. [CrossRef]
25. Ibrahim, O.M.O.; Tayeh, B.A. Combined effect of lightweight fine aggregate and micro rubber ash on the properties of cement mortar. *Adv. Concr. Constr.* **2020**, *10*, 537–546.
26. Çakır, Ö. Experimental analysis of properties of recycled coarse aggregate (RCA) concrete with mineral additives. *Constr. Build. Mater.* **2014**, *68*, 17–25. [CrossRef]

27. Nili, M.; Afroughsabet, V. Combined effect of silica fume and steel fibers on the impact resistance and mechanical properties of concrete. *Int. J. Impact Eng.* **2010**, *37*, 879–886. [CrossRef]
28. Siddique, R. Utilization of silica fume in concrete: Review of hardened properties. *Resour. Conserv. Recycl.* **2011**, *55*, 923–932. [CrossRef]
29. Xie, J.; Fang, C.; Lu, Z.; Li, Z.; Li, L. Effects of the addition of silica fume and rubber particles on the compressive behaviour of recycled aggregate concrete with steel fibres. *J. Clean. Prod.* **2018**, *197*, 656–667. [CrossRef]
30. Onuaguluchi, O.; Panesar, D.K. Hardened properties of concrete mixtures containing pre-coated crumb rubber and silica fume. *J. Clean. Prod.* **2014**, *82*, 125–131. [CrossRef]
31. Güneyisi, E.; Gesoğlu, M.; Özturan, T. Properties of rubberized concretes containing silica fume. *Cem. Concr. Res.* **2004**, *34*, 2309–2317. [CrossRef]
32. Mohammed, B.S.; Khed, V.C.; Nuruddin, M.F. Rubbercrete mixture optimization using response surface methodology. *J. Clean. Prod.* **2018**, *171*, 1605–1621. [CrossRef]
33. C293/C293M-16, A. *Standard Test Method for Flexural Strength of Concrete (Using Simple Beam with Center-Point Loading)*; ASTM International: West Conshohocken, PA, USA, 2016.
34. European Federation for Specialist Construction Chemicals and Concrete Systems. *Guidelines for Self-Compacting Concrete*, English ed.; European Federation for Specialist Construction Chemicals and Concrete Systems: Norfolk, UK, February 2002.
35. Topçu, I.B.; Bilir, T. Experimental investigation of some fresh and hardened properties of rubberized self-compacting concrete. *Mater. Des.* **2009**, *30*, 3056–3065. [CrossRef]
36. Wang, X.; Subramaniam, K.V. Ultrasonic monitoring of capillary porosity and elastic properties in hydrating cement paste. *Cem. Concr. Compos.* **2011**, *33*, 389–401. [CrossRef]
37. Mohammed, B.S.; Xian, L.W.; Haruna, S.; Liew, M.S.; Abdulkadir, I.; Zawawi, N.A.W.A. Deformation Properties of Rubberized Engineered Cementitious Composites Using Response Surface Methodology. *Iran. J. Sci. Technol. Trans. Civ. Eng.* **2020**, 1–12. [CrossRef]
38. Mohammed, B.S.; Haruna, S.; Wahab, M.M.B.A.; Liew, M. Optimization and characterization of cast in-situ alkali-activated pastes by response surface methodology. *Constr. Build. Mater.* **2019**, *225*, 776–787. [CrossRef]
39. Ganesan, N.; Raj, J.B.; Shashikala, A. Flexural fatigue behavior of self compacting rubberized concrete. *Constr. Build. Mater.* **2013**, *44*, 7–14. [CrossRef]
40. Khed, V.C.; Mohammed, B.S.; Nuruddin, M.F. Effects of different crumb rubber sizes on the flowability and compressive strength of hybrid fibre reinforced ECC. In *IOP Conference Series: Earth and Environmental Science*; IOP Publishing: Bristol, UK, 2018.
41. Haruna, S.; Mohammed, B.S.; Wahab, M.M.A.; Al-Fakih, A. Compressive strength and workability of High Calcium One-Part alkali activated mortars using response surface methodology. In *IOP Conference Series: Earth and Environmental Science 2nd International Conference on Civil and Environmental Engineering*; IOP Science: Langkawi, Malaysia, 2020.

*sustainability*

MDPI

Article

# An Experimental Study on the Performance of Corrugated Cardboard as a Sustainable Sound-Absorbing and Insulating Material

**Chun-Won Kang [1], Mina K. Kim [2] and Eun-Suk Jang [1,3,*]**

[1] Department of Housing Environmental Design and Research Institute of Human Ecology, College of Human Ecology, Jeonbuk National University, Jeonju 54896, Korea; kcwon@jbnu.ac.kr
[2] Department of Food Science & Human Nutrition and Research Institute of Human Ecology, College of Human Ecology, Jeonbuk National University, Jeonju 54896, Korea; minakim@jbnu.ac.kr
[3] R&D Center, Sambo Scientific Co., Ltd., Seoul 07528, Korea
* Correspondence: esjang@sambosc.com

**Abstract:** The continuing development of industrialization and increasing population density has led to the emergence of noise as an increasingly common problem, requiring various types of sound absorption and insulation methods to address it. Meanwhile, the recycling of resources to ensure a sustainable future for the planet and mankind is also required. Therefore, this study investigates the potential of corrugated cardboard as a resource for noise reduction. The sound absorption and insulation performance of non-perforated corrugated cardboard (NPCC) were measured, and modified corrugated boards were fabricated by drilling holes either through the surface of the corrugated board alone or through the corrugated board in its entirety. The sound-absorption/insulation performance both of perforated corrugated cardboard (PCC) and perforated corrugated cardboard with multi-frequency resonators (PCCM) were measured using the transfer function method and the transmission matrix method. To determine the effectiveness of NPCC, PCC, and PCCM in noise reduction, the sound pressure level was analyzed by applying it to a home blender. The results showed PCCM's sound absorption and insulation performance to be excellent. On the basis of these findings, we propose the use of PMMC as an eco-friendly noise-reduction material.

**Keywords:** eco-friendly sound-absorbing material; corrugated cardboard; perforated corrugated cardboard; sound-absorption coefficient; sound transmission loss; transfer function method; transfer matrix method; multi-frequency resonator

**Citation:** Kang, C.-W.; Kim, M.K.; Jang, E.-S. An Experimental Study on the Performance of Corrugated Cardboard as a Sustainable Sound-Absorbing and Insulating Material. *Sustainability* **2021**, *13*, 5546. https://doi.org/10.3390/su13105546

Academic Editor: Shervin Hashemi

Received: 8 April 2021
Accepted: 14 May 2021
Published: 16 May 2021



## 1. Introduction

Because Republic of Korea is a country with small land mass and high population density, apartment housing is common [1]. Noise complaints in these apartment complexes are an unavoidable reality [2]. According to the National Noise Information System (NOISEINFO), a government agency that monitors noise problems in the country, the number of reported noise-related problems has increased exponentially, from 1829 cases in 2012 to 10,142 cases in 2019 [3].

Overall noise level reduction is necessary for maintaining an agreeable sound environment [4]. Blocking or reducing noise through sound absorption and sound insulation is a mainstream method [5]. Accordingly, a significant amount of research has been conducted in order to identify materials with excellent sound absorption and sound insulation properties [6,7]. Sound-absorbing materials typically include one or more components from porous materials [8], plate or membrane vibration materials [9], or resonators [10,11].

Traditionally, the most important metric by which a potential sound-absorbing material is assessed is its absorption ability. However, while the materials typically relied on for indoor acoustic control are typically derived from petroleum [12], recent years have seen a new focus on the utilization of sustainable green materials, including agricultural by-products, to fulfill this function [13–15]. In this spirit, a wide range of studies have been conducted of eco-friendly sound-absorption materials such as rice straw [16], rice husks [17], palm fibers [18], giant reeds [19], egg cartons [20], and wood paper [21].

Corrugated cardboard is a bio-degradable, eco-friendly paper material that is inexpensive and robust in relation to its weight [22]. Its thickness and empty middle space make it a useful sound insulating material [23]. It can be used as a resonance sound-absorbing material either by perforating only the surface of the corrugated board or by penetrating the entire thickness of the corrugated board. Its sound-absorption properties for specific frequency bands can be enhanced by modifying hole size and depth [24]. Corrugated cardboard can also be used as an interior building material and, when discarded, can be reused for pulp or paper [25]. Corrugated cardboard is known for its utility as a building material [25,26], and some studies have suggested the potential of hydrophobic treatment to avoid moisture absorption and further expand its versatility [27,28]. Moreover, it may act as a flame retardant, improving building safety [29].

With an eye toward these benefits, we set out to determine whether corrugated cardboard could be utilized as a sustainable noise-reducing building material.

Berardi and Iannace [30] measured cardboard's sound-absorption coefficient by inserting its veins in a direction parallel to the impedance tube. This resulted in excellent sound-absorption performance at medium and high frequencies but poor performance at low frequencies below 400 Hz. In short, while the material performed well in the veins of the cardboard direction, as a practical matter, it is not easy to use the material in this way.

Kang and Seo [31] investigated changes to the resonant frequency of the cardboard as a function of changes to the aperture ratio. They found no significant changes but reported that sound absorption at a specific frequency was significantly increased by perforations of a certain depth and size. Kang et al. [32] reported that applying porous polyurethane foam attached to a corrugated cardboard to a household blend reduced the sound-absorption coefficient and sound transmission loss. Polyurethane foam, however, is not an eco-friendly sound-absorbing material, and sound-absorbing performance may be improved with additional research into the corrugated board itself.

This study developed a natural sound absorber using triple-layer-corrugated cardboard whose inner surface layers were pierced with holes to enhance its resonance sound-absorbing properties. Three types of corrugated cardboard were prepared: non-perforated corrugated cardboard (NPCC), perforated corrugated cardboard (PCC), and perforated corrugated cardboard with multi-frequency resonator (PCCM). The sound absorption and insulation properties of each of these types were then measured using the transfer function method and the transmission matrix method.

To test these corrugated sound-absorbing materials, they were applied to the use case of home blenders. A home blender was selected as a noise generator since it is one of the most common noise-producing household appliances [33].

The noise level of the blender's rotor was first analyzed. After this, a Helmholtz resonator actively formed in the thickness direction corresponding to the frequency characteristics of the noise source was created on the surface of the corrugated cardboard.

An additional cavity layer was installed between the single-resonator corrugated cardboard and the NPCC, while part of the surface perforation was connected to the rear layer in order to create a multi-resonator with multiple frequencies. The noise reduction effect on the blender was evaluated by measuring and comparing the sound-absorption rate, acoustic transmission loss, and noise level of the fabricated single-resonator and multi-resonator-perforations in relation to hole diameter and perforation ratio.

## 2. Materials and Methods

### 2.1. Sample Preparation

As shown in Figure 1, triple-walled, seven-layer corrugated cardboard with 1800 g/m$^2$ base weight and 15.0 mm thickness was sourced from a Korean market (Daeyoung Packaging Co, Ltd., Ansan-si, Korea).

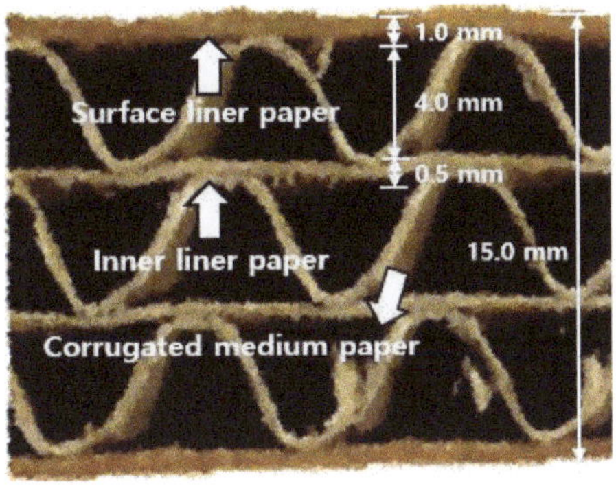


**Figure 1.** Triple-wall corrugated cardboard structure.

In this study, three types of corrugated cardboard were used as sound absorbers. Figure 2 shows the respective structures of the three corrugated cardboard types, while Figure 3 shows application of the cardboard to the blender. NPCC denotes non-perforated corrugated cardboard (Figure 3a). PCC denotes single-resonance sound-absorbing corrugated cardboard whose surface liner paper was pierced with 2.3 mm diameter holes at 14 mm intervals (Figure 3b).

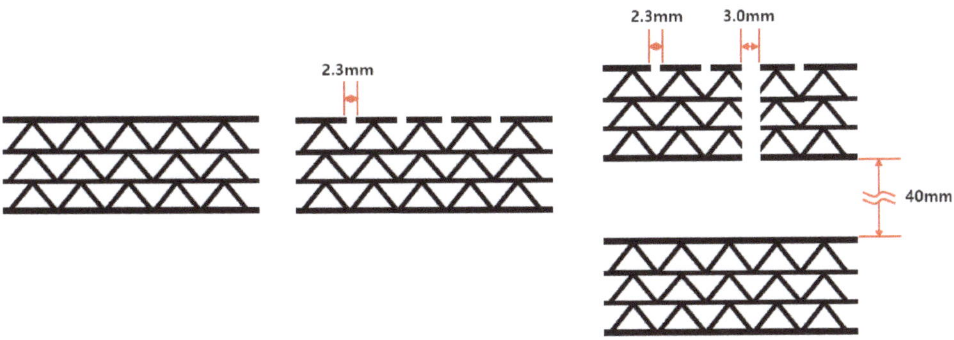


**Figure 2.** Three corrugated cardboard type structures. (**a**): NPCC, (**b**): PCC and, (**c**): PCCM.

PCCM denotes PCC with additional 3.0 mm diameter holes (1/4 of the number of 2.3 mm diameter holes). The 3.0 mm diameter holes pierced all 7 layers of the corrugated cardboard. There was a 4 cm air cavity at the back where we added NPCC (Figure 3c).

Resonance occurs at different frequencies depending on perforated hole diameter, surface liner paper thickness, perforated hole area, and distance from the inner liner paper.

The resonance sound-absorption frequency was calculated as follows:

$$f_0 = c/2\pi(G/V)^{0.5} \quad (1)$$

$$G = s/le \quad (2)$$

$$le = l + \delta \quad (3)$$

where *c*: sound velocity; *G*: neck conductivity; *V*: cavity volume; *s*: neck surface area; *le*: effective neck length; *l*: neck length; and *δ*: end correction (≒ 0.8d)

Therefore,

$$f_0 = c/2\pi(s/V(1 + \delta))^{0.5} \quad (4)$$

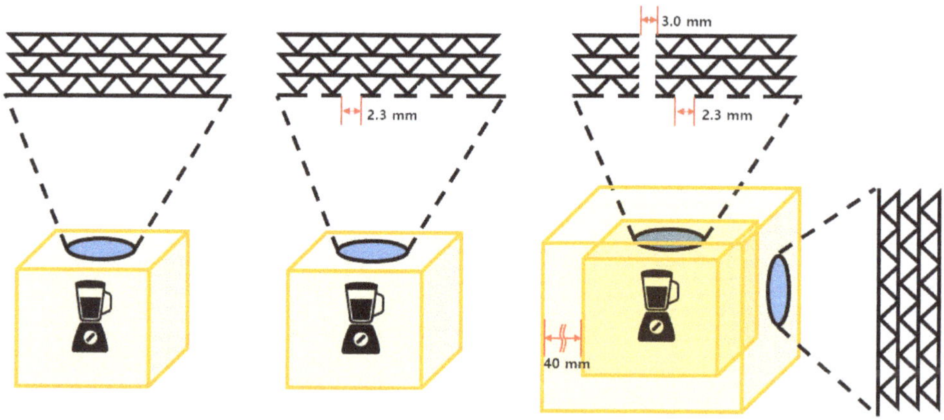


**Figure 3.** Three types of corrugated cardboard types for sound-absorbing materials applied to the blender. (**a**): NPCC, (**b**): PCC, and (**c**): PCCM.

*2.2. Measurement of Sound-Absorption Coefficient Using Transfer Function Method*

The sound-absorption coefficient (SAC) of the NPCC, PCC, and PCCM was measured using a B&K type 4206 impedance tube according to ISO 10534-2 [34] (Figure 4). We additionally calculated the noise-reduction coefficient (NRC) as the average of the sound-absorption rates of 200, 500, 1000, and 2000 Hz.

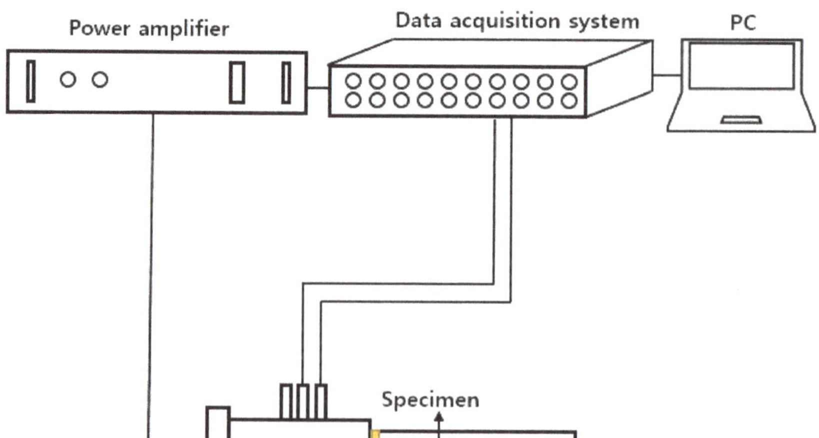


**Figure 4.** Schematic for B&K type 4206 impedance tube used to measure SAC.

Specimens were cut into 29 mm diameter pieces and inserted into an impedance tube. We added silicon O-rings to prevent experimental errors due to gaps between the sample and the wall of the impedance tube. SAC was measured in the 100–6400 Hz frequency range. Temperature, relative humidity, and air pressure were 25.8 °C, 53%, and 1012.00 hPa, respectively. Sound velocity, air density, and acoustic impedance were 346.62 m/s, 1.177 kg/m$^3$, and 408.0 Pa/(m/s), respectively.

### 2.3. STL Measurements Using Transmission Matrix Method

Sound transmission loss (STL), the ratio of the difference between incident sound energy to the material and transmitted sound energy to the material, represents the material's sound insulation performance. In this study, STL was measured in the 100–6400 Hz frequency band by the transmission matrix method according to ASTM E-2611 [35], using a B&K type 4206-T impedance tube to measure acoustic transmission loss (Figure 5). Temperature, relative humidity, and air pressure during measurement were 26.3 °C, 50%, and 1010.0 hPa, respectively.

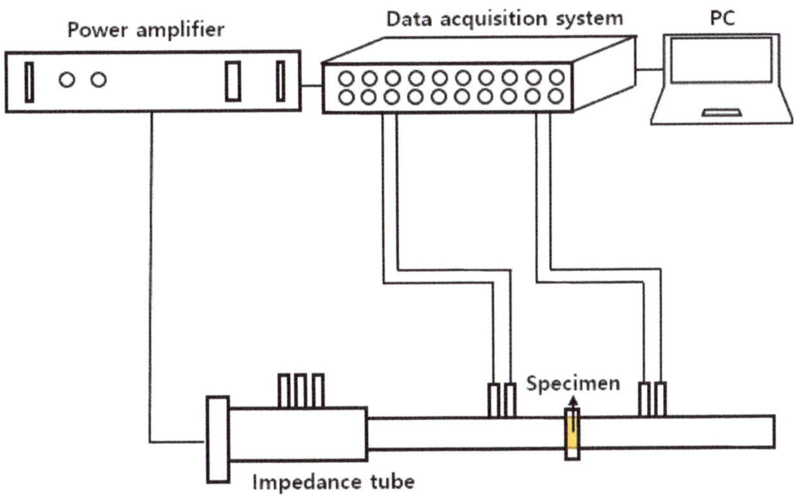


**Figure 5.** Schematic for B&K 4206-T impedance used to measure STL.

### 2.4. Sound Pressure Level Analysis

To verify the actual noise reduction effect, we fabricated a cover using NPCC, PCC, and PCCM and applied this to the blender. First, we performed a sound pressure level analysis. Sound pressure level was measured using a B&K type 2250 handheld sound analyzer in the 63–16,000 Hz frequency range with a 1/3 octave analyzer. Sound pressure level can be measured from 63 Hz. However, there is an experimental error due to noise generation below 50 Hz; therefore, it was measured at frequencies above 100 Hz [33,34]. Sound pressure was measured approximately 1 m from the top of the blender, and the sound pressure level is given as the average of the measured values over 20 s. At the blender's maximum power, the maximum noise peak was in the 1–2 kHz frequency band. Using this information, the resonance cover was optimized for the 1–2 kHz frequency band. The blender was enclosed with the NPCC, PCC, and PCCM covers, and noise levels measured. Temperature, relative humidity, and air pressure during measurements were 28.7 °C, 54%, and 1017.2 hPa, respectively.

## 3. Results and Discussion

### *3.1. SAC Results from Transfer Function Method*

Figure 6 shows the SAC for NPCC, PCC, and PCCM in the 100–6400 Hz frequency range measured by impedance tube. The average SAC and NRC of NPCC were 0.062 (SD 0.010) and 0.055 (SD 0.012), respectively. This indicates almost no sound absorption. The average SAC and NRC of PCC were 0.331 (SD 0.009) and 0.346 (SD 0.007), respectively. The average SAC and NRC of PCCM were 0.362 (SD 0.017) and 0.423 (0.009), respectively.

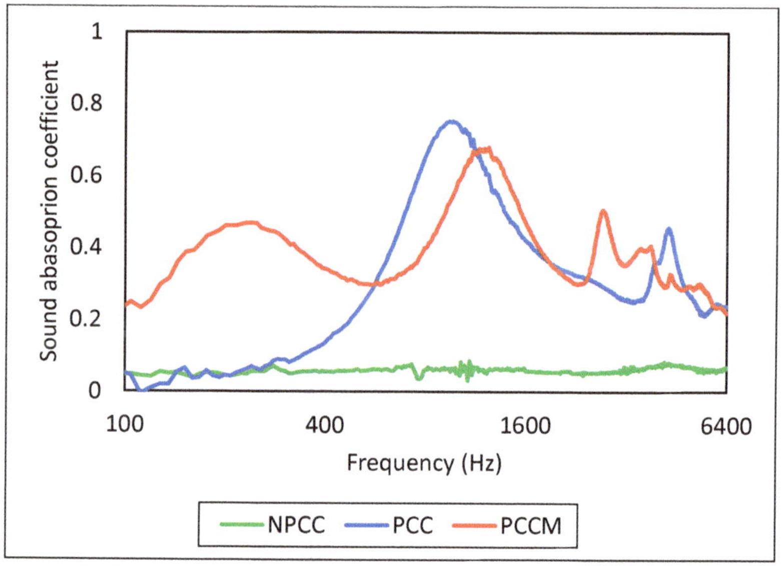


**Figure 6.** SAC results for NPCC, PCC, and PCCM.

As shown in Figure 2b, the PCC's void volume between the perforated surface liner paper and the inner liner paper becomes a single resonator. This is why it can resonate at specific frequencies.

The theoretical resonance frequency of PCC calculated from Equation 4 was 1102 Hz. As shown in Figure 6, the resonance frequency of the PCC obtained experimentally was 936 Hz. As Figure 2c shows, the PCCM's void volume is equal to that of the PCC, plus additional void volume between the corrugated cardboard and the rear space. This means that resonance occurs in two places, at 380 and 1050 Hz. The theoretically calculated resonance frequency and experimentally measured resonance frequency were little bit different. The cause of such an error is that the corrugated cardboard does not only absorb sound by resonance, but an effect due to weak plate vibration may be added. In addition, in this study, since the hole was drilled by the experimenter, not by a machine such as CNC, the hole size may not be constant; therefore, the frequency mismatch can be regarded as an experimental error.

As a result, the average SAC of the PCC increased approximately 5 times more than that of the NPCC, while the NRC of the PCC increased approximately 6 times more than that of the NPCC. In addition, the PCC had peak values of SAC, which were 0.754 at 936 Hz and 0.457 at 4264 Hz. The SAC of the PCC was significantly increased at a specific frequency.

The average SAC of the PCCM was similar to that of the PCC, but the NRC of the PCCM increased approximately 1.2 times more than that of the PCC. The SAC peak values for the PCCM were 0.680 at 1232 Hz, 0.628 at 2704 Hz, and 0.469 at 240 Hz. The PCCM is a multi-resonator and showed peak SAC values at various frequencies.

Compared with other NRC natural fiber composite board (Bagasse: 0.32, Bamboo: 0.35, Banana 0.40, Coir of 0.29, Corn: 0.36) [36], the PCCM demonstrated higher sound-absorption capabilities.

We also compared the sound-absorption performance against Wooden MPP (micro-perforated panels), which are an eco-friendly sound-absorbing material in wide use. We extracted raw data for the sound-absorbing graph result of wooden MPP with holes of 2 mm in diameter and 10 mm intervals in a 5 mm wooden board and the rear air cavity set to 50 mm from the previous study [37] using Engauge Digitzer software [38], and compared the performance of this material with our cardboard. In addition, we added to compare with parallel direction of NPCC from Beradi et al. [27].

As shown Figure 7, PCC's SAC absorbed sound better over 680 Hz than Wooden MPP, and PCCM generally performed better at sound absorption than Wooden MPP, with the exception of the 400–600 Hz range. The wooden MPP had a rear space of 50 mm, while that of the PMCC was 40 mm. Were the space behind the PCC and PCCM to be further increased, the SAC at low frequencies might increase even more.

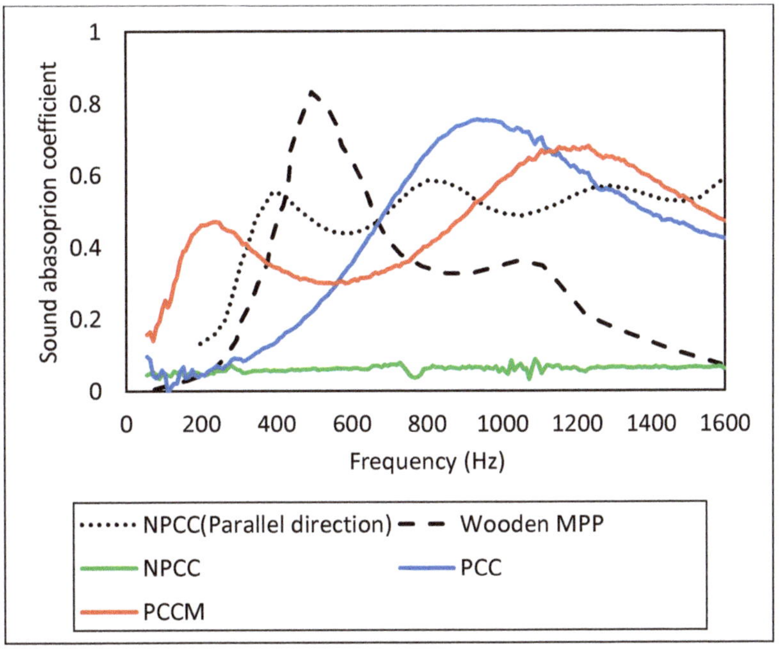


**Figure 7.** Compared SAC results for Wooden MPP from Song et al. [37], NPCC (Parallel direction) from Berardi et al. [27], NPCC, PCC, and PCCM from this work.

PMCC generally performed better at sound absorption than parallel direction of NPCC from Beradi et al. [27], with the exception of the 320–940 Hz range.

In sum, PCC and PMCC did not perform worse than Wooden MPP and parallel direction of NPCC. Corrugated cardboard is cheaper than wooden boards, lighter, easier to install, and easier to recycle. In addition, parallel direction of NPCC is not easy to use practically as sound-absorption material. Therefore, PCC's and PMCC's many advantages make them ideal environmentally friendly sound-absorbing materials.

In sum, PCC and PCCM did not perform worse than Wooden MPP. Corrugated cardboard is cheaper than wooden boards, lighter, easier to install, and easier to recycle; PCC's and PMCC's many advantages make them ideal environmentally friendly sound-absorbing materials.

### *3.2. STL Results from Transmission Matrix Method*

Figure 8 shows the SLT of NPCC, PCC, and PCCM in the 100–6400 Hz frequency range.

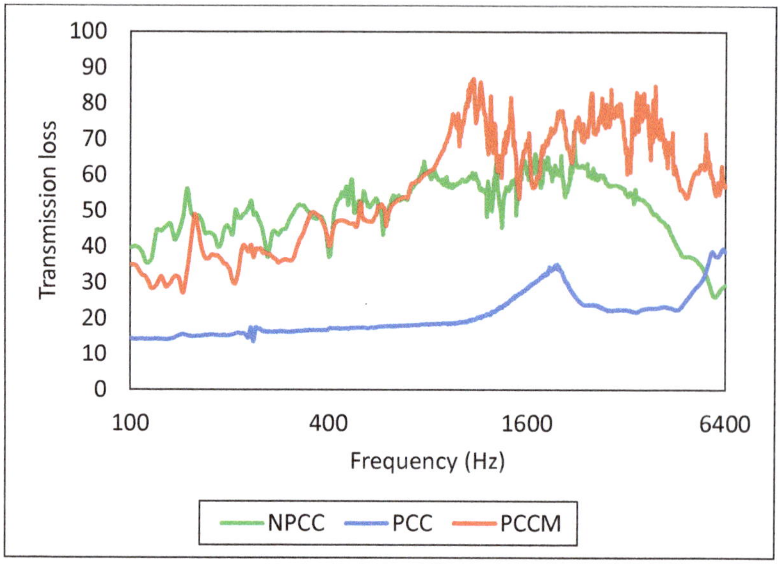


**Figure 8.** STL for CC, PCC, and PCCM.

The average STL of NPCC, PCC, and PCCM were 48.246 (SD 4.683) dB, 25.590 (SD 1.839) dB, and 65.011 (0.878) dB, respectively.

NPCC is a good sound insulation material in itself. The PCC's STL was significantly lower than that of the NPCC due to the surface liner paper perforations. However, the PCCM also showed good sound-insulation performance because the multi-perforated corrugated cardboard on the front side and the NPCC on the rear-side block the sound energy.

Corrugated cardboard has a low specific gravity and is thick; therefore, transmission loss is generally high. As frequency increases, transmission loss decreases. Below 1000 Hz, corrugated cardboard is a good sound-insulation material.

### *3.3. Sound Pressure Level Analysis*

The average sound pressure level of the blender without a cover was 86.267 (SD 1.840) dB, while the levels using NPCC, PCC, and PCCM were 75.500 (SD 0.432) dB, 72.133 (SD 1.096) dB, and 64.367 (SD 0.573) dB, respectively (Figure 9).

NPCC application already reduced the blender's sound level pressure by 11 dB solely on account of its sound insulation effect. PCC application lowered the noise reduction rate by 13 dB, and the PCCM reduced the noise by 22 dB. There was no difference in the average SAC between the PCC and the PCCM, but the PCCM's blender noise reduction effect was greater than that of the PCC. This is because the PCCM's sound-absorption peak frequency range was similar to the blender's noise frequency range.

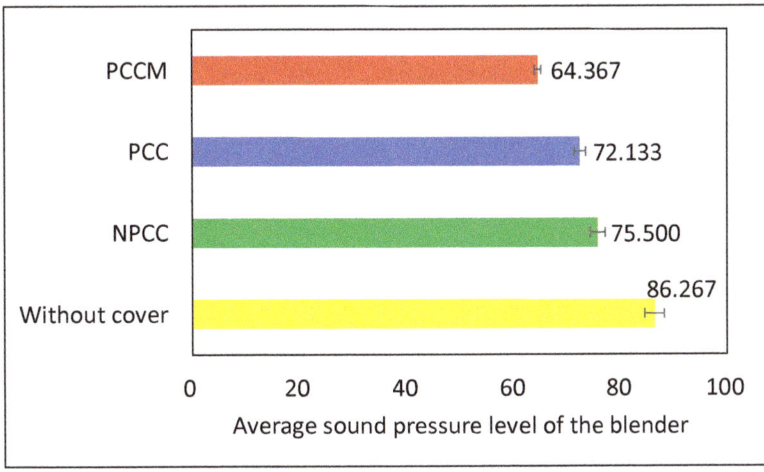


**Figure 9.** Sound level results of blender without cover, blender with NPCC cover, blender with PCC cover, and blender with PCCM cover.

## 4. Conclusions

The possibility of using corrugated cardboard as an eco-friendly sound-absorbing and insulating material was investigated. The material was applied to a blender to evaluate its noise reduction effect. The results of the study are as follows:

1. Corrugated cardboard itself had a sound insulation effect.
2. The NRC of PCC and PCCM were 0.346 (SD 0.007) and 0.423 (0.009), respectively. The average sound pressure level of the blender using NPCC, PCC, and PCCM were 75.500 (SD 0.432) dB, 72.133 (SD 1.096) dB, and 64.367 (SD 0.573) dB, respectively.
3. Compared with other NRC natural fiber composite board, Wooden MPP, and NPCC (parallel direction), and the PCCM demonstrated higher sound-absorption capabilities
4. PCCM shows considerable promise as a sustainable, eco-friendly sound-absorbing and insulating material.

On the basis of the findings in this study, it will be possible to develop a variety of sound-absorbing and sound-insulating materials using the method of corrugated cardboard perforation. In the low-carbon era, in which the recycling of resources has become a necessity for the sustainable future of the earth and humanity, corrugated cardboard is likely to become an increasingly valuable resource as an eco-friendly sound-absorbing material.

**Author Contributions:** C.-W.K.: conceptualization, investigation, methodology, writing—original draft, writing—review and editing. M.K.K.: conceptualization, methodology, supervision, writing—review and editing. E.-S.J.: experiment, data analysis, supervision, writing—original draft, writing—review and editing. All authors have read and agreed to the published version of the manuscript.

**Funding:** This research was supported by the Basic Science Research Program through the National Research Foundation of Korea (NRF) funded by the Ministry of Education (NRF-2019R1I1A3A02059471). It was supported under the international cooperation program framework managed by the NRF of Korea (NRF-2020K2A9A2A08000181).

**Institutional Review Board Statement:** Not applicable.

**Informed Consent Statement:** Not applicable.

**Data Availability Statement:** Not applicable.

**Conflicts of Interest:** The authors declare no conflict of interest.

## References


1. Kim, E.-S. Sound and the Korean public: Sonic citizenship in the governance of apartment floor noise conflicts. *Sci. Cult.* **2016**, *25*, 538–559. [CrossRef]
2. Szopińska, K.; Krajewska, M. Prices of apartments in relation to noise level in Poland. *J. Civ. Eng. Archit.* **2013**, *7*, 1189–1195. [CrossRef]
3. NOISEINFO. Field Diagnosis and Measurement Service of Noise. Available online: http://www.noiseinfo.or.kr/about/stats/fieldDiagnosisSttus_02.jsp (accessed on 6 December 2020).
4. Moebus, S.; Gruehn, D.; Poppen, J.; Sutcliffe, R.; Haselhoff, T.; Lawrence, B. Acoustic quality and urban health-more than just noise and silence. *Bundesgesundh. Gesundh. Gesundh.* **2020**, *63*, 997–1003. [CrossRef]
5. Hassan, T.; Jamshaid, H.; Mishra, R.; Khan, M.Q.; Petru, M.; Novak, J.; Choteborsky, R.; Hromasova, M. Acoustic, Mechanical and Thermal Properties of Green Composites Reinforced with Natural Fibers Waste. *Polymers* **2020**, *12*, 654. [CrossRef] [PubMed]
6. Atienzar-Navarro, R.; Del Rey, R.; Jesus, A.; Sanchez-Morcillo, V.J.; Pico, R. Sound Absorption Properties of Perforated Recycled Polyurethane Foams Reinforced with Woven Fabric. *Polymers* **2020**, *12*, 401. [CrossRef]
7. Leiva, C.; Arenas, C.; Vilches, L.F.; Alonso-Farinas, B.; Rodriguez-Galan, M. Development of fly ash boards with thermal, acoustic and fire insulation properties. *Waste Manag.* **2015**, *46*, 298–303. [CrossRef]
8. Feng, L. Enhancement of low frequency sound absorption by placing thin plates on surface or between layers of porous materials. *J. Acoust. Soc. Am.* **2019**, *146*, EL141. [CrossRef]
9. Gardonio, P.; Zilletti, M. Integrated tuned vibration absorbers: A theoretical study. *J. Acoust. Soc. Am.* **2013**, *134*, 3631–3644. [CrossRef]
10. Long, H.; Cheng, Y.; Liu, X. Reconfigurable sound anomalous absorptions in transparent waveguide with modularized multi-order Helmholtz resonator. *Sci. Rep.* **2018**, *8*, 15678. [CrossRef]
11. Kim, S.; Kim, Y.H.; Jang, J.H. A theoretical model to predict the low-frequency sound absorption of a helmholtz resonator array. *J. Acoust. Soc. Am.* **2006**, *119*, 1933–1936. [CrossRef] [PubMed]
12. Iannace, G.; Trematerra, A. Acoustic measurements and correction of a council room. *Noise Vib. Worldw.* **2014**, *45*, 12–16. [CrossRef]
13. Iannace, G.; Ciaburro, G. Modelling sound absorption properties for recycled polyethylene terephthalate-based material using Gaussian regression. *Build. Acoust.* **2020**. [CrossRef]
14. Asdrubali, F.; Schiavoni, S.; Horoshenkov, K. A review of sustainable materials for acoustic applications. *Build. Acoust.* **2012**, *19*, 283–311. [CrossRef]
15. Iannace, G.; Bravo-Moncayo, L.; Ciaburro, G.; Puyana-Romero, V.; Trematerra, A. The use of green materials for the acoustic correction of rooms. In Proceedings of the INTER-NOISE and NOISE-CON Congress and Conference Proceedings, Madrid, Spain, 30 September 2019; pp. 2589–2597.
16. Yang, H.S.; Kim, D.J.; Kim, H.J. Rice straw-wood particle composite for sound absorbing wooden construction materials. *Bioresour. Technol.* **2003**, *86*, 117–121. [CrossRef]
17. António, J.; Tadeu, A.; Marques, B.; Almeida, J.A.; Pinto, V. Application of rice husk in the development of new composite boards. *Constr. Build. Mater.* **2018**, *176*, 432–439. [CrossRef]
18. Chen, C.; Wang, Z.; Zhang, Y.; Bi, M.; Nie, K.; Wang, G. Investigation of the hydrophobic and acoustic properties of bio windmill palm materials. *Sci. Rep.* **2018**, *8*, 13419. [CrossRef] [PubMed]
19. Iannace, G.; Trematerra, A.; Trematerra, P. Acoustic correction using green material in classrooms located in historical buildings. *Acoust. Aust.* **2013**, *41*, 213–218.
20. Iannace, G.; Berardi, U.; Ciaburro, G.; Trematerra, A. Egg cartons used as sound absorbing systems. In Proceedings of the INTER-NOISE and NOISE-CON Congress and Conference Proceedings, Seoul, Korea, 12 October 2020; pp. 405–412.
21. Jang, E.-S.; Kang, C.-W.; Kang, H.-Y.; Jang, S.-S. Sound Absorption Property of Traditional Korean Natural Wallpaper (Hanji). *J. Korean Wood Sci. Technol.* **2018**, *46*, 703–712. [CrossRef]
22. Lin, L.; Yang, J.; Ni, S.; Wang, X.; Bian, H.; Dai, H. Resource utilization and ionization modification of waste starch from the recycling process of old corrugated cardboard paper. *J. Environ. Manag.* **2020**, *271*, 111031. [CrossRef]
23. Asdrubali, F.; Pisello, A.; D'Alessandro, F.; Bianchi, F.; Cornicchia, M.; Fabiani, C. Innovative cardboard based panels with recycled materials from the packaging industry: Thermal and acoustic performance analysis. *Energy Procedia* **2015**, *78*, 321–326. [CrossRef]
24. Langfeldt, F.; Hoppen, H.; Gleine, W. Resonance frequencies and sound absorption of Helmholtz resonators with multiple necks. *Appl. Acoust.* **2019**, *145*, 314–319. [CrossRef]
25. Diarte, J.; Vazquez, E.; Shaffer, M. Tooling Cardboard for Smart Reuse: A Digital and Analog Workflow for Upcycling Waste Corrugated Cardboard as a Building Material. In Proceedings of the International Conference on Computer-Aided Architectural Design Futures, Daejeon, Korea, 26–28 June 2019; pp. 384–398.
26. Schonwalder, J.; Rots, J.G. Cardboard: An innovative construction material. In *Sustainable Construction Materials and Technologies*; Chun, Y., Claisse, P., Naik, T.R., Ganjian, E., Eds.; Taylor & Francis: Leiden, The Netherlands, 2007; pp. 731–740.
27. Xia, X.; Dong, Z.; Zhag, X. Study on a New Type of Hydrophobic and Moisture Proof Packaging Board and Its Performance. *Packag. Eng.* **2010**, *31*, 13–16.

28. Cataldi, P.; Profaizer, M.; Bayer, I.S. Preventing water-induced mechanical deterioration of cardboard by a sequential polymer treatment. *Ind. Eng. Chem. Res.* **2019**, *58*, 6456–6465. [CrossRef]
29. Yang, G.; Liu, J.; Xu, B.; Liu, Z.; Ma, F.; Zhang, Q. Effect of silicon-containing nitrogen and phosphorus flame-retardant system on the mechanical properties and thermal and flame-retardant behaviors of corrugated cardboard. *J. Anal. Calorim.* **2020**. [CrossRef]
30. Berardi, U.; Iannace, G. Acoustic characterization of natural fibers for sound absorption applications. *Build. Environ.* **2015**, *94*, 840–852. [CrossRef]
31. Kang, C.; Seo, Y. Sound absorption and sound transmission loss of perforated corrugated board. *J. Korea TAPPI* **2018**, *50*, 32–39. [CrossRef]
32. Kang, C.-W.; Kim, M.K.; JANG, E.S.; Lee, Y.-H.; Jang, S.-S. Sound Absorption Coefficient and Sound Transmission Loss of Porous Sponge Attached Corrugated Cardboard of Noise Insulation Cover. *J. Korea TAPPI* **2020**, *52*, 38–44. [CrossRef]
33. Epp, S.; Konz, S. Home appliance noise: Annoyance and speech interference. *Home Econ. Res. J.* **1975**, 3, 205–209. [CrossRef]
34. ISO11534-2. *Acoustics-Determination of Sound Absorption Coefficient and Impedance in Impedance Tubes-Part 2 Transfer-Function Method*; International Organization for Standardization (ISO): Geneva, Switzerland, 2001.
35. ASTM E2611-19. *Standard Test Method for Normal Incidence Determination of Porous Material Acoustical Properties Based on the Transfer Matrix Method*; ASTM International: West Conshohocken, PA, USA, 2019.
36. Sharma, S.; Shukla, S.; Sethy, A. Acoustical behaviour of natural fibres-based composite boards as sound-absorbing materials. *J. Indian Acad. Wood Sci.* **2020**, *17*, 66–72. [CrossRef]
37. Song, B.; Peng, L.; Fu, F.; Liu, M.; Zhang, H. Experimental and theoretical analysis of sound absorption properties of finely perforated wooden panels. *Materials* **2016**, *9*, 942. [CrossRef]
38. Mitchell, M.; Muftakhidinov, B.; Winchen, T.; Jedrzejewski-Szmek, Z. Engauge Digitizer Software. Available online: https://markummitchell.github.io/engauge-digitizer/ (accessed on 8 April 2021).

*sustainability*

MDPI

*Article*

# Digital Technologies, Circular Economy Practices and Environmental Policies in the Era of COVID-19

**Syed Abdul Rehman Khan [1,2], Pablo Ponce [3], George Thomas [4], Zhang Yu [5,6], Mohammad Saad Al-Ahmadi [7] and Muhammad Tanveer [8,*]**

1 School of Management and Engineering, Xuzhou University of Technology, Xuzhou 221018, China; khan_sar@xzit.edu.cn
2 Beijing Key Laboratory of Urban Spatial Information Engineering, Beijing 100084, China
3 Carrera de Economía and Centro de Investigaciones Sociales y Económicas, Loja 110150, Ecuador; Pablo.ponce@unl.edu.ec
4 Department of Marketing, College of Business Administration, Prince Sultan University, Rafah Street, Riyadh 11586, Saudi Arabia; gthomas@psu.edu.sa
5 School of Economics and Management, Chang'an University, Xi'an 710054, China; 2020023001@chd.edu.cn
6 Department of Business Administration, ILMA University, Karachi 75190, Pakistan
7 Information Systems & Operations Management Department, KFUPM Business School, King Fahd University of Petroleum and Minerals, Dhahran 31261, Saudi Arabia; alahmadi@kfupm.edu.sa
8 Prince Sultan University, Rafah Street, Riyadh 11586, Saudi Arabia
* Correspondence: mtanveer@psu.edu.sa

**Citation:** Khan, S.A.R.; Ponce, P.; Thomas, G.; Yu, Z.; Al-Ahmadi, M.S.; Tanveer, M. Digital Technologies, Circular Economy Practices and Environmental Policies in the Era of COVID-19. *Sustainability* **2021**, *13*, 12790. https://doi.org/10.3390/su132212790

Academic Editor: Shervin Hashemi

Received: 15 October 2021
Accepted: 16 November 2021
Published: 19 November 2021



**Abstract:** The degradation of the environment is associated with economic activity, particularly with the linear way in which the economy does not make efficient use of resources. However, the circular economy is opposed to this linear paradigm, since it makes the most of the resources in trying to achieve zero waste. In this context, this study investigates the relationship between industry 4.0 technologies, COVID-19 outbreak, environmental regulation policies and circular economy practices. A questionnaire is designed to collect information from 214 big and private manufacturing firms in Ecuador, and subsequently, through CB-SEM, the information is processed, and the study paths are validated. The results suggest that industry 4.0 technologies and environmental regulation policies are driving circular economy practices during the pandemic. The study finds no evidence favoring COVID-19 being a determining factor in the adoption of the circular economy. The results provide a policy framework for the adoption of a circular economy.

**Keywords:** industry 4.0; circular economy; environmental regulations; manufacturing supply chains; COVID-19

## 1. Introduction

The outbreak of COVID-19 has been the cause of the current economic crisis, stemming from the global supply chain's decline [1]. In this regard, the pandemic has shown the worst deficiencies of firms and consumers and their vulnerability to risk situations [2]. Despite this, not everything has been discouraging, since it has become clear that environmental sustainability is the way to face a situation of risk and uncertainty, which the supply chain faces [3]; however, the economy faces an economic system of linear production, in which a real challenge to achieve environmental sustainability is presented, since resources are not used efficiently, and too much waste is generated—that is, it is a model that is based on taking that makes waste [4]. On the contrary, circular economy (CE) practices focus on the efficient use of resources based on the 10Rs (reject, rethink, reduce, reuse, repair, recondition, remanufacture, reclaim, recycle and recover) [5]. The decision to adopt cleaner production systems in which the CE is accentuated has taken on a greater force since COVID-19 [6]. Consequently, the CE captures locals' and strangers' attention, since it is crucial for the future to achieve a cleaner and more environmentally-friendly production, based on a production system with zero waste [7]. Therefore, reaching a CE is an arduous process in which economic agents' mentality must be changed to design ecological products in which cleaner production methods are used through the sustainable management of the supply chain from front to back [8]. It should be noted that entrepreneurs are agents of change that adapt to circumstances and can contribute eloquently in the transition towards a sustainable CE [9,10].

Although the CE process is very understandable in theory, the CE establishes a radical change from a linear economic model to a circular one, in which each phase of production represents a systemic change that is very complex to put into practice [11]. Generally, the implementation CE is difficult due to barriers such as lack of resources, ignorance of the innovation of the process, fear of failure, and return on investment, among others [12,13]. Otherwise, the adoption of emerging technologies from industry 4.0 (I4.0) are poised to be determinants to overcome these barriers [14,15] and to implement CE during the COVID-19 pandemic [2]. Some of the I4.0 technologies are artificial intelligence and big data, which allow firms to make better decisions when making sustainable decisions around the choice of resources [16,17].

Therefore, digital transformation is increasingly used in manufacturing industries, allowing stronger, more resilient, and intelligent production processes, which allow firms to apply CE [16]. Conversely, institutional regulation is a factor that determines the adoption of CE in the supply chain, which is why this research constitutes an element of analysis [18].

Even though emerging technologies are varied and used to achieve environmental sustainability, in developing countries, there is little evidence on the role of I4.0 in the adoption of CE during the pandemic [5]. Therefore, this document is one of the pioneers to be developed in Ecuador. In this country, the impulse for the adoption of a CE is recent; at the end of 2019, the government authorities signed the "Pact for the Circular Economy," which seeks to promote the industrialization of waste, use of renewable energy, sustained use of resources among others [19]. However, according to the careful review of the literature carried out, there are no formal investigations that examine the determinants of CE in the country, much less during the COVID-19 pandemic.

In this sense, this research aims to examine the relationship between I4.0, COVID-19, institutional regulation, and CE in Ecuador during the pandemic. Information is collected from various firms through a questionnaire which contains several questions representing each examined construct, with a response option according to the Likert Scale. Subsequently, covariance-based SEM (CB-SEM) is used to process the information collected in 214 Ecuadorian manufacturing firms that are big and private. The research findings reveal unpublished results and validate the hypotheses of the study, which serve to formulate policy measures to guarantee the application of CE during and after the pandemic. In addition, the motivation for carrying out this study is due to its contribution to the current state of the literature and to provide empirical evidence for policymakers' correct definition of policy aimed at achieving environmental sustainability.

Consequently, the contributions of the study are varied, which are indicated below. First, primary information sources are used from CE firms, which allows for obtaining direct information to assess the analysis situation better. In addition, it allows the investigation to be designed effectively, considering all the aspects to be examined. Second, the study uses stylized methods to test the hypotheses raised and maintain their scientific rigor. The applied methodology, CB-SEM, allows for obtaining efficient and unbiased statistics, which support the study's conclusions. Third, no studies have examined the relationship studied in the country of analysis, which makes it an unpublished investigation. Additionally, the study contributes to understanding how COVID-19 has influenced the decision to implement CE in a developing country. Fourth, there is little empirical evidence on the implementation of CE in developing countries [20]. Therefore, this research contributes to the literature by providing evidence on how to adopt CE in developing countries. These aspects show signs of concrete actions to achieve the environmental sustainability that the planet requires. After the introduction, the research structure is as follows: Section 2 describes the recent literature and the definition of hypotheses; Section 3 describes the data and methodological approach; Section 4 discusses the data and results; Consequently, Section 5 discusses the results obtained; furthermore, Section 6 contains the conclusions and policy implications.

## 2. Literature Review

### 2.1. Industry 4.0 and Circular Economy Practices

The CE focuses on the use of resources in the 10Rs to generate economic and environmental benefits [5]; however, some obstacles are related to the lack of advanced technologies to apply this circular flow [21], as well as the uncertainty of the economic benefit to be obtained from the investment made [22,23]; however, the emerging technologies of I4.0 are dynamizing the unlocking of CE and the collaborative economy, respectively [4].

Industry 4.0, known for the adoption of technology to maintain the efficiency of the firm [14], uses systems such as "Adoption of smart factory components", "Integration of digital and physical systems", "Environmental Product Design and life cycle analysis", and Adoption of advanced machine learning algorithms "in order to improve the performance of resources in the production process" [24]. These improvements include saving time in the processing of a product, reducing the cost of the final product, integrating the value chain, the resilience of the production process, making processes more flexible, and focusing on the efficiency of useful resources [25]. Consequently, the application of I4.0 technologies has gained great space in industrial transformation due to their great benefits for the firm and the environment [26]. Hence, I4.0 technologies contribute to applying the CE 10Rs; for example, they promote the reuse of discarded products, restore old products and update them, or design remanufactured products [27].

Indeed, blockchain technology contributes to the firm's CE since its application reduces transaction costs, improves performance and communication in the supply chain, generates a waste reduction, and, consequently, reduces carbon footprint [28]. In Indonesia, Ref. [29] states that the collection, transport, and processing of commercial waste presents CE processes efficiently achieved through the internet of things (IoT). Moreover, in South Africa, Ref. [30] mentions that the analysis of big data, information technologies and human capital are essential factors of I4.0, which are positively related to sustainable production constitutes a determinant for the CE.

Besides, Ref. [31], through structural equation modeling (SEM) in South Africa, find that the use of cloud computing, IoT, smart objects, GPS, radio-frequency identifications devices (RFID), among others, contribute to the application of CE, such as the design of products for reuse and recycling and reduction of solid waste and wastewater management costs, among others. In the same direction, Ref. [16] mention that the various tools of I4.0 allow improving the configuration of the production process, obtaining; as a result, the decrease in the total cost of production and the energy consumption of the machinery. On the other hand, Ref. [32] carried out a study in Estonia through semi-structured interviews; their results reveal that additive manufacturing and Big Data and Advance Analysis in all CE practices.

In addition, from a global perspective, the authors of Ref. [33] carry out an analysis with the heads of CE projects in Europe, America, Asia, and Africa. Their results show that I4.0 directly influences the EC, especially reducing the consumption of materials, energy, waste generation, and emissions. They also indicate that additive manufacturing and robotics are the technologies with the greatest impact on the EC. Finally, Ref. [34], through an in-depth review of the literature, indicate that I4.0 are the drivers to implement CE processes and achieve environmental sustainability. Therefore, we developed the following hypothesis:

**Hypothesis 1 (H1).** *Industry 4.0 technologies contribute to circular economy practices.*

### 2.2. COVID-19 and Circular Economy Practices

Before the outbreak of COVID-19, the CE was considered to be a strong economic approach for the future, but the crisis modified this approach, showing initiatives to incubate CE ideas to challenge the economy's linear model [6]. The COVID-19 outbreak has shown that the global economic model has been developing in a linear way in which the obtaining of economic benefits has prevailed without taking energy consumption into account, which is why the pandemic has become a trigger to bet towards a low carbon economy based on the CE [22,35]. Ref. [31] mentions that the COVID-19 outbreak has demonstrated the importance of implementing CE in production processes to mitigate the economic losses generated by a pandemic. Similarly, in Ref. [36], it is affirmed that the COVID-19 outbreak generated economic losses in companies. However, COVID-19 was an opportunity to apply CE, which allowed them to improve the product delivery service using fewer resources.

Ref. [2] mentions that the COVID-19 outbreak has elucidated the existing deficiencies in the supply chains' consumption and production processes. The current pandemic becomes an opportunity to improve them through CE, enabled by blockchain technology. In this way, sustainability becomes a competitive advantage for firms that seek to remain in the market in the current crisis since the COVID-19 outbreak has prompted firms to design new goods and services through the efficient use of resources and adequate waste management [9,10]. In the same sense, Ref. [32] indicates that due to the lockdown measures implemented during the COVID-19 outbreak, the energy sector has experienced a significant contraction, except for renewable energy consumption, which increased by 3%; specifically, this has been seen in renewable bioenergy, which is has a circular focus with zero waste.

In contrast, lockdown and mobility restriction measures have led to disruption in the shipping recycling supply chain in Bangladesh, India, and Pakistan, due to a lack of workforce to recycle ship components at the end of their useful life, representing a loss of around USD 20 billion [33]. Ref. [34] establishes that the pandemic has increased the need to take care of oneself with the use of single-use products (plastic water bottles, gloves, face shields, among others), which puts environmental sustainability at risk due to the large amount of waste that is generated, especially in low-income economies where there are linear economies.

Conversely, the pandemic has increased the demand for single-use plastic personal protective equipment to prevent the spread of COVID-19; however, they have become a polluting factor due to indiscriminate waste and added to the absence of CE for waste reduction, recycling, and recovery [36,37]. Similarly, [38] mention that the demand for plastic personal protective equipment, food, and groceries packaged in plastic has increased due to the COVID-19 outbreak; thus, CE, through recycling, constitutes one of the solutions to counteract the effect of plastic on the environment. Consequently, considering previous studies, Hypothesis 2 is defined:

**Hypothesis 2 (H2).** *The COVID-19 outbreak does not drive circular economy practices.*

### 2.3. Institutional Regulation and Circular Economy Practices

Concerns about environmental degradation have led governments to the establishment of policies that force firms to adopt clean, sustainable, and circular production systems [39].

Therefore, the government's regulatory role is an essential element when evaluating the adoption of CE motivated by institutional change [40]. However, there is a diversity of findings and criteria on institutional regulation's role in adopting the CE.

For this reason, Ref. [41], in a study of 60 countries, finds that the definition of government policies for the adoption of CE is making notable progress; however, the policies' efficiency is not according to expectations. For example, in a study conducted in Ref. [42] found that entrepreneurs do not trust the government to lead environmental development towards an CE; however, they believe that business incubators would help in this process. In Tanzania, Ref. [43] mentions that environmental laws and regulations are not aimed at promoting CE; on the contrary, they exacerbate resource shortages, and moreover, the private sector and NGOs have contributed initiatives to launch the CE.

Likewise, through SEM processes information collected from Brazilian industrial firms, in Ref. [17] the main findings affirm that the adoption of CE depend to a great extent on the decisions of owners/shareholders, the institutional regulatory pressure of the government is incipient in determining the practices of CE, and ISO 14001 and ISO 9001 certification contribute to better CE adoption. Moreover, Ref. [13] identifies that institutional regulation constitutes a barrier for adopting a CE, given that the countries' institutional infrastructure favors the linear economy. Besides, international supply chains are highly coordinated, so that each country's heterogeneous policies become barriers to the adoption of CE practices [13].

Conversely, through SEM in the European Union (EU) Ref. [11] affirms that the institutional entrepreneurial role is key to adopting the CE; however, this varies according to the legislation of each EU country. Similarly, Ref. [44] found that the re-manufacturing industry in China has developed significantly, which has been achieved satisfactorily due to institutional and legal reforms and government policies implemented in recent years, and have contributed to green economic development, enacting a CE. Ref. [45] mentions that the reducing, reusing, recycling and repurposing processes are not borne of an initiative of the firm to achieve a CE, but as a product of institutional regulations; moreover, they assert that start-ups motivate the definition and modification of institutional regulations and legal regulations for the adoption of CE. Ref. [46] examines the role of institutional quality in adopting CE in EU countries using a structural equation model. Their results find that institutional promotion is a driving factor for the adoption of CE. Likewise, Ref. [47] examines the implementation of the Chinese CE through institutional policies through a systematic review of the literature. Their results show that the EC was notably boosted when China entered the Year Plan period (2016–2020). Finally, Ref. [48] examines how EU policies contribute to the adoption of CE. Their findings show that this type of institutional regulation has contributed notably to the CE, such as reducing waste. According to the empirical evidence described, Hypothesis 3 is defined as follows:

**Hypothesis 3 (H3).** *Institutional regulation contributes to the adoption of a circular economy practices.*

Consequently, Figure 1 synthesizes the hypotheses defined between constructs, which will be examined in the present investigation.

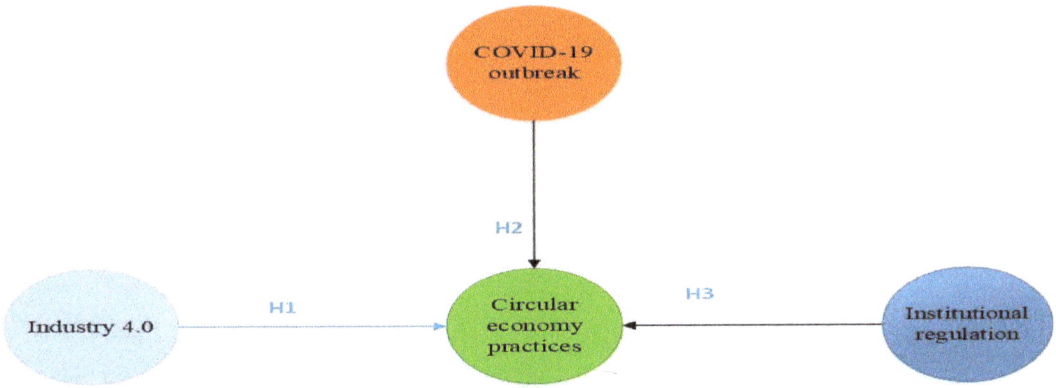


**Figure 1.** Proposed model.

## 3. Data and Methodological Approach

Through a questionnaire, information was collected from 214 manufacturing firms from different cantons of the four provinces contributing the highest percentage to the country's national production. The firms are from the private sector, considered as big firms. According to official statistics published by Ref. [49], 27.1% of the Gross Value Added (GVA) of manufacturing comes from the province of Guayas, 24.2% from Pichincha, and to a lesser extent it attributes Manabí with 5.5% and Azuay with 4.9%. Table 1 shows the distribution of the information collected.

**Table 1.** Demographical data collection.

| Provinces | Numbers | Cities | Numbers |
|---|---|---|---|
| Guayas | 88 | Guayaquil | 18 |
| | | Durán | 15 |
| | | Yaguachi | 10 |
| | | Coronel Marcelino Maridueña | 08 |
| | | Milagro | 11 |
| | | Sanborondón | 05 |
| | | Daule | 07 |
| | | Naranajal | 06 |
| | | Nobol | 03 |
| | | Santa Lucía | 05 |
| Pichincha | 45 | Quito | 20 |
| | | Rumiñahui | 10 |
| | | Mejía | 06 |
| | | Cayambe | 07 |
| | | Pedro Moncayo | 02 |
| Manabí | 36 | Montecristi | 25 |
| | | Manta | 09 |
| | | Jaramijó | 02 |
| Azuay | 30 | Cuenca | 12 |
| | | Gualaceo | 08 |
| | | Chordeleg | 06 |
| | | Sigsig | 04 |
| Total | 214 | Total | 214 |

A questionnaire was designed with several questions according to the examined constructs (see Appendix A) to gather the information from firms. The questionnaire is based on the Likert Scale (1 = strongly disagree, and 5 = strongly agree). The questionnaire was then applied online through Google forms. Consequently, the questionnaire was sent to the firms to be filled in. The information was collected from August to November 2020. After obtaining the information, the covariance-based structural equation modeling (CB-SEM) econometric approach was used in order to examine the relationship between constructs according to the information provided by their factors; likewise, it allows us to capture the effect of latent interaction and moderation [50]. One of the advantages of the CB-SEM approach is its particularity in investigating various relationships between interrelated constructs simultaneously [51]. Another advantage is that it improves the robustness of the estimators, optimizes the results for the interpretation of the interaction effects and efficiently controls the measurement error [52]. Conversely, the proposed model meets the minimum criteria for applying CB-SEM. CB-SEM can be used with samples greater than 200 observations [53,54], and likewise, CB-SEM offers the availability to use several constructs and with any number of constructed latent interaction indicators [55].

Similarly, to determine the suitability of the constructs to be used in the research, reliability and validity tests are performed to obtain robust and unbiased results using exploratory factor analysis (EFA) and confirmatory factor analysis (CFA) [56].

## 4. Analysis of Data and Results

After collecting the information of firms using the online questionnaire, the reliability and validity of the data must be confirmed. The questionnaire is designed with several questions (items) that represent each construct. The "items" column of Table 2 indicates the number of questions that each construct contains, which were sent to the firms for their answer. Therefore, the study uses the exploratory factor analysis (EFA) and confirmatory factor analysis (CFA). First, the EFA is applied to examine the unidimensionality of the constructs and, consequently, define the factors responsible for representing a thing in common, in other words, a set of diverse factors that explain a single concept [57]. Thus, Table 2 shows the results of factor loadings that are higher than 0.6, which supports the unidimensionality of the constructs of circular economy (CE) practices, industry 4.0 (I4.0) technologies, coronavirus disease 2019 (COVID-19) outbreak and institutional regulations for environment (IRE). Additionally, Cronbach's Alpha values are between 0.81 and 0.92, which are higher than 0.7 and provide an argument favoring internal consistency between the measurement elements within each construct.

**Table 2.** Instrument reliability and validity.

| Constructs | Symbol | Items | Factor Loading Ranges | Cronbach's Alpha | AVE | CR |
|---|---|---|---|---|---|---|
| Circular economy practices | CE | 12 | 0.741–0.892 | 0.91 | 0.637 | 0.864 |
| Industry 4.0 technologies | I4.0 | 11 | 0.759–0.934 | 0.92 | 0.696 | 0.905 |
| Coronavirus disease 2019 | COVID-19 | 5 | 0.68–0.873 | 0.87 | 0.671 | 0.833 |
| Institutional regulations for environment | IRE | 6 | 0.582–0.816 | 0.81 | 0.568 | 0.739 |

Subsequently, the CFA examines the factors included in the dimensions of the study; first, Composite Reliability (CR) is a more efficient measure of internal consistency; the values are higher than 0.7 and vary between 0.739 and 0.905 [58]. In addition, the average variance extracted (AVE) values are more significant than 0.5, which guarantees that the variables meet the convergent criteria, respectively. Consequently, the Fornell and Larcker criterion confirms the discriminant validity, as reported in Table 3. Then, results of the CFA that are displayed in Table 4.

**Table 3.** Discriminant validity analysis.

| Constructs | COVID-19 | CE | IRE | I4.0 | Mean | S.D. |
|---|---|---|---|---|---|---|
| COVID-19 | 0.822 | | | | 3.52 | 0.621 |
| CE | 0.515 | 0.798 | | | 2.98 | 0.495 |
| IRE | 0.379 | 0.261 | 0.753 | | 2.38 | 0.337 |
| I4.0 | 0.497 | 0.507 | 0.479 | 0.834 | 4.55 | 0.754 |

**Table 4.** Model fitness results.

| Fit induces | NNFI | NFI | CFI | GFI | AGFI | TLI | $\chi 2/df$ | RMSEA | SRMR |
|---|---|---|---|---|---|---|---|---|---|
| Criteria | $\geq$0.90 | $\geq$0.90 | $\geq$0.90 | $\geq$0.90 | $\geq$0.90 | $\geq$0.90 | $\leq$3 | $\leq$0.08 | $\leq$0.08 |
| Measurement model | 0.913 | 0.910 | 0.911 | 0.907 | 0.929 | 0.938 | 1.892 | 0.037 | 0.027 |
| Structural model | 0.922 | 0.922 | 0.934 | 0.912 | 0.932 | 0.941 | 1.282 | 0.038 | 0.026 |

Table 4 shows model fitness, whose values of the nine indicators show an adequate specification of the measurements of the variables as well as the model [59]. The indicators used are the non-normed fit Index (NNFI), normative fit index (NFI), comparative fit index (CFI), the goodness of fit index (GFI), adjusted goodness of fit index (AGFI), Tucker–Lewis index (TLI), Chi-square to the degree of freedom ($\chi 2/df$), root-mean-square error of approximation (RMSEA) and standardized root mean squared residual (SRMR).

The values of NNFI, NFI, CFI, GFI, AGFI, and TLI are greater than the limit established for each indicator following [60]. Besides, the value of $\chi 2/df$ (1.892) is less than three and that of RMSEA (0.037) less than 0.08, which agrees with the criteria established by [61]. Additionally, the approximate fit, which measures the difference between the observed correlation matrix and the implicit correlation matrix of the model, is determined by SRMR, whose value (0.027) is less than 0.08 [62]. Based on what was previously analyzed, it is confirmed that the proposed model has a great fit with the data; therefore, the next step is to analyze the results of the hypotheses.

In this scenario, Figure 2 and Table 5 present the standardized regression estimates of variables in the structural model and the significance of path weights. The regression carried out was of the ordinal type, since each question has a response option from 1 to 5, according to the Likert Scale. Therefore, ordinal regression allows for the processing of responses of polytomous variables. I4.0 maintains a positive and significant relationship with CE ($\alpha = 0.716$, $p = 0.000$) as expected. Then, the effect of the COVID-19 outbreak is examined, in which it is identified that there is no strong or statistically significant association with CE ($\alpha = 0.871$, $p = 0.541$). This result is novel and was shown to be unknown due to the few precedents associated with the COVID-19 outbreak in the industry. Finally, the relationship between IRE and CE were examined, in which evidence is found in favor of the fulfilment of the hypothesis, IRE has a positive and statistically significant effect on CE ($\alpha = 0.834$, $p < 0.01$), whose results are found according to expectations. The results obtained provide important information that supports the fulfilment of the study hypotheses H1–H3.

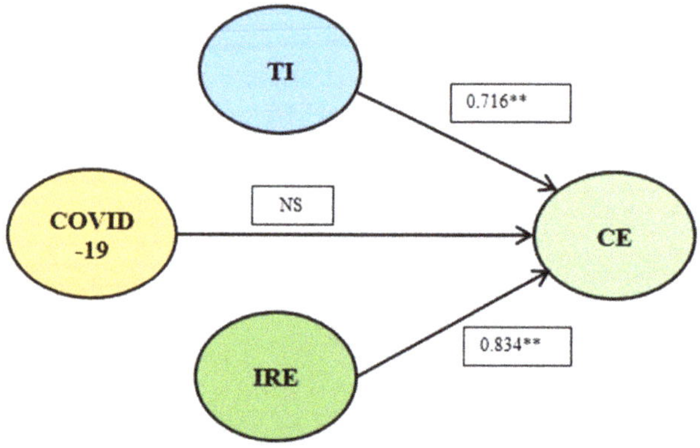


**Figure 2.** The results of the structural model. Note: ** indicates the significance.

**Table 5.** Standardized parameter estimates for structural model.

| Hypothesis | Paths | Standardized Estimate | *p*-Value | Results |
|---|---|---|---|---|
| H1 | I4.0 → CE | 0.716 ** | 0.000 | Supported |
| H2 | COVID-19 → CE | 0.871 | 0.541 | Not supported |
| H3 | IR → CE | 0.834 ** | 0.006 | Supported |

Note: ** indicates significance at 1% and 5% respectively.

## 5. Discussion of Results

The COVID-19 outbreak has directly impacted various economic activities worldwide due to the lockdown and mobility restriction measures that have been defined to mitigate the spread of the pandemic. Likewise, the it has generated a favorable scenario to promote the CE as an alternative to improve the efficiency of using resources, which brings with it a cleaner and more environmentally-friendly approach to production [6]. Consequently, this research examines the relationship between the technologies of industry 4.0, the COVID-

19 outbreak, institutional regulations for environment and circular economy practices in manufacturing firms in Ecuador. This fact constitutes a valuable contribution to the country of study due to the few studies that examine the determinants of CE.

The results shown in the previous section show the direct effects of the explanatory variables of the study on CE, which were described in Hypotheses 1–3. In the first place, I4.0 has a direct effect on the adoption of CE in manufacturing industries, from which it should be understood that those firms that have modern information systems, with state-of-the-art technology in their production processes, intelligent processing of data, among others, are easier to adopt and apply CE. This fact could be explained by the fact that a production process with a high component of I4.0 generates information in each phase for the adequate decision-making of the firm on the efficient use of resources, such as reducing waste, saving electricity and water, among others. Moreover, the reuse or re-manufacturing of inputs and products is more efficient with modern technological processes of I4.0. These findings coincide with the study by the authors Ref. [16], who affirm that the adoption of intelligent technologies, such as those of I4.0, contributes to increasing the efficiency of resources, reducing waste and saving energy in machinery. Similar to our findings, Ref. [26] affirms that the use of I4.0, such as IoT, GPS, or RFID drives the design of products for reuse or recycling, as well as improving waste management.

Next, the COVID-19 outbreak shows a positive sign but not statistically significant. In other words, the pandemic was able to corroborate that a linear system of the economy does not contribute to environmental sustainability due to the excessive amount of waste and inefficiency of resources used in the production of goods and services. Thus, several firms have been inclined to carry out CEs before the onset of the COVID-19 outbreak. However, the research does not find evidence in favor of COVID-19 having caused the implementation of CE in Ecuadorian firms, which could be explained for two main reasons. First, firms that adopt CE do so in an organized and planned way, considering the firm's performance, not because of a risk situation due to mitigation measures for the spread of COVID-19. Thus, for example, manufacturing firms did not have the necessary response capacity to adapt to this situation, and on the contrary, their planning towards a circular economy becomes uncertain [63]. Second, COVID-19 generated a situation of uncertainty in the firm performance, which causes firm managers to have greater doubts when it comes to adapting the firm to apply CE in its production processes since it is required of a significant investment which becomes riskier in the middle of a pandemic as suggested by [64]. Thus, the results obtained in this study are contrary to those found by the author of Ref. [9], who mentions that the COVID-19 outbreak has been the driving force for firms to design new products and improve waste management by adopting CE.

Conversely, IRE presents a positive and significant effect on CE. This result must be associated with the important role that the government plays, through its environmental regulation policies, in order for the production processes of manufacturing firms to be clean and not to degrade the environment. However, government intervention may not achieve the environmental solutions that an economy requires [17]. These types of regulations, such as ISO 4001 or ISO 9001, aim to promote the reduction of the environmental impact resulting from industrial activity, for which it is recommended to achieve environmental sustainability through the adoption of CE. Similar findings to those found by Ref. [65] in China indicate that the Chinese re-manufacturing industry has achieved sustainable development with CE, thanks to the institutional and legal reforms of the country in recent years. On the contrary, Refs. [13,66] state that IRE becomes a public barrier to adopt the CE because the manufacturing supply chains are internationally coordinated, and environmental policies are different for each country.

## 6. Concluding Remarks and Policy Implications

CE is a commitment to sustainable development to change the linear model of the economy, causing the deterioration of the environment. However, the adoption of the CE is an arduous task that requires all economic agents' effort, especially government leadership, to apply measures that force firms to adopt cleaner production systems. The adoption of CE has taken on great importance in the current pandemic that the world is going through since it is necessary to make the most of productive resources to deplete the supply chain of the manufacturing sector. In this sense, this study has found essential and unpublished findings on this particular topic by collecting primary information from firms in the Ecuadorian manufacturing sector, which was subsequently processed with CB-SEM. The results find evidence favoring the existence of direct effects of I4.0 and IRE on the CE. However, there is no evidence favoring COVID-19 being a determinant of CE adoption in manufacturing firms in the country. Based on the results obtained from the study, the following policy implications emerge:

- The Central Government should lead the implementation of I4.0 in manufacturing firms with a legal project that is progressively achieved, especially in industries where they have a mostly marked linear economy approach.
- The Central Government must provide facilities for firms to acquire I4.0 technological equipment, such as 0% tariffs for importing this type of equipment. Similarly, the State must guarantee loans with low-interest rates so that firms can finance such technological acquisitions.
- The manufacturing supply chain must be promoted by adopting I4.0 so that the CE is not only an isolated practice of specific firms, but on the contrary, it is a green practice of all the agents that participate in the chain of supply.
- Tax incentives or subsidies must be generated to those firms in which their production has been achieved through CE and generate public-private alliances to open local and international markets to commercialize these products. In the same way, incentives must be generated in consumers to direct their choice of consumption to the goods that come from a production process through CE.
- The government must give a more leading role and increase the powers of the institutions of environmental regulation and progressively increase the penalties for firms that continuously degrade the environment, due to the obsolete production process.
- Likewise, there should be a greater presence of these institutions to improve the efficiency of regulation and learn about each industry's particularities to improve the transition to manufacturing firms with CE.
- The COVID-19 outbreak does not represent a determining factor for the adoption of CE; however, firms should have management plans to face that presents uncertainty and risk to be able to adapt the production process in these scenarios and be ready to apply the 10R of CE.

Like most studies, one of the limitations is that the level of detail of the information is limited. However, despite the study's limitations, the contribution is unprecedented and constitutes one of the significant pioneering studies on the subject in the country and the region.

**Author Contributions:** Conceptualization, S.A.R.K., P.P. and M.T.; methodology, M.S.A.-A., Z.Y. and G.T.; software, Z.Y. and P.P.; validation, S.A.R.K., G.T. and M.S.A.-A.; formal analysis, P.P. and Z.Y.; investigation, S.A.R.K. and M.T.; resources, P.P. and M.S.A.-A.; data curation, P.P. and Z.Y.; writing—original draft preparation, S.A.R.K., M.T. and P.P.; writing—review and editing, G.T., M.S.A.-A. and Z.Y.; visualization, P.P. and Z.Y.; supervision, S.A.R.K., G.T. and M.T.; project administration, S.A.R.K., M.T.; funding acquisition, S.A.R.K. and G.T. All authors have read and agreed to the published version of the manuscript.

**Funding:** This research is supported by the Beijing Key Laboratory of Urban Spatial Information Engineering (No. 20210218) and the APC funded by Prince Sultan University.

**Acknowledgments:** This research is supported by the Beijing Key Laboratory of Urban Spatial Information Engineering (No. 20210218). Furthermore, all authors of this article would like to thank the Prince Sultan University for its financial and academic support to publish this paper in "*Sustainability*".

**Conflicts of Interest:** The authors declare no conflict of interest.

## Appendix A. Questionnaire

| *Constructs and Questions* |
|---|
| **Industry 4.0** |
| Advanced technology contributes to inefficient decision-making. |
| Artificial intelligence applications help to reduce polluting activities. |
| Advanced technologies help in the processing of large amounts of data. |
| Technology allows obtaining reliable information on the production process. |
| The adoption of smart technology improves the use of resources. |
| The use of technology makes it possible to improve circular economy practices. |
| The use of technology improves the efficiency of the firm. |
| Advanced technology allows to redistribute product delivery and reduce waste. |
| Technology contributes to the green design of products. |
| Technology allows for improving the firm's response to adverse natural events. |
| Technology allows flexibility and improvement of the operational processes of the firm. |
| **Circular Economy Practices** |
| Design of Products for Reduced Consumption of Material/Energy. |
| Design of Products for Reuse, Recycle, Recovery of Material, or Components Parts. |
| Design of Products to Avoid or Reduce Use of Hazardous Products & Their Manufacturing Process. |
| Ensure suppliers meet their environmental objectives. |
| Requires suppliers to have certified EMS like ISO 14001. |
| Ensure purchased materials contain green attributes. |
| Requires suppliers to develop and maintain an EMS. |
| Decreasing toxic and hazardous chemicals in manufacturing processes. |
| Reducing fossil fuel energy consumption. |
| Using green materials in manufacturing. |
| The firm's production process prioritizes the consumption of raw materials and energy. |
| The firm's initiative improves the energy efficiency of production equipment. |
| **Institutional regulations for the environment** |
| Implementation of carbon-taxation. |
| Heavy penalties and fines due to violation of environmental policies. |
| Safety training to the employees. |
| Environmental awareness training for the labor-force. |
| Zero and/or low-interest-rate loans for the circular economy projects. |
| Tax exemption policies for green projects. |
| **COVID-19 Pandemic** |
| Lockdown creates pressure on firms to adopt eco-friendly practices. |
| During the pandemic, online delivery of products improves long-lasting sustainability. |
| Due to the COVID-19 pandemic, firms efficiently consume energy and water to increase sustainability. |
| Lockdown policies are a solution to increased environmental performance. |
| During the pandemic, firms adopting teleworking patterns in the context of sustainability. |
| *Note: This questionnaire is used to collect the data from the manufacturing firms. The questionnaire is based on the Likert Scale (1 = strongly disagree, and 5 = strongly agree)* |

## References


1. Chowdhury, P.; Paul, S.K.; Kaisar, S.; Moktadir, M.A. COVID-19 pandemic related supply chain studies: A systematic review. *Transp. Res. Part E* **2021**, *148*, 102271. [CrossRef]
2. Nandi, S.; Sarkis, J.; Hervani, A.; Helms, M.M. Redesigning supply chains using blockchain-enabled circular economy and COVID-19 experiences. *Sustain. Prod. Consum.* **2021**, *27*, 10–22. [CrossRef] [PubMed]
3. Lahcen, B.; Brusselaers, J.; Vrancken, K.; Dams, Y.; Paes, C.D.S.; Eyckmans, J.; Rousseau, S. Green Recovery policies for the COVID-19 crisis: Modelling the impact on the economy and greenhouse gas emissions. *Environ. Reso. Econ.* **2020**, *76*, 731–750. [CrossRef]
4. Jabbour, C.J.C.; Fiorini, P.D.C.; Wong, C.W.; Jugend, D.; Jabbour, A.B.L.D.S.; Seles, B.M.R.P.; Pinheiro, M.A.P.; da Silva, H.M.R. First-mover firms in the transition towards the sharing economy in metallic natural resource-intensive industries: Implications for the circular economy and emerging industry 4.0 technologies. *Res. Policy* **2020**, *66*, 101596. [CrossRef]

5. Bag, S.; Gupta, S.; Kumar, S. Industry 4.0 adoption and 10R advance manufacturing capabilities for sustainable development. *Int. J. Prod. Econ.* **2021**, *231*, 107844. [CrossRef]
6. Wuyts, W.; Marin, J.; Brusselaers, J.; Vrancken, K. Circular economy as a COVID-19 cure? *Resour. Conserv. Recycl.* **2020**, *162*, 105016. [CrossRef]
7. Levänen, J.; Lyytinen, T.; Gatica, S. Modelling the interplay between institutions and circular economy business models: A case study of battery recycling in Finland and Chile. *Ecol. Econ.* **2018**, *154*, 373–382. [CrossRef]
8. Bag, S.; Yadav, G.; Wood, L.C.; Dhamija, P.; Joshi, S. Industry 4.0 and the circular economy: Resource melioration in logistics. *Resour. Policy* **2020**, *68*, 101776. [CrossRef]
9. Neumeyer, X.; Ashton, W.S.; Dentchev, N. Addressing resource and waste management challenges imposed by COVID-19: An entrepreneurship perspective. *Resour. Conserv. Recycl.* **2020**, *162*, 105058. [CrossRef] [PubMed]
10. Tanveer, M.; Hassan, S.; Bhaumik, A. Covid-19 quarantine and consumer behavior that change the trends of business sustainability & development. *Acad. Strateg. Manag. J.* **2020**, *23*, 1–14.
11. Alonso-Almeida, M.; Rodríguez-Anton, J.; Bagur-Femenías, L.; Perramon, J. Institutional entrepreneurship enablers to promote circular economy in the European Union: Impacts on transition towards a more circular economy. *J. Clean. Prod.* **2021**, *281*, 124841. [CrossRef]
12. Agyemang-Duah, W.; Peprah, C.; Peprah, P. "Let's talk about money": How do poor older people finance their healthcare in rural Ghana? A qualitative study. *Int. J. Equity Health* **2019**, *18*, 1–12. [CrossRef] [PubMed]
13. Grafström, J.; Aasma, S. Breaking circular economy barriers. *J. Clean. Prod.* **2021**, *292*, 126002. [CrossRef]
14. Yadav, G.; Luthra, S.; Kumar, S.; Kumar, S.; Rai, D.P. A framework to overcome sustainable supply chain challenges through solution measures of industry 4.0 and circular economy: An automotive case. *J. Clean. Prod.* **2020**, *254*, 120112. [CrossRef]
15. Di Nardo, M. Developing a conceptual framework model of Industry 4.0 for industrial management. *Ind. Eng. Manag. Syst.* **2020**, *19*, 551–560. [CrossRef]
16. Kumar, R.; Kr, R.; Kr, Y. Application of industry 4.0 technologies in SMEs for ethical and sustainable operations: Analysis of challenges. *J. Clean. Prod.* **2020**, *275*, 124063. [CrossRef]
17. Boccella, A.R.; Centobelli, P.; Cerchione, R.; Murino, T.; Riedel, R. Evaluating centralized and heterarchical control of smart manufacturing systems in the era of Industry 4.0. *Appl. Sci.* **2020**, *10*, 755. [CrossRef]
18. Rajput, S.; Singh, S.P. Industry 4.0 Model for circular economy and cleaner production. *J. Clean. Prod.* **2020**, *277*, 123853. [CrossRef]
19. Ministry of the Environment. Pact for the Circular Economy. 2019. Available online: https://www.ambiente.gob.ec/ (accessed on 15 November 2021).
20. Halog, A.; Anieke, S. A review of circular economy studies in developed countries and its potential adoption in developing countries. *Circ. Econ. Sustain.* **2021**, *1*, 1–22. [CrossRef]
21. Jabbour, C.; Seuring, S.; de Sousa Jabbour, A.B.L.; Jugend, D.; Fiorini, P.; Latan, H.; Izeppi, W.C. Stakeholders, innovative business models for the circular economy and sustainable performance of firms in an emerging economy facing institutional voids. *J. Environ. Manag.* **2020**, *264*, 110416. [CrossRef]
22. Piscitelli, G.; Ferazzoli, A.; Petrillo, A.; Cioffi, R.; Parmentola, A.; Travaglioni, M. Circular economy models in the industry 4.0 era: A review of the last decade. *Proc. Manuf.* **2020**, *42*, 227–234. [CrossRef]
23. Abdul-hamid, A.; Helmi, M.; Tseng, M.; Lan, S.; Kumar, M. Impeding challenges on industry 4.0 in circular economy: Palm oil industry in Malaysia. *Comput. Oper. Res.* **2020**, *123*, 105052. [CrossRef]
24. Tanveer, M.; Bhaumik, A.; Hassan, S.; Haq, I.U. Covid-19 pandemic, outbreak educational sector and students online learning in Saudi Arabia. *J. Entrep. Educ.* **2020**, *23*, 1–14.
25. Fatorachian, H.; Kazemi, H. A critical investigation of Industry 4.0 in manufacturing: Theoretical operationalisation framework. *Prod. Plan. Control.* **2018**, *29*, 633–644. [CrossRef]
26. Fettermann, D.C.; Cavalcante, C.G.S.; de Almeida, T.D.; Tortorella, G.L. How does Industry 4.0 contribute to operations management? *J. Ind. Prod. Eng.* **2018**, *35*, 255–268. [CrossRef]
27. Sun, M.; Li, J.; Yang, C.; Schmidt, G.A.; Bambacus, M.; Cahalan, R.; Huang, Q.; Xu, C.; Noble, E.U.; Li, Z. A web-based geovisual analytical system for climate studies. *Future Internet* **2012**, *4*, 1069–1085. [CrossRef]
28. Upadhyay, A.; Mukhuty, S.; Kumar, V.; Kazancoglu, Y. Blockchain technology and the circular economy: Implications for sustainability and social responsibility. *J. Clean. Prod.* **2021**, *293*, 126130. [CrossRef]
29. Arifatul, Y.; Govindan, K.; Murniningsih, R.; Setiawan, A. Industry 4.0 based sustainable circular economy approach for smart waste management system to achieve sustainable development goals: A case study of Indonesia. *J. Clean. Prod.* **2020**, *269*, 122263. [CrossRef]
30. Bag, S.; Yadav, G.; Dhamija, P.; Kumar, K. Key resources for industry 4.0 adoption and its effect on sustainable production and circular economy: An empirical study. *J. Clean. Prod.* **2021**, *281*, 125233. [CrossRef]
31. Ibn-Mohammed, T.; Mustapha, K.B.; Godsell, J.; Adamu, Z.; Babatunde, K.A.; Akintade, D.D.; Acquaye, A.; Fujii, H.; Ndiaye, M.M.; Yamoah, F.A.; et al. A critical analysis of the impacts of COVID-19 on the global economy and ecosystems and opportunities for circular economy strategies. *Resour. Conserv. Recycl.* **2021**, *164*, 105169. [CrossRef]
32. Prause, G.; Atari, S. On sustainable production networks for Industry 4.0. *Entrep. Sustain. Issues* **2017**, *4*, 421–431. [CrossRef]
33. Laskurain-Iturbe, I.; Arana-Landín, G.; Landeta-Manzano, B.; Uriarte-Gallastegi, N. Exploring the influence of industry 4.0 technologies on the circular economy. *J. Clean. Prod.* **2021**, *321*, 128944. [CrossRef]

34. De Sousa Jabbour, A.B.L.; Jabbour, C.J.C.; Godinho Filho, M.; Roubaud, D. Industry 4.0 and the circular economy: A proposed research agenda and original roadmap for sustainable operations. *Ann. Oper. Res.* **2018**, *270*, 273–286. [CrossRef]
35. Wicker, R.J.; Kumar, G.; Khan, E.; Bhatnagar, A. Emergent green technologies for cost-effective valorization of microalgal biomass to renewable fuel products under a biorefinery scheme. *Chem. Eng. J.* **2021**, *415*, 128932. [CrossRef]
36. Hossain, M. The effect of the COVID-19 on sharing economy activities. *J. Clean. Prod.* **2021**, *280*, 12478. [CrossRef]
37. Rahman, S.M.M.; Kim, J.; Laratte, B. Disruption in circularity? Impact analysis of COVID-19 on ship recycling using Weibull tonnage estimation and scenario analysis method. *Resour. Conserv. Recycl.* **2021**, *164*, 105139. [CrossRef]
38. Carenbauer, M.G. Essential or dismissible? Exploring the challenges of waste pickers in relation to COVID-19. *Geoforum* **2021**, *120*, 79–81. [CrossRef]
39. Parashar, N.; Hait, S. Plastics in the time of COVID-19 pandemic: Protector or polluter? *Sci. Total Environ.* **2021**, *759*, 144274. [CrossRef]
40. Vanapalli, K.; Sharma, H.; Ranjan, V.; Samal, B.; Bhattacharya, J.; Dubey, B.; Goel, S. Challenges and strategies for effective plastic waste management during and post COVID-19 pandemic. *Sci. Total Environ.* **2021**, *750*, 141514. [CrossRef]
41. Bag, S.; Ham, J.; Pretorius, C.; Gupta, S.; Dwivedi, Y.K. Role of institutional pressures and resources in the adoption of big data analytics powered artificial intelligence, sustainable manufacturing practices and circular economy capabilities. *Technol. Forecast. Soc. Chang.* **2021**, *163*, 120420. [CrossRef]
42. Schulz, C.; Hjaltadóttir, R.E.; Hild, P. Practising circles: Studying institutional change and circular economy practices. *J. Clean. Prod.* **2019**, *237*, 1–10. [CrossRef]
43. Fitch-Roy, O.; Benson, D.; Monciardini, D. All around the world: Assessing optimality in comparative circular economy policy packages. *J. Clean. Prod.* **2021**, *286*, 125493. [CrossRef]
44. Hull, C.; Millette, S.; Williams, E. Challenges and opportunities in building circular-economy incubators: Stakeholder perspectives in Trinidad and Tobago. *J. Clean. Prod.* **2021**, *296*, 126412. [CrossRef]
45. Rweyendela, A.G.; Kombe, G.G. Institutional influences on circular economy: A Tanzanian perspective. *Sustain. Prod. Consum.* **2021**, *26*, 1062–1073. [CrossRef]
46. Alonso-Almeida, M.D.M.; Rodríguez-Antón, J.M.; Bagur-Femenías, L.; Perramon, J. Sustainable development and circular economy: The role of institutional promotion on circular consumption and market competitiveness from a multistakeholder engagement approach. *Bus. Strategy Environ.* **2020**, *29*, 2803–2814. [CrossRef]
47. Cui, T.; Zhang, J. Bibliometric and review of the research on circular economy through the evolution of Chinese public policy. *Scientometrics* **2018**, *116*, 1013–1037. [CrossRef]
48. Camilleri, M.A. European environment policy for the circular economy: Implications for business and industry stakeholders. *Sustain. Dev.* **2020**, *28*, 1804–1812. [CrossRef]
49. Central Bank of Ecuador. Regional Indicators. 2019. Available online: https://www.bce.fin.ec/ (accessed on 15 November 2021).
50. Yuan, X.; Liu, M.; Yuan, Q.; Fan, X.; Teng, Y.; Fu, J.; Ma, Q.; Wang, Q.; Zuo, J. Transitioning China to a circular economy through remanufacturing: A comprehensive review of the management institutions and policy system. *Resour. Conserv. Recycl.* **2020**, *161*, 104920. [CrossRef]
51. Närvänen, E.; Mattila, M.; Mesiranta, N. Institutional work in food waste reduction: Start-ups' role in moving towards a circular economy. *Ind. Mark. Manag.* **2020**, *93*, 605–616. [CrossRef]
52. Marsh, H.W.; Craven, R.G. Reciprocal effects of self-concept and performance from a multidimensional perspective: Beyond seductive pleasure and unidimensional perspectives. *Perspect. Psychol. Sci.* **2006**, *1*, 133–163. [CrossRef]
53. Marsh, H.W.; Hau, K.-T.; Wen, Z. Search of golden rules: Comment on hypothesis-testing approaches to setting cutoff values for fit indexes and dangers in overgeneralizing hu and bentler's (1999) findings. *Struct. Equ. Model. Multidiscip. J.* **2004**, *11*, 320–341. [CrossRef]
54. Cheah, J.H.; Memon, M.A.; Richard, J.E.; Ting, H.; Cham, T.H. CB-SEM latent interaction: Unconstrained and orthogonalized approaches. *Australas. Mark. J.* **2020**, *28*, 218–234. [CrossRef]
55. Prasojo, L.; Habibi, A.; Faiz, M.; Yaakob, M. Dataset relating to the relationship between teacher self-concept and teacher efficacy as the predictors of burnout: A survey in Indonesian education. *Data Brief* **2020**, *30*, 4–8. [CrossRef] [PubMed]
56. Chandra, D.; Kumar, D. Evaluating the effect of key performance indicators of vaccine supply chain on sustainable development of mission indradhanush: A structural equation modeling approach. *Omega* **2020**, *101*, 102258. [CrossRef]
57. Khan, S.A.R.; Ponce, P.; Tanveer, M.; Aguirre-Padilla, N.; Mahmood, H.; Shah, S.A.A. Technological innovation and circular economy practices: Business strategies to mitigate the effects of COVID-19. *Sustainability* **2021**, *13*, 8479. [CrossRef]
58. Kenny, D.A.; Judd, C.M. Estimating the nonlinear and interactive effects of latent variables. *Psychol. Bull.* **1984**, *96*, 201–210. [CrossRef]
59. Yasri, Y.; Susanto, P.; Hoque, M.; Gusti, M. Price perception and price appearance on repurchase intention of Gen Y: Do brand experience and brand preference mediate? *Heliyon* **2020**, *6*, e05532. [CrossRef]
60. Hattie, J. Methodology Review: Assessing unidimensionality of tests and ltenls. *Appl. Psychol. Meas.* **1985**, *9*, 139–164. [CrossRef]
61. Hair, J.F.; Black, W.C.; Babin, B.J.; Roleph, A. *Multivariate Data Analysis: A Global Perspective*, 7th ed.; Pearson: London, UK, 2010; ISBN 0135153093.
62. Byrne, B.M. *Structural Equation Modeling with AMOS: Basic Concepts, Applications, and Programming*; Lawrence Erlbaum Associates Publishers: Mahwah, NJ, USA, 2001.

63. Browne, M.W.; Cudeck, R. Alternative ways of assessing model fit. In *Testing Structural Equation Models*; Bollen, K.A., Long, J.S., Eds.; Sage: Newbury Park, CA, USA, 1993; pp. 136–162.
64. Hu, L.-T.; Bentler, P.M. Fit indices in covariance structure modeling: Sensitivity to underparameterized model misspecification. *Psychol. Methods* **1998**, *3*, 424–453. [CrossRef]
65. Djekic, I.; Nikolić, A.; Uzunović, M.; Marijke, A.; Liu, A.; Han, J.; Brnčić, M.; Knežević, N.; Papademas, P.; Lemoniati, K.; et al. COVID-19 pandemic effects on food safety—Multi-country survey study. *Food Control* **2021**, *122*, 107800. [CrossRef]
66. Chowdhury, T.; Sarkar, A.; Paul, S.K. A case study on strategies to deal with the impacts of COVID-19 pandemic in the food and beverage industry. *Oper. Manag. Res.* **2020**, 1–13. [CrossRef]

*sustainability*

MDPI

*Article*

# Development of Rice Husk Power Plants Based on Clean Development Mechanism: A Case Study in Mekong River Delta, Vietnam

**Nguyen Van Song [1], Thai Van Ha [2,*], Tran Duc Thuan [3], Nguyen Van Hanh [4], Dinh Van Tien [2], Nguyen Cong Tiep [1], Nguyen Thi Minh Phuong [5], Phan Anh Tu [6] and Tran Ba Uan [7]**

1 Faculty of Economics and Rural Development, Vietnam National University of Agriculture (VNUA), Ha Noi 10000, Vietnam; nguyensonghua@gmail.com (N.V.S.); nctiep@vnua.edu.vn (N.C.T.)
2 Faculty of Business Administration, Ha Noi University of Business and Technology (HUBT), Ha Noi 10000, Vietnam; dvtien.napa@yahoo.com
3 Faculty of Economics and Administration, Dong Nai Technology University (DNTU), Dong Nai 76000, Vietnam; tranducthuan@dntu.edu.vn
4 Institute of Energy of Viet Nam (IEVN), Ton That Tung, Dong Da, Ha Noi 10000, Vietnam; nguyenvanhanh53@gmail.com
5 Faculty of Economics, Vinh University (VU), Nghe An 43000, Vietnam; minhphuongn78@yahoo.com
6 International Business Department, School of Economics, Can Tho University, Can Tho 94000, Vietnam; patu@ctu.edu.vn
7 Faculty of Economic and Finance, Dien Bien Technical Economic College, Dien Bien 32000, Vietnam; bauandb@gmail.com
* Correspondence: thaivanha.hubt@gmail.com

**Citation:** Song, N.V.; Ha, T.V.; Thuan, T.D.; Hanh, N.V.; Tien, D.V.; Tiep, N.C.; Phuong, N.T.M.; Tu, P.A.; Uan, T.B. Development of Rice Husk Power Plants Based on Clean Development Mechanism: A Case Study in Mekong River Delta, Vietnam. *Sustainability* **2021**, *13*, 6950. https://doi.org/10.3390/su13126950

Academic Editor: Shervin Hashemi

Received: 18 May 2021
Accepted: 17 June 2021
Published: 21 June 2021



**Abstract:** In this research, we planned and conducted estimations for developing a pilot-scale Clean Development Mechanism (CDM) project for group plant activities in the Vietnam electricity/energy sector. The overall aim of this paper is to assess the power generation potential of rice husk power plants in the Mekong Delta. We intend to set up a rice husk energy balance flowchart for the whole Mekong River Delta in the year 2021 and suggest policies that can be used for the power generation of unused rice husk, to avoid having them pollute rivers and canals. We put forward a safe and environmentally friendly solution to thoroughly minimize the current serious pollution of rivers and canals in the Mekong River Delta caused by the increasing quantity of unused rice husk. The results of this paper are based on the estimation of electricity potential of a group of rice husk power development plants in the Mekong River Delta with a capacity of 11 MW per plant, including carbon dioxide emission reductions (CERs) and CER credits, along with estimations of their economic criteria (NPV, B/C, IRR), both W/CDM and W/O CDM.

**Keywords:** rice husk; power plants; $CO_2$; emission reductions; Clean Development Mechanism

## 1. Introduction

Vietnam has an impressive economic growth rate, and it has succeeded in transforming itself from a command economy to a market economy, especially in transforming and developing its agricultural sector. With a major impact on employment, GDP, and export, the importance of the agricultural sector in Vietnam is profound. Having both a continuing agricultural development in general and a rapid paddy production growth in particular is very necessary. The Mekong River Delta is an important agricultural area amongst the agricultural areas in Viet Nam.

Vietnam's renewable energy report of 2018 [1] highlighted some future plans and key points, including an electricity growth rate demand of about 9% per year, in which the demand of renewable energy demand growth rate is around 10% per year. The report pointed out that the growth rate of renewable energy supply is likely to increase the fastest at 23.2%, compared with hydropower at 2.6% and coal gas fire at 7.8%, in the period between 2020 and 2030.

The intensive paddy farming and rapid growth of rice production in the Mekong River Delta has led to dumping and discharging a large amount of rice husk from the local dense milling center networks. Currently, rice husk discharged from milling centers can be used for fueling brick kilns, porcelain furnaces, and rural household cooking (under 20% total), for open-air burning to fertilize the planted areas (not considerable; under 20% total), or it can be dumped (uncontrollable at over 70%) [2].

Most of paddy milling plants in the Mekong River Delta are located on the banks of canals and two major rivers. About 1.4–1.5 million tons of rice husk discharged from dense milling center network into rivers cause serious negative environmental impact (See Figure 1). Amongst the three aforementioned rice husk disposal modes (i.e., used as fuel source, open-air burning of rice husk for fertilizing the planted areas, uncontrollable dumping of unused rice husk into rivers), power generation using rice husk is considered to be the only mode with an environmentally friendly context.

**Figure 1.** Rice husks from rice-milling plants/factories pollute the environment in the Mekong Delta.

The study in [3] provided a comprehensive overview of three main types of renewable energy in Vietnam: solar, mini-hydro power, and biomass energy. In this present study, the use of rice husk and bagasse to fuel the bioelectricity generation plants is first considered in details on both a qualitative and quantitative basis. The preliminary data and analysis of the study on the rice husk potential in the Mekong River Delta (South Vietnam) are very useful for preparing the CDM-PDD of a 11 MW rice husk fueled biopower plant and for achieving the policy recommendations for using the rice husk potential of provinces in the Mekong River Delta [4].

The objectives of the study are to assess the CDM-based potential of rice husk power development plants and to recommend a regional strategy for developing a group of rice husk power generation plants with a 11 MW installed capacity per plant for minimizing the uncontrollable dumping of unused rice husk produced from paddy milling plants to rivers.

## 2. Literature Review

Cheewaphongphan et al. [5] studied the rice straw in Thailand and calculated its potential to serve as a fuel source. The study results of Ji and Nananukul [6] assisted in the decision making of biomass projects based on the study of supply chains for sustainable renewable energy from biomass. Weldekidan et al. [7] concluded that "gases have a high concentration of combustible products and as fuels in engines". Another fuel source is industrial wastewater and livestock manure resources, which have a potential of 7800 to 13,000 TJ/year (terajoule per year) [8]. Kinoshita et al. [9] found that biomass production is an important target of the Japanese government. In the study by Beagle and Belmont [10], they considered the power plants near beetle kill mortality to be ideal candidates for co-firing. Jasiulewicz [11] described the conditions when making a decision on replacing hard coal with local biomass. The results of the study by Luk et al. [12] showed that the overall efficiency with proper drying and heat integration is improved by about 5% when compared to a process without drying. In the study by Botelho et al. [13], they concentrated on the importance of performing an equity analysis; they also found that while the sulfur content of coal can reach 4%, the biomass sulfur content is between 0 and approximately 1%. Tokarski [14] showed that the most widespread method of producing electricity from renewable sources in power plants involves the co-firing of biomass with fossil fuels. Cereals were found to have a major contribution (about 74.67%) in the surplus biomass [15]. The results of Zhang et al. [16] showed that the proposed feedstock supply pattern is able to significantly increase the profits of biomass plants. The study by Gao et al. [17] encouraged building wind power plants in desert areas where possible. In the study by Moretti et al. [18], they compared benchmarks of biomass-fueled combined heat and power systems (CHPs) with conventional separate production technologies; they also identified the main sources of environmental impacts and assessed the potential environmental performance.

The study by Wang and Watanabe [19] on straw-based biomass power generation showed that risk changing in the biomass supply chain is one of the reasons why farmers are unwilling to supply straw. Visser et al. [20] showed the details of the cost of biomass power plants in South Africa. Yang et al. [21] pointed out that a pulverized biomass/coal co-firing power plant with carbon capture and stores (CCS) can achieve near-zero emissions. Thakur et al. [22] found that the bundled and chipped forest harvest residues used at a power plant ranges from 21.4 to 21 g $CO_2$ eq/kWh. Ferreira et al. [23] pointed out that the total potential estimated for various sectors of Portugal is 42.5 GWh/year. The economic and environmental results given by Mohamed et al. [24] showed the efficiencies of the carbon capture and stores and non-CCS plants. Singh [25] examined the cereal crops, sugarcane, and cotton contribution in the production of surplus biomass. In the study by Song et al. [26] on hydro power plants, they concluded that the electricity price would have to be increased to 5.7 US cents/kWh in order to cover the full costs of the Yali hydro power plant. In the environmental analysis of Roy et al. [27], they found an environmental benefit value of about 430,014 USD/year of using biomass power plants.

## 3. Research Methods

### 3.1. Data Collection

We determined rice husk availability based on estimating the rice husk potential of milling centers located alongside the Tien Giang River in the Mekong Delta. We also considered the capability needed to transport the rice husk that is needed by not only the considered pilot rice husk power plant but also similar ones planned at the Mekong Delta for future use, and we found that waterways are the most economical method. We also found the current local rice husk using and pricing by interviewing the relevant companies and stakeholders. In the data collection process, we asked them questions (Appendix A) in a number of areas, such as their willingness to participate in the pilot plant of the current local milling centers in the capacity of the plant developers; their willingness to sell the stored rice husk; the rice husk selling capability, and the acceptable rice husk pricing level of current rice milling centers. The steady rice husk availability and procurement for bioelectricity generation was considered for provinces of the Mekong River Delta (South Vietnam).

### 3.2. Estimation of GHG (Greenhouse Gas) Emissions by Sources

In this section, we present the estimation of GHG emissions in the project. All equations were created based on the Clean Development Mechanism and GHG emission reduction requirement.

#### 3.2.1. Project Emissions

We calculated the $CO_2$ on- and off-site transportation and the $CO_2$ from start-up/auxiliary fuel use.

| | | | | | | | | | | | | |
|---|---|---|---|---|---|---|---|---|---|---|---|---|
| (a) Biomass electricity generation | | | | | | | | | | | | |
| Annual $CH_4$ released | = | Heat value of rice husk used by power plant | × | Methane emission factor for rice husk combustion | × | Global warming potential (GWP) of $CH_4$ | | | | | | |
| ($tCO_2e$/year) | | (TJ (tera-joule)/year) | | ($tCH_4$/TJ (terajoule)) | | ($tCO_2e/tCH_4$) | | | | | | |
| (b) Transportation of biomass | | | | | | | | | | | | |
| Distance traveled | = | Total rice husk consumed by plant | ÷ | Truck capacity | × | Return trip distance to supply site | | | | | | |
| (km/year) | | (t/year) | | (t) | | (km) | | | | | | |
| Emission factor | = | $CO_2$ emission factor | ÷ | $CH_4$ emission factor | × | Global warming potential (GWP) of $CH_4$ | + | $N_2O$ emission factor | × | Global warming potential (GWP) of $N_2O$ | | |
| ($tCO_2e$/km) | | ($tCO_2$/km) | | ($tCH_4$/km) | | ($tCO_2e/tCH_4$) | | ($tN_2O$/km) | | ($tCO_2e/tN_2O$) | | |
| Annual emission | = | Emission factor | × | Distance traveled | | | | | | | | |
| ($tCO_2e$/year) | | ($tCO_2e$/km) | | (km/year) | | | | | | | | |
| (c) Start-up/auxiliary fuel use | | | | | | | | | | | | |
| • For residual oil: | | | | | | | | | | | | |
| CO2 emission factor | = | Carbon emission factor | × | Fraction of Carbon oxidized | × | Mass conversion factor | | | | | | |
| ($tCO_2$/TJ (terajoule)) | | (tC/TJ (terajoule)) | | - | | ($tCO_2$/tC) | | | | | | |
| • For $CH_4$ and $N_2O$ | | | | | | | | | | | | |
| Emission factor | = | $CO_2$ emission factor | + | $CH_4$ emission factor | × | GWP of $CH_4$ | + | $CO_2$ emission factor | + | $N_2O$ emission factor | × | GWP of $N_2O$ |
| ($tCO_2e$/TJ) | | ($tCO_2$/TJ) | | ($tCH_4$/TJ) | | ($tCO_2e$/tCH4) | | ($tCO_2$/TJ) | | ($tN_2O$/TJ) | | ($tCO_2e/tN_2O$) |
| • For fuel consumption in energy equivalent | | | | | | | | | | | | |
| Fuel consumption in energy equivalent | = | Fuel oil (FO) consumption | × | Net calorific value of FO | × | Density of FO | | | | | | |
| (TJ/year) | | (L/year) | | (TJ/$10^3$t) | | (t/L) | | | | | | |
| Annual emission | = | Emission factor | × | Fuel consumption in energy | | | | | | | | |
| ($tCO_2e$/year) | | ($tCO_2e$/TJ) | | (TJ/year) | | | | | | | | |

(d) Estimate anthropogenic emissions by sources:

$$\mathbf{E\ (ton\ CO_2/year) = \sum_j E_j\ (ton\ CO_2/year)} \quad (1)$$

where $E_j$ is the $CO_2$ emissions per year of the generation mode j, calculated as:

$$\mathbf{E_j\ (ton\ CO_2/year) = PG_j\ (MWh/year) \times EF_j\ (ton\ C/TJ) \times OF_j \times CF/TE_j\ (\%)} \quad (2)$$

where $\mathbf{PG_j}$ is the electricity generation of power plant **j**; $\mathbf{EF_j}$ is the emission capacity of the fuel-fired power plant **j**; $\mathbf{OF_j}$ is the oxidation factor; **CF** is the unit conversion factor, i.e., 44/12 ($\mathbf{C-CO_2}$) × 0.36 (**TJ−MWh**); and $\mathbf{TE_j}$ is the thermal efficiency of the electric generation mode **j**.

The weighted average emission (**E**), representing the emission intensity, is given by:

$$\textbf{(E) (ton CO}_2\textbf{/MWh) = E(ton CO}_2\textbf{/year)} / \text{(Power Generation (MWh per Year)} \textbf{ (PG) (MWh/year)} \quad (3)$$

where **E** is given by Equation (1); PG (**MWh/year**) is $\sum_j \mathbf{PG_j}$ (**MWh/year**). The emission intensity coefficient, $\mathbf{(E)_{baseline}}$, is thus obtained as:

$$\underset{\text{(ton CO}_2\text{/MWh)}}{(\text{E})_{\text{baseline}}\ (\text{ton CO}_2/\text{MWh}) = \{(\text{E})_{\text{operating margin}}} + \underset{\text{(ton CO}_2\text{/MWh)}}{(\text{E})_{\text{build margin}}}\}/2 \quad (4)$$

Finally, the baseline emissions are given by:

$$\mathbf{E_{baseline}\ (ton\ CO_2/MWh) = (E)_{baseline}\ (ton\ CO_2/MWh) \times CG\ (MWh/year)} \quad (5)$$

### 3.2.2. Estimating the Anthropogenic Emissions by GHG Sources of Baseline

| (a) Grid electricity generation | | | | | | | | | | |
|---|---|---|---|---|---|---|---|---|---|---|
| $CO_2$ emission from grid | = | Grid fuel consumption | × | Net calorific value | × | Carbon emission factor | × | Fraction of carbon oxidized | × | Mass conversion factor |
| ($tCO_2$) | | ($10^3$t) | | (TJ/$10^3$t) | | (tC/TJ) | | – | | ($tCO_2$/tC) |
| $CO_2$ emission factor | = | Sum of all $CO_2$ emission from grid | ÷ | Grid electricity generated | | | | | | |
| ($tCO_2$/MWh) | | ($tCO_2$) | | (MWh) | | | | | | |
| $CO_2$ emission displaced by plant | = | Electricity exported by plant | × | $CO_2$ emission factor | | | | | | |
| ($tCO_2$/year) | | (MWh/year) | | ($tCO_2$/MWh) | | | | | | |
| (b) Open-air burning for biomass disposal | | | | | | | | | | |
| Carbon released | = | Rice husk used as fuel by the biopower plant | × | Carbon fraction of biomass | | | | | | |
| (tC/year) | | (t biomass/year) | | (tC/t biomass) | | | | | | |
| Annual CH4 released | = | Carbon released in total | × | Carbon released as $CH_4$ in open-air | × | Mass conversion factor | × | GWP of $CH_4$ | | |
| ($tCO_2$e/year) | | (tC/year) | | (%) | | ($tCH_4$/tC) | | ($tCO_2$e/$tCH_4$) | | |
| (c) Baseline emissions summary | | | | | | | | | | |
| $CO_2$ emission from grid electricity | + | $CH_4$ emission from open-air burning of rice husk | = | Total baseline emissions | | | | | | |
| ($tCO_2$/year) | | ($tCO_2$e/year) | | ($tCO_2$e/year) | | | | | | |

### 3.2.3. Representing the Emission Reductions of Plant Activity

| Emission reduction | = | Emission from grid electricity generation | + | Emission from open-air burning for rice husk disposal | − | Emission from biomass-fueled electricity generation | − | Emission from transportation of rice husk for the plant | − | Emission from fuel oil used for the plant (start-up) |
|---|---|---|---|---|---|---|---|---|---|---|

3.2.4. Emission Reductions

| Total baseline emissions | – | Total plant emissions | = | Emission reductions |
|---|---|---|---|---|
| ($tCO_2$/year) | | ($tCO_2e$/year) | | ($tCO_2e$/year) |

### 3.3. *Benefit–Cost Analysis*

3.3.1. Total Cost

The total cost is calculated as follows:

$$Ct = Ct\ inv. + Ct\ O\&M + Ct\ fuel\ (RH)$$

where Ct inv. is the investment cost; Ct O&M is the operation and maintenance cost; and Ct fuel (RH) is the fuel rice husk cost (including rice husk transport and storage costs).

3.3.2. Total Benefit

The total benefit is calculated as follows:

$$Bt = Bte + BtCER + Bash$$

where Bte is the benefit given by rice husk electricity sale = peWt; BtCER is the benefit given by CER sale = $pCO_2CER$; Bt ask is the benefit given by rice husk ash sale = pashWt; Pe = rice husk electricity sale price; $pCO_2$ is the CER sale price; pash is the rice husk ash sale price; and Wt is the rice husk electricity sale to the Vietnam Electricity (EVN) grid in year t.

## 4. Results and Discussion

### 4.1. *Assessment of the $CO_2$ Emission Reductions (CERs) and CER Credits Determined by Different Assumed $CO_2$ Prices*

Assessment of the $CO_2$ (CERs) and CER credits determined by different assumed $CO_2$ prices was realized for a group of five similar pilot grid-connected rice husk power development plants 5 × 11 MW installed capacity. As presented in Section 3, these five identified and recommended power plants are similar in relation to their size and employed technology. Although they are originally presented as a single CDM plant, this single plant in actuality comprises five similar rice husk power plants with the installed capacity of 11 MW per plant. The assessment of their CERs and CER credits is carried out only for an individual rice husk power plant, and then its assessed CER and CER credit is multiplied by 5 to make the CER and CER credit account for the whole CDM power plant.

### 4.2. *IRR, NPV, BCR Power Plant of the Rice Husk Power Plants with and without CDM*

4.2.1. Calculation and Comparison of IRR, NPV and B/C—With and without CDM

The unit investment costs of the proposed rice husk power plant are 1350, 1570, and 1700 USD/KW. The electricity sale prices of the proposed rice husk power plant are 0.04, 0.05, 0.06, and 0.07 USD/KWh. The $CO_2$ sale prices of the proposed rice husk power plant are namely: (W/O CDM), 3 (W CDM), 9 (W CDM), and 15 (W CDM) USD/ton of $CO_2e$. The rice husk ash price of proposed rice husk power plant, which is assumed to be at a constant pricing level, is 0.02 USD/t of ash. The calculations are given in Table 1.

**Table 1.** Benefit–cost analysis of the rice husk-fueled biopower plants with and without CDM.

| Unit Investment Cost (USD/KW) | Electricity Sale Price (USD/KWh) | IRR (%) By $CO_2$ Prices (USD/t$CO_2$) of: | | | | NPV (1000 USD) By $CO_2$ Prices (USD/t$CO_2$) of: | | | |
|---|---|---|---|---|---|---|---|---|---|
| | | 0 (w/o CDM) | 3 (w/CDM) | 9 (w/CDM) | 15 (w/CDM) | 0 (w/o CDM) | 3 (w/CDM) | 9 (w/CDM) | 15 (w/CDM) |
| 1350 | 0.040 | <12 (8.99) | <12 (10) | <12 (10.63) | <12 (11.67) | −874.23 | −395.82 | 561.00 | 1517.81 |
| | 0.045 | <12 | <12 | =12 | >12 | 716.82 | - | 2152.05 | - |
| | 0.050 | <12 (8.47) | <12 (9.04) | >12 (13.95) | >12 (14.88) | −130.14 | −826.73 | 3743.10 | 4699.24 |
| 1570 | 0.040 | <12 (6.52) | <12 (7.06) | <12 (8.09) | <12 (9.08) | −3318.66 | −2840.25 | −1883.43 | −926.61 |
| | 0.045 | <12 | <12 | <12 | <12 (10.64) | - | - | - | - |
| | 0.050 | <12 (6.53) | <12 (10.33) | <12 (11.24) | >12 (12.11) | −3312.03 | 341.85 | 1298.67 | 2255.49 |

#### 4.2.2. Calculation and Comparing of IRR, NPV and B/C ratios—With and without CDM

We made calculations using the maximal running number of days (332 days/year) as given above, and the average running day number (200 days/year), which is the realistic case, based on realistic input parameters (Table 1).

We took into consideration the current serious pollution of Mekong Delta's rivers and canals caused by unused rice husk, as river pollution is a threat to the health of local communities and their livelihood, especially their traditional aquaculture and pisciculture. This region-wide environmental threat is expected to rapidly increase with the following contexts:

- Mekong River Delta, which leads to a rapid increase in the local rice husk generation;
- Basic change in rice husk end-uses of local communities from using rice husk fuel to using the commercial energy types, leading to the rapid reduction of the local rice husk consumption and the increase of the local unused rice husk dumping;
- Lack of region-wide cooperation in looking for an environmentally friendly and effective solution to thoroughly minimize the pollution of the Mekong Delta's rivers and canals with rice husk pollution by paddy milling centers.

From 2004, the search for a thorough solution to eliminate the increasing pollution of rivers in the Mekong River Delta became an urgent task faced by local authorities, administrators, agriculture, and energy development planners. Safe and environmentally friendly disposal of 3.7 million tons of rice husk per year with over 70% of that (2.5 million tons per year) to be dumped is one of the major problems of the Mekong Delta's sustainable development.

In this context, the development of a group of 5 rice husk power plants with an installed capacity of 5 × 11 MW was selected as the most thorough and sustainable solution to solve this problem.

## 5. Conclusions and Recommendations

In this study, we investigated the potential of rice husk power plants using secondary data and direct survey data in the study area and applying methods of project analysis along with emission reduction estimations based on the Clean Development Mechanism. The prices of electricity generated by rice husk power plants and sold to Vietnam Electricity (EVN) through national power grids should be considered by the government and EVN with the concession of electricity pricing to small renewable (rice husk) power producers so that EVN could agree to purchase rice husk electricity with the electricity pricing level from 0.045 to 0.050 USD per KWh.

During a plant's projected lifespan of 20 years (2020–2040), the average annual CER of a proposed rice husk power plant is calculated to be 26,700 tons of $CO_{2e}$ with a time of use (TOU) of 4800 h/year (or 200 operating days per year), and its average annual CER credits by $CO_2$ prices of 9 and 15 USD per ton of $CO_{2e}$ is expected to be from 240 to 400 thousand USD, respectively. For the whole group of five similar rice husk power plants with a 5 × 11 MW installed capacity, these figures are 5 × 26,700 = 133,500 tons of $CO_{2e}$ per year, and 1200–2000 thousand USD per year, respectively.

Initial construction and installation costs are still high compared to other types of electricity power sources. Currently, the cost of rice husk is almost zero, and the only costs involved are shipping costs. In the future, if rice husk power plants are developed in the area, rice husk prices are likely to increase, and so further studies are needed to ensure the sustainable development of these rice husk power plants.

The research recommends developing in the Mekong River Delta a group of five similar pilot rice husk power plants having a total installed capacity of $5 \times 11$ MW at five locations, namely An Hoa (An Giang province), Thoi Hoa (limitrophe area of three provinces: An Giang, Dong Thap, and Can Tho), Thoi Lai (Can Tho province), Cai Lay (Tien Giang province), and Tan An (Long An province) (Figure 2). Besides these five locations, a reserved location in Tan Chau (limitrophe area of two provinces such as An Giang and Dong Thap, and the Kingdom of Cambodia) was selected for the future development of a paddy milling center network as well as rice husk power centers.

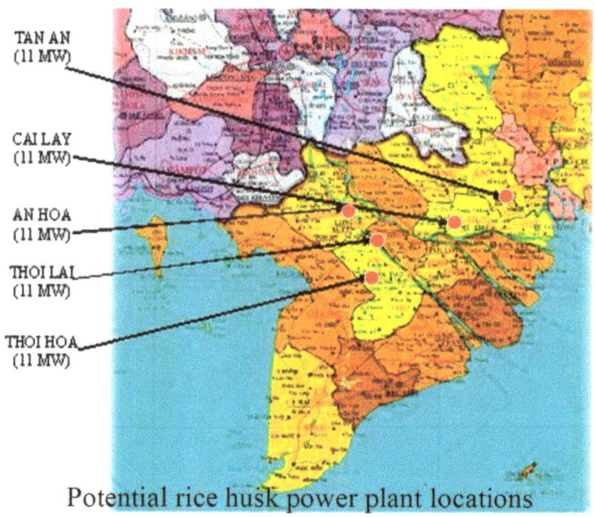


**Figure 2.** Locations of potential rice husk power plants in Vietnam.

**Author Contributions:** Conceptualization, original draft writing—review and editing, N.V.S., N.V.H.; N.V.S., and N.V.S.; data curation, T.D.T., N.V.H., N.T.M.P., P.A.T., and T.B.U.: formal analysis, N.V.H., N.V.S., N.C.T., and P.A.T.; investigation: D.V.T., N.T.M.P., P.A.T., and T.B.U.; methodology: N.V.S., T.V.H., and N.V.H.; project administration: T.V.H. and D.V.T.; resources: N.V.S.; software: T.D.T., N.T.M.P., and P.A.T.; supervision, N.V.S. and N.V.H.; validation, N.C.T.; visualization, T.V.H. and D.V.T. All authors have read and agreed to the published version of the manuscript.

**Funding:** This research received no external funding.

**Institutional Review Board Statement:** Not applicable.

**Informed Consent Statement:** Not applicable.

**Data Availability Statement:** Data are available as MDPI Research Data Policies.

**Conflicts of Interest:** All authors declare that there are no conflicts of interest.

## Appendix A

Annex of the paper: **Questionnaires**

Four kinds of questionnaires needed for this study:

**A1. Questionnaire Implemented by the Project Team for Field Trips in Typical Provinces of Mekong Delta River (An Giang and Can Tho Provinces)**

1. Milling centers; dense system of canal and fluvial transport in these provinces; high transportability of these uncontrollable and free fluvial trans port system, where the local small-scale milling centers can discharge their rice husk into the water, and cause the pollutions to harm the aquaculture livelihood (cage fishing) and the health (for local population)
2. Questions about estimating the rice husk generation of local milling centers that could be discharged into the canal systems.
3. Questions about estimating the willingness to pay for pricing the rice husk for rice husk fueled thermal power plants in the future, etc.

**A2. Questionnaire Implemented as the Following:**

1. People Committee of Provinces in Mekong River Delta (through Provincial Industrial Service):

- What is the scale of milling centers? The average scale of a milling center in the average milling center(at average level: milling 5–10 tons/day of rice);
- What is the quantity of rice husk to be generated per month and per milling center?
- Willingness to answer: How is transportability of rice husk generated and discharged by the milling center: free and directly into the local canal system, paying for transport the rice husk from silos to the boats.
- Willingness to pay for the rice husk to be used for fueling the rice husk thermal power plants in the future? Answer: If in the future, there will be such a thermal power plant, their rice husk fuel could be sold with the price of 200 VND per kg rice husk-fuel.

**A3. Questionnaire Implemented for an Average Milling Center (15 Tons Per Day) to Be Surveyed as a Typical Milling Owner in the Province (Young, Dynamic, Relatively Rich, etc.):**

- What is the scale of milling centers? the mill owner self-proclaimed that he is rich, he has private boats, logistics, etc.
- What is the quantity of rice husk to be generated per month/per milling center? He answer: about 500 tons of rice per month.
- Willingness to answer: How is the transportability of rice husk generated and discharged by the milling center: free and directly into the local canal system, paying for transport the rice husk? He answers: Directly and freely discharge into the local canal system, is unique way for dumping the rice husk of his milling center; today he has pay for transport the rice husk from silos to the boats. The use of rice husk for fueling the thermal power is welcomed, but he is not possible to invest in such a plant. The investment of this rice husk = fueled thermal power plant is quite high, about 2000 USD per KW installed capacity.

**A4. General Questionnaire**

How is the rice husk energy balance of Vietnam?

Through the field trip in An Giang, it is possible to identify the different criteria and standards of economics and energy relating to rice husk. The general questionnaire that would be installed will be that: How is the Rice Husk Energy Balance of Vietnam:

Based on this balance, it is possible to identify the uniqueness of Mekong River Delta in capacity of a region of rice husk fueled thermal power plant in Vietnam with 5 promising rice husk thermal power plants.

## References

1. Vietnam Investment Review. *Vietnam Renewable Energy Report 2018*; Institute of Energy of Viet Nam: Hanoi, Vietnam, 2018.
2. Vietnam General Statistics Office. *Vietnam Statistical Yearbook 2019*; Statistical Publishing House Hanoi: Hanoi, Vietnam, 2020.
3. Toan, P.K.; Hanh, N.V.; Cuong, N.D. *Quantitative Feasibility Study for Using the Solar Energy, Mini-Hydropower and Biomass Energy in Vietnam*; Institute of Energy of Viet Nam: Hanoi, Vietnam, 2005.
4. Institute of Energy-EVN-MOI. *The Vietnam Power Development Master Plan. for the Period. of 2005–2015 with Perspective up to 2025*; Institute of Energy of Viet Nam: Hanoi, Vietnam, 2005.

5. Cheewaphongphan, P.; Junpen, A.; Kamnoet, O.; Garivait, S. Study on the potential of rice straws as a supplementary fuel in very small power plants in Thailand. *Energies* **2018**, *11*, 270. [CrossRef]
6. Ji, J.; Nananukul, N. Supply chain for sustainable renewable energy from biomass. *Int. J. Logist. Syst. Manag.* **2019**, *33*, 568–590. [CrossRef]
7. Weldekidan, H.; Strezov, V.; Li, R.; Kan, T.; Town, G.; Kumar, R.; He, J.; Flamant, G. Distribution of solar pyrolysis products and product gas composition produced from agricultural residues and animal wastes at different operating parameters. *Renew. Energy* **2020**, *151*, 1102–1109. [CrossRef]
8. Prasertsan, S.; Sajjakulnukit, B. Biomass and biogas energy in Thailand: Potential, opportunity and barriers. *Renew. Energy* **2006**, *31*, 599–610. [CrossRef]
9. Kinoshita, T.; Ohki, T.; Yamagata, Y. Woody biomass supply potential for thermal power plants in Japan. *Appl. Energy* **2010**, *87*, 2923–2927. [CrossRef]
10. Beagle, E.; Belmont, E. Technoeconomic assessment of beetle kill biomass co-firing in existing coal fired power plants in the Western United States. *Energy Policy* **2016**, *97*, 429–438. [CrossRef]
11. Jasiulewicz, M. The possibilities of meeting energy demands in system thermal power plants by using local solid biomass. *Roczniki* **2019**, *2019*, 1230-2020-745. [CrossRef]
12. Luk, H.T.; Lam, T.Y.G.; Oyedun, A.O.; Gebreegziabher, T.; Hui, C.W. Drying of biomass for power generation: A case study on power generation from empty fruit bunch. *Energy* **2013**, *63*, 205–215. [CrossRef]
13. Botelho, A.; Lourenço-Gomes, L.; Pinto, L.; Sousa, S.; Valente, M. Using stated preference methods to assess environmental impacts of forest biomass power plants in Portugal. *Environ. Dev. Sustain.* **2016**, *18*, 1323–1337. [CrossRef]
14. Tokarski, S.; Głód, K.; Ściążko, M.; Zuwała, J. Comparative assessment of the energy effects of biomass combustion and co-firing in selected technologies. *Energy* **2015**, *92*, 24–32. [CrossRef]
15. Singh, J. Overview of electric power potential of surplus agricultural biomass from economic, social, environmental and technical perspective—A case study of Punjab. *Renew. Sustain. Energy Rev.* **2015**, *42*, 286–297. [CrossRef]
16. Zhang, X.; Luo, K.; Tan, Q. A feedstock supply model integrating the official organization for China's biomass generation plants. *Energy Policy* **2016**, *97*, 276–290. [CrossRef]
17. Gao, C.-K.; Na, H.-M.; Song, K.-H.; Dyer, N.; Tian, F.; Xu, Q.-J.; Xing, Y.-H. Environmental impact analysis of power generation from biomass and wind farms in different locations. *Renew. Sustain. Energy Rev.* **2019**, *102*, 307–317. [CrossRef]
18. Moretti, C.; Corona, B.; Rühlin, V.; Götz, T.; Junginger, M.; Brunner, T.; Obernberger, I.; Shen, L. Combining biomass gasification and solid oxid fuel cell for heat and power generation: An early-stage life cycle assessment. *Energies* **2020**, *13*, 2773. [CrossRef]
19. Wang, L.; Watanabe, T. The development of straw-based biomass power generation in rural area in Northeast China—An institutional analysis grounded in a risk management perspective. *Sustainability* **2020**, *12*, 1973. [CrossRef]
20. Visser, H.; Thopil, G.A.; Brent, A. Life cycle cost profitability of biomass power plants in South Africa within the international context. *Renew. Energy* **2019**, *139*, 9–21. [CrossRef]
21. Yang, B.; Wei, Y.-M.; Hou, Y.; Li, H.; Wang, P. Life cycle environmental impact assessment of fuel mix-based biomass co-firing plants with $CO_2$ capture and storage. *Appl. Energy* **2019**, *252*, 113483. [CrossRef]
22. Thakur, A.; Canter, C.E.; Kumar, A. Life-cycle energy and emission analysis of power generation from forest biomass. *Appl. Energy* **2014**, *128*, 246–253. [CrossRef]
23. Ferreira, S.; Monteiro, E.; Brito, P.; Vilarinho, C. Biomass resources in Portugal: Current status and prospects. *Renew. Sustain. Energy Rev.* **2017**, *78*, 1221–1235. [CrossRef]
24. Mohamed, U.; Zhao, Y.; Huang, Y.; Cui, Y.; Shi, L.; Li, C.; Pourkashanian, M.; Wei, G.; Yi, Q.; Nimmo, W. Sustainability evaluation of biomass direct gasification using chemical looping technology for power generation with and w/o $CO_2$ capture. *Energy* **2020**, *205*, 117904. [CrossRef]
25. Singh, J. A roadmap for production of sustainable, consistent and reliable electric power from agricultural biomass—An Indian perspective. *Energy Policy* **2016**, *92*, 246–254. [CrossRef]
26. Van Song, N.; Huyen, V.N.; Van Hanh, N.; Diep, N.X.; Huu, N.X.; Lan, P.T.; Cuong, H.N.; Trang, T.T.; Phuong, N.T. Environmental and External Costs of Yali Hydropower Plant and Policy Recommendations in Vietnam. *J. Environ. Prot.* **2020**, *11*, 344–358. [CrossRef]
27. Roy, D.; Samanta, S.; Ghosh, S. Performance assessment of a biomass fuelled advanced hybrid power generation system. *Renew. Energy* **2020**, *162*, 639–661. [CrossRef]

sustainability

Review

# The Recent Progress of Natural Sources and Manufacturing Process of Biodiesel: A Review

**Eko Supriyanto [1], Jayan Sentanuhady [2,*], Ariyana Dwiputra [1], Ari Permana [2] and Muhammad Akhsin Muflikhun [2,*]**

1 PLN Research Institute, Jakarta 12760, Indonesia; email4ecko@gmail.com (E.S.); ariyana@pln.co.id (A.D.)
2 Mechanical and Industrial Engineering Department, Gadjah Mada University, Yogyakarta 12760, Indonesia; aripermana@mail.ugm.ac.id
* Correspondence: jayan@ugm.ac.id (J.S.); akhsin.muflikhun@ugm.ac.id (M.A.M.)

**Abstract:** Biodiesel has caught the attention of many researchers because it has great potential to be a sustainable fossil fuel substitute. Biodiesel has a non-toxic and renewable nature and has been proven to emit less environmentally harmful emissions such as hydrocarbons (HC), and carbon monoxide (CO) as smoke particles during combustion. Problems related to global warming caused by greenhouse gas (GHG) emissions could also be solved by utilizing biodiesel as a daily energy source. However, the expensive cost of biodiesel production, mainly because of the cost of natural feedstock, hinders the potential of biodiesel commercialization. The selection of natural sources of biodiesel should be made with observations from economic, agricultural, and technical perspectives to obtain one feasible biodiesel with superior characteristics. This review paper presents a detailed overview of various natural sources, their physicochemical properties, the performance, emission, and combustion characteristics of biodiesel when used in a diesel engine. The recent progress in studies about natural feedstocks and manufacturing methods used in biodiesel production were evaluated in detail. Finally, the findings of the present work reveal that transesterification is currently the most superior and commonly used biodiesel production method compared to other methods available.

**Keywords:** biodiesel; engine performance; emissions; natural feedstocks; production method

**Citation:** Supriyanto, E.; Sentanuhady, J.; Dwiputra, A.; Permana, A.; Muflikhun, M.A. The Recent Progress of Natural Sources and Manufacturing Process of Biodiesel: A Review. *Sustainability* **2021**, *13*, 5599. https://doi.org/10.3390/su13105599

Academic Editor: Elio Dinuccio

Received: 29 March 2021
Accepted: 11 May 2021
Published: 17 May 2021

**Publisher's Note:** MDPI stays neutral with regard to jurisdictional claims in published maps and institutional affiliations.



## 1. Introduction

Nowadays, as we live in the modern era where various aspects of human activity have been automated and run by fueled machines, the need for energy will continue to increase over time. According to the International Energy Outlook 2019, it has been predicted that world energy consumption will rise nearly 50% from 2018 to 2050 [1–5]. As of today, global energy consumption highly depends on fossil resources such as crude oil, natural gas, and coal, which make up about 80% of the total consumption. Oil is the most widely used type of fossil fuel and the demand will continue to increase and is projected to reach 109.1 million barrels per day by 2045 [6,7].

The problems caused by the use of fossil fuels are their finite resources and also the environmental pollution resulted from their combustion. Fossil resources have a very slow regeneration rate. Based on current daily fossil-usage data, it is only a matter of time until the fossil resources are completely depleted on earth [7]. Combined with global warming issues partly caused by greenhouse gases (GHG) emissions from fossil-fuel combustion, renewable energy development becomes a key and challenge for the world society to reduce fossil-fuel usage [8]. Bioenergy and biofuels could solve this problem due to their renewable nature. Carbon dioxide in the atmosphere absorbed by plants will be released back during biofuel combustion, which means there is no extra carbon emission accumulated in the atmosphere from the use of biofuels [9]. Biodiesel is one of the widely used biofuels due to its considerably similar properties to fossil diesel [10–12].

Biodiesel is defined as a non-toxic, biodegradable, and renewable fuel made of vegetable oil, animal fat, and waste cooking oil that can be produced by different techniques [13–17].

Biodiesel is considered the most carbon neutral fuel with the ability to reduce carbon dioxide emissions by 78% compared to diesel fuel. Moreover, the biodegradability of biodiesel has been proven to be really high, ranging from 80.4% to 91.2% after 30 days while fossil diesel only has 24.5% biodegradability [9,18]. Another advantage using biodiesel is that its emission content is considered less harmful to the environment. Although it widely varies according to study parameters in engine types and operating conditions, the dominant trend from some studies shows that combustion of biodiesel emits less content of CO, Particulate Matters (PM), Hydrocarbons (HC), and produces almost no sulfate emissions. However, an increase in NOx emission is usually found from the usage of biodiesel [12,19,20].

The use of vegetable-based biodiesel (sunflower, soybean, grapeseed, corn, rice bran, olive) in the semi-industrial boiler has been studied by Bazooyar et al. [21]. The combustion performance, as well as emission characteristics of biodiesel is comparable to diesel fuel, especially for the lower air fuel ratio (AFR) and higher energy rates. Biodiesel application in boiler results in an increment of combustion efficiency and significantly reduces environmentally harmful emissions except for NOx. Another comparative study of sunflower and soybean-based biodiesel in an experimental boiler showed different result patterns where the combustion of biodiesel was found to be more efficient in lower energy rates [22]. Combustion performance increased along with the content of vegetable oil in the fuel, but their high viscosity would be a problem when used in high percentage blends. Therefore, it is recommended to blend vegetable oil up to 40% composition with diesel to prevent problems caused by their high viscosity as well as the need to modify the engine [23].

Palm oil biodiesel with B100 composition was tested on a 14-hp Kubota RT140 DI diesel engine for 800 h of engine operation under high load and low-speed conditions, ran 12 h a day to aerate a fish pond [24]. The experiment was conducted to determine the long-term mechanical durability and reliability of the usage of pure biodiesel on a small agricultural diesel engine. Based on the ferrographic analysis and visual inspection, the engine was found to develop the usual rate of wear. Therefore, it can be concluded that biodiesel B100 can be used as an alternate fuel for a small diesel engine without any serious mechanical durability problems. In an experimental study conducted by Shahir et al. [25], a turbocharged CRDI engine was operated using several blends of animal fat-based biodiesel (B10, B20, B30, B40, and B50) at a constant speed of 2800 rpm. B30 animal fat biodiesel was found to have optimum performance and emission characteristics, even better than diesel. Higher composition of biodiesel resulted in higher BSFC and lower thermal efficiency due to lower calorific value and higher viscosity. It also increased the emissions of $CO_2$ and NOx due to higher oxygen content in biodiesel.

Economically speaking, some types of biodiesel are found to be feasible and suitable for commercial-scale production [26]. Bazooyar et al. [27] conducted a study in order to compare the annual cost needed to utilize a boiler power plant fueled with vegetable oil, biodiesel, diesel fuel, and their blends (B5 and B20). They simulated and calculated both internal costs associated with fuel prices and external costs associated with gaseous emissions needed to run a boiler power plant within a yearlong period. The results showed that vegetable oil and pure biodiesel were not economically feasible compared to conventional diesel fuel, but the blends B5 and B20 indicated the opposite results. Although the fuel prices of B5 and B20 are higher in the market, their external costs in the boiler are much lower. Thus, replacing diesel fuel with B5 and B20 could reduce the total costs of the power plant up to USD 1452 and USD 1878, respectively, in a year.

However, to be widely used, biodiesel must be able to be produced at a lower cost. The biodiesel industry currently depends strongly on the cost of feedstock as it accounts for most of the biodiesel production cost. Some feedstock types such as non-edible oil plants and waste cooking oil can provide cheaper cost, but these low-cost feedstock types are used to have a higher amount of impurities. Additional treatments are needed in the manufacture to produce standard quality biodiesel with low-cost feedstock to increase manufacturing cost such as by using recent technology in machine learning and computational analysis [28–30].

## 2. Methods of Screening Paper

We searched for literature in the Google Scholar and ScienceDirect databases from December 2020 to January 2021 using the following criteria and boundaries: (1) available as open access literature and free to download, and (2) related to ("biodiesel characteristics" OR "biodiesel emission" OR "biodiesel properties" OR "biodiesel performance") AND ("biodiesel production" OR "biodiesel method" OR "production cost"). Filters for the access type have been set to "open access", while for the article type have been set to "review articles" and "research articles". Subsequently, the literature obtained from the original search was screened manually by reading the abstract. To include as much relevant literature as possible, the literature was further expanded by reading the references of the articles encountered when reviewing other studies.

## 3. Biodiesel Sources

In this part, various natural sources that have potential as biodiesel feedstock are discussed, especially related to previous studies about their engine performances and emission characteristics compared to diesel. The characteristic data were obtained by carrying out an engine test using biodiesel and blends as fuel. Although the procedure of the test varies, they are usually with the general scheme of engine test as shown in Figure 1.

The characteristic performance of biodiesel based on the material that used to produce biodiesel (natural sources) can be seen in Table 1. It is shown that performance and emission characteristics of biodiesel-fueled engine is influenced by various input parameters such as biodiesel blend composition, fuel injection pressure, and injection timing. Each parameter affects the characteristics differently, so it is important to investigate the best parameter for biodiesel with different natural feedstocks [31]. The effect of injection parameter on thermal performance and emission characteristics of an oil burner fueled with B20 palm oil has been studied by Abu-Hamdeh et al. [32]. An increase in injection pressure gives several advantages as it has been proven to be able to enhance the mixing rate and complete combustion and significantly reduce CO and soot particle emissions. However, NOx emission was found to increase with injection pressure. The comprehensive emission characteristic of the various sources of biodiesel can be seen in Table 2.

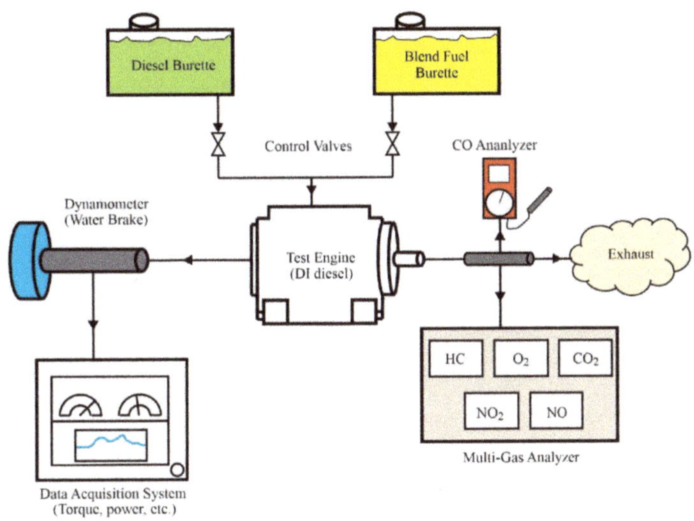


**Figure 1.** The basic scheme for engine performance and emission test [33].

**Table 1.** Performance characteristics of biodiesel from various natural sources.

| Fuel Type | Blend Composition | Engine Type | Test Condition | BSFC | BTE | Exhaust Gas Temperature | Ref |
|---|---|---|---|---|---|---|---|
| Palm | B5 | Mitsubishi 4D68 SOHC DI 4-stroke, 4-cylinder engine with EGR | variable speeds from 1000 to 3000 rpm | ↑ 3.80% (1000 rpm)<br>↑ 4.22% (3000 rpm) | | | [34] |
| | | single cylinder air | variable loads at a | ↑ 4.2% B20 | ↓ 2.61% B20 | | |

**Table 1.** *Cont.*

| Fuel Type | Blend Composition | Engine Type | Test Condition | BSFC | BTE | Exhaust Gas Temperature | Ref |
|---|---|---|---|---|---|---|---|
| Hemp | B10, B20, B30, B50, B100 | Kirloskar TV1, single cylinder, 4-stroke, DI engine | variable loads at a constant speed of 1500 rpm | ↓ 2.56% B10 & B20, ↑ 2.56% B30, ↑ 5.13% B50 & B100 | ↓ 0.35% B10, ↓ 1.06% B20, ↓ 10.28% B30, ↓ 14.54% B50, ↓ 16.31% B100 | ↑ (B30 > B20 > B50 > B10 > B100) | [51] |
| Camelina | B20, B100 | 4-cylinder, 4-stroke, DI diesel engine | variable speeds between 1200 and 2600 rpm (interval 200 rpm) at full load | ↑ 0.86% B20, ↑ 3.84% B100 | | | [52] |
| Grease | B20, B100 | 4-cylinder, 4-stroke, DI Fiat diesel engine | variable loads at a constant speed of 1500 rpm | ↑ significantly (B100 > B20) | ↑ slightly (B100 > B20) | | [53] |
| Waste cooking oil | B20 | single cylinder, 4-stroke, DI Kirloskar diesel engine | variable loads at a constant speed of 1500 rpm | ↑ 2.96% (at full load) | ↑ 10.79% (at full load) | | [54] |
| Beef (beef bone marrow) | B50, B100 | 4-stroke, 4-cylinder, direct-injection diesel generator engine | variable loads (3.6, 7.2, and 10.8 kW) | ↑ (B100 > B50) at all loads | ↓ (B100 > B50) at 3.6 and 10.8 kW ↓ (B50 > B100) at 7.2 kW | ↑ (B100 > B50) at all loads | [55] |
| Sheep (sheep fat) | B25, B50, B75, B100 | Kirloskar TV1, single cylinder, WC diesel engine | variable loads at a constant speed of 1500 rpm | | ↓ from 1.8% to 2.3% (B100 < B75 < B50 < B25) | | [56] |
| Pork (pork lard) | B100 | single cylinder, WC, DI diesel engine | variable loads at a constant speed of 1500 rpm | | ↓ | | [57] |
| Algae (*Scenedesmus obliquus*) | B50 | single cylinder, 4-stroke, WC diesel engine (CR 20:1) | variable loads at a constant speed of 1400 rpm | ↑ 14.55% | ↓ 4.75% | ↓ 5.21% | [58] |
| Algae (*Spirulina platensis*) | B10, B20, B100 | 1-cylinder, 4-stroke, compression ignition engine | variable loads at a constant speed of 1500 rpm | | ↓ 1.48% B10, ↓ 4.49% B20, ↓ 2.50% B100 (full load) | | [59] |
| Fish | B20, B40, B60, B80, B100 | single cylinder, 4-stroke, air-cooled diesel engine | variable loads and speeds | | ↓ up to 12.68% (B100 < B80 < B60 < B40 < B20) | | [60] |
| Rapeseed oil | B10, B20, B30, B50 | Four cylinder, 4-stroke, diesel engine | 2000 rpm | ↑ 30% (2000 rpm) compared to diesel fuel | | | [61] |

Note: ↑ More than, Increase. ↓ Less than, Decrease.

**Table 2.** Emission characteristics of biodiesel from various natural sources.

| Fuel Type | Blend Composition | Engine Type | Test Condition | HC | CO | $CO_2$ | NOx | Smoke Opacity | Ref |
|---|---|---|---|---|---|---|---|---|---|
| Palm | B5 | Mitsubishi 4D68 SOHC DI, 4-stroke, 4-cylinder engine with EGR | variable speeds from 1000 to 3000 rpm | ↓ insignificantly | ↓ (↑ at 1500 rpm) | ↓ insignificantly (↓ significantly at 3000 rpm) | ↑ significantly | | [34] |
| Soy | B20, B40, B100 | single cylinder, air cooled, vertical, DI diesel engine | variable loads at a constant speed of 1500 rpm | ↓ 15% B20, ↓ 27%, B40, ↓ 38.4%, B100 | ↓ 11.36 B20, ↓ 29% B40, ↓ 41.7% B100 | | ↑ 7.5%, B20 (B100 > B40 > B20) | ↓ 20.5% B20, ↓ 33.41% B40, ↓ 48.23% B100 | [35] |
| Corn | B10, B20, B30 | single cylinder, 4-stroke, WC diesel engine | variable loads with maximum speed of 1500 rpm | ↓ 7.69% B10, ↑ 15.38% for B20, ↑ 30.77% for B30 | | | ↓ 1.48% B10, ↓ 2.79% B20, ↓ 4.07% B30 | | [36] |
| Canola | B10 | DI, CI diesel engine | variable loads at maximum speed of 2200 rpm | ↓ | ↓ | ↑ | ↑ | ↓ | [37] |
| Jatropha | B100 | double cylinder, DI, CI diesel engine | variable loads at a constant speed of 1500 rpm | | | | ↑ 37.70% at full load | | [38] |

**Table 2.** *Cont.*

| Fuel Type | Blend Composition | Engine Type | Test Condition | HC | CO | $CO_2$ | NOx | Smoke Opacity | Ref |
|---|---|---|---|---|---|---|---|---|---|
| Jojoba | B20, B40 | 4-stroke, single cylinder, CI engine | variable compression ratios (17:1, 17.5:1, and 18:1), variable loads at a constant speed | B40 > diesel > B20l (CR 17:1) Diesel > B20 > B40 (CR 17,5:1) B20 > Diesel > B40 (CR 18) | Various for different CR and loads, but generally Diesel > B20 > B40 | - | B20 > diesel > B40 (CR 17,5:1 and 18:1) B20 > B40 > diesel (CR 17:1) | B40 > diesel > B20 at full loads | [39] |
| Sun flower (hydrotreated refined) | B25, B100 | 4-stroke, stationary DI diesel engine | variable loads at a constant speed of 1500 rpm | ↓ 42% for B25, ↓ 55% for B100 (at full load) | ↓ 9% for B25, ↓ 37% for B100 (at full load) | | ↓ 10% B25, ↓ 18.18%. B100 | | [40] |
| Peanut | B5, B20, B50, B100 | 3-cylinder Yanmar 3009D diesel engine | variable engine speeds (rated power = 14.2 kW) | ↑ for B20 and B100 (↑ 30%, B100), d for B5 and B50 | ↓ up to 29%, B50 (B50 < B100 < B5 < B20) | ↑ up to 18%, B100 (B100 > B50 > B5 > B20) | ↑ up to 30%, B100 (B100 > B50 > B20 > B5) | | [41] |
| Flax | B10, B20, B30 | an inline 4-cylinder, WC Mitsubishi Pajero engine | variable loads at a constant speed of 2000 rpm | ↑ up to 15.8% at full load, B30 (B30 > B20 > B10) | ↓ up to 27.7% at full load, B20 (B20 < B30 < B10) | | ↑ up to 14.4% at full load, B30 (B30 > B20 > B10) | | [42] |
| Safflower | B100 | a single cylinder, 4-stroke, CI engine | variable loads at a constant speed of 1500 rpm | ↓ 3.85% (full load) | ↓ 4.21% (full load) | | ↑ 1.27% (full load) | ↓ 2.57% (full load) | [43] |
| Castor Seed | B20, B40, B60, B80, B100 | Techno-mate, TNM-TDE-700 machine | fixed load and speed | ↓ 20% B20, ↓ 37% B40, ↓49% B60, ↓ 59% B80, ↓ 67% B100 | ↓ 13% B20, ↓ 24% B40, ↓ 33% B60, ↓ 40% B80, ↓ 48% B100 | | ↑ 2% B20, ↑ 4% B40, ↑ 6% B60, ↑ 8% B80, ↑ 10% B100 | | [44] |
| Cotton | B20 | Kirloskar, single cylinder, 4-stroke, diesel engine (CR 18:1) | variable loads at a constant speed of 1500 rpm | ↓ 3.86% | ↓ 18.4% | ↑ 14.0% | ↑ 8.0% | | [45] |
| Avocado | B20, B40, B50, B60, B80, B100 | single cylinder, 4-stroke, air cooled direct injection-compression ignition engine (CR 20:1) | variable loads at a constant speed of 3600 rpm | ↓ (B100 < B80 < B60 < B50 < B40 < B20) | | ↓ (B100 < B80 < B60 < B50 < B40 < B20) | ↑ (B100 > B80 > B60 > B50 > B40 > B20) | ↓ (B100 < B80 < B60 < B50 < B40 < B20) | [46] |
| Mahua | B5, B10, B15, B20 | Kirloskar, twin cylinder diesel engine | variable loads at a constant speed of 1500 rpm | ↓ 6.56% B5, ↓ 11.48% B10, ↓ 16.39% B15, ↓ 21.31% B20 (full load) | ↓ 21.05% B5, ↓ 31.58% B10, ↓ 36.84% B15, ↓ 42.11% B20 (full load) | | ↑ 11.11% B5, ↑ 15.24% B10, ↑ 22.33% B15, ↑ 26.98% B20 (full load) | ↓ 12.5% B5, ↓ 22.92% B10, ↓ 37.5% B15, ↓ 50% B20 (full load) | [47] |
| *Pongamia pinnata* | B20, B40, B60, B80, B100 | Kirloskar, single cylinder 4-stroke, WC, CI engine. | variable loads at a constant speed of 1500 rpm | ↓ B20, almost zero for B40, B60, B80, and B100 | ↓ B80 and B100, zero for B20, B40, B60 (75% load) | ↓ except for B20 (B60 < B40 < B100 < B80 < B20) | ↓ (B60 < B40 < B80 < B100 < B20) | | [48] |
| Mustard | B10, B20 | an inline 4-cylinder, WC, Mitsubishi Pajero engine (CR 21:1) | variable loads and speeds ranging from 1000 to 4000 rpm | ↓ 24% B10, 42% B20 | ↓ significantly (19-40% lower) (B20 < B10 < B0) | | ↑ 9% B10, ↑ 12% B20 | | [49] |
| Coconut | B5, B15 | a one-cylinder, 4-stroke, DI diesel engine | variable loads at a constant speed of 2200 rpm | ↓ 13.89% for B5 and ↓ 22.88% for B15 (full load) | ↓ 13.38% for B5 and ↓ 21.51% for B15 (full load) | ↑ 2.54% for B5 and ↑ 4.64% for B15 (full load) | ↑1.42% for B5 and ↑ 3.19% for B15 (full load) | | [50] |
| Hemp | B10, B20, B30, B50, B100 | Kirloskar TV1, single cylinder, 4-stroke, DI engine | variable loads at a constant speed of 1500 rpm | ↓ 11.11% B10, ↓ 2.22% B20, ↑ 2.22% B30, ↓ 8.89% B50, ↓ 4.44% B100 | ↓ 0.13% B10, ↓ 0.14% B20, ↓ 0.17% B30, ↓ 0.18% B50 ↓ 0.21% B100 | ↑ 2.08% B10 and B50, unchanged for B20, ↓ 5.00% B30, ↑ 6.25% B100 | ↑ 4.17% B10, ↑ 10.42% B20, ↑ 19.79% B30 and B50, ↑ 20.83% B100 | ↓ 18.79% B10, ↓ 15.03% B20, ↓ 12.72% B30, ↓ 4.34% B50, ↑ 6.36% B100 | [51] |
| Camelina | B20, B100 | 4-cylinder, 4-stroke, DI diesel engine | variable speeds between 1200 and 2600 rpm (interval 200 rpm) at full load | | ↓ 15.4% B100 (B100 < B20) | ↓ 13.8% B100 (B100 < B20) | ↑ 9.6% B100 (B100 > B20) | | [52] |

**Table 2.** *Cont.*

| Fuel Type | Blend COM-POSITION | Engine Type | Test Condition | HC | CO | $CO_2$ | NOx | Smoke Opacity | Ref |
|---|---|---|---|---|---|---|---|---|---|
| Grease | B20, B100 | 4-cylinder, 4-stroke, DI Fiat diesel engine | variable loads at a constant speed of 1500 rpm | ↓ slightly B20, ↓ significantly B100 | ↓ 11.26% B20 ↓ 43% B100 (full load) | ↑ 1.8% B20, ↑ 0.7% B100 (full load) | ↑ 12.28% B20 ↑ 20.52% B100 (full load) | | [53] |
| Waste cooking oil | B20 | single cylinder, 4-stroke, DI Kirloskar diesel engine | variable loads at a constant speed of 1500 rpm | ↓ 2.86% (at full load) | ↓ 29.07% (at full load) | ↑ 3.19% (at full load) | ↑ 5.33% (at full load) | | [54] |
| Beef (beef bone marrow) | B50, B100 | 4-stroke, 4-cylinder, direct-injection diesel generator engine | variable loads (3.6, 7.2, and 10.8 kW) | ↓ 24% B50 ↓ 12% B100 | Diesel > B100 > B50 at all loads | ↑ slightly for both B50 and B100 | ↓ slightly (B50 > B100) | - | [55] |
| Sheep (sheep fat) | B25, B50, B75, B100 | Kirloskar TV1, single cylinder, WC diesel engine | variable loads at a constant speed of 1500 rpm | ↑ slightly except for B100 (B100 > B75 > B50 > B25) | ↑ significantly (B100 > B75 > B50 > B25) | | ↓ significantly at full load (B100 < B75 < B50 < B25) | ↑ (B100 > B75 > B50 > B25) | [56] |
| Pork (pork lard) | B100 | single cylinder, WC, DI diesel engine | variable loads at a constant speed of 1500 rpm | ↓ slightly (less than 7.5%) | ↓ 12.32% | | ↓ 3.74% (full load) | ↓ 7.69% | [57] |
| Algae (*Scenedesmus obliquus*) | B50 | single cylinder, 4-stroke, WC diesel engine (CR 20:1) | variable loads at a constant speed of 1400 rpm | ↓ 2.84% | ↓ 4.63% | ↓ 4.46% | ↑ 4.63% | | [58] |
| Algae (*Spirulina platensis*) | B10, B20, B100 | 1-cyllinder, 4-stroke, compression ignition engine | variable loads at a constant speed of 1500 rpm | ↓ up to 33.33% (B100 < B20 < B10) | ↓ 17.42% B10, ↓ 24.84% B20, ↓ 61.29% B100 (at full load) | | ↑ 1.75% B10, ↑ 6.25% B20, ↑ 23.31% B100 | ↓ (B100 < B20 < B10) | [59] |
| Fish | B20, B40, B60, B80, B100 | single cylinder, 4-stroke, air-cooled diesel engine | variable loads and speeds | ↓ up to 20.45% (B100 < B80 < B60 < B40 < B20) | ↓ up to 43.94% (B100 < B80 < B60 < B40 < B20) | | ↑ up to 55.03%, (B100) | | [60] |
| Rapeseed oil | B100 | Single cylinder, 4-stroke, diesel engine | variable speeds from 900 to 1800 rpm | ↓ insignificantly | ↓ insignificantly | ↓ insignificantly | ↑ significantly | | [62] |

Note: ↑ More than, Increase. ↓ Less than, Decrease.

## 3.1. Plant Oils

### 3.1.1. Palm

Palm oil, like other vegetable oils, can be used to produce biodiesel for an internal combustion engine. An experimental study using biodiesel from palm oil fuel with a composition of 5:95 (B5) was conducted on a Mitsubishi 4D68 SOHC DI, 4-stroke, 4-cylinder diesel engine. The results on engine performance showed an increase in BSFC up to 4.22% compared to the data obtained when using straight diesel at an engine speed of 3000 rpm. It was also found that the utilization of palm oil B5 could reduce the CO emission level and insignificant amount of HC and $CO_2$ emissions, while NOx emission was found to be higher compared to that of diesel [34].

### 3.1.2. Soybean

Soybean oil is a vegetable oil extracted from the seeds of soybean. It is commonly used as cooking oil and natural feedstock in biofuel production. Biodiesel from soybean blends with 20% (B20), 40% (B40), and 100% (B100) proportions of biodiesel were tested on a single cylinder, air-cooled, vertical, DI diesel engine [35]. Several performances and emission parameters of each blend were noted when running a diesel engine under variable loads and constant speed of 1500 rpm condition, then the results were compared to that of diesel. Soybean blends was indicated having poor performance compared to diesel as it had a higher value of BSFC by 4.2%, 8.7%, and 14.65% and lower brake thermal efficiency by 2.61%, 4.95%, and 8.07% for B20, B40, and B100 blends, respectively. On the other side, biodiesel showed better characteristics in terms of emission as the use of these blends could significantly reduce HC, CO, and smoke emission levels. The greater amount of emission cut gained from the blend, the higher the composition of soy biodiesel. However, up to a 7.5% increase in NOx emission was observed from the use of biodiesel blends sourced from soybean.

#### 3.1.3. Corn

Corn seed methyl ester biodiesel was mixed with diesel fuel into B10, B20, and B30 biodiesel fuel blends that were used to run a single-cylinder, 4-stroke diesel engine in an experimental test conducted by several researchers [36,63]. As shown in Table 2, an increase in both BSFC and BTE values was observed. Hydrocarbon emissions were reported higher than when the engine was fueled by neat diesel (up to 30.77%), while nitrogen emissions were slightly lower. The result of NOx emission level emitted by corn seed methyl ester is consistent with the results of similar study conducted by Reddy et al. where at 100% engine load condition, NOx levels are observed as (in ppm) 1930 for B10, 1950 for B15, 2030 for B25, and 2040 for diesel fuel [64]. These prove that corn seed methyl ester biodiesel could reduce NOx emission to some degree. These results shown that in most cases, biodiesel emits more NOx than diesel, with the exception of low loads at low and medium speeds [63].

#### 3.1.4. Canola

Canola is a crop with plants from three to five feet tall that produce pods from which seeds are harvested and crushed to produce canola oil and food. A recent study conducted by Öztürk et al. investigated the engine performance and emission of B10 canola biodiesel fuel on a DI diesel engine under various load conditions. A reduction in engine performance was observed as an increase in BSFC and a decrease in BTE was found from B10 canola biodiesel fuel. However, the use of canola biodiesel gave a significant improvement in emission characteristics such as lower amounts of HC, CO, and smoke emissions. However, NOx and $CO_2$ emissions were found to be higher compared to diesel [37].

#### 3.1.5. Jatropha

*Jatropha curcas* seeds contain 27–40% oil that can be processed to produce high-quality biodiesel fuel that is usable in a standard diesel engine. Shehu et al. reviewed and evaluated *Jatropha* as natural feedstock for biodiesel related to its ecological requirements and land suitability in Nigeria. It was found that the choice of *Jatropha* for biodiesel production in Nigeria is seen to be beneficial considering that suitable ecological conditions exist to support the cultivation of the crops in most parts of the country. *Jatropha* has less rigid ecological requirements that make it easy and cheap to be produced [65]. Paul et al. [38] ran an experimental test on a double cylinder, DI, CI diesel engine using pure *Jatropha* biodiesel (B100) as fuel, and also tested straight diesel fuel as a comparison. The test was carried out at variable engine loads and a constant speed of 1500 rpm conditions. The use of B100 *Jatropha* biodiesel has resulted in a significant loss of engine performance as BSFC value was found 56.55% higher and BTE was 26.70% lower at full load than that obtained from straight diesel. It was also found that combustion of *Jatropha* pure biodiesel in diesel engine emitted a 37.70% higher amount of NOx emission.

#### 3.1.6. Jojoba

Jojoba has promising potential as an alternative biofuel source. Oil extracted from jojoba seeds contains a long chain of ester compounds and has a very high boiling point of nearly 400 °C that makes it suitable to be used as fuel. Hariram et al. compared performance and emission characteristics between straight diesel and jojoba biodiesel (B20 and B40) when used as fuel on a single-cylinder, 4-stroke, CI engine [39]. At full load condition, the BSFC of jojoba biodiesel blends was higher and the BTE was lower compared to diesel, except at the compression ratio of 18:1 where the BTE was found to be higher. The emissions from B20 and B40 jojoba biodiesel at full engine load condition were generally lower for HC, CO, and smoke particles and higher for NOx parameters, but it was noted that B40 fuel emitted less NOx at a compression ratio of 17.5:1 and 18:1.

#### 3.1.7. Sunflower

Sunflower oil is a vegetable oil extracted from the seeds of sunflowers. Refined sunflower oil which has been hydrotreated was tested on 4-stroke, stationary DI diesel engine by Hemanandh et al. [40] The engine performance and emission parameters obtained from utilizing the hydrotreated refined sunflower oil and its blend by 25% proportion with diesel was indicated as a favorable mix. From the engine performance improvement observed, there were 10% and 38% increase in BTE and 25% and 12.5% decrease in BSFC for B25 and B100 fuels, respectively. Decreases in all emission parameters were also observed at full load conditions with a maximum reduction of 55% found in HC emission for B100 fuel. However, $CO_2$ and smoke emissions were not determined in this experiment.

#### 3.1.8. Peanut

Santos et al. [41] carried out an experimental study to investigate the performance and emission parameters of peanut oil biodiesel and its blends when performed on a 3-cylinder Yanmar 3009D diesel engine, Japan. Different blends based on the proportion of biodiesel in sample fuels were made consist of B5, B20, B50, and B100 biodiesel fuels. From the experiment, relatively unsatisfactory parameters were obtained from peanut oil biodiesel fuels as it indicated up to 9% higher of BSFC and significantly increased $CO_2$ and NOx emissions by 18% and 30% maximum, respectively, for B100 fuel and straight diesel. HC emission levels were found to be increased on B20 and B100, while it was decreased on B5 and B50. Finally, a maximum 29% decrease in CO emission obtained by B50 fuel was also observed from the experiment.

#### 3.1.9. Flax

An experimental study using B10, B20, and B30 flax oil biodiesel was carried out on an inline 4-cylinder, Water Cooled Mitsubishi Pajero engine, Japan. The engine was operated under variable loads at a constant speed of 2000 rpm conditions and data related to engine performance and emission parameters were recorded. Compared to parameters obtained when the engine ran by diesel fuel, higher BSFC (by up to 13%, B30) and lower BTE (by up to 7.47%, B20) values were gained from flax oil biodiesel blend samples when the engine was running at maximum load. Less favorable parameters were also found in terms of emission. The uses of flax biodiesel blend fuels resulted in a higher amount of HC and NOx emissions by up to 15.8% and 14.4%, respectively, for B30 fuel at full load condition. The only positive parameter came from CO emission as the combustion of all biodiesel blend fuels emitted significantly less amount of CO emission (by up to 27.7% for B20 at full load) compared to that of diesel [42].

#### 3.1.10. Safflower

Balasubramanian et al. [43] ran an experimental test utilizing pure safflower biodiesel (B100) and straight diesel as fuel to run a single-cylinder, 4-stroke, CI engine. The test was carried out in variable engine loads and a constant speed of 1500 rpm conditions. A slight decrease in engine performance was observed when running the engine with B100 safflower biodiesel as BSFC value was found 16.87% higher and BTE was 15.09% lower than that obtained from straight diesel at full load condition. An insignificant decrease in HC, CO, and smoke emission levels, as well as a slight increase in NOx emission, were observed when using B100 safflower biodiesel fuel with less than 5% change in every parameter compared to emission from straight diesel. This result could make safflower biodiesel one of the most comparable fuels with diesel, even utilizing safflower biodiesel in a lower proportion of biodiesel blend could possibly result in a renewable fuel with better characteristics than diesel.

#### 3.1.11. Castor Seed

Biodiesel fuel derived from castor seed was produced using the transesterification method, they were then blended with diesel into B10, B20, B30, and B40 fuels. An engine test was carried out on a Techno-mate, TNM-TDE-700 machine to obtain performance and emission characteristics resulted from utilizing these biodiesel blended fuels in the diesel engine. In terms of BSFC, the castor biodiesel blends were found to have slightly higher yet acceptable values compared to that of diesel with the BSFC values, which increased along with biodiesel composition in the fuel blend. Promising results were also found from the investigation on emission characteristics using different variations of biodiesel blends (B20, B40, B60, B80, and B100). The utilization of castor biodiesel in a diesel engine could reduce HC and CO emission by up to 67% and 48%, respectively, compared to diesel. However, the slight increase in NOx emission by less than 10% was observed, making it one challenge to be solved in the future [44].

#### 3.1.12. Cotton

Sundar et al. [45] carried out an engine test using cotton oil biodiesel fuel with a volume ratio of 20:80 (B20) on a Kirloskar single-cylinder, 4-stroke diesel engine equipped with 18:1 compression ratio. The engine was run at a constant speed of 1500 rpm and some engine performance and emission parameters were noted. The BSFC of B20 cotton biodiesel was 17.1% higher and the BTE was 4.13% lower compared to that of diesel. In terms of emissions, combustion of B20 cotton oil biodiesel on the diesel engine emitted higher levels of $CO_2$ and NOx by 14% and 8%, respectively, and lower amounts of HC and CO by 3.86% and 18.4%, respectively.

#### 3.1.13. Avocado

Anawe et al. [46] studied performance and emissions on a single-cylinder, compression ignition engine fueled with six different compositions of avocado biodiesel blends (B20, B40, B50, B60, B80, and B100). They analyzed engine performance parameters such as BSFC, BTE, and exhaust gas temperature as well as parameters for emission. The results showed that BSFC and NOx emission were higher, whereas BTE, exhaust gas temperature as well as HC, $CO_2$, and smoke emissions were lower than pure diesel. It was also noted that the amount of the increase or decrease in all parameters observed was proportional to the amount of biodiesel in the blend composition.

#### 3.1.14. Mahua

Raman et al. [47] tested a Kirloskar twin cylinder diesel engine with B5, B10, B15, and B20 compositions of mahua biodiesel blends. They reported a slight increase in Brake Specific Fuel Consumption (BSFC) except for B5 fuel that was noted to have a 9.42% lower BSFC than diesel. An increase of 26.98% in NOx emission and reduction of 21.31% in HC, 42.11% in CO, and 50% in smoke particles during engine operation using B20 fuel were observed as the maximum or minimum value of the mentioned emission parameters compared to diesel and other biodiesel blends.

#### 3.1.15. Pongamia pinnata

*Pongamia pinnata* or also known as Karanja is a non-edible plant species capable of growing in almost all types of lands. It can even grow in an extreme environment such as saltwater and withstand extreme weather conditions. A single tree of *Pongamia pinnata* could yield about 25 to 100 kg of seeds containing around 27% to 50% of oil annually. Raw *Pongamia* oil has relatively lower kinematic viscosity (37.12 $mm^2/s$ at 40 °C) compared to other raw vegetable oils, which should give advantages to its fuel properties. Sureshkumar et al. [48] experimented on the engine performance and emission parameters of a single-cylinder, 4-stroke, water-cooled diesel engine running with blends of Karanja biodiesel (B20, B40, B60, B80, and B100). They found out that B20 blend gave the best results in terms of engine performance. It has lower BSFC even when compared to diesel.

In emission characterization, all blends of Karanja biodiesel were found to emit a lower amount or similar amount of harmful emissions such as HC, CO, $CO_2$, and even NOx with an exception for the B20 blend that shows an increase in $CO_2$ emission compared to diesel.

#### 3.1.16. Mustard

Engine performance and emission characteristics of mustard biodiesel blends of B10 and B20 compositions were evaluated in an experimental study using an inline 4-cylinder, Water Cooled Mitsubishi Pajero engine, Japan with a compression ratio of 21 as the experimental engine. The value of BSFC, as well as emission parameters such as HC, CO, and NOx were analyzed in the experiment. The results from the test of mustard biodiesel blends showed that BSFC and NOx emissions were 8.5% and 9% higher for B10, respectively. Meanwhile, for B20, BSFC and NOx emissions were 13.4% and 12% higher, respectively. However, HC and CO emissions were reduced by up to 42% and 40%, respectively, compared to that of diesel [49].

#### 3.1.17. Coconut

B5 and B15 blends of coconut oil biodiesel were utilized on a single-cylinder, 4-stroke, DI diesel engine under variable engine speeds. Performance and emission parameters of these samples were recorded and compared to that of diesel. The BSFC and exhaust gas temperature were noted as slightly higher than diesel by less than a 3.33% difference, which makes these blends comparable to diesel fuel. The engine emissions when using the B15 blend were 22.88% and 21.51% lower for HC and CO emissions, respectively, but 4.64% and 3.19% higher for $CO_2$ and NOx emissions. The emission characteristics of the B5 blend were noted as similar to B15, with higher emission in B15, but both were lower than that of diesel [50].

#### 3.1.18. *Cannabis sativa*

*Cannabis sativa* or hemp is a crop plant grown in temperate zones cultivated annually from seeds. Hemp seeds have high oil content ranging from 26% to 42% which makes them suitable as a biofuel feedstock. Different biodiesel blends consisting of diesel and biodiesel produced by alkali base transesterification of *C. sativa* oil in B10, B20, B30, B50, and B100 compositions were tested on Kirloskar TV1 single-cylinder, 4-stroke DI diesel engine. The BSFC of B10 and B20 fuels were noted lower than diesel, while an acceptable increase was observed for other blends. The results also showed that, for emission characteristics, the use of hemp biodiesel promotes lower emission levels of HC, CO, and smoke particles in general, comparable emission of $CO_2$, and higher emission of NOx compared to that of diesel [51].

#### 3.1.19. Camelina

*Camelina sativa*, also known as false flax, is a plant that is commonly cultivated as an oilseed crop in Europe or Northern America. Engine performance and emission characteristics of false flax biodiesel (B100), diesel, and their blend (B20) were evaluated in an experimental study using a 4-cylinder, 4-stroke, direct injection diesel engine as an experimental setup. The value of BSFC, as well as emission parameters such as CO, $CO_2$, and NOx were determined from the experiment. The results from the test using B100 false flax biodiesel showed that BSFC and NOx emissions were 3.84% and 9.6% higher, respectively, but CO and $CO_2$ emissions were reduced by 15.4% and 13.8%, respectively, compared to those using diesel [52].

#### 3.1.20. Algae

Algae have become a popular biofuel feedstock as it is known to store a high amount of lipid and oil contents. Algae could also produce biomass very rapidly; some species even have a doubling time of only six hours. The engine performance and emission when running with algae biodiesel have been studied utilizing various species of algae. B50 fuel derived from *Scenedesmus obliquus* mixed with diesel fuel has been tested on a single-cylinder, 4-stroke, water-cooled diesel engine [58]. The B50 fuel has a 14.55% higher BSFC, and 4.75% and 5.21% lower BTE and exhaust gas temperature, respectively. Reduction of HC, CO, and $CO_2$ emissions by 2.84%, 4.63%, and 4.46% was also observed, while NOx emission was found higher by 4.63% compared to diesel. In the other study, another algae species, *Spirulina platensis*, was converted into biodiesel blends of B10 and B20 [59]. The engine test on a diesel engine has been done, and the results showed a similar trend on BTE and emission parameters to what has been found in *S. obliquus* biodiesel. For B100 *Spirulina platensis* biodiesel, the BTE, HC, and CO emission levels were found lower by 2.5%, 33.33%, 61.29%, respectively, while NOx emission was increased by 23.31% compared to diesel.

### 3.2. Waste Cooking Oils

#### 3.2.1. Grease

Grease is thickened oil having a solid or semi-solid phase, which is also a dispersion of thickening agent in liquid lubricant oil. There are several types of grease such as brown and yellow grease, both sourced from the waste of food industry but having different oil content and characteristics. A study of biodiesel fuel produced from yellow grease collected from restaurant waste was conducted by Chaichan et al. [53] Yellow grease biodiesel was prepared using the base transesterification method, then the product was blended with diesel into B20 fuel. Diesel fuel (B0), pure yellow grease biodiesel (B100), and B20 blend were tested on a single-cylinder, water-cooled, DI diesel engine to investigate its performance and emission characteristics. They reported an increase in both BSFC and BTE parameters, significant decrease in HC and CO emissions, slight increase in $CO_2$ emission, and significant increase in NOx emission from the use of B20 and pure yellow grease biodiesel.

#### 3.2.2. Waste Cooking Oil

Waste cooking oil (WCO) is a potential source of relatively cheap feedstock for biodiesel production, but it has high free fatty acid (FFA) contents that can be a serious bottleneck for the process of transesterification. Avase et al. [54] in their study, made biodiesel fuel from the transesterification of waste cooking oil (WCO) mixed with diesel in the proportion of 2:8 (B20). The results show that the production process of B20 WCO biodiesel could decreased the amount of HC and CO emissions by 2.86% and 29.07%, respectively, but slightly increased the amount of NOx and $CO_2$ emissions by 5.33% and 3.19% compared to that of diesel. An increase in both BSFC and BTE parameters by 2.96% and 10.79%, respectively, at full load was also observed from the experiment.

### 3.3. Animal Fats

#### 3.3.1. Beef

Beef bone marrow contains fat that could be potentially converted into biodiesel fuel. An experimental study has been done by Erdoğan et al. [55] to test engine performance and emission run with beef bone marrow fat biodiesel and its blend with straight diesel in the proportion of 1:1 (B50). The engine used in the study was a 4-cylinder, 4-stroke, DI diesel generator engine, and the test was carried out under various load conditions. The results showed that the use of beef bone marrow biodiesel and its blend caused a reduction in engine performance due to higher BSFC and lower thermal efficiency. However, positive effects were shown in the form of a significant amount of reduction in HC, CO, and NOx emissions. Only a slight increase in $CO_2$ emission was reported.

#### 3.3.2. Sheep

Oil and fat contained in sheep meat could also be used as a source of biodiesel. Ultrasonic-assisted transesterification process was carried out to convert sheep fat into biodiesel and then several proportions of biodiesel were blended with diesel to obtain B25, B50, and B75 fuels. Sheep fat biodiesel (B100) and its blends were tested on a Kirloskar TV1, single-cylinder, water-cooled diesel engine at a constant engine speed of 1500 rpm.

The brake thermal efficiency of sheep fat biodiesel and biodiesel blend fuels was slightly lower compared to diesel, ranging from 1.8% to 2.3%. The results also shown that the NOx emission in biodiesel (B100) has a higher value compared to diesel fuel (B0) [56,63].

3.3.3. Pork

Pork lard, like other animal fats, has a high cetane number, rich oxygen content, and very close lower heating values when compared to standard diesel. An experimental study utilizing pure biodiesel (B100) derived from waste pork lard on a single-cylinder, water-cooled, DI diesel engine has been done to observe the performance and emission characteristics of pork lard biodiesel. Although using pork lard biodiesel in the engine indicated a reduction in brake thermal efficiency, the emission parameters obtained were satisfactory. Using pork lard biodiesel has been proven to be able to reduce all emission parameters observed in this study, including NOx by a maximum 12.32% reduction found in CO emission compared to diesel [57].

3.3.4. Fish

Some species of fish considered as oily fish store high content of oil in their tissues. Fish oil is commonly used by people as a supplement, but it is also possible to use fish oil as a material for biodiesel production. Performance and emission tests of Fish Oil Methyl Ester (FOME) and their blends (80%, 60%, 40%, and 20%) with diesel were carried out on a single-cylinder, 4-stroke, CI engine at variable load conditions. The results indicated a reduction in engine performance as up to 12.68% decreases in BTE were observed from B100 fuel. The HC and CO emissions were reduced by up to 20.45% and 43.94%, respectively, while NOx emission was increased by up to 55.03% compared to that of diesel [60].

## 4. Manufacturing Process of Biodiesel

The manufacturing process of biodiesel is one of the most important factors in biodiesel development. The common methods of biodiesel production that are currently used to yield biodiesel from crude oil are direct use and blending, microemulsion, pyrolysis, and transesterification. The choice of production method determines not only the quality of biodiesel yielded but also the cost and duration of the manufacturing process. In this section, the recent studies about biodiesel production related to their efficiency, stability, cost, factors that affected the process, and quality of the products are evaluated in detail.

Recently, waste from biodiesel production and combustion also has been reused to produce other forms of bioenergy or useful materials, making biodiesel even more environmentally friendly. A study conducted by Yin et al. [66] showed huge potential from ash produced from the combustion of oil palm biomass waste to be reused for several purposes. Its composition and toxicity were assessed by observation with electron microscope and toxicity characteristic leaching procedure (TCLP). It is concluded that oil palm ash should not be considered as toxic material and have the potential to be used as crude fertilizer and cement mixture. Other waste from oil palm production such as empty fruit bunches (EFBs) also have reusability potential as a fuel for an oil palm mill boiler after being processed through the torrefaction process to reduce its moisture content [67]. The use of torrefied EFBs in the boiler as alternative fuel offers cleaner combustion and additional revenue for palm oil processing.

### *4.1. Direct Use and Blending*

The direct use of natural oils, especially those sourced from vegetables to run diesel engines, has been applied since the 1900s. Vegetable oil offers numerous advantages over diesel fuel such as liquid nature, heat content, which is about 80% of diesel fuel's, availability, and renewability. Higher viscosity, lower volatility, and the reactivity of unsaturated hydrocarbon chains are some disadvantages of vegetable oil that cause many inherent problems when they are used as fuel on diesel engines. Thus, crude vegetable oils are often mixed directly or diluted with diesel fuel to improve the viscosity [68,69].

Experiments related to the direct use of vegetable oil as fuel for diesel engines and their performance and emission characteristics have been done by many researchers. Almeida et al. [70] tested pure palm oil to run a diesel generator engine (4-stroke, NA, DI) for 350 h of operation. Palm oil was heated at 50 °C and 100 °C before the fuel pump to promote smooth flow, then the effect of oil temperature was observed. A high level of deposits was found in the combustion chamber when it was operated with palm oil heated at 50 °C due to incomplete combustion, but the use of oil heated at 100 °C successfully reduced the deposits to an acceptable level and presented better combustion. In terms of engine performance, diesel engine fueled by pure palm oil has higher ignition delay and specific fuel consumption compared to that of diesel. Investigation of the emission characteristics also showed unsatisfactory results where there were higher amount of CO, $CO_2$, and HC emissions. However, the emission of NOx was noted to be significantly lower than diesel. This result made a good agreement with the experiment using other sources of vegetable oils (raw sunflower oil, raw cottonseed oil, raw soybean oil, refined corn oil, distilled opium poppy oil, and refined rapeseed oil [62] as well as Karanja oil investigated in another study [71]), except the fact that some of these oils emitted less $CO_2$ emission than diesel.

Rapeseed oil (RSO) and its blend with standard low sulfur diesel with various compositions were tested using 2.0 L, 4-cylinder, 16-valve, direct injection diesel engine [61]. The NOx emission for RSO and its blend was lower than diesel, but they produced a much higher amount of soot emissions. Further reduction of NOx emission by 22% was achieved on 30% RSO blend by retarding the injection timing up to 3° bTDC and increasing the injection pressures up to 1200 bar, but the concentration of soot particles was still higher compared to diesel under this condition. Reduction of NOx emission at retarded injection timing also occurred when lemongrass oil (LGO) and its blends were run in diesel engines [72]. However, advanced injection timing (27° bTDC) is considered to be an optimal condition for LGO-diesel blends as it indicated better performance and emitted a lower amount of HC and smoke emissions.

A mixture of sesame oil and diesel fuel in the ratio of 1:1 was tested in Lombardini 6 LD 400, one cylinder, 4-stroke, air-cooled, direct injection diesel engine to investigate its performance and emission characteristics [73]. The results were satisfactory as this mixture could successfully produce and maintain power and torque close to that produced by diesel, also emitted less amount of CO and NOx. The only disadvantage reported was the higher number of BSFC due to the lower heating value of the blend compared to diesel. The direct use and blending of *Jatropha* as biofuel was studied by Forson et al. [74] Diesel fuel, *Jatropha* oil, and biodiesel with the proportion of 20:80%, 50:50%, and 2.6:97.4% were used to operate an air-cooled, direct-injection, single-cylinder, 4-stroke diesel engine. The unmodified engine ran well on all fuels tested. The results on investigation of performance and emission characteristics put 2.6:97.4% of *Jatropha* oil blend as the best even when it was compared to straight diesel, making *Jatropha* oil recommended for use as an additive for diesel fuel in the low amount.

A study conducted by Roy et al. [33] compared emission characteristics between the blends of canola oil–diesel and transesterified canola biodiesel as fuels on a 4-stroke, 2-cylinder, naturally aspirated DI diesel engine. The canola biofuel had a higher viscosity and a lower oxygen content compared to canola biofuel fuel at the same proportion which contributed to the lower fuel conversion efficiency of canola biodiesel fuel. It was concluded that canola oil–diesel fuel generally emitted a higher amount of CO, HC, and NOx emissions compared to biodiesel, except for CO emission at canola oil–biodiesel proportion of up to 5% and HC emission at canola biodiesel proportion of up to 5% and low-speed condition.

Although for a short-term use with low ratios of vegetable biodiesel blends have been found to be successful, the direct use of vegetable oil in diesel engines has generally been considered to be not satisfactory and impractical for both direct and indirect diesel engines. The high viscosity, acid composition, free fatty acid content, as well as gum formation due to oxidation and polymerization during storage and combustion, carbon deposits, and lubricating oil thickening are obvious problems that cannot be ignored. Some significant engine modifications are required for enabling vegetable oil to be used as diesel fuel, including changing of piping and injector construction materials. Otherwise, engine running times are decreased, maintenance costs are increased due to higher wear, and the danger of engine failure is increased [68,69].

### 4.2. Microemulsion

Micro-emulsification is a potential method to solve the problem of the high viscosity of crude natural oil. A microemulsion is defined as a colloidal equilibrium dispersion of optically isotropic fluid microstructures with dimensions generally in the 1–150 nm range formed spontaneously from two normally immiscible liquids and one or more ionic or non-ionic amphiphiles. The three components that make up a microemulsion are usually acted as an oil phase, an aqueous phase, and a surfactant [68,69].

Patidar et al. [75] studied phase behavior and physicochemical properties of Karanja oil-ethanol microemulsion with span 80 (sorbitan monooleate) and span 85 (sorbitan trioleate) as surfactants. Karanja oil was a mixture with up to 15% ethanol. Variable amounts of the span were added when the concentration of ethanol reached above 15% under 500 rpm of continuous stirring condition. It can be seen from the phase diagram that span 80 made better solubilization of ethanol in the oil phase due to its higher content of OH group. The physicochemical analysis showed that the ethanol with diesel ratios of 65:35, 70:30, 75:25, 80:20, 85:15, 90:10, and 95:5 with span 80 had kinematic viscosity which meet the ASTM standard of B100 biodiesel (1.9–6 $mm^2s^{-1}$) at 313.15 K. Therefore, Karanja oil–ethanol-span 80 microemulsion could be used as alternative diesel fuel and is comparable to that made by transesterification process.

Span 80 as a hydrophilic surfactant was mixed with Tween 80 as a hydrophobic surfactant in different mixing ratios to obtain higher solubility and better stability for bio-oil-in-diesel microemulsion (BDM). GC-MS and other analysis methods were performed to investigate properties of BDM resulted, then the prepared fuels were stored over 90 days at indoor temperature to determine their stability. The results showed that mixed surfactant of 7:3 of Span 80-Tween 80 with the help of 2% n-hexanol as cosurfactant is the optimum formula to get BDM with better properties. This BDM formula was tested in a single-cylinder, 4-stroke, DI diesel engine and found to have well performance characteristics (higher BSFC and BTE compared to diesel), while reduced CO, $CO_2$, and NOx emissions by 21.4–66.7%, 7.1–27.3%, and 1.5–14.7%, respectively [76].

A non-ionic sunflower oil mixed with ethanol microemulsion that have a composition of 53.3% sunflower oil, 13.3% 190-proof ethanol, and 33.4% 1-butanol by volume was tested in in 4-cylinder, turbocharged, DI diesel engine, and the effects were evaluated. Although the engine fueled by microemulsion completed the 200 h of EMA screening test cycle and was observed to reduce BSFC by 4% and smoke emission significantly, the use of this microemulsion still has so many disadvantages and is not recommended for long-term use in DI diesel engine. The major problems found during the screening test were incomplete combustion during engine start and low load conditions, premature injection nozzle deterioration, and excessive deposits on the internal parts of the fuel injection pump [77]. In another study, coconut oil-based hybrid fuel which was coconut oil–aqueous ethanol microemulsion with butan-1-ol as a surfactant in various compositions was prepared and performed on PowerTec 170FG, 4-stroke, single-cylinder, air-cooled, DI diesel engine. Positive results were recorded from the test as micro emulsification of coconut oil was successful to obtain diesel-like fuel with relatively low viscosity close to the viscosity value of diesel. The engine efficiency gained from the use of coconut oil–ethanol microemulsion is almost similar to diesel and the significantly lower NO, $SO_2$, and $CO_2$ emissions level were noted. However, the specific fuel consumption and CO emission level of hybrid fuel were higher compared to that of diesel due to lower gross calorific value and incomplete combustion, respectively [78].

The experiment conducted by Charoensaeng et al. [79] used butanol, octanol, and decanol as cosurfactant in microemulsion biofuel. The effect of that cosurfactant on fuel consumption and emission was compared as they vary in the carbon chain length. All microemulsion biofuel samples indicated better emission characteristics for CO, NOx, and $CO_2$ with an increase in BSFC. Moreover, it was found that the use of the longer carbon chain alcohol as cosurfactant resulted in lower fuel consumption and CO emission, but NOx and $CO_2$ emissions were gradually increased.

Some vegetable oils have been used as blend materials for microemulsions. Engine test and physicochemical analysis of those microemulsions have been done in the past studies. Table 3 presents data collected from several studies related to physicochemical properties and engine performance and emission characteristics obtained from vegetable oil-based microemulsions.

**Table 3.** The physicochemical properties and engine test parameters of several biodiesel microemulsions.

| Emulsion | 80% B30 Tung Oil-Diesel—20% Ethanol | 80% Soybean Biodiesel—20% Ethanol | 81% B20 Soybean Biodiesel—15% NP5EO—4% Water | 52% Crude *Pongamia* Oil (CPO)—23% Ethanol—25% Buthanol | 95% B15 Palm Biodiesel—5% Ethanol |
|---|---|---|---|---|---|
| Reference | [80] | [81] | [82] | [83] | [84] |
| Density (g/mL) | 0.843 | 0.8552 | 0.871 | 0.864 | 0.833 |
| Kinematic viscosity (mm2/s) | 4.8 (at 40 °C) | | | 13.51 (at 21 °C) | 3.23 (at 40 °C) |
| Lower heating value (kJ/kg) | 37,990 | 31,283 | | | |
| Flash point (°C) | | | 46.2 | | 84.5 |
| Stoichiotric air-fuel ratio (AFR) (kg/kg) | 12.87 | 11.249 | | | |
| Engine | 6-cylinder, 4-stroke, turbocharged, inter-cooled, CRDI diesel engine | single cylinder, 4-stroke, WC, CR 16.5:1, DI diesel engine | single-cylinder, 4-stroke, AC, DI diesel engine | single-cylinder, 4-stroke, DI diesel engine | single cylinder, 4-stroke, DI diesel engine |
| Performance | BSFC ↑, BTE ↑ slightly | BSFC ↑ slightly | | BSFC ↑, BTE ↓ 2.75% | BSFC ↑, BTE ↓ |

Note: ↑ More than, Increase. ↓ Less than, Decrease.

### 4.3. Pyrolysis

Vegetable oil is known to have a problem with high viscosity and density that affects the fuel atomization in the engine, so it could not be used directly as fuel in a diesel engine without modification. Pyrolysis or thermal cracking offers a solution to this problem since this process is able to convert the complex structure of hydrocarbons into its simplest structure by means of heat, with or without a catalyst. Pyrolysis will occur in high-temperature conditions ranging between 250 °C and 350 °C [85]. The flow of biodiesel production process using the pyrolysis method is shown in Figure 2.

A study conducted by Du et al. [86] assessed biofuel production from microalgae biomass named *Chlorella* sp. using the microwave-assisted pyrolysis method. The pyrolysis was carried out in a microwave oven at a frequency of 2450 MHz for 20 min of reaction time. The maximum bio-oil yield of 28.6% was produced under the microwave power of 750 W. The algae bio-oil produced in this experiment was characterized and found to have a density of 0.98 kg/L, a viscosity of 61.2 cSt, and a higher heating value (HHV) of 30.7 MJ/kg.

Pyrolysis of soybean oil and high oleic safflower oil was carried out in air and nitrogen sparge by Schwab et al. [87] The distillation method and equipment used were based on the ASTM standard. Gas chromatography-mass spectrometry (GC-MS) analysis was conducted to determine the composition of pyrolyzed materials. Biofuels with properties of two-third lower viscosity and higher cetane number were successfully obtained by distillation. The main compositions of the products observed were alkanes, alkenes, aromatics, and carboxylic acids with carbon numbers ranging from four to more than 20 by percentages as shown in Table 4.

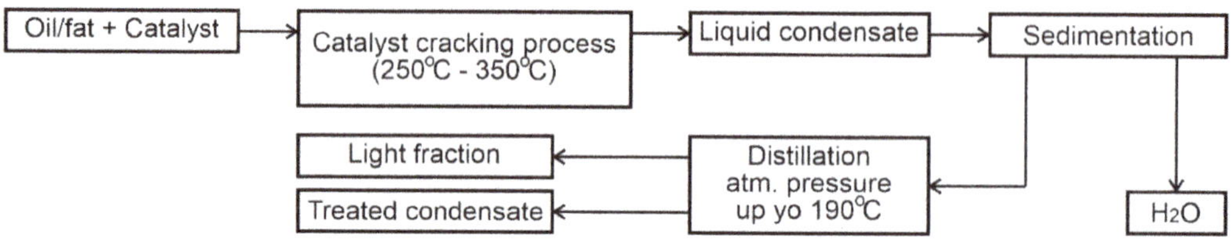


**Figure 2.** The flow of biodiesel production process using the pyrolysis method [85].

**Table 4.** Data of pyrolyzed oil compositions [87].

| | Percent by Weight | | | |
|---|---|---|---|---|
| | High Oleic Safflower | | Soy | |
| | N2 Sparge | Air | N2 Sparge | Air |
| Alkanes | 37.5 | 40.9 | 31.3 | 29.9 |
| Alkenes | 22.2 | 22.0 | 28.3 | 24.9 |
| Alkadienes | 8.1 | 13.0 | 9.4 | 10.9 |
| Aromatics | 2.3 | 2.2 | 2.3 | 1.9 |
| Unresolved unsaturates | 9.7 | 10.1 | 5.5 | 5.1 |
| Carboxylic acids | 11.5 | 16.1 | 12.2 | 9.6 |
| Unidentified | 8.7 | 12.7 | 10.9 | 12.6 |

A study by Lima et al. [88] investigated pyrolysis reactions and products of soybean, palm tree, and castor oils. The pyrolysis oil distillates were separated into four fractions based on the distillation temperature (<80 °C, 80–140 °C, 140–200 °C, and >200 °C) and the fraction gained from distillation temperature >200 °C that contains 60–75% of pyrolyzed products was observed by GC, FTIR, and ASTM analysis methods for biodiesel. Some compounds such as hydrocarbons alkanes, alkenes, alkadienes, and carboxylic acids were identified from this fraction. However, different from the results of the previous study, there was no sign of aromatic product detected. Physicochemical properties of pyrolysis products from soybean and palm tree oils were found comparable with standard diesel fuel used in Brazil. Furthermore, the preliminary test of catalytic deoxygenating of soybean pyrolyzed product using HZSM-5 zeolite at 400 °C was successful to obtain an enriched hydrocarbon diesel-like fuel.

Wood pyrolysis oil (WPO) obtained from the vacuum pyrolysis process of pine wood was mixed with diesel fuel by volume ratio of 1:9 and some percentages of diethyl ether (DEE). An experiment was carried out using a single-cylinder, four-stroke, air-cooled, DI diesel engine fueled by WPO emulsion in order to investigate its combustion, performance, and emission parameters. The WPO emulsion was found exhibiting longer ignition delay and combustion duration due to the poor ignition quality of the WPO. The BTE of WPO emulsion increased by 6.35%, but the exhaust gas temperature was observed to be lower than that of diesel. In terms of emission, the WPO emulsion emitted 19.21% lower NO, 14.28% higher HC, and slightly higher CO. Addition of DEE on WPO emulsion by percentages of 2% and 4% has been proven to cause improvement in all parameters observed [89]. In another study by the same author, the WPO of 5%, 10%, and 15% by volume was emulsified with *Jatropha* methyl ester (JME) and tested using the same diesel engine. Based on the experimental results, JME and its emulsion with WPO indicated better combustion at high load, higher BTE, and lower smoke emissions compared to diesel [90].

Thermal catalytic cracking or pyrolysis with catalyst process was used in past studies to produce diesel-like fuel from rice bran and palm oils [91,92]. The thermal cracking process of rice bran oil was carried out at 450 °C using calcium oxide (CaO) in a load ranging between 0.5% and 3% to the weight of oil. The process using a catalyst with more CaO content was observed to produce more product up to 71.5% with a faster process. Meanwhile, crude palm oil was thermally cracked at 450 °C using sodium carbonate (Na2CO3) of 20% proportion to the weight of oil. The rice bran and palm oils gained from the thermal cracking process were then distilled and physical-chemical characterized using standard ASTM and other analysis methods. The properties of diesel-like fuels produced from rice bran and palm oils can be seen in Tables 5 and 6, respectively.

**Table 5.** Physicochemical properties of the cracked samples of rice bran oil compared to diesel. Adapted from [91].

| | | Catalyst Load (kg/100 kg Oil) | | | |
|---|---|---|---|---|---|
| | **Diesel** | **0.5** | **1.0** | **2.0** | **3.0** |
| API gravity | 31–41 | 32.08 | 35.75 | 33.80 | 33.99 |
| Specific gravity | 0.82–0.87 | 0.865 | 0.846 | 0.856 | 0.855 |
| Pour point (°C) | +4.5–15 | +12 | +6 | 0.0 | +6 |
| Kinematic viscosity ($m^2S^{-1} \times 10^6$) | ≤7 | 8.61 | 7.19 | 9.42 | 9.39 |
| Calorific Value (MJ/kg) | 44.3 | 40.437 | 38.199 | 39.675 | 35.860 |
| Heating value compared to diesel | 1 | 0.91 | 0.86 | 0.89 | 0.81 |
| Flash point (°C) | ≥55 | 54 | 51 | 56 | 51 |
| Cetane number | ≥55 | 59.69 | 52.24 | 52.26 | 59.68 |
| Cetane index | - | 51 | 54 | 52 | 54 |

**Table 6.** Physicochemical properties of the cracked samples of palm oil (green diesel) compared to diesel. Adapted from [92].

| Properties | Unit | Green Diesel | Diesel |
|---|---|---|---|
| Density | $kg/m^3$ | 790 | 820–880 |
| Kinematic viscosity | $mm^2/s$ | 1.48 | 2–5 |
| Flash point, min | °C | 10 | 38 |
| Copper strip corrosion, 3 h 50 °C, max | - | 1a | 1a |
| Carbon residue, max | wt% | 0.02 | 0.25 |
| Acid value | mg KOH/g | 1.68 | 0.5 |
| Saponification value | mg KOH/g | 7.93 | - |
| Refraction index | - | 1.44 | - |
| Ester value | mg KOH/g | 6.25 | - |

### 4.4. Transesterification

Transesterification is the reaction of fat or oil with an alcohol to form esters (biodiesel) as the main product and glycerol as a by-product. This displacement of an alcohol from an ester by another alcohol occurred in transesterification is similar to hydrolysis process, except that the hydrolysis process employs water instead of alcohol. Thus, transesterification is also known as alcoholysis. The process of transesterification is basically a sequence of three step reversible reactions. The first step is the conversion of triglycerides to diglycerides, followed by the conversion of diglycerides to monoglycerides and of monoglycerides to glycerol. One methyl ester molecule is yielded from each glyceride at each step [68,69,93–95]. The reaction is shown in Figure 3. Where the basic scheme of biodiesel production with transesterification is shown in Figure 4.

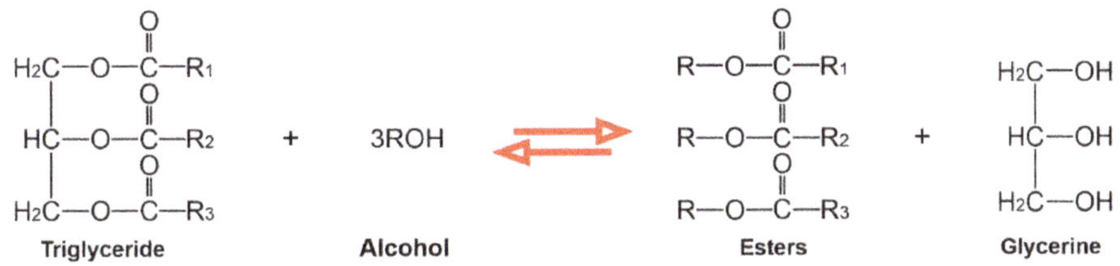


**Figure 3.** Transesterification of triglycerides with alcohol [52].

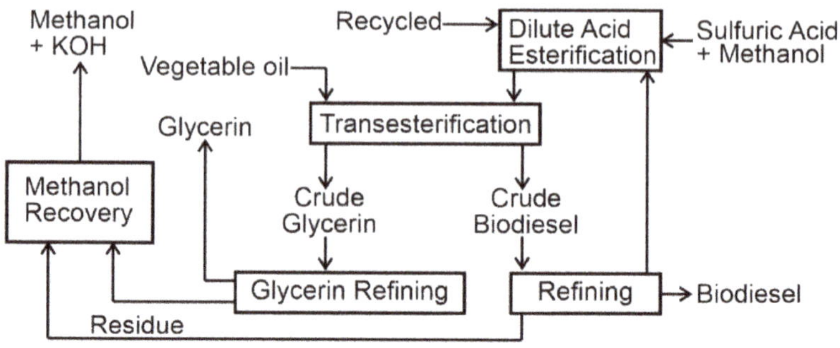


**Figure 4.** The basic scheme of biodiesel production with transesterification [94].

Moisture or water content is strongly affected the efficiency of transesterification, especially for a method called in situ transesterification where the process is carried out on very raw biomass rather than oil. A study by Sathish et al. [96] about the effect of moisture on in situ transesterification of microalgae clearly supported this statement. They reported a significant reduction in biodiesel yield occurred when the moisture content exceeded around 15 to 20 wt% of algae dry mass. They also reported that an increase in the amount of sulfuric acid as a catalyst could overcome this problem and improve the biodiesel yield to greater than 80% even when using algae biomass with 84% moisture content. However, this addition resulted in a higher cost of production, which is not economically sustainable. A study by Haas et al. [97] also concluded a similar result about the comparison of the biodiesel production cost when using low-moisture soy flakes and full-moisture soy flakes. They found that the use of low-moisture soy flakes with 2.6 wt% of moisture content could substantially reduce the reagent requirements for high-efficiency transesterification. It required 40% less methanol and 33% less NaOH compared to that required by transesterification of full-moisture flakes, making the biodiesel production significantly cheaper.

Many researchers have studied the kinetics of the transesterification process to reveal how parameters such as the molar ratio of alcohol to oil, reaction temperature, and reaction time affect the production of biodiesel. Kusdiana et al. [98] studied the kinetics of free catalyst transesterification of rapeseed oil made in subcritical and supercritical methanol. They found the optimum condition for the free-catalyst process of biodiesel fuel production from rapeseed oil was at a temperature of 350 °C with a 42:1 of the molar ratio of methanol and oil. Another study by Kapilakarn et al. [99] used simulation software to find the optimum operating condition of transesterification of triglycerides based on the percentage of purity of the products and predicted cost. The molar ratio of methanol and oil at 6:1, reaction temperature of 70 °C, and 20-min reaction time were the optimum conditions of transesterification proposed from the simulation analysis.

The effect of various catalysts on the transesterification process has been studied by many researchers in the past. The efficiency of two-dimensional zeolites (Na-zeolite) as a basic catalyst of the transesterification process was studied by Pang et al. [100] The prepared Na-zeolite samples were used as catalysts in the base transesterification reaction of triolein (soybean oil) with methanol at a mass ratio of methanol:triolein:catalyst = 100:10:1 and 60 °C reaction temperature. Na/zeolite catalysts were able to increase the yield of biodiesel up to 80% and 2D Na/ITQ-2 was found to be the best catalyst compared to other samples. 2D Na/ITQ-2 is a two-dimensional basic zeolite contains natrium and ITQ-2 zeolite which possess large external surface areas, hierarchical characteristics, stable active sites, and a high concentration of strong basic sites that contribute to enhancing the mass transfer of bulky molecules and providing more accessible and stable active sites in the transesterification of bio-derived oil to produce biodiesel. A similar investigation conducted by Mansir et al. [101] using bimetallic tungsten zirconia (W-Zr) samples to catalyze simultaneous esterification and transesterification of unrefined palm-derived waste oil (PDWO) to biodiesel. The catalyst with a 7 wt% concentration of W-Zr loaded on CaO support (7WZC) was able to achieve a maximum biodiesel yield of 94% in one-hour reaction time under the optimum reaction condition (methanol and oil molar ratio of 15:1, 2 wt% catalyst loading, 80 °C reaction temperature). The high catalytic activity of 7WZC was attributed to its moderate basicity and acidity composition that could easily convert both FFA and triglycerides present in PDWO to biodiesel.

Recently, some studies also have been conducted related to the utilization of biodiesel production waste from the transesterification process. Production of 100 kg of biodiesel via the transesterification process generates 10 kg of crude glycerol as a by-product. This crude glycerol derived from the biodiesel transesterification process can be used as a material to produce useful chemicals such as 1,3-propanediol and 2,3-butanediol. The production of those chemicals by a co-fermentation process with help from isolated strain *Enterobacter* sp. MU-01 bacteria has been proven efficient [102].

Compared to other biodiesel production methods available, transesterification of natural oils and fat is currently the best method due to its high efficiency and superior characteristics indicated by its products. Transesterification is affected by several factors such as the molar ratio of glycerides and alcohol, the catalyst, reaction temperature and time, and the content of moisture [68,69].

## 5. Future Trends

Based on findings from all the papers reviewed in this study, there are two main challenges in the development of biodiesel as renewable alternative energy that could possibly become a trend in future studies related to biodiesel. First, mass production of biodiesel is still considered expensive compared to that of diesel fuel, which causes less biodiesel use on a daily basis. Researchers are encouraged to conduct studies and observations on various potential biomasses until one feasible feedstock is found. Second, it can be seen from Table 2 that using biodiesel as fuel results in an increase in NOx emissions for almost all types of biodiesel. There should be some efforts to reduce NOx emissions from biodiesel fuel combustion in the future. Formation of NO in the combustion of biodiesel and diesel fuels in a non-pressurized, water-cooled combustion chamber were observed by Bazooyar et al. [103]. The results showed that while the amount of thermal NO formed in biodiesel and diesel fuel was comparable, biodiesel produced significantly higher prompt NO than diesel fuel, making biodiesel emiting more NO in total than diesel. This higher formation of prompt NO in biodiesel fuel combustion is attributed to its higher number of bis-allylic sites. Sun et al. [104] found that the use of exhaust gas recirculation (EGR) system in diesel engines could effectively suppress NOx formation on biodiesel combustion, proven by a decrease of NOx emissions as the EGR rate increased. However, it has the opposite effects on CO and HC emissions and some disadvantages on engine combustion and performance characteristics. As of today, there has not been found a beneficial solution to reduce NOx emission in biodiesel combustion without raising any other problems. Thus, solving these issues should be the main focus for future researchers or other people interested in this topic. Moreover, the metals and chemical compounds contaminants in biodiesel became the concern of researchers since in the real time operation, the time and long term operational have affect to the emissions and the engine performance [105,106].

## 6. Conclusions

In this review, various types of biodiesel and its production methods were comprehensively discussed. It is expected to assist the researchers, biodiesel producers, and other people who might be interested in this topic. This paper can ensure the information regarding natural sources of biodiesel, biodiesel production methods, and properties of biodiesel yielded from each source and method observed. Finally, the engine performance and exhaust emission characteristics of different types of biodiesel as well as their blends on the compression ignition engines have been evaluated in detail. Overall, the following critical findings could be drawn from the present study according to the published literature.

1. With increasing concerns about global warming partly caused by greenhouse gas (GHG) emissions from fossil-fuel combustion and the limited availability of fossil-fuels, sustainable, environmentally friendly, and renewable fuels such as biodiesel are important to be developed and applied immediately on a daily basis.
2. The cost of raw material takes up most of the biodiesel production cost. Economic, agricultural, and technical evaluations are needed to choose the most feasible natural sources in biodiesel production without having to sacrifice the quality.
3. The use of biodiesel is very beneficial from the environmental perspective since they generally reduce the amount of HC, CO, and smoke emissions. However, running biodiesel fuel in diesel engines usually causes an increase in NOx emission compared to straight diesel.
4. Biodiesel has some poor properties such as higher viscosity, higher density, and lower calorific value that cause a reduction in engine performance parameters when they are used as fuel on a diesel engine.
5. Performance and emission characteristics of biodiesel fuel and blends vary, depending on their natural sources, physicochemical properties, and engine operating conditions.
6. Of the various biodiesel production methods available, transesterification is currently the most superior method since it can efficiently yield biodiesel products that have superior characteristics compared to other biodiesel produced with other methods.

**Author Contributions:** Conceptualization, J.S., M.A.M. and A.D.; methodology, J.S., M.A.M. and A.D.; formal analysis, J.S., M.A.M., E.S., A.P. and A.D.; investigation, J.S., M.A.M., E.S., A.P. and A.D.; data curation, J.S., M.A.M., E.S., A.P. and A.D.; writing—original draft preparation, J.S., M.A.M., E.S., A.P. and A.D.; writing—review and editing, J.S., M.A.M., E.S., A.P. and A.D.; visualization, J.S., M.A.M., E.S., A.P. and A.D.; supervision, M.A.M.; project administration, J.S., M.A.M. and A.D.; funding acquisition, J.S., M.A.M. All authors have read and agreed to the published version of the manuscript.

**Funding:** This research received no external funding.

**Acknowledgments:** The author would like to thank PT PLN and the Innovation Center for Automotive (ICA) Universitas Gadjah Mada for the funding and support of this study.

**Conflicts of Interest:** The authors declare no conflict of interest.

## Nomenclature

| | |
|---|---|
| AC | air-cooled |
| ASTM | American society for testing and materials |
| B0 | diesel fuel |
| B10 | blended diesel fuel (90%) and biodiesel (10%) |
| B20 | blended diesel fuel (80%) and biodiesel (20%) |
| B50 | blended diesel fuel (50%) and biodiesel (50%) |
| B100 | pure biodiesel fuel |
| BDM | bio-oil-in-diesel microemulsion |
| BSFC | brake specific fuel consumption |
| BTE | brake thermal efficiency |
| CI | compression ignition |
| CO | carbon monoxide |
| $CO_2$ | carbon dioxide |
| CR | compression ratio |
| CRDI | common rail direct injection |
| DEE | diethyl ether |
| DI | direct injection |
| EFBs | empty fruit bunches |

| | |
|---|---|
| EGR | exhaust gas recirculation |
| EMA | engine manufacturers association |
| FOME | fish oil methyl ester |
| FTIR | Fourier transform infrared spectroscopy |
| GC | gas chromatography |
| GC-MS | gas chromatography-mass spectrometry |
| GHG | greenhouse gas |
| HC | hydrocarbons |
| JME | jatropha methyl ester |
| LGO | lemongrass oil |
| NA | naturally aspirated |
| NOx | nitrogen oxides |
| PDWO | palm-derived waste oil |
| PM | particulate matters |
| rpm | rotation per minute (unit) |
| RSO | rapeseed oil |
| TCLP | toxicity characteristic leaching procedure |
| WC | water-cooled |
| WCO | waste cooking oil |
| WPO | wood pyrolysis oil |

## References


1. Administration, U.S.E.I. Energy Information Administration. *Choice Rev. Online* **2007**, *44*, 44–3624. [CrossRef]
2. Li, L.; Sun, W.; Hu, W.; Sun, Y. Impact of natural and social environmental factors on building energy consumption: Based on bibliometrics. *J. Build. Eng.* **2021**, *37*, 102136. [CrossRef]
3. Chong, C.T.; Loe, T.Y.; Wong, K.Y.; Ashokkumar, V.; Lam, S.S.; Chong, W.T.; Borrion, A.; Tian, B.; Ng, J.-H. Biodiesel sustainability: The global impact of potential biodiesel production on the energy–water–food (EWF) nexus. *Environ. Technol. Innov.* **2021**, *22*, 101408. [CrossRef]
4. Soliman, M.N.; Guen, F.Z.; Ahmed, S.A.; Saleem, H.; Khalil, M.J.; Zaidi, S.J. Energy consumption and environmental impact assessment of desalination plants and brine disposal strategies. *Process Saf. Environ. Prot.* **2021**, *147*, 589–608. [CrossRef]
5. Bourdeau, M.; Zhai, X.Q.; Nefzaoui, E.; Guo, X.; Chatellier, P. Modeling and forecasting building energy consumption: A review of data-driven techniques. *Sustain. Cities Soc.* **2019**, *48*, 101533. [CrossRef]
6. OPEC. *OPEC 2020 World Oil Outlook 2045*; OPEC: Vienna, Austria, 2020; Volume 14, ISBN 9783950489002.
7. Luque, R.; Lin, C.S.K.; Wilson, K.; Clark, J. *Handbook of Biofuels Production: Processes and Technologies*, 2nd ed.; Woodhead Publishing: Cambridge, UK, 2016; ISBN 9780081004562.
8. Schmidt, J.; De Rosa, M. Certified palm oil reduces greenhouse gas emissions compared to non-certified. *J. Clean. Prod.* **2020**, *277*, 124045. [CrossRef]
9. Lee, S.; Shah, Y.T. *Biofuels and Bioenergy Processes and Technologies*; CRC Press: Florida, FL, USA, 2013; ISBN 9781420089561.
10. Mohiddin, M.N.B.; Tan, Y.H.; Seow, Y.X.; Kansedo, J.; Mubarak, N.; Abdullah, M.O.; Chan, Y.S.; Khalid, M. Evaluation on feedstock, technologies, catalyst and reactor for sustainable biodiesel production: A review Mohd. *Build. Environ.* **2020**, *184*, 107229. [CrossRef]
11. Raheem, I.; Mohiddin, M.N.B.; Tan, Y.H.; Kansedo, J.; Mubarak, N.M.; Abdullah, M.O.; Ibrahim, M.L. A review on influence of reactor technologies and kinetic studies for biodiesel application. *J. Ind. Eng. Chem.* **2020**, *91*, 54–68. [CrossRef]
12. Aydın, S. Comprehensive analysis of combustion, performance and emissions of power generator diesel engine fueled with different source of biodiesel blends. *Energy* **2020**, *205*. [CrossRef]
13. Bibin, C.; Gopinath, S.; Aravindraj, R.; Devaraj, A.; Gokula Krishnan, S.; Jeevaananthan, J.K.S. The production of biodiesel from castor oil as a potential feedstock and its usage in compression ignition Engine: A comprehensive review. *Mater. Today Proc.* **2019**, *33*, 84–92. [CrossRef]
14. Jayakumar, M.; Karmegam, N.; Gundupalli, M.P.; Gebeyehu, K.B.; Asfaw, B.T.; Chang, S.W.; Balasubramani, R.; Awasthi, M.K. Heterogeneous base catalysts: Synthesis and application for biodiesel production—A review. *Bioresour. Technol.* **2021**, 125054. [CrossRef] [PubMed]
15. Alagumalai, A.; Mahian, O.; Hollmann, F.; Zhang, W. Environmentally benign solid catalysts for sustainable biodiesel production: A critical review. *Sci. Total Environ.* **2021**, *768*, 144856. [CrossRef]
16. Vignesh, P.; Kumar, A.R.P.; Ganesh, N.S.; Jayaseelan, V.; Sudhakar, K. A review of conventional and renewable biodiesel production. *Chin. J. Chem. Eng.* **2020**. [CrossRef]
17. Chozhavendhan, S.; Singh, M.V.P.; Fransila, B.; Kumar, R.P.; Devi, G.K. A review on influencing parameters of biodiesel production and purification processes. *Curr. Res. Green Sustain. Chem.* **2020**, *1*, 1–6. [CrossRef]
18. Syafiuddin, A.; Chong, J.H.; Yuniarto, A.; Hadibarata, T. The current scenario and challenges of biodiesel production in Asian countries: A review. *Bioresour. Technol. Rep.* **2020**, *12*, 100608. [CrossRef]

19. Pullen, J.; Saeed, K. Factors affecting biodiesel engine performance and exhaust emissions—Part I: Review. *Energy* **2014**, *72*, 1–16. [CrossRef]
20. Macor, A.; Pavanello, P. Performance and emissions of biodiesel in a boiler for residential heating. *Energy* **2009**, *34*, 2025–2032. [CrossRef]
21. Bazooyar, B.; Ghorbani, A.; Shariati, A. Combustion performance and emissions of petrodiesel and biodiesels based on various vegetable oils in a semi industrial boiler. *Fuel* **2011**, *90*, 3078–3092. [CrossRef]
22. Ghorbani, A.; Bazooyar, B.; Shariati, A.; Jokar, S.M.; Ajami, H.; Naderi, A. A comparative study of combustion performance and emission of biodiesel blends and diesel in an experimental boiler. *Appl. Energy* **2011**, *88*, 4725–4732. [CrossRef]
23. Alonso, J.S.J.; Sastre, J.A.L.; Romero-Ávila, C.; López, E. A note on the combustion of blends of diesel and soya, sunflower and rapeseed vegetable oils in a light boiler. *Biomass Bioenergy* **2008**, *32*, 880–886. [CrossRef]
24. Suthisripok, T.; Semsamran, P. The impact of biodiesel B100 on a small agricultural diesel engine. *Tribol. Int.* **2018**, *128*, 397–409. [CrossRef]
25. Shahir, V.K.; Jawahar, C.P.; Suresh, P.R.; Vinod, V. Experimental Investigation on Performance and Emission Characteristics of a Common Rail Direct InjectionEngine Using Animal Fat Biodiesel Blends. *Energy Procedia* **2017**, *117*, 283–290. [CrossRef]
26. Ishola, F.; Adelekan, D.; Mamudu, A.; Abodunrin, T.; Aworinde, A.; Olatunji, O.; Akinlabi, S. Biodiesel production from palm olein: A sustainable bioresource for Nigeria. *Heliyon* **2020**, *6*, e03725. [CrossRef] [PubMed]
27. Bazooyar, B.; Shariati, A.; Hashemabadi, S.H. Economy of a utility boiler power plant fueled with vegetable oil, biodiesel, petrodiesel and their prevalent blends. *Sustain. Prod. Consum.* **2015**, *3*, 1–7. [CrossRef]
28. Aghbashlo, M.; Peng, W.; Tabatabaei, M.; Kalogirou, S.A.; Soltanian, S.; Hosseinzadeh-Bandbafha, H.; Mahian, O.; Lam, S.S. Machine learning technology in biodiesel research: A review. *Prog. Energy Combust. Sci.* **2021**, *85*, 100904. [CrossRef]
29. Gebremariam, S.N.; Marchetti, J.M. Economics of biodiesel production: Review. *Energy Convers. Manag.* **2018**, *168*, 74–84. [CrossRef]
30. Rochelle, D.; Najafi, H. A review of the effect of biodiesel on gas turbine emissions and performance. *Renew. Sustain. Energy Rev.* **2019**, *105*, 129–137. [CrossRef]
31. Sharma, A.; Maurya, N.K.; Singh, Y.; Singh, N.K.; Gupta, S.K. Effect of design parameters on performance and emissions of DI diesel engine running on biodiesel-diesel blends: Taguchi and utility theory. *Fuel* **2020**, *281*. [CrossRef]
32. Abu-Hamdeh, N.H.; Bantan, R.A.R.; Alimoradi, A.; Pourhoseini, S.H. The effect of injection pressure on the thermal performance and emission characteristics of an oil burner operating on B20 palm oil biodiesel-diesel blend fuel. *Fuel* **2020**, *278*, 118174. [CrossRef]
33. Roy, M.M.; Wang, W.; Bujold, J. Biodiesel production and comparison of emissions of a DI diesel engine fueled by biodiesel-diesel and canola oil-diesel blends at high idling operations. *Appl. Energy* **2013**, *106*, 198–208. [CrossRef]
34. Yasin, M.H.M.; Paruka, P.; Mamat, R.; Yusop, A.F.; Najafi, G.; Alias, A. Effect of Low Proportion Palm Biodiesel Blend on Performance, Combustion and Emission Characteristics of a Diesel Engine. *Energy Procedia* **2015**, *75*, 92–98. [CrossRef]
35. Al-Dawody, M.F.; Bhatti, S.K. Experimental and computational investigations for combustion, performance and emission parameters of a diesel engine fueled with soybean biodiesel-diesel blends. *Energy Procedia* **2014**, *52*, 421–430. [CrossRef]
36. Balaji, D.; Pillai, T.M.; Gnanasekaran, K.; Balachandar, M.; Ravikumar, T.S.; Sathish, S.; Sathyamurthy, R. Dataset for compression ignition engine fuelled with corn oil methyl ester biodiesel. *Data Br.* **2019**, *27*, 104683. [CrossRef] [PubMed]
37. Öztürk, E.; Can, Ö.; Usta, N.; Yücesu, H.S. Effects of retarded fuel injection timing on combustion and emissions of a diesel engine fueled with canola biodiesel. *Eng. Sci. Technol. Int. J.* **2020**, *23*, 1466–1475. [CrossRef]
38. Paul, G.; Datta, A.; Mandal, B.K. An experimental and numerical investigation of the performance, combustion and emission characteristics of a diesel engine fueled with jatropha biodiesel. *Energy Procedia* **2014**, *54*, 455–467. [CrossRef]
39. Hariram, V.; Solomon, G.R.; Raj, D.S.; Dev, M.J.; Kumar, U.N.; Gokulakesavan, M.; Premkumar, T.M.; Seralathan, S. Impact of compression ratio in the emission and performance phenomenon of a CI engine fuelled with jojoba biodiesel blends. *Mater. Today Proc.* **2020**. [CrossRef]
40. Hemanandh, J.; Narayanan, K.V. Emission and Performance analysis of hydrotreated refined sunflower oil as alternate fuel. *Alex. Eng. J.* **2015**, *54*, 389–393. [CrossRef]
41. Santos, B.S.; Capareda, S.C.; Capunitan, J.A. Engine Performance and Exhaust Emissions of Peanut Oil Biodiesel. *J. Sustain. Bioenergy Syst.* **2013**, *03*, 272–286. [CrossRef]
42. Rashedul, H.K.; Masjuki, H.H.; Kalam, M.A.; Ashraful, A.M.; Rashed, M.M.; Sanchita, I.; Shaon, T. Performance and emission characteristics of a compression ignition engine running with linseed biodiesel. *RSC Adv.* **2014**, *4*, 64791–64797. [CrossRef]
43. Balasubramanian, K.; Purushothaman, K. Performance, emission and combustion characteristics of safflower, neem and corn biodiesels fuelled in a CI engine. *Nat. Environ. Pollut. Technol.* **2019**, *18*, 1265–1273.
44. Islam, S.; Ahmed, A.S.; Islam, A.; Aziz, S.A.; Xian, L.C.; Mridha, M. Castor Biodiesel. *J. Chem.* **2014**, *2014*, 1–8. [CrossRef]
45. Sundar, K.; Udayakumar, R. Comparative evaluation of the performance of rice bran and cotton seed biodiesel blends in VCR diesel engine. *Energy Rep.* **2020**, *6*, 795–801. [CrossRef]
46. Anawe, P.A.L.; Folayan, J.A. Data on physico-chemical, performance, combustion and emission characteristics of Persea Americana Biodiesel and its blends on direct-injection, compression-ignition engines. *Data Br.* **2018**, *21*, 1533–1540. [CrossRef] [PubMed]

47. Raman, R.; Kumar, N. Performance and emission characteristics of twin cylinder diesel engine fueled with mahua biodiesel and DEE. *Transp. Eng.* **2020**, *2*, 100024. [CrossRef]
48. Sureshkumar, K. Performance and Characteristics Study of the Use of Environment Friendly Pongamia Pinnata Methyl Ester in CI Engines. *J. Energy Environ.* **2007**, *5*, 60–71.
49. Sanjid, A.; Masjuki, H.H.; Kalam, M.A.; Abedin, M.J.; Rahman, S.M.A. Experimental Investigation of Mustard Biodiesel Blend Properties, Performance, Exhaust Emission and Noise in an Unmodified Diesel Engine. *APCBEE Procedia* **2014**, *10*, 149–153. [CrossRef]
50. Liaquat, A.M.; Masjuki, H.H.; Kalam, M.A.; Fattah, I.M.R.; Hazrat, M.A.; Varman, M.; Mofijur, M.; Shahabuddin, M. Effect of coconut biodiesel blended fuels on engine performance and emission characteristics. *Procedia Eng.* **2013**, *56*, 583–590. [CrossRef]
51. Afif, M.K.; Biradar, C.H. Production of Biodiesel from Cannabis sativa (Hemp) Seed oil and its Performance and Emission Characteristics on DI Engine Fueled with Biodiesel Blends. *Int. Res. J. Eng. Technol.* **2019**, *6*, 246–253.
52. Akar, M.A. Performance and emission characteristics of compression ignition engine operating with false flax biodiesel and butanol blends. *Adv. Mech. Eng.* **2016**, *8*, 1–7. [CrossRef]
53. Chaichan, M.T.; Sabah, P.; Ahmed, T. Evaluation of Performance and Emissions Characteristics for Compression Ignition Engine Operated with Disposal Yellow Grease. *Int. J. Eng. Sci.* **2013**, *2*, 111–122.
54. Avase, S.A.; Srivastava, S.; Vishal, K.; Ashok, H.V.; Varghese, G. Effect of Pyrogallol as an Antioxidant on the Performance and Emission Characteristics of Biodiesel Derived from Waste Cooking Oil. *Procedia Earth Planet. Sci.* **2015**, *11*, 437–444. [CrossRef]
55. Erdoğan, S.; Balki, M.K.; Aydın, S.; Sayın, C. Performance, emission and combustion characteristic assessment of biodiesels derived from beef bone marrow in a diesel generator. *Energy* **2020**, *207*. [CrossRef]
56. Selvam, M.; Srinivasan, C. An investigation Performances, Emission and Combustion Characteristics on Sheep Fat oil as Biodiesel with help of Ultrasonic assisted Transesterification Process. *Int. J. Eng. Appl. Sci.* **2017**, *4*, 257385.
57. John Panneer Selvam, D.; Vadivel, K. The effects of ethanol addition with waste pork lard methyl ester on performance, emission, and combustion characteristics of a diesel engine. *Therm. Sci.* **2014**, *18*, 217–228. [CrossRef]
58. Elkelawy, M.; Alm-Eldin Bastawissi, H.; El Shenawy, E.A.; Taha, M.; Panchal, H.; Sadasivuni, K.K. Study of performance, combustion, and emissions parameters of DI-diesel engine fueled with algae biodiesel/diesel/n-pentane blends. *Energy Convers. Manag. X* **2020**, 100058. [CrossRef]
59. Nautiyal, P.; Subramanian, K.A.; Dastidar, M.G.; Kumar, A. Experimental assessment of performance, combustion and emissions of a compression ignition engine fuelled with Spirulina platensis biodiesel. *Energy* **2020**, *193*. [CrossRef]
60. Prakash, S.; Prabhahar, M.; Sendilvelan, S.; Venkatesh, R.; Singh, S.; Bhaskar, K. Experimental studies on the performance and emission characteristics of an automobile engine fueled with fish oil methyl ester to reduce environmental pollution. *Energy Procedia* **2019**, *160*, 412–419. [CrossRef]
61. Labecki, L.; Cairns, A.; Xia, J.; Megaritis, A.; Zhao, H.; Ganippa, L.C. Combustion and emission of rapeseed oil blends in diesel engine. *Appl. Energy* **2012**, *95*, 139–146. [CrossRef]
62. Altin, R.; Çetinkaya, S.; Yücesu, H.S. Potential of using vegetable oil fuels as fuel for diesel engines. *Energy Convers. Manag.* **2001**, *42*, 529–538. [CrossRef]
63. Chen, H.; Xie, B.; Ma, J.; Chen, Y. NOx emission of biodiesel compared to diesel: Higher or lower? *Appl. Therm. Eng.* **2018**, *137*, 584–593. [CrossRef]
64. Reddy, E.R.K.; Subbalakshmi, Y.; Raju, V.D.; Rao, K.A.; Kumar, M.H.; Reddy, S.R.; Sai, P.T.; Subbalakshmi, Y.; Raju, V.D.; Rao, K.A.; et al. Assessment of performance, combustion and emission characteristics of the Assessment of performance, combustion and emission characteristics of the diesel engine powered with corn biodiesel blends. *Int. J. Ambient Energy* **2019**, 1–9. [CrossRef]
65. Shehu, B.G.; Clarke, M.L. Successful and sustainable crop based biodiesel programme in Nigeria through ecological optimisation and intersectoral policy realignment. *Renew. Sustain. Energy Rev.* **2020**, *134*. [CrossRef]
66. Yin, C.Y.; Kadir, S.A.S.A.; Lim, Y.P.; Syed-Ariffin, S.N.; Zamzuri, Z. An investigation into physicochemical characteristics of ash produced from combustion of oil palm biomass wastein a boiler. *Fuel Process. Technol.* **2008**, *89*, 693–696. [CrossRef]
67. Sukiran, M.A.; Abnisa, F.; Syafiie, S.; Daud, W.M.A.W.; Nasrin, A.B.; Aziz, A.A.; Loh, S.K. Experimental and modelling study of the torrefaction of empty fruit bunches as a potential fuel for palm oil mill boilers. *Biomass Bioenergy* **2020**, *136*, 105530. [CrossRef]
68. Hanna, M.; Fangrui, M. Biodiesel production: A review. *Bioresour. Technol.* **1999**, *70*, 1–15.
69. Abbaszaadeh, A.; Ghobadian, B.; Omidkhah, M.R.; Najafi, G. Current biodiesel production technologies: A comparative review. *Energy Convers. Manag.* **2012**, *63*, 138–148. [CrossRef]
70. De Almeida, S.C.A.; Belchior, C.R.; Nascimento, M.V.G.; Vieira, L.D.S.R.; Fleury, G. Performance of a diesel generator fuelled with palm oil. *Fuel* **2002**, *81*, 2097–2102. [CrossRef]
71. Agarwal, A.K.; Rajamanoharan, K. Experimental investigations of performance and emissions of Karanja oil and its blends in a single cylinder agricultural diesel engine. *Appl. Energy* **2009**, *86*, 106–112. [CrossRef]
72. Sathiyamoorthi, R.; Sankaranarayan, G. Fuel injection timings of a direct injection diesel engine running on neat lemongrass oil-diesel blends. *Int. J. Automot. Mech. Eng.* **2015**, *11*, 2348. [CrossRef]
73. Altun, Ş.; Bulut, H.; Öner, C. The comparison of engine performance and exhaust emission characteristics of sesame oil-diesel fuel mixture with diesel fuel in a direct injection diesel engine. *Renew. Energy* **2008**, *33*, 1791–1795. [CrossRef]
74. Forson, F.K.; Oduro, E.K.; Hammond-Donkoh, E. Performance of jatropha oil blends in a diesel engine. *Renew. Energy* **2004**, *29*, 1135–1145. [CrossRef]

75. Patidar, V.; Chandra, A.; Singh, M.; Kale, R.K. Phase behaviour and physicochemical study of Karanj oil-ethanol microemulsion as alternative renewable biofuel. *J. Sci. Ind. Res.* **2014**, *73*, 461–464.
76. Liang, J.; Qian, Y.; Yuan, X.; Leng, L.; Zeng, G.; Jiang, L.; Shao, J.; Luo, Y.; Ding, X.; Yang, Z.; et al. Span80/Tween80 stabilized bio-oil-in-diesel microemulsion: Formation and combustion. *Renew. Energy* **2018**, *126*, 774–782. [CrossRef]
77. Ziejewski, M.; Kaufman, K.R.; Schwab, A.W.; Pryde, E.H. Diesel engine evaluation of a nonionic sunflower oil-aqueous ethanol microemulsion. *J. Am. Oil Chem. Soc.* **1984**, *61*, 1620–1626. [CrossRef]
78. Singh, P.J.; Khurma, J.; Singh, A. Preparation, characterisation, engine performance and emission characteristics of coconut oil based hybrid fuels. *Renew. Energy* **2010**, *35*, 2065–2070. [CrossRef]
79. Charoensaeng, A.; Khaodhiar, S.; Sabatini, D.A.; Arpornpong, N. Exhaust emissions of a diesel engine using ethanol-in-palm oil/diesel microemulsion-based biofuels. *Environ. Eng. Res.* **2018**, *23*, 242–249. [CrossRef]
80. Qi, D.H.; Yang, K.; Zhang, D.; Chen, B.; Wei, Q.; Zhang, C.H. Experimental investigation of a turbocharged CRDI diesel engine fueled with Tung oil-diesel-ethanol microemulsion fuel. *Renew. Energy* **2017**, *113*, 1201–1207. [CrossRef]
81. Qi, D.H.; Chen, H.; Matthews, R.D.; Bian, Y.Z. Combustion and emission characteristics of ethanol-biodiesel-water micro-emulsions used in a direct injection compression ignition engine. *Fuel* **2010**, *89*, 958–964. [CrossRef]
82. Neto, A.A.D.; Fernandes, M.R.; Neto, E.L.B.; Dantas, T.N.C.; Moura, M.C.P.A. Effect of Biodiesel/Diesel-Based Microemulsions on the Exhaust Emissions of a Diesel Engine. *Braz. J. Pet. Gas* **2014**, *7*, 141–153. [CrossRef]
83. Prasad, S.S.; Singh, A.; Prasad, S. Degummed Pongamia oil—Ethanol microemulsions as novel alternative CI engine fuels for remote Small Island Developing States: Preparation, characterization, engine performance and emissions characteristics. *Renew. Energy* **2020**, *150*, 401–411. [CrossRef]
84. Imtenan, S.; Masjuki, H.H.; Varman, M.; Arbab, M.I.; Sajjad, H.; Fattah, I.M.R.; Abedin, M.J.; Abu, A.S. Emission and performance improvement analysis of biodiesel-diesel blends with additives. *Procedia Eng.* **2014**, *90*, 472–477. [CrossRef]
85. Rajalingam, A.; Jani, S.P.; Kumar, A.S.; Khan, M.A. Production methods of biodiesel. *J. Chem. Pharm. Res.* **2016**, *8*, 170–173.
86. Du, Z.; Li, Y.; Wang, X.; Wan, Y.; Chen, Q.; Wang, C.; Lin, X.; Liu, Y.; Chen, P.; Ruan, R. Microwave-assisted pyrolysis of microalgae for biofuel production. *Bioresour. Technol.* **2011**, *102*, 4890–4896. [CrossRef] [PubMed]
87. Schwab, A.W.; Dykstra, G.J.; Selke, E.; Sorenson, S.C.; Pryde, E.H. Diesel fuel from thermal decomposition of soybean oil. *J. Am. Oil Chem. Soc.* **1988**, *65*, 1781–1786. [CrossRef]
88. Lima, D.G.; Soares, V.C.D.; Ribeiro, E.B.; Carvalho, D.A.; Cardoso, É.C.V.; Rassi, F.C.; Mundim, K.C.; Rubim, J.C.; Suarez, P.A.Z. Diesel-like fuel obtained by pyrolysis of vegetable oils. *J. Anal. Appl. Pyrolysis* **2004**, *71*, 987–996. [CrossRef]
89. Prakash, R.; Singh, R.K.; Murugan, S. Experimental Studies on a Diesel Engine Fueled withWood Pyrolysis Oil Diesel Emulsions. *Int. J. Chem. Eng. Appl.* **2011**, *2*, 395–399. [CrossRef]
90. Prakash, R.; Singh, R.K.; Murugan, S. Experimental investigation on a diesel engine fueled with bio-oil derived from waste wood-biodiesel emulsions. *Energy* **2013**, *55*, 610–618. [CrossRef]
91. Megahed, O.A.; Keera, S.T.; Abdallah, R.I.; Zaher, F.A. Thermally decomposed ricebran oil as a diesel fuel. *Grasas Aceites* **1998**, *49*, 165–169. [CrossRef]
92. Da Mota, S.D.P.; Mancio, A.A.; Lhamas, D.E.L.; De Abreu, D.H.; Da Silva, M.S.; Dos Santos, W.G.; De Castro, D.A.R.; De Oliveira, R.M.; Araújo, M.E.; Borges, L.E.P.; et al. Production of green diesel by thermal catalytic cracking of crude palm oil (Elaeis guineensis Jacq) in a pilot plant. *J. Anal. Appl. Pyrolysis* **2014**, *110*, 1–11. [CrossRef]
93. Barnwal, B.K.; Sharma, M.P. Prospects of biodiesel production from vegetable oils in India. *Renew. Sustain. Energy Rev.* **2005**, *9*, 363–378. [CrossRef]
94. Marchetti, J.M.; Miguel, V.U.; Errazu, A.F. Possible methods for biodiesel production. *Renew. Sustain. Energy Rev.* **2007**, *11*, 1300–1311. [CrossRef]
95. Fukuda, H.; Kondo, A.; Noda, H. Biodiesel fuel production by transesterification of oils. *J. Biosci. Bioeng.* **2001**, *92*, 405–416. [CrossRef]
96. Sathish, A.; Smith, B.R.; Sims, R.C. Effect of moisture on in situ transesterification of microalgae for biodiesel production. *J. Chem. Technol. Biotechnol.* **2014**, *89*, 137–142. [CrossRef]
97. Haas, M.J.; Scott, K.M. Moisture removal substantially improves the efficiency of in situ biodiesel production from soybeans. *J. Am. Oil Chem. Soc.* **2007**, *84*, 197–204. [CrossRef]
98. Kusdiana, D.; Saka, S. Kinetics of transesterification in rapeseed oil to biodiesel fuel as treated in supercritical methanol. *Fuel* **2001**, *80*, 693–698. [CrossRef]
99. Kapilakarn, K.; Peugtong, A. A comparison of costs of biodiesel production from transesterication. *Int. Energy J.* **2007**, *8*, 1–6.
100. Pang, H.; Yang, G.; Li, L.; Yu, J. Efficient transesterification over two-dimensional zeolites for sustainable biodiesel production. *Green Energy Environ.* **2020**. [CrossRef]
101. Mansir, N.; Teo, S.H.; Mijan, N.A.; Taufiq-Yap, Y.H. Efficient reaction for biodiesel manufacturing using bi-functional oxide catalyst. *Catal. Commun.* **2021**, *149*, 106201. [CrossRef]
102. Kongjan, P.; Jariyaboon, R.; Reungsang, A.; Sittijunda, S. Co-fermentation of 1,3-propanediol and 2,3-butanediol from crude glycerol derived from the biodiesel production process by newly isolated *Enterobacter* sp.: Optimization factors affecting. *Bioresour. Technol. Rep.* **2021**, *13*, 100616. [CrossRef]
103. Bazooyar, B.; Hashemabadi, S.H.; Shariati, A. NOX formation of biodiesel in utility power plant boilers; Part B. Comparison of NO between biodiesel and petrodiesel. *Fuel* **2016**, *182*, 323–332. [CrossRef]

104. Sun, C.; Liu, Y.; Qiao, X.; Ju, D.; Tang, Q.; Fang, X.; Zhou, F. Experimental study of effects of exhaust gas recirculation on combustion, performance, and emissions of DME-biodiesel fueled engine. *Energy* **2020**, *197*. [CrossRef]
105. Sentanuhady, J.; Majid, A.I.; Prasidha, W.; Saputro, W.; Gunawan, N.P.; Raditya, T.Y.; Muflikhun, M.A. Analysis of the Effect of Biodiesel B20 and B100 on the Degradation of Viscosity and Total Base Number of Lubricating Oil in Diesel Engines with Long-Term Operation Using ASTM D2896 and ASTM D445-06 Methods. *TEKNIK* **2020**, *41*, 269–274. [CrossRef]
106. Sentanuhady, J.; Saputro, W.; Muflikhun, M.A. Metals and chemical compounds contaminants in diesel engine lubricant with B20 and B100 biofuels for long term operation. *Sustain. Energy Technol. Assess.* **2021**, *45*, 101161. [CrossRef]

*sustainability* MDPI

*Review*

# A Review of Groundwater Management Models with a Focus on IoT-Based Systems

**Banjo Ayoade Aderemi [1,*], Thomas Otieno Olwal [1], Julius Musyoka Ndambuki [2] and Sophia Sudi Rwanga [3]**

1 Department of Electrical Engineering, Tshwane University of Technology, Pretoria 0183, South Africa; OlwalTO@tut.ac.za
2 Department of Civil Engineering, Tshwane University of Technology, Pretoria 0183, South Africa; NdambukiJM@tut.ac.za
3 Department of Civil Engineering, Vaal University of Technology, Vanderbijlpark 1900, South Africa; sophiar@vut.ac.za
* Correspondence: ayoadebanjo93@hotmail.com; Tel.: +27-78-354-2284

**Abstract:** Globally, groundwater is the largest distributed storage of freshwater and plays an important role in an ecosystem's sustainability in addition to aiding human adaptation to both climatic change and variability. However, groundwater resources are dynamic and often change as a result of land usage, abstraction, as well as variation in climate. To solve these challenges, many conventional solutions, such as certain numerical techniques, have been proffered for groundwater modelling. The global evolution of the Internet of Things (IoT) has enhanced the culture of data gathering for the management of groundwater resources. In addition, efficient data-driven groundwater resource management relies hugely on information relating to changes in groundwater resources as well as their availability. At the moment, some studies in the literature reveal that groundwater managers lack an efficient and real-time groundwater management system which is needed to gather the required data. Additionally, the literature reveals that the existing methods of collecting data lack the required efficiency to meet computational model requirements and meet management objectives. Unlike previous surveys, which solely focussed on particular groundwater issues related to simulation and optimisation management methods, this paper seeks to highlight the current groundwater management models as well as the IoT contributions.

**Keywords:** Internet of Things (IoT); groundwater level; groundwater resource; groundwater management models; groundwater monitoring system; wireless sensor network

**Citation:** Aderemi, B.A.; Olwal, T.O.; Ndambuki, J.M.; Rwanga, S.S. A Review of Groundwater Management Models with a Focus on IoT-Based Systems. *Sustainability* **2022**, *14*, 148. https://doi.org/10.3390/su14010148

Academic Editor: Shervin Hashemi

Received: 17 November 2021
Accepted: 21 December 2021
Published: 23 December 2021

**Publisher's Note:** MDPI stays neutral with regard to jurisdictional claims in published maps and institutional affiliations.



## 1. Introduction

About approximately one-third of global freshwater consumption depends on groundwater resources; thus, it has become an important source of freshwater globally [1]. In the water cycle, the freshwater resource accounts for approximately 4% of the total water available on earth, while the remaining 96% is salty, found within seas and oceans [2]. Meanwhile, only about 0.001% of water is available as a groundwater resource that is hidden underground, while 75% is ice and about 25% is liquid water [3]. Therefore, groundwater resources constitute approximately 98% of all fresh liquid water available on earth [4]. Since both plants and animals depend on water, the interaction between surface water systems and groundwater resources is essential for basic life on the earth. This makes groundwater the largest distributable storage of freshwater which plays an important role in an ecosystem's sustainability as well as in aiding human adaptation to both climatic change and variability [5]. Human beings largely depend on groundwater resources as a major supply of their drinking water. Thus, the efficient measurement, monitoring, and management of groundwater resources are crucial to ensure future sustainability. Nonetheless, human activities and a lack of planning for these activities have led to groundwater quantity shortage in the aquifer, land subsidence, saltwater intrusion, high cost of water production, overexploitation, and unhygienic and quality contamination of water [6–16]. Additionally, the management and sustainability of groundwater resources is a global critical issue. This is in line with goal six of the Sustainable Development Goals (SDGs) adopted by the United Nations General Assembly which seeks to achieve a universal reduction in water scarcity, sustainable water management, and to substantially increase the efficiency of water usage by 2030 [17].

In the aquifer, the system of groundwater is slow and dynamic. This is because it responds to variation in hydrological stresses like climate change, land use, and abstraction among other things in two ways [18,19]. The first way is to balance out the re-charge and discharge of the aquifer system in a process known as an equilibrium state, thus, allowing the aquifer to exist indefinitely, or secondly, when the hydrological stress is so large and beyond the equilibrium state it leads to a finite aquifer's life span [20,21]. Depending on the region, the replenishment of a utilised groundwater resource is typically slow, thus leading to a reduction in the level of resources. In addition to usage, as a result of the high non-linear and non-stationary nature of groundwater resources in time series, its efficient management depends on various complex factors such as precipitation and other environmental factors [22]. Consequently, groundwater models are essential management tools that are used to locate these adverse effects and the ideal balance of groundwater abstraction on the aquifer [23]. Furthermore, groundwater modelling assists in determining an aquifer's response to hydrological conditions as well as promoting the sustainable management of groundwater resources [24]. Therefore, groundwater models are crucial to ensuring a sustainable future for water scarcity reduction and overexploitation as well as for the improvement in the efficient use of water resources. Consequently, it is of vital importance that an effective management model for accurate groundwater-level management models be developed [25,26].

The measurement of groundwater levels is important to avoid depletion of groundwater resources as earlier stated [2]. Groundwater level measurement assists in determining the hydrological stress acting in an aquifer and provides data for efficient management. For long-term forecast and management, groundwater level measurements supply data to develop a groundwater model [27,28]. This has made groundwater management models become a standard instrument used by water managers and professionals to solve most groundwater-related problems. Thus, there are many models, such as data-driven models, numerical models, as well as nonlinear models, which are used for groundwater modelling [29–31]. However, it is important to understand the nature and characteristics of groundwater levels to be able to efficiently measure or model groundwater resources. Additionally, the nature of groundwater level measurement in time series is described to be non-stationary and non-linear which means it depends on various complex surrounding agents such as precipitation, intervening aquitards, groundwater aquifers, and other hydrogeological characteristics of the aquifers involved [32]. Additionally, the groundwater aquifers systems' interaction with the surface water at different temporal-spatial scales is described as intrinsic in heterogeneous systems that are strained by a knotty hydrogeological state [33,34]. Thus, the processes and the physical attributes that form groundwater flow within an aquifer are very heterogeneous [35]. To effectively take into consideration all these properties of groundwater resources, various groundwater management models warrant complex and highly distributed models in a time series. Previously, the tendency in groundwater resource management through distributed modelling was to increase the numerical resolution, include many physical attributes, and to enlarge the model's domain using either a finite difference approximation or a finite element method [36–39]. Although the application of either the finite element or finite difference methods to groundwater resource management has enabled real-world and complex systems to be modelled, this has led to an increase in the model's running time.

In many instances, groundwater levels are manually measured using a data logger coupled with a pressure transducer and electrical probes or a graduated steel tape dipper [2,40]. Although these techniques are simple and easy to use, they are prone to human error, are unreliable, and are inefficient. Since the development of super-fast-speed computers as well as user-friendly software, the field of hydrogeology has frequently used physical-based numerical techniques to model groundwater. However, it is difficult to define the needed information accurately because spatial data on hydrological as well as geological properties of the aquifer are needed [18]. Hence, data gathering is essential to maintaining groundwater resource management and to perform crucial water-related functions, such as mitigating against drought [40]. Likewise, many researchers have combined many physical numerical techniques such as both simulation and optimisation modelling methods to manage groundwater resources [28,41–47]. Additionally, different groundwater resource management models have been developed to minimise groundwater resource scarcity [18,22,29,32–34,48–65].

Chang et al. [34], developed a numerical groundwater resource model from a conceptual model but failed to capture the computational complexities in the nature of the groundwater aquifer system while focusing only on the fundamental and main principles [34]. This is because it is difficult to formulate both groundwater flow equations and prove the hydrogeological parameters for conceptual models due to computational complexity. Additionally, there are uncertainties and difficulties in obtaining long-term series data for groundwater levels using numerical modelling processes [32,50,66]. Thus, making long-term series models time-consuming, laborious, unscalable, and costly [54]. However, it is important to achieve an acceptable groundwater resources model for performance efficiency. However, past information was not often available because most government agencies in charge of water often collect this data just once or twice a year during the agricultural season [64]. Additionally, the majority of traditional groundwater models are process-based [18]. This means much supplementary spatial data on the aquifer's hydrological and geological properties is mandatory. Further challenges occur in groundwater modelling due to the complexity of hydrogeology sub-surfaces as well as a spatiotemporally variability in societal pumping activities [65].

As identified in most pieces of literature, there have been many efforts developed previously to proffer solutions to groundwater modelling. Over the years, the use of machine learning as an alternative solution for the modelling of groundwater has gained more attraction [34]. In contrast with numerical models, machine learning models are data-driven and require learning from historically measured data from the aquifer system. These models do not need to comprehend the physical processes or the internal framework of an aquifer [23,67]. Other solutions such as linear and nonlinear programming techniques which are driven solely by data acquisition have also been proffered [9,14,15,41,43,46,47,54–56,68]. Data-driven classification models such as Genetic Programming (GP), Artificial Neutral Networks (ANNs), Fuzzy Logic, as well as Support Vector Machines (SVMs), and time-series techniques such as Autoregressive Integrated Moving Average (ARIMA) and Autoregressive Moving Average (ARMA), are proven alternatives to conceptual models [22,29,50,51]. However, they are not scalable enough when there is a change in dynamic groundwater levels over time as well as when available data are insufficient to give accurate results [18]. Furthermore, Husna et al. and Kenda et al. observed that the majority of groundwater resource management models have failed to capture computation efficiency and scalability [18,52]. Despite the huge improvements in the existing groundwater resource management models, researchers pointed out lack of efficiency and scalability as a result of considerable uncertainty, over-dependent on unavailable additional datasheets and potential substantial errors [58,69–75]. Both data-driven machine learning models and numerical techniques have been used extensively in modelling groundwater resources, however, the numerical techniques have physical limitations such as the aquifer's state as well as it having extensive physical properties qualification, being laborious, and being expensive [76]. Therefore, it is necessary to develop a model that can identify these limitations and use the information for groundwater management and policymaking [77].

In 1990, John Romkey created the first world's Internet of Things (IoT) Interop Internet Toaster device. However, the term IoT was first coined out in 1999 by Kevin Ashton [78]. This was a toasting machine that was operated from a computer system that was accessing the management information base using a simple network management protocol and the internet protocol [79]. Since then, over 500 million devices have been connected to the internet. Furthermore, about 20 to 30 billion IoT devices are estimated to be connected to the internet by 2021 [80]. Over the years, the concept of the IoT has been embraced in areas including smart water, smart healthcare, smart agriculture, smart climate management, and so on [81,82]. This concept of the IoT refers to a phenomenon in which there is a connection of many smart things such as sensors, mobile phones, utility devices and industrial components through networks that possess data analysis capabilities. Thus, smart sensors and pressure transducers have found their usage in groundwater resource management, using the traditional IoT architecture for continuous data acquisition, analysis, and processing. In traditional IoT architecture, the IoT data generated by the smart devices are transmitted into the remote cloud through the internet [83]. As a result of huge data transfers from IoT devices into the cloud, the traditional IoT internet architecture has become inefficient for analysing and processing the collected data [82]. This is because this computation process exerts pressure on the cloud computing linked network, thereby causing untold stress. Furthermore, as identified in most pieces of literature, the cloud is not efficient and lacks the scalability to sustain large sums of IoT data in real-time due to its communication latency and network bandwidth [81–85].

Furthermore, the use of IoT in the management of groundwater levels as well as in the development of groundwater data-driven models has been instrumental in detecting areas of groundwater resource quantity reduction. However, the sensing instruments deployed into the observation wells have a low resolution and are too indirect to be of use for local regional assessments in geodetic methods [69,71,74,75]. Consequently, improvements in IoT technology in combination with improved geophysical modelling and data-assimilation are needed to meet computation that is efficient and scalable for the IoT architecture needs for groundwater resource management models.

While previous surveys have solely focussed on particular groundwater issues related to mathematical modelling and simulation models, this current review seeks to provide an IoT based inclusion perspective as it relates to groundwater resources management. Therefore, this paper aims to present a review of existing groundwater resource management models and IoT-based monitoring systems.

The arrangement of this paper is as follows; Section 2 discusses the overview and organisation of groundwater-level management models. Groundwater management and IoT are presented in Section 3. Section 4 contains the conclusion and, thereafter, the references and acknowledgements. The list of all acronyms and abbreviations used in this article is given in Table 1.

**Table 1.** Acronyms.

| Abbreviation | Description |
|---|---|
| ANFIS | Adaptive Neuro-Fuzzy Inference System |
| ANN | Artificial Neural Network |
| ARIMA | Autoregressive Integrated Moving Average |
| ARMA | Autoregressive Moving Average |
| EEMD | Ensemble Empirical Mode Decomposition |
| EMD | Empirical Mode Decomposition |
| EMI | Environmental data Management Interface technique |
| FDMT | Finite-Difference Model Technique |

Table 1. Cont.

| Abbreviation | Description |
| --- | --- |
| FEMT | Finite Element Modelling Technique |
| FFNN | Feed-Forward Neural Networks |
| GB | Gradient Boosting |
| GIS | Geographic Information Systems |
| GP | Genetic Programming |
| GPU | Graphics Processing Unit |
| GRG | Generalized Reduced Gradient |
| GWFM | Groundwater Flow Model |
| HEM | Hydro-Economic Model |
| HYDRUS-1D | Heat, Water, and Solute Movement Simulator in One-Dimensional Variably Saturated Media |
| ICT | Information and Communication Technologies |
| IDE | Integrated Development Enviroment |
| IGRAC's GGIS | International Groundwater Resources Assessment Centre's Global Groundwater Information Service |
| IMFs | Intrinsic Mode Functions |
| IoT | Internet of Things |
| ISOQUAD | Groundwater flow and Contaminants Transport Simulation Model |
| LCSNs | Low-Cost Wireless Sensors Networks |
| LoRa | Long Range |
| LR | Linear Regression |
| MAR | Managed Aquifer Recharge |
| MLPN | Multilayer Perception Network |
| MODFLOW | Modular Three-Dimensional Finite-Difference Groundwater Flow Model |
| MODPATH | Modular Particle-Tracking Post Processing Package |
| MMA | Multimodel Application |
| MQTT | Message Queuing Telemetry Transport server |
| NLP | Non-Linear Programming |
| NoSQL | No relational Structured Query Language |
| PLASM | Prickett-Lonnquist Aquifer Simulation Model |
| RS | Remote Sensing |
| RTs | Regression Trees |
| SEAWAT | A computer program for simulation of 3-Dimensional, Variable-Density, Transient Groundwater Flow |
| SDGs | Sustainable Development Goals |
| SGMP | Soil and Groundwater Management Plans |
| SGRA | Sequential Gradient Restoration Algorithm |
| SIMGRO | Simulation of Groundwater and Surface Water Levels |
| SMO | Sequential Minimal Optimisation |
| SVAT | Soil-Vegetation-Atmosphere Modelling |
| SVM | Support Vector Machine |
| SVMs | Support Vector Machines |
| SVM-QPSO | Support Vector Machine- Quantum-Behaved Particles Swarm Optimisation |
| SWAP | Soil-Water-Atmosphere-Plant Package |
| SWMS-2D | Water and Solute Movement Simulator in Two-Dimensional Variably Saturated Media |
| UCODE | A Universal Inverse Code |
| WBM | Water Balance Model |
| WebGIS | Web-based Geographic Information Systems |
| WOFOST | World Food Studies |

## 2. Overview and Organisation of Groundwater-Level Management Models

This section aims to provide an overview and organisation of the groundwater-level management model hierarchy.

### 2.1. Sources of Groundwater Recharge for an Aquifer

Naturally, the main source of groundwater aquifer recharge is the rainfall from the climate. Therefore, any upset in the climate conditions, such as less-intensive rainfalls, can lead to an alteration in the quantity of recharge for an unconfined aquifer [86]. Therefore, in a groundwater aquifer system, there is an assumption that the discharge is cancelled by the recharge, hence, there is no prolonged period of shortage in storage [20]. Although, before an abstraction occurs, there may be daily occurrences of fluctuations in the groundwater's aquifer system. However, the system remains in a dynamic equilibrium state [20]. Moreover, hydrological stress on the groundwater system is a result of fluctuations between the discharge and recharge due to climatic changes [21]. The quantity of water required to recharge an aquifer is also affected as a result of an evaporation increase [87]. Consequently, the concept of water balance or budget follows the law of the conservation of mass, it delineates the condition of an aquifer in its natural balance before abstraction over a prolonged duration [21].

After an abstraction has occurred, the groundwater aquifer system can be artificially recharged via a pumped borehole. Therefore, a groundwater aquifer system is balanced by water from any of a combination of three methods. The methods are increasing recharge to the borehole via pumping, reducing storage in the aquifer, and/or decreasing the discharge from the aquifer caused via pumping [20,88]. During the artificial pumping of groundwater from a borehole, a capture is created. A capture principal as a result of pumping can be described as a decrease in discharge as well as an increase in the recharge of water that will not voluntarily flow into the groundwater system [89,90]. Hence, the new equilibrium or water balance will be achieved after enough capture is achieved [90].

### 2.2. Overview of Groundwater Management Models

A groundwater management model is a powerful mathematical aquifer management tool that utilises numerical methods such as linear and quadratic programming and machine learning with a combination of groundwater governing flow and transport equations to solve groundwater management problems [91–99]. Therefore, a traditional simulation model is utilised to answer 'what if' related questions while an optimisation model provides a solution to 'what is best' under the given limiting boundaries [28]. Over the years, groundwater hydrogeologists have attempted to evaluate groundwater resources using numerical simulation models. The application of numerical simulation models by the researchers in the field of groundwater hydrology has assisted them to increase their understanding of regional aquifers' functions within a particular facet of groundwater systems as well as testing the hypotheses [13,28,35,41,43,59,60,100]. Additionally, the modelling of real and complex groundwater systems has been made possible via the use of finite difference and finite element simulation models. Furthermore, this has enabled the idea for evaluation and conception frameworks towards the effective management of groundwater resources, including impacts of chemical contaminants and surface water interaction and its withdrawal. In a real-world groundwater resource management scenario, the groundwater management problems are multi-dimensional and single objective simulation management methods are not adequate.

Although simulation as a tool is often utilised by groundwater managers, it fails to capture critical functional and physical restrictions while also sideling management goals. Therefore, the use of only the simulation method will not be able to achieve the optimal management objectives because determining the proper objective function of a groundwater system is very challenging and cannot be ignored. Consequently, a combination of simulation and optimisation management models is needed. An optimisation-based groundwater management model aims to provide a solution to a specified goal in the best possible manner within the various limiting boundaries. These limiting boundaries emanate from both the physical patterns of the groundwater system and the manager's requirements. In addition, optimisation as a simple tool employs the capabilities of both nonlinear and linear formulations to solve a given complex mathematical problem. Therefore, simulation-optimisation groundwater management models in optimisation formulation merge aquifer simulation models in exchange for certain constraints. Thus, optimisation and simulation groundwater models have been developed for various areas of utilisation, like the management of groundwater resource policies, the restoration of surface and groundwater resources, the control of aquifer flows, and aquifer simulation simultaneously [28,101–105].

### *2.3. The Organisation of Groundwater Management Models*

According to SM Gorelick [100], studies on groundwater management models may be organised into two classes; groundwater hydraulic management models, and groundwater policy and allocation models. The main distinction between these classes is that the hydraulic management models are primarily concerned with the management of groundwater withdrawal rate in relation to its recharge rate known as stress within a given aquifer. Consequently, these models use both hydraulic heads and stress directly as decision-making variables. Two approaches, embedding and response matrix methods, are under this model class. The embedding method for the groundwater hydraulic management model uses the Finite Difference Management Technique (FDMT) element or finite element approximations as members of the constraint set of linear programming formulation. The response matrix method uses an external groundwater simulation model to formulate unit responses [100–113]. The groundwater policy evaluation and allocation models class may be further divided into hydraulic economic response model, linked simulation–optimization models and hierarchical models. This class can be used to perform complex economic interactions analysis and policy evaluation of surface-groundwater quota challenges [102]. This class is symbolized by the use of multiple simulations and optimization techniques. These classes are discussed extensively by Remson et al., Pinder and Bredehoeft, and Gorelick, S.M. [28,32,100,102,106,111–113]. Figure 1 shows the organisation of groundwater management models into groups.

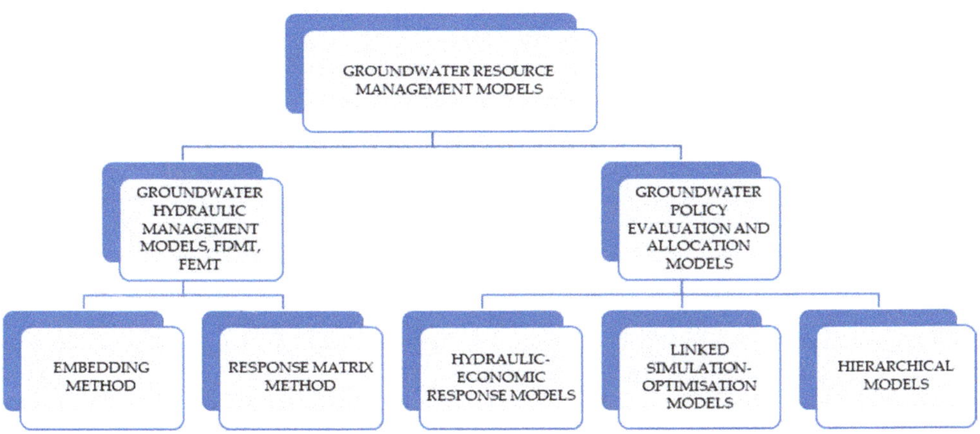


**Figure 1.** The classification of groundwater resource management models [100].

According to Kisi et al., studies on groundwater management models may be divided into two groups. These are the physical or process-based approach/model and data-driven-based approach [56]. The physical-based models are based on the use of the physical parameters of the groundwater bed to determine any changes in the water level. However, these models are difficult to execute, expensive, and need to be partitioned to obtain numerical information [62]. Furthermore, although groundwater management models are essential tools to examining the negative effects of human activities on the dynamic nature of groundwater resources, the physical and the hydrogeological surroundings require reliable data [107,108]. The aquifer of the physical groundwater management model's aquifer surrounding includes the topography, soil, climate, land use, agricultural practice, demand for groundwater usage, drainage, and canal ditches. Whereas the hydrogeological surrounding refers to the groundwater dynamism, the hydraulic parameters for each of the aquifer layers, aquifer system parameters, and aquifer boundary conditions. This hydrogeological surrounding varies in time and space [49,109]. Above all, to achieve accurate modelling results from a physical model, it is essential to have a calibration of the model using accurate data. Thus, achieving more efficient and accurate computation must be explored deeper.

Data-driven models' groups are differentiated on the objective function where the decision is based solely on groundwater hydraulic functions and the other whose management decisions are based on the evaluation of policy as well as the allocation of groundwater economy. Thus, the former is aimed at the management of groundwater stresses, like recharging and discharging, well costs, and aquifer hydraulics. These groundwater hydraulic management model groups treat both the hydraulic heads and stresses as direct management objective function decision variables. On the other hand, the groundwater policy evaluation and allocation model group aims to solve groundwater management problems by examining the complex relationship between groundwater and surface water and their economic interactions. These evaluations form the basis of regional groundwater policy [69,101,102]. The process-based/physical-based groundwater model in its primitive structure, possess four basic components; it is nonlinear concerning its decision variables; requires the solution of nonlinear partial differential equations to describe the transport as well as the flow of groundwater; it is stochastic as its primary uncertain source is associated with the aquifer simulation mode; and it is a mixed-integer programming decision because it contains discrete and continuous objective functions [14,110]. Therefore, generally these models are identified by their multiple objectives optimisation characteristics and possess robust economic management units.

In all these groups of optimisation and simulation groundwater management models, a set of different differential mathematical equations are used to describe the groundwater flow. Thus, a mathematical model is formed as a result of the combination of these different differential equations with their boundaries as well as initial conditions. The basis of groundwater modelling is governed by the combination of Darcy's law and the conservation of mass law through an anisotropic, nonhomogeneous, porous medium [111–113]. The combined flow equation is represented by Equation (1):

$$\frac{\partial}{\partial x}\left(T_x\frac{\partial h}{\partial x}\right)+\frac{\partial}{\partial y}\left(T_y\frac{\partial h}{\partial y}\right)+W = S\frac{\partial h}{\partial x} \tag{1}$$

where h is the hydraulic potential (L), x and y are cartesian coordinates (L), $T_x$ and $T_y$ are the components of the transmissivity tensor ($L^2T^{-1}$), W is a general source/sink term ($LT^{-1}$), while S is the storage coefficient, and t is time (T). The efficient measurement of groundwater level (head) is important to improving the model's usage. Hence, various non-linear time-series mathematical models have been proposed to measure groundwater levels in one as well as two-dimensional flow. However, many physical management model parameters should be treated in three dimensions. This three-dimensional management application method is an open method for groundwater resource management.

### 2.4. Data-Driven Process-Based Management Models

The latest machine learning data-driven classification models, such as Artificial Neural Network (ANN) technique, Genetic programming (GP), Adaptive Neuro-Fuzzy Inference System (ANFIS), and support vector machine (SVM) as well as time series methods such as Autoregressive Integrated Moving Average (ARIMA), Multiobjective Function Approach, and Autoregressive Moving Average (ARMA), are proven alternatives to the management of groundwater resources. They are treated as nonlinear standard estimators that can overcome the difficulties associated with physical models and are less costly. However, they are less accurate with insufficient availability of data and when the focus of research is not based on a physical mechanism [51]. Furthermore, there are numerical groundwater models developed from a conceptual model. However, these models often neglect complexities while focusing only on the main fundamental principles of the groundwater systems [34]. Meanwhile, finding long time-series data for a numerical model is exceedingly challenging during the modelling processes.

*2.5. Data-Driven Mathematical Models for Groundwater Resources Management*

The groundwater resources prediction model is described as a regression problem [18,114,115]. This means the goal is to forecast, predict, and manage any changes in groundwater resources on any particular day according to the available data. This makes the availability of accurate data important for efficient groundwater management. Thus, data-driven mathematical models use previous data based on the underlying process behind a particular phenomenon to learn the best forecast result. Therefore, the goal of this section is to provide insight into the mathematically formulation for the management of groundwater resources. Although data-driven models are discussed, they can also be used as process-based management models in order to manage groundwater resources.

2.5.1. Sequential Gradient Restoration Algorithm (SGRA) Model

The Sequential Gradient Restoration Algorithm (SGRA) is a dual-phase path to solving Non-Linear Programming (NLP) problems. Beginning from a feasible design point of view, the gradient phase decreases the value of the objective function while at the same time fulfilling the obligation of the linearised active constraints. This will lead to constraints violation of the nonlinear active constraints. Consequently, the restoration phase catalyses the design towards feasibility, which may produce a new as well as a distinct set of constraints. This process of a two-phase cycle is repeated until the optimum value is achieved [116]. This technique incorporates a decent property with each cycle. The SGRA uses an active constraint strategy just like other algorithms. This method has been favourably compared to the generalised reduced gradient (GRG) and the gradient projection method. However, in comparison, the SGRA combines the two phases [116,117].

The SGRA assumes the inequality constraints of the form by Equation (2):

$$g_j(X) \geq 0; \quad j = 1, 2, \ldots, m \tag{2}$$

The general problem is defined as:
Minimise

$$f(X), \quad [X]_n \tag{3}$$

Subject to:

$$[h(X)]_t = 0 \tag{4}$$

$$[g(X)]_m \leq 0 \tag{5}$$

$$X^{low} \leq X \leq X^{up} \tag{6}$$

Only active inequality constraints are of interest in the SGRA. Although, equality constraints are always active. Active inequality constraints also include violated constraints. If "v" indicates the set of active constraints, Equations (4) and (5) can be combined into a vector of active constraints ($\Phi$);

$$\Phi(X) = \begin{bmatrix} h(X) \\ g(X) \geq 0 \end{bmatrix}_v \tag{7}$$

The number of active inequality is v−1. The Lagrangian for the problem can be expressed in terms of the active constraints alone (since the multipliers for the $g_f(X) < 0$ will be set to zero as part of Kuhn-Tucker (KT) conditions). In optimization, the KT conditions for a solution in NLP are described as the first optimal test derivation that must be satisfied. The KT conditions are then expressed by Equations (8) to (10).

$$\nabla_X F(X, \lambda^v) = \nabla_x f(X) + [\nabla_X \Phi]\lambda^v \tag{8}$$

$$\Phi(X) = 0 \tag{9}$$

$$\lambda_{v-t} \geq 0 \tag{10}$$

where $[\nabla_x \Phi] = [\nabla_x \Phi_1\ \nabla_x \Phi_2 \ldots \nabla_x \Phi_v]$; $\lambda^v = [\lambda_1\ \lambda_2\ \ldots\ \lambda_v]^T$.

Gradient phase: Given a feasible design $X^i$, the neighbouring gradient point can be expressed as the following where δf < 0 and δΦ = 0:

$$\tilde{X}_g = X^i + \Delta X \tag{11}$$

By imposing a quadratic constraint on the displacement ΔK in Equation (9), the problem can be set up as an optimisation subproblem in which the KT conditions are determined as given in Equation (12).

$$\Delta X = -a\nabla F_x\left(X^i, \lambda^v\right) = aS \tag{12}$$

In this model, the search direction is directly proportional to the gradient of the Lagrangian, which is an improved and model feature. To solve the Lagrangian, the Lagrange multipliers must be computed. This is solved by solving a system of $v$ linear equations, given in Equation (13).

$$\left[\nabla_X \Phi\left(X^i\right)\right]^T \nabla_X f\left(X^i\right) + \left[\nabla_X \Phi\left(X^i\right)\right]^T \left[\nabla_X \Phi\left(X^i\right)\right]\lambda^v = 0 \tag{13}$$

Stepsize for gradient phase: The stepsize calculation is based on driving the optimality conditions in (8) to zero. Therefore, if

$$\tilde{X}_g = X^i - \alpha \nabla F_X\left(X^i, \lambda^v\right) \nabla_X F(X, \lambda^v) = \nabla_x f(X) + [\nabla_X \Phi]\lambda^v \tag{14}$$

The optimum α^* is solved by cubic interpolation while trying to satisfy the condition in Equation (15).

$$\nabla_X F(\tilde{X}_g, \lambda^v)^T \nabla_X F\left(X^i, \lambda^v\right) = 0 \tag{15}$$

However, an adequate step must be taken to show that this stepsize does not cause significant constraint violation. This can be ensured by capping the squared error in the constraints by a suitable upper bound which is set up as:

$$\Phi(\tilde{X}_g)^T \Phi \tilde{X}_g \leq P_{max} \tag{16}$$

Furthermore, the $P_{max}$ is related to another performance index Q, which is the error in the optimality conditions [97]. Thus,

$$Q = \nabla F_X(\tilde{X}_g, \lambda^v)^T \nabla F_X\left(X^i, \lambda^v\right) \tag{17}$$

Restoration Phase: After the end of the gradient phase, it is assumed the function will have decreased but there would have been some infeasible constraints (supposing at the initial stage of the gradient phase, there was at least more than one nonlinear active constraint). In the restoration phase, a feasible solution within the neighbour would be established. This is achieved by making sure that the linearised constraints are feasible. However, before this the active constraint set has to be updated ($\tilde{v}$) since the initial feasible constraints could have become infeasible and previously infeasible constraints could have become feasible as a result of the change in design caused by the gradient phase. Thus, the design vector and the changes in design for this restoration phase can be written as:

$$\tilde{X}_r = \tilde{X}_g + \Delta\tilde{X}_r \quad (18)$$

The change in design for this restoration phase $\Delta\tilde{X}_r$ can be obtained as a least square error in the design changes, subject to the satisfaction of the linear constraints. Thus, Equation (19) is used to calculate this change in the design using NLP.

$$\Delta\tilde{X}_r = -\nabla_X \Phi(\tilde{X}_g)\sigma^{-v} \quad (19)$$

The above is valid where $\sigma^{-v}$ is the $\bar{v}$ vector Lagrange multiplier of the quadrantic subproblem. Furthermore, the values for the multipliers are established through the linear equation in Equation (20), where the factor $\mu$ is a user-controlled parameter to discourage large design changes.

$$\mu\Phi(\tilde{X}_g) - \nabla_X \Phi(\tilde{X}_g)^T \nabla_X \Phi(\tilde{X}_g)\sigma^{-v} = 0 \quad (20)$$

The Restoration phase is iteratively applied until Equation (21) is feasible, where $\varepsilon_1$ represent a small number.

$$\Phi(\tilde{X}_r)^T \Phi(\tilde{X}_r) \leq \varepsilon_1 \quad (21)$$

After the restoration phase, the constraints are feasible, and the next cycle of the Gradient-Restoration phase can be applied again.

### 2.5.2. Linear Regression

Linear Regression (LR) is one of the oldest and most widely used groundwater management models. Therefore, LR can be defined by Equation (22) [118,119].

$$y = X\beta + \varepsilon \quad (22)$$

This is valid where y is the vector of the groundwater level changes values from the observation well, X is the aquifer's independent parameters such as the historic weather data in a two-row matrix, $\beta$ is the parameter of forecasting vector that is under the investigation, while $\varepsilon$ is the errors' vector. In LR, the goal of applying this model technique is to minimise the validation set errors as well as to learn the vector parameter $\beta$. Many parameter estimation model techniques are available; however, the least-squares estimation techniques are the most popular [18].

### 2.5.3. Regression Tree and Gradient Boosting

Decision making is an important aspect of the linear regression groundwater management model [120,121]. Therefore, a decision-tree-based algorithm known as Regression Trees (RTs) was developed. RTs operate by diving each groundwater resource value into smaller subspaces that are represented by a tree leaf. Thus, the learning samples are obtainable by averaging all the samples as well as by introducing another LR model at the node. The accuracy of this model is a function of introduction ensembles of regression. In environmental data-driven modelling applications, the RTs algorithm is very important as it is easy, fast, and successfully deployable. Gradient Boosting (GB) is the most used method in various fields. This is because GB makes practical and effective use of ensemble weaker trees from the learning set to provide final predictable results. Nonetheless, it stacks them additively [122,123]. Thus, the loss function differentials are approximated in each succeeding stage. Equation (23) is an example of a loss function.

$$L(y, F_m(x)) = \frac{1}{2}(y - F_m(x))^2 \quad (23)$$

From (21), the true value is represented by $y$. $F_m(x)$ is the model's prediction after the m-th stage. As stated earlier, The RTs model prediction $F_m(x)$ combines all the weaker tree's results.

2.5.4. Artificial Neural Network (ANN) Model

An artificial neural network (ANN) is a black box tool with a resemblance to a human brain's biological neural networks in certain performance characteristics which consist of an enormous equidistant distribution processing system [25,124,125]. However, Feed-Forward Neural Network (FFNN) models are the most commonly used and employed in modelling [32,126,127]. A normal ANN model is made up of a three-layer FFNN model with an input layer, hidden layer, and output layer as well as artificial neurons with each layer interwoven with those in the next layer known as Multilayer Perception Network (MLPN). As such, the output of a node layer determines the weight as well as the type of transform of the sole function of the input it receives from the former layer. Therefore, previous research trained the ANN with the Bayesian regularisation algorithm and tan-sigmoid transfer function because it performs better [18,29,32,51]. This is mathematically expressed by Equation (24), where the input vector is $x_i$, the output is $y_j$, the bias is $b_i$, the weight connecting $x_i$ and $x_i$ is $x_{ji}$ the number of nodes is represented by $N$, while the activation function is $f$ within the presentation layer

$$y_j = f\left(\sum_{i=1}^{N} w_{ji}x_i + b_j\right) \tag{24}$$

The application of ANN by many researchers has proven to be a success. The ANN time series data trained model was used successfully to predict the principal factors affecting algal blooms in the man-made Lake Juam reservoir [128]. Furthermore, ANN models have been applied in various scenarios such as the estimation of a regional index flood, in an ungauged catchment of the Chindwin River in Myanmar, to evaluate an extreme daily precipitation potential in Athens, Greece, and for river flow rate and sediment load from Rantau Panjang Station on Johor River [129–132]. The results showed that ANN models are superior to the conventional regression method.

2.5.5. Adaptive Neuro-Fuzzy Inference System (ANFIS) Model

An Adaptive Neuro-Fuzzy Inference System (ANFIS) model can be described as an adaptive neural hybrid algorithm that is based on a fuzzy inference system. This model was first inaugurated in 1993 by Jang et al. [133,134]. Furthermore, ANFIS has been found to be capable of approximating any continuous and real function in a compact set to an acceptable degree of accuracy universally. In the ANFIS model, it is assumed that the fuzzy inference system has two inputs, x and y, as well as one output f. Thus, it can be expressed by the set of rules in Equations (25) and (26).

$$Rule\,1: If\ x\ is\ A_1\ and\ y\ is\ B_1; then\ f_1 = p_1x + q_1y + r_1 \tag{25}$$

$$Rule\,2: If\ x\ is\ A_2\ and\ y\ is\ B_2; then\ f_2 = p_2x + q_2y + r_2 \tag{26}$$

This is expressed where $x$ and $y$ are the crisp inputs to the node $i$, the low, and medium or high characterised linguistic labels by the convenient membership functions are represented by $A_i$ and $B_i$. Parameters of the first-order fuzzy model are represented by $p_i$, $q_i$, and $r_i$ $(i = 1, 2)$. The ANFIS models have a higher capability for modelling non-linear dynamic hydrology and diverse water resources among all the models for effective water management. In terms of efficiency and accuracy, the ANFIS models were utilised with machine learning for the estimation of daily pan evaporation as well as long-term dam inflow water [135–137]. The result showed that the ANFIS model performed better than any traditional empirical techniques. The ANFIS model has been used for rainfall-runoff modelling, groundwater modelling, as well as evaporation modelling [32,138–140].

### 2.5.6. Support Vector Machine (SVM) Model

The support vector machine (SVM) is a machine learning approach characterised by the statistical learning principle [141,142]. Therefore, its solution is obtainable via an optimisation algorithm using a regression hyperplane. Although, in a regression SVM model, an insensitivity loss function, as well as a regression hyperplane, is a convex dual optimisation problem. In addition, the Sequential Minimal Optimisation (SMO) algorithm is often used to provide a solution to problems involving the dual optimisation of the SVM. Mathematically, the SVM deterministic function is expressed as shown in Equation (27).

$$f(x) = w \cdot \phi(x) + b \tag{27}$$

In this equation $w_i$ is a weight vector, while $b$ is a bias, and the high-dimensional feature space $x$ is mapped by a non-linear transfer function $\phi$. This SVM model has been deployed by many researchers in the field of engineering to solve hydrogeology and hydrology drawbacks. The results show that the SVM model performed better than the ANN model. Limited climatic data were used to evaluate daily evapotranspiration using the SVM model in an extremely arid region [143]. Of the four models used, the SVM model proved to be the best. Additionally, the SVM model assembled with Quantum behaved Particles Swarm Optimisation SVM-QPSO model was used to forecast streamflow every month [144]. The result shows that the SVM model ensures a high prediction degree of streamflow reliability and accuracy. In addition, several researchers applied the SVM model to predict, estimate, and evaluate the streamflow as well as the water level of Lake Van in Turkey [66,145–147]. The results showed that the SVM model outperformed the regression and ANN models.

### 2.5.7. Empirical Mode Decomposition (EMD)

Empirical Mode Decomposition (EMD) is a data-adaptive full algorithm technique used for analysing signals that are non-linear as well as non-stationary [148]. Thus, for EMD to perform its basic function of decomposing an original signal to a various number of intrinsic mode functions (IMFs), two conditions must be met: (a) number of extrema and zero crossings must be equal or differ at most by one, and (b) functions are symmetric and the mean value of the upper as well as lower envelopes should be zero. However, mode mixing has been identified as the major shortcoming of EMD's implementation [149]. Mathematically, the EMD can be obtained using Equation (28).

$$x[x] = \sum_{k=1}^{K} IMF_k[n] + R[n] \tag{28}$$

This equation works where the corresponding residue to the signal approximation at the lowest resolution is represented by $R[n] = m_K[n]\ and\ IMF_k[n]$ is the $k$-th IMF. It is, however, widely claimed that the EMD, as well as the Ensemble EMD (EEMD), are computationally intensive to run. Over the past few years, both the EMD and EEMD have demonstrated higher effectiveness in comparison to the Fourier techniques. Thus, it has been extensively deployed in various disciplines such as image analysis, the health sector for diagnosis, biomedicine, and big data logging.

Table 2 shows a summary of the latest data-driven groundwater resource management methods of modelling.

**Table 2.** Summary of data-driven groundwater resource management modelling methods.

| Authors | Methods | Objectives | Shortcoming |
| --- | --- | --- | --- |
| Valipour et al. [63,64] | Improved Random Forest Regression With A Combination Of Random Features | These authors provide the application of simulation modelling methods in groundwater resource management. | The effectiveness of these methods and solutions could not be established by these authors with a field experiment. This is due to costs and time constraints, increase in computational complexity, as well as needs partitioning |
| Xuanhui et al. [56,64] | Canonical correlation forest algorithm with a combination of the random features simulation model and neuro-particle swarm optimisation and neuro-differential evolution methods | These authors proposed this method to solve the problems of data scarcity at a site and low-dimensional data | The accuracy of this method is instantaneous. Hence, this makes it unsuitable for groundwater resource prediction over a long period. |
| Emamgholizadeh et al. [29,50–52,61] | Artificial Neural Network (ANN) based | The objective of this method was to provide accurate predictions without an increase in costly computational time. | ANN models are prone to incur the problems of local minima and overfitting. |
| Yoon et al. [54,61,64,66,150–152] | Super Vector Machine Method (SVM) | The objective of this method is to overcome the variation in groundwater level predictions | Although SVM is robust for groundwater resource prediction, as highlighted by these authors, it is sensitive to redundant and outlier data. Additionally, it is not scalable enough and requires more time because of the trial and error it requires. SVM also has high parametrisation complexity. |
| Mustafa et al. [28,57,58,61,65] | Groundwater flow model and multiobjective method | The objective of this model was to evaluate the effects of human activities on groundwater dynamics | This model is negatively hampered due to a lack of data in arid and semi-arid regions. Furthermore, this model requires good-quality evidence-based data for it to be reliable |
| Kisi et al. [48,56,57,61,63,140] | Adaptive Neuro-Fuzzy Inference System (ANFIS) | The objective of this model is to overcome challenges with both the ANN and SVM models | Although ANFIS has performance acceptability in modelling many environmental and hydrological phenomena, it has many weaknesses, such as probable entrapment at local minima and slow convergence, making it ineffective in modelling. |

With the advancements in data mining for modelling, optimisation, and simulation techniques for groundwater resources management, the use of finite difference and finite elements have increased exponentially. Although, the use of the Finite Element Modelling Technique (FEMT) was first instigated by Lee and Cheng in 1974, for seawater encroachment in a coastal aquifer [153]. Likewise, Tyson and Weber, in 1963, advocated the use of electronic computers in the simulation of the dynamic behaviour of groundwater basins using the Finite-Difference Model Technique (FDMT) [154]. Consequently, both the FEMT and FDMT have been used extensively for the Groundwater Flow Model (GWFM), Hydro-Economic Model (HEM), Calibration (C), Sensitivity Analysis, (SA), Water Balance Model (WBM), as well as Validation/Verification (V) [104]. However, the efficacy of these modelling solutions depends largely on their adequate verification. Table 3 is a summary of some of the relevant researchers that have applied these two modelling techniques in groundwater resources management [104].

**Table 3.** Summary of relevant application of FEMT and FDMT and their purposes.

| Study | Application/Code | Modelling Technique | Objectives | Scheme |
|---|---|---|---|---|
| [155–158] | • GWFM<br>• SWAP<br>• WOFOST<br>• SEBAL<br>• SLURP<br>• SWAP-SWATRE<br>• SIMGRO | • FEMT | The authors combined these applications/codes with the FEMT to solve a regional groundwater flow challenge. The simulation result shows a better performance in comparison to using only MODFLOW. | • Calibration<br>• Validation/Verification<br>• Simulation |
| [159,160] | • MODFLOW<br>• GFLOW<br>• MODPATH | • FDMT | The objective of this method was to use cross-correlation analyses on the field data in order to achieve a theoretical understanding of an aquifer. | • Calibration<br>• Simulation |
| [20,161–164] | • HYDRUS-1D<br>• SWMS-2D<br>• PLASM<br>• MODFLOW<br>• AQUIFEMM-1<br>• ISOQUAD<br>• SVAT<br>• SIMGRO | • FEMT<br>• FDMT | The objective was to combine the digital groundwater models to access water and dissolved solutes in critical basins and tile-drained soil aquifer systems. The simulated result indicated the possibility of simulating the seasonal behaviour of the aquifer system for future planning and management. | • Calibration<br>• Validation/Verification<br>• Simulation |
| [165] | • SGMP | • FDMT | The objective of this model was to achieve groundwater balance analysis of different regions. The result shows that these models can be used for the analysis of future groundwater behaviour | • Calibration<br>• Simulation<br>• Sensitivity Analysis |
| [162,163,166–176] | • MODFLOW<br>• SVAT<br>• SIMGRO<br>• SEAWAT<br>• UCODE<br>• ISOQUAD<br>• MT-3D<br>• UPFLOW<br>• HEM<br>• WBM<br>• MMA<br>• GLUE | • FDMT | The objective of these models is to simulate the effect of water flow in saturated and unsaturated zones in the surface water in an integrated approach. The simulated result shows that this approach can show the effects and impacts of groundwater usage, drainage, and irrigation on the evapotranspiration of various crops. | • Calibration<br>• Simulation<br>• Validation/Verification |

## 3. Groundwater Management and Internet of Things

This section aims to highlight the relationship and importance of the Internet of Things, which comprises Remote Sensing (RS) and Geographic Information Systems (GIS), among many other approaches [176,177], as it relates to the measurement, monitoring, management, forecasting, and modelling of groundwater resources.

### *3.1. Introduction*

Effective groundwater resource management, as well as modelling, is a function of the availability of good quality data pertaining to the information of the observation well. This is because the groundwater resource measurements are the most basis supporting existence for evaluating the information of the quantity of groundwater resources stored within the aquifer [27]. The information about the aquifer's properties may include and is not limited to changes in groundwater level, storage, flow rate, and recharge as well as discharge rates. However, data are not collected automatically. Furthermore, Calderwood et al. state that the negative impact of overexploitation, as well as reduction in the recharge of groundwater resources, is often unknown, even after years have passed, due to data limitations [40]. In addition, groundwater resource information is strenuous to collect and use due to the lack of proper integration between the equipment deployed, irrelevant and inconsistent data as a result of lack of stationary large-scale flow rate hindrances, un-automated groundwater analysis processes, and a lack of interoperability in previous systems [177–182]. Consequently, these limitations are causing untold strains in groundwater resource management. Additionally, most groundwater resource management models cannot provide any reliable decisions or support without the required data input. Hence, there is a need for a contemporary, scalable, and real-time IoT-based management system solution for groundwater resources management. Therefore, via the advancements recorded in the field of Information and Communication Technologies (ICT), there are many opportunities available to manage the groundwater resource crises.

There are various groundwater level monitoring systems. These systems vary in technology, monitoring, and management tasks, scalability, the solutions they render, and the cost implications. Historically, the traditional technique of measuring groundwater levels involves the use of a manual tape measure. Furthermore, there are emerging threats that the majority of groundwater level measuring networks are being regularly abandoned due to a decline in global groundwater monitoring [183–185]. Consequently, groundwater resources are often inadequately monitored, despite the fact that they are needed for calibration and the validation of groundwater resources models [185,186]. These aforementioned challenges can be solved through the use of the Internet of Things (IoTs) techniques [177].

The application and deployment of low-cost IoT-enabled Wireless Sensors Networks (WSN) technologies in surface water management have gained tremendous progress. However, the contrary is the case in groundwater-level management [187–190]. Although there are few application cases, they are synonymous with the use of high power and cost networks; most deployed are commercial pressure transducers without outlining their energy implications, power consumption analysis, or the overall cost implications [187]. Consequently, the use of the Internet of Things based Low-Cost Wireless Sensors Networks (LCSNs) has emerged as an alternative to solve these challenges. Contrarily, according to Mao et al. and Chan et al., the growing deployment of LCSNs in groundwater resource management has produced a limited quantity of studies with corresponding open data as well as them having failed to provide much-needed information to the water managers, scientists, and other policymakers [188–200]. Rather, non-technical issues like socio-economic contexts, operational and financial mechanisms, and stakeholder engagements, which are hindering the advantages of deploying these LCSNs, are mostly concentrated on [189]. Therefore, there is an urgent need to have a better understanding of these emerging challenges through the existence of more research papers, the development of scientific tools, as well as the assessment of possible available future opportunities for the deployment of IoT-based technologies for groundwater resources management.

### 3.2. Architecture of IoT Based Groundwater Monitoring System

Generally, the groundwater level IoT-based system architecture can be subdivided into three layers. These are the physical layers, the service layer, and the presentation layer [200–225]. In the physical layer, the communication equipment, as well as various sensor nodes for basic data acquisition of relevant groundwater level and aquifer information, are built-in. The raw data measured and collected are deployed to the service layer. At the service layer, there are various tools for data analytics. This layer also stores the received data from the physical layer. Both the application and business logic implementation are implemented at this layer, thus, making it an important part of the architecture. The presentation layer is the visualisation layer where the users are allowed to interact as well as view the displayed information on a screen monitor.

Figure 2 shows an overview and working representation of the workflow of groundwater level IoT based monitoring system. This shows the interconnection of the sensors deployed into the well, the nodes, gateway, and the end-user platform. The arrows indicate the direction of the flow of data. The solid line represents the direction of the data received, while the broken lines represent the request sent for data. In this system, the IoT is combined with the GIS to ensure the real-time uploading of groundwater level data, this improves the quality of the database and provides geographical information about the system. The sensors deployed into the observation wells acquire and send the corresponding measurements of groundwater level parameters to their nodes over the Modbus protocol employed. From the nodes, the data is sent via the Long Range (LoRa) wireless network to the gateway. The data is saved into the Message Queuing Telemetry Transport (MQTT) server platform, which is connected to the No relational Structured Query Language (NoSQL) database or any other Graphics Processing Unit (GPU) platform for real-time display. Examples of these web-based client GPUs are ArcGIS, API, Google Maps API, ThingSpeak, and WebGL [215,226–231].

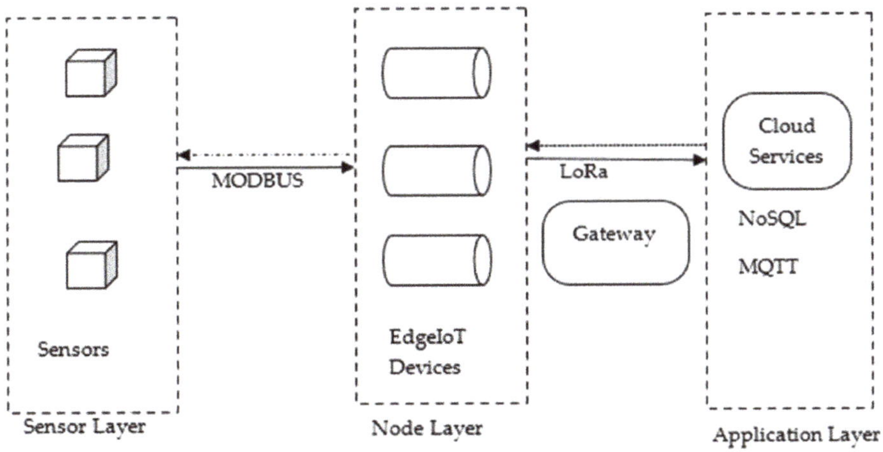


**Figure 2.** The architecture of Groundwater resource IoTs based monitoring system.

### 3.3. *IoT and Groundwater Level Measurement Techniques*

In recent years, the internet has changed the way human lives. Therefore, the Internet of Things (IoT) as a facet of the internet has become the ultimate layer on which several things such as smart gadgets are interconnected. This concept of the IoT has been embraced in many areas of human endeavours, including smart water and groundwater-level management. Thus, IoT techniques are used to collect, transfer, and analyse the required groundwater level data. The major advantage of IoT's deployment is its ability to be able to combine with several technologies, such as wireless sensors, cloud computing, ubiquitous computing, RFIP, and software [191]. Thus, groundwater level data management in an IoT environment involves the combination of smart technologies such as sensors to collect data and transmit it over a network area, with a combination of Integrated Development Environment (IDE) software into the cloud server.

Remote sensing (RS) is an example of a classical way of obtaining the much-needed hydrological data for groundwater-level measurements using the internet [192]. However, this can only provide point data. Thus, the challenge remains on how to navigate using point data alone to determine regionally distributed data. Although the RS can be used to obtain certain groundwater resources parameters, these parameters are not often useful for groundwater management modelling. Consequently, another model will be required to manipulate the acquired data into usable or verifiable data as an input in spatially distributed models [193]. Furthermore, the data acquired via the RS are prone to noise. The substantial and most relevant data for groundwater resource management modelling are the recharge and discharge information. Of these two, recharge is very crucial for the sustainability of groundwater-level management. The application of RS for groundwater level modelling was carried out by Dams et al., for mapping out impervious surface changes for hydrological modelling [194]. More researchers also applied the RS technique for the management of groundwater resources [192,195–197]. To solve these challenges, the IoT and machine learning techniques can be employed. Apart from RS, groundwater-level management is using in situ networks in many aquifer systems around the world. However, the difference in each monitoring well depends on the technology used as well as the frequency of the measuring data [198–200]. The in situ technique assists in the management of groundwater level by estimating the changes in (groundwater) storage, providing the user's information, flow model calibration, and general up to date system information [200–202].

The IoT and machine learning techniques involve the acquisition, processing, transfer, and analysis of groundwater level data interconnected using intelligent nodes. This combined with the evolution of web techniques has been used in many studies in the literature. The application of the IoT for monitoring the groundwater level's daily variations and safety quality in the mining environment was carried out by Reddy et al. using sensor technologies [203]. Additionally, Neyens et al. monitored groundwater quality and quantity from a desktop using the IoT-enabled Environmental data Management Interface (EMI) technique [204]. This same technique was also applied to saltwater intrusion monitoring. A low cost IoT-based real-time groundwater resource management system which was built for monitoring a community-based network consisting of eleven wells sited in Nova Scotia, Canada [205]. Additionally, an automated low-cost sensor network for monitoring groundwater levels was designed at the South American Subbasin Groundwater Observatory (GWO), in a real-time scenario [40]. Botta et al. presented the IoTs based sensor network review for both quality and quantity measurement of water in a smart city [206]. It should be noted that with the large amount of groundwater level data collected from many deployed sensors, there comes a new challenge of how to process, store, and deal with it. This is an open research bottleneck that needs to be unravelled.

Starting from the evolution of the web technique in the year 1993, several database managers have started to develop web-based Geographic Information Systems (WebGIS) to store the collected real-time and fast data streams [207,208]. The GIS as an example of Database Management Systems (DBMS) is embedded with the topography, geology, geometry, as well as coordinate data to assist in the storage, explanation, location, and manipulation of inputs as well as the corresponding output data information [209]. Therefore, the WebGIS technique performs better in terms of user's quality of service (QoS), can be used by multiple users, providing the benefits of cost reduction, global reach, as well as cross-platform compatibility [210].

There has been exponential growth in the development of groundwater level measurement IoT based management techniques, with a combination of relevant WebGIS. The GIS software known as ARCVIEW, and the groundwater model (MODFLOW) were combined for the numerical modelling of groundwater resources by Chenini and Mammou [211]. This combination was used in the central region of Tunisia. Similarly, the combination of the Managed Aquifer Recharge (MAR) and International Groundwater Resources Assessment Centre's Global Groundwater Information Service (IGRAC's GGIS) was successfully implemented using enhanced historical data from approximately 1200 site studies in about 62 nations [212]. The results show an increase in groundwater resource storage, good recharging levels, and improved water distribution management. In Table 4 a few other existing groundwater level IoT-based management techniques are presented. Most of the highlighted studies are combined both web browser and GIS software and support multiple users as well as tasks.

**Table 4.** Summary of relevant existing groundwater level WebGIS technique.

| Technique | Numerical Model | Spatial Interpolation | Limitation |
| --- | --- | --- | --- |
| Cloud-based MODFLOW ArcGIS [213,214] | None | Several methods | This is based on the simulation modules only. This limitation does not permit extensive interpolation due to the lack of a numerical model. |
| Collaborative Geographic Information Systems (CGIS) [177,215] | Multi hydraulic | Several methods | It is based on the computer's Random-Access Memory (RAM) size. This causes delays or freezing up during processing. |
| Delft-FEWS [216,217] | Multi hydraulic | Several methods | It is based on the computer's Random-Access Memory (RAM) size. As a result of this limitation, the model experiences low computation memory during its operation. |
| FREEWAT was developed as a plug in GIS desktop software QGIS (QGIS Development Team 2017) [218–220] | Multi hydraulic | QGIS | The interpolation speed is based on the computer's RAM. This causes a delay in operations. |
| HydroShare [221,222] | It is based on the user's developed numerical model | Based on the user's upload | It is based on the developed model and lacks space and speed. |
| Tethy [223] | It is based on multi hydraulic models | Several methods | Based on a developed model |
| MAGNET [224] | It is based on multi hydraulic models | Several methods | It is limited to a 10,000-mesh number. Anything above this number is not possible. |

### 3.4. Groundwater Quantity Monitoring System

Now, to properly manage, monitor, or model the groundwater resource, the availability of data is very crucial. Furthermore, all the data-driven machine learning techniques could only be used based on the availability of historical data. Hence, an IoT-based and online groundwater quantity monitoring system is necessary for data acquisition, timely access to processed information, to assess the impact of drought on groundwater resources, as well as to format decision-making tools and to derive novel drought indicators [230]. The application of a groundwater quantity monitoring system encompasses a large distribution network of monitoring wells with corresponding distributed monitoring sensors utilising different architecture systems [231,232]. Thus, this network of observation wells provides information about the effects of hydrologic stress on the groundwater system [27]. The conventional method of obtaining data from these monitoring wells is by visiting the site once or twice a year to manually collect said data at different times. Thus, a lot of laborious manpower, money, and time is spent to monitor the groundwater level [232]. Over the years, this conventional manual method has proven to be ineffective because the human in charge may fail to carry out the monitoring. Additionally, the method lacks real-time monitoring information which is critical for making an urgent decision by a water manager. Thus, Tauro et al. suggest that the biggest challenge in groundwater management is monitoring groundwater level and flow through diverse channels. This makes hydrological data loggers expensive. Thus, the application of IoT-based low-cost and open-source instrumentation to monitor and manage groundwater quantity systems is one of the strongly needed tools. This also allows for the adjustment and tailoring of sensors to a specific hydrological need [233].

Several studies have been carried out to examine groundwater level monitoring in a network of wells. Prinos et al. used groundwater level data to design a real-time groundwater level monitoring network and portrayal of hydrologic data in Southern

Florida [234]. The researchers noticed the balance between recharge and discharge is mostly tenuous during droughts. Thus, they analysed 26 years of historical data from three groundwater aquifer systems in South Florida using regression analysis. Through their development, they were able to transmit groundwater information to the groundwater managers and the public via the created web portal in a timely fashion.

Anumalla et al. developed a groundwater monitoring system based on a network of wirelessly linked field-programmable arrays and pressure sensors in Western Nebraska aquifers in the United States of America [230]. These researchers used components such as the Data Acquisition Unit (DAU), the Data Transfer Unit (DTU), and the Data Processing Unit (DPU) to design this monitoring system. Each unit performs various functions. The DAU is responsible for collecting the data from the sensor and responds to data requests from the DTU. The DTU will transport the data from the DAU to the DPU reliably. Finally, the DPU, after gathering the data from sensors, will analyse and store the data for end-use. The system operated on the researchers' argument that timely and quick groundwater data is required to make critical decisions regarding the impact of droughts on groundwater resources.

Two low-cost groundwater level sensors based on Plastic Optical Fibres (POF) were proposed by Esequiel Mesquita et al. In their work, they presented the operation as well as the examination of the impact of using POF sensor on the monitoring of groundwater levels [234]. This work was based on the experimental principle that there is a decrease in signal between the fibre grooves and water in comparison to air. The introduction of grooves in this work increased optical signal strength as the water level also rises as well and vice and versa. The advantages of this experiment are based in its simplicity, low cost, and suitable sensitivity.

A low-cost, low-powered, IoT enabled approach integrated with energy harvesting groundwater resource monitoring system was designed by Kombo et al. [187]. In this study, the authors developed this groundwater monitoring system to produce real-time data to aid decision making in groundwater resource management in Bandamaji station situated at Zanzibar Tanzania. The brain behind this monitoring system was the Arduino UNO ATmega328P based microcontroller platform which embed with MS5803-14BA and MB280 sensors. The advantage of this prototype lies in its ability to overcome energy barriers by inculcating an automatic energy harvesting technique. This prolonged the life span of this prototype's battery.

Calderwood et al. developed a low-cost, open-source wireless sensor network for real-time, scalable groundwater monitoring utilising cellular telemetry in the South American Subbasin groundwater observatory, located in California, USA [40]. This prototype consists of sensors for groundwater level data acquisition, cellular telemetry units that transfer the acquired data via an open-source data mining pipeline, as well as a visualisation web dashboard. Additionally, the amount of water used was estimated and managed using IoT by Robles et al. at a region in Zaragoza, Spain [188]. Through this technique, efficient water monitoring, usage, and management were achieved. Furthermore, the latest water level data was continuously received by a wireless sensor network developed by Wadekar et al. [232]. They were able to visualise and manage an urban area water usage via a smartphone. This system was able to assist the consumers to plan as well as predict their water usage. Table 5 summarises a few of the related groundwater quantity monitoring systems.

**Table 5.** Summary of relevant existing groundwater quantity monitoring systems.

| Monitoring System and Sensor Type | Network Technologies | Processing Board | Limitation |
|---|---|---|---|
| Unidata Pressure Sensor [230] | 802.11 based WLAN | PIC12F675 microcontroller on Altera Nios FPGA board | The details of the power consumption and energy harvesting were not reported. However, the cost analysis was reported. |
| Plastic Optical Fibres Sensor [234] | MiWiTM-P2 P wireless module from Microchip Technologies | PIC24FJ256DA206 16-bit microcontroller from Microchips Technologies | Salty water affects its performance as well as the temperature calibration not being precise. The details of power consumption analysis were not reported. |
| MS5803-02BA and NXP MPX5010 DP [190] | Not Specified | Arduino Pro Mini or Nano | Except for the reported cost analysis, the power consumption analysis was not given. |
| Redesigned MS58030-14BA and MBE280 [187] | LoRa and GSM | Arduino Uno or Mega | It requires one or more gateways because the single-channel gateway used limited the number of nodes that can communicate with the LoRa-enabled gateway simultaneously. |
| A low-cost ultrasonic Sensor [235] | Texas Instruments CC1200-DK and Semetech LoRa iM880A | TI MSP432P$01R 32-bit ARM core M4 with 64 kB RAM | Due to inertia, the piezoelectric transducer takes some time to build up to maximum amplitude; High power radios; Water tank reservoirs. |
| Ultrasonic Sensor [236] | GSM/GPRS | Arduino UNO | Commercial ultrasonic sensors are inflexible, have a short maximum distance from the target, and there was no power saving analysis. |
| Solinst Leveloggers [40] | GPRS/GSM | Solinst in-built board | Apart from the reported detailed cost analysis, the power consumption analysis was not given. |

## 4. Conclusions and Future Directions

This review presented and discussed the existing trend in groundwater resources management. In the past, most of the existing groundwater resource management models have been combined with optimisation and simulation techniques using the appropriate mathematical programming to proffer solutions to challenges within aquifers. Consequently, the previous surveys have presented a narrow review involving simulation and optimisation management models; however, this review presented a much broader IoT-enabled management perspective. This is because, for any of the management modelling of groundwater resources to be achieved, the measurement of resource data is important. However, part of the general limitations are the uncertainties, from the input parameters to the system modelling. Furthermore, despite huge research attention towards solving such problems, little research evidence exists concerning achieving computationally efficient and scalable models for groundwater resource management in real-time operations. Although there are various modelling tools, as presented in this review, their field applications must be ascertained.

Furthermore, there has been an unprecedented increase in the amount of data generated by electrical sensors in the Internet of Things (IoT) over the years. The application of groundwater resource IoT-based techniques is a very useful tool in data acquisition, monitoring, and manipulation in the management of groundwater resources. This technique combined with GIS has a huge potential in the field of water management. However, IoT data are mostly processed via a computing resource situated in a data centre location far away using either the internet or cloud computing. Consequently, this has led to insecurity in the privacy of users, low latency, and scalability problems. Since the IoT large data sums are transmitted into the cloud in high volumes, it is necessary to have an efficient and scalable IoT platform to extract valuable information in real-time for groundwater resource management. Additionally, the existing groundwater-level management systems lack large scale applications in situ management techniques.

For future research directions, the challenges of high computational inefficiency and scalability must be addressed. This will enable the groundwater-level management model to achieve computationally efficient and scalability. Additionally, the current IoT-enabled automated data processing systems for transmitting the generated data from IoT sensors into the centralised cloud are not scalable and efficient. Therefore, there is a need to develop an alternative model for the IoT-enabled groundwater-level management model. These are open research directions that should be explored. Although this is a review of existing management models for groundwater resources, it is not all-inclusive. Thus, there is a possibility of missing some other publications because it is impractical to review them all. Therefore, these gaps could also be filled with further review publications.

**Author Contributions:** Conceptualization, B.A.A., T.O.O., J.M.N., S.S.R.; methodology, B.A.A., T.O.O., J.M.N., S.S.R.; software, B.A.A., T.O.O., J.M.N., S.S.R.; formal analysis, B.A.A.; investigation, B.A.A., T.O.O., J.M.N., S.S.R.; resources, B.A.A., T.O.O., J.M.N., S.S.R.; data curation, B.A.A.; writing—original draft preparation, B.A.A.; writing—review and editing, B.A.A., T.O.O., J.M.N., S.S.R.; visualization, B.A.A.; supervisions, T.O.O., J.M.N. and S.S.R.; project administration, B.A.A., T.O.O., J.M.N., S.S.R.; funding acquisition, T.O.O. and J.M.N. All authors have read and agreed to the published version of the manuscript.

**Funding:** This research received no external funding.

**Institutional Review Board Statement:** Not applicable.

**Informed Consent Statement:** Not applicable.

**Data Availability Statement:** Not applicable.

**Acknowledgments:** The authors acknowledge the research supports received from the Nichea Area of Climate, Water Security and Management of the Tshwanne University of Technology, Tshwane University of Technology, Pretoria, and the Water Research Council (WRC), Pretoria, South Africa.

**Conflicts of Interest:** The authors declare no conflict of interest.

## References

1. Rejani, R.; Jha, M.K.; Panda, S.N. Simulation-Optimization Modelling for Sustainable Groundwater Management in a Coastal Basin of Orissa, India. *Water Resour. Manag.* **2009**, *23*, 235–263. [CrossRef]
2. Mary, X.A.; Rose, L.; Rajasekaran, K. Continuous and Remote Monitoring of Ground Water Level Measurement in a Well. *Int. J. Water* **2018**, *12*, 356–369. [CrossRef]
3. Aziz, S.Q.; Saleh, S.M.; Omar, I.A. Essential Treatment Processes for Industrial Wastewaters and Reusing for Irrigation. *Zanco J. Pure Appl. Sci.* **2019**, *31*, 269–275.
4. Chambel, A. The Role of Groundwater in the Management of Water Resources in the World. *Proc. Int. Assoc. Hydrol. Sci.* **2015**, *366*, 107. [CrossRef]
5. Taylor, R.; Scanlon, B.; Döll, P.; Rodell, M.; Van Beek, R.; Wada, Y.; Longuevergne, L.; Leblanc, M.; Famiglietti, J.S.; Edmunds, M. Ground Water and Climate Change. *Nat. Clim. Change* **2013**, *3*, 322–329. [CrossRef]
6. Yin, J.; Tsai, F.T.-C. Saltwater Scavenging Optimization under Surrogate Uncertainty for a Multi-Aquifer System. *J. Hydrol.* **2018**, *565*, 698–710. [CrossRef]
7. Yin, J.; Tsai, F.T.-C. Steady-State Approximate Freshwater–Saltwater Interface in a Two-Horizontal-Well Scavenging System. *J. Hydrol. Eng.* **2019**, *24*, 06019008. [CrossRef]
8. Motevalli, A.; Moradi, H.R.; Javadi, S. A Comprehensive Evaluation of Groundwater Vulnerability to Saltwater Up-Coning and Sea Water Intrusion in a Coastal Aquifer (Case Study: Ghaemshahr-Juybar Aquifer). *J. Hydrol.* **2018**, *557*, 753–773. [CrossRef]
9. Mani, A.; Tsai, F.T.-C.; Paudel, K.P. Mixed Integer Linear Fractional Programming for Conjunctive Use of Surface Water and Groundwater. *J. Water Resour. Plan. Manag.* **2016**, *142*, 04016045. [CrossRef]

10. Conway, B.D. Land Subsidence and Earth Fissures in South-Central and Southern Arizona, USA. *Hydrogeol. J.* **2016**, *24*, 649–655. [CrossRef]
11. Castellazzi, P.; Martel, R.; Rivera, A.; Huang, J.; Pavlic, G.; Calderhead, A.I.; Chaussard, E.; Garfias, J.; Salas, J. Groundwater Depletion in Central Mexico: Use of GRACE and inSAR to Support Water Resources Management. *Water Resour. Res.* **2016**, *52*, 5985–6003. [CrossRef]
12. Erban, L.E.; Gorelick, S.M.; Zebker, H.A.; Fendorf, S. Release of Arsenic to Deep Groundwater in the Mekong Delta, Vietnam, Linked to Pumping-Induced Land Subsidence. *Proc. Natl. Acad. Sci. USA* **2013**, *110*, 13751–13756. [CrossRef]
13. Yang, Y.; Song, X.F.; Zheng, F.D.; Liu, L.C.; Qiao, X.J. Simulation of Fully Coupled Finite Element Analysis of Nonlinear Hydraulic Properties in Land Subsidence Due to Groundwater Pumping. *Environ. Earth Sci.* **2015**, *73*, 4191–4199. [CrossRef]
14. Yeh, W.W. Optimization Methods for Groundwater Modeling and Management. *Hydrogeol. J.* **2015**, *23*, 1051–1065. [CrossRef]
15. Liu, Z.; Merwade, V.; Jafarzadegan, K. Investigating the Role of Model Structure and Surface Roughness in Generating Flood Inundation Extents Using One- and Two-Dimensional Hydraulic Models. *J. Flood Risk Manag.* **2019**, *12*, e12347. [CrossRef]
16. Raju, K.R.S.R.; Varma, G.H.K. Knowledge-Based Real-Time Monitoring System for Aquaculture Using IoT. In Proceedings of the 2017 IEEE 7th International Advance Computing Conference (IACC), Hyderabad, India, 5–7 January 2017; pp. 318–321.
17. United Nations General Assembly. *Transforming Our World: The 2030 Agenda for Sustainable Development*; Resolution Adopted by the General Assembly on 25 September 2015; United Nations: New York, NY, USA, 2015.
18. Kenda, K.; Čerin, M.; Bogataj, M.; Senožetnik, M.; Klemen, K.; Pergar, P.; Laspidou, C.; Mladenić, D. Groundwater Modeling with Machine Learning Techniques: Ljubljana polje Aquifer. *Proceedings* **2018**, *2*, 697. [CrossRef]
19. Alley, W.M.; Leake, S.A. The Journey from Safe Yield to Sustainability. *Groundwater* **2004**, *42*, 12–16. [CrossRef] [PubMed]
20. Konikow, L.F.; Bredehoeft, J.D. Ground-Water Models Cannot Be Validated. *Adv. Water Resour.* **1992**, *15*, 75–83. [CrossRef]
21. Bredehoeft, J.; Durbin, T. Groundwater Development—The Time to Full Capture Problem. *Groundwater* **2009**, *47*, 506–514. [CrossRef]
22. Khaki, M.; Yusoff, I.; Islami, N. Simulation of Groundwater Level through Artificial Intelligence System. *Environ. Earth Sci.* **2015**, *73*, 8357–8367. [CrossRef]
23. Coppola, E.; Poulton, M.; Charles, E.; Dustman, J.; Szidarovszky, F. Application of Artificial Neural Networks to Complex Groundwater Management Problems. *Nat. Resour. Res.* **2003**, *12*, 303–320. [CrossRef]
24. Saatsaz, M.; Chitsazan, M.; Eslamian, S.; Sulaiman, W.N.A. The Application of Groundwater Modelling to Simulate the Behaviour of Groundwater Resources in the Ramhormooz Aquifer, Iran. *Int. J. Water* **2011**, *6*, 29–42. [CrossRef]
25. Gong, Y.; Zhang, Y.; Lan, S.; Wang, H. A Comparative Study of Artificial Neural Networks, Support Vector Machines and Adaptive Neuro-Fuzzy Inference System for Forecasting Groundwater Levels near Lake Okeechobee, Florida. *Water Resour. Manag.* **2016**, *30*, 375–391. [CrossRef]
26. Verma, A.; Singh, T. Prediction of Water Quality from Simple Field Parameters. *Environ. Earth Sci.* **2013**, *69*, 821–829. [CrossRef]
27. Taylor, C.J.; Alley, W.M. *Ground-Water-Level Monitoring and the Importance of Long-Term Water-Level Data (no. 1217-2002)*; US Geological Survey: Reston, VA, USA, 2002.
28. Singh, A. Simulation and Optimization Modeling for the Management of Groundwater Resources. II: Combined Applications. *J. Irrig. Drain. Eng.* **2014**, *140*, 04014002. [CrossRef]
29. Emamgholizadeh, S.; Moslemi, K.; Karami, G. Prediction the Groundwater Level of Bastam Plain (Iran) by Artificial Neural Network (ANN) and Adaptive Neuro-Fuzzy Inference System (ANFIS). *Water Resour. Manag.* **2014**, *28*, 5433–5446. [CrossRef]
30. Sun, S.; Zhang, C.; Yu, G. A Bayesian Network Approach to Traffic Flow Forecasting. *IEEE Trans. Intell. Transp. Syst.* **2006**, *7*, 124–132. [CrossRef]
31. Sun, S.; Xu, X. Variational Inference for Infinite Mixtures of Gaussian Processes with Applications to Traffic Flow Prediction. *IEEE Trans. Intell. Transp. Syst.* **2011**, *12*, 466–475. [CrossRef]
32. Gong, Y.; Wang, Z.; Xu, G.; Zhang, Z. A Comparative Study of Groundwater Level Forecasting Using Data-Driven Models Based on Ensemble Empirical Mode Decomposition. *Water* **2018**, *10*, 730. [CrossRef]
33. Chang, J.; Wang, G.; Mao, T. Simulation and Prediction of Suprapermafrost Groundwater Level Variation in Response to Climate Change Using a Neural Network Model. *J. Hydrol.* **2015**, *529*, 1211–1220. [CrossRef]
34. Chang, F.-J.; Chang, L.-C.; Huang, C.-W.; Kao, I.-F. Prediction of Monthly Regional Groundwater Levels through Hybrid Soft-Computing Techniques. *J. Hydrol.* **2016**, *541*, 965–976. [CrossRef]
35. Asher, M.J.; Croke, B.F.; Jakeman, A.J.; Peeters, L.J. A Review of Surrogate Models and Their Application to Groundwater Modeling. *Water Resour. Res.* **2015**, *51*, 5957–5973. [CrossRef]
36. Diersch, H. *FEFLOW Finite Element Subsurface Flow and Transport Simulation System Reference Manual*; WASY Institute for Water Resources Planning and Systems Research: Berlin, Germany, 2005; p. 292.
37. Harbaugh, A. MODFLOW-2005, The US Geological Survey Modular Groundwater Model: The Groundwater Flow Process. In *U.S. Geological Survey Technology Methods*; US Geological Survey: Reston, VA, USA, 2005; Volume 6-A16, pp. 1–99.
38. Doherty, J.; Simmons, C.T. Groundwater Modelling in Decision Support: Reflections on a Unified Conceptual Framework. *Hydrogeol. J.* **2013**, *21*, 1531–1537. [CrossRef]
39. Leube, P.; Nowak, W.; Schneider, G. Temporal Moments Revisited: Why There Is No Better Way for Physically Based Model Reduction in Time. *Water Resour. Res.* **2012**, *48*, W11527. [CrossRef]

40. Calderwood, A.J.; Pauloo, R.A.; Yoder, A.M.; Fogg, G.E. Low-Cost, Open Source Wireless Sensor Network for Real-Time, Scalable Groundwater Monitoring. *Water* **2020**, *12*, 1066. [CrossRef]
41. Ahlfeld, D.P.; Baro-Montes, G. Solving Unconfined Groundwater Flow Management Problems with Successive Linear Programming. *J. Water Resour. Plan. Manag.* **2008**, *134*, 404–412. [CrossRef]
42. Peralta, R.; Timani, B.; Das, R. Optimizing Safe Yield Policy Implementation. *Water Resour. Manag.* **2011**, *25*, 483–508. [CrossRef]
43. Xu, J.; Tu, Y.; Zeng, Z. Bilevel Optimization of Regional Water Resources Allocation Problem under Fuzzy Random Environment. *J. Water Resour. Plan. Manag.* **2013**, *139*, 246–264. [CrossRef]
44. Tsai, F.T.-C.; Katiyar, V.; Toy, D.; Goff, R.A. Conjunctive Management of Large-Scale Pressurized Water Distribution and Groundwater Systems in Semi-Arid Area with Parallel Genetic Algorithm. *Water Resour. Manag.* **2009**, *23*, 1497. [CrossRef]
45. Ayvaz, M.; Elci, A. A Groundwater Management Tool for Solving the Pumping Cost Minimization. *J. Hydrol.* **2013**, *478*, 63–76. [CrossRef]
46. Kifanyi, G.E.; Ndambuki, J.M.; Odai, S.N. A Quantitative Groundwater Resource Management under Uncertainty Using a Retrospective Optimization Framework. *Sustainability* **2017**, *9*, 2. [CrossRef]
47. Li, M.; Guo, P.; Ren, C. Water Resources Management Models Based on Two-Level Linear Fractional Programming Method Under Uncertainty. *J. Water Resour. Plan. Manag.* **2015**, *141*, 05015001. [CrossRef]
48. Fallah-Mehdipour, E.; Haddad, O.B.; Mariño, M. Prediction and Simulation of Monthly Groundwater Levels by Genetic Programming. *J. Hydro-Environ. Res.* **2013**, *7*, 253–260. [CrossRef]
49. Fu, G.; Crosbie, R.S.; Barron, O.; Charles, S.; Dawes, W.; Shi, X.; Van Niel, T.; Li, C. Attributing Variations of Temporal and Spatial Groundwater Recharge: A Statistical Analysis of Climatic and Non-Climatic Factors. *J. Hydrol.* **2019**, *568*, 816–834. [CrossRef]
50. He, Z.; Zhang, Y.; Guo, Q.; Zhao, X. Comparative Study of Artificial Neural Networks and Wavelet Artificial Neural Networks for Groundwater Depth Data Forecasting with Various Curve Fractal Dimensions. *Water Resour. Manag.* **2014**, *28*, 5297–5317. [CrossRef]
51. Hosseini, S.M.; Mahjouri, N. Integrating Support Vector Regression and a Geomorphologic Artificial Neural Network for Daily Rainfall-Runoff Modeling. *Appl. Soft Comput.* **2016**, *38*, 329–345. [CrossRef]
52. Mao, X.; Shang, S.; Liu, X. Groundwater Level Predictions Using Artificial Neural Networks. *Tsinghua Sci. Technol.* **2002**, *7*, 574–579.
53. Husna, N.-e.-a.; Bari, S.H.; Hussain, M.M.; Urrahman, M.T.; Rahman, M. Ground Water Level Prediction Using Artificial Neural Network. *Int. J. Hydrol. Sci. Technol.* **2016**, *6*, 371–381. [CrossRef]
54. Khosravi, K.; Panahi, M.; Bui, D.T. Spatial Prediction of Groundwater Spring Potential Mapping Based on an Adaptive Neuro-Fuzzy Inference System and Metaheuristic Optimization. *Hydrol. Earth Syst. Sci.* **2018**, *22*, 4771–4792. [CrossRef]
55. Kisi, O.; Azad, A.; Kashi, H.; Saeedian, A.; Hashemi, S.A.A.; Ghorbani, S. Modeling Groundwater Quality Parameters Using Hybrid Neuro-Fuzzy Methods. *Water Resour. Manag.* **2019**, *33*, 847–861. [CrossRef]
56. Kisi, O.; Keshavarzi, A.; Shiri, J.; Zounemat-Kermani, M.; Omran, E.-S.E. Groundwater Quality Modeling Using Neuro-Particle Swarm Optimization and Neuro-Differential Evolution Techniques. *Hydrol. Res.* **2017**, *48*, 1508–1519. [CrossRef]
57. Moosavi, V.; Vafakhah, M.; Shirmohammadi, B.; Behnia, N. A Wavelet-ANFIS Hybrid Model for Groundwater Level Forecasting for Different Prediction Periods. *Water Resour. Manag.* **2013**, *27*, 1301–1321. [CrossRef]
58. Mustafa, S.M.T.; Hasan, M.M.; Saha, A.K.; Rannu, R.P.; Van Uytven, E.; Willems, P.; Huysmans, M. Multi-Model Approach to Quantify Groundwater-Level Prediction Uncertainty Using an Ensemble of Global Climate Models and Multiple Abstraction Scenarios. *Hydrol. Earth Syst. Sci.* **2019**, *23*, 2279–2303. [CrossRef]
59. Ndambuki, J.; Otieno, F.; Stroet, C.; Veling, E. Groundwater Management Under Uncertainty: A Multi-Objective Approach. *Water SA* **2000**, *26*, 35–42.
60. Ndambuki, J.M. Multi-Objective Groundwater Quantity Management: A Stochastic Approach. Ph.D. Thesis, Delft University, Delft, The Netherlands, 2001.
61. Shirmohammadi, B.; Vafakhah, M.; Moosavi, V.; Moghaddamnia, A. Application of Several Data-Driven Techniques for Predicting Groundwater Level. *Water Resour. Manag.* **2013**, *27*, 419–432. [CrossRef]
62. Taormina, R.; Chau, K.; Sethi, R. Artificial Neural Network Simulation of Hourly Groundwater Levels in a Coastal Aquifer System of the Venice Lagoon. *Eng. Appl. Artif. Intell.* **2015**, *25*, 1670–1676. [CrossRef]
63. Valipour, M.; Banihabib, M.E.; Behbahani, S.M.R. Comparison of the ARMA, ARIMA, and the Autoregressive Artificial Neural Network Models in Forecasting the Monthly Inflow of Dez Dam Reservoir. *J. Hydrol.* **2013**, *476*, 433–441. [CrossRef]
64. Xuanhui, W.; Tailian, L.; Xilai, Z.; Hui, P.; Jia, X.; Bo, Z. Short-Term Prediction of Groundwater Level Using Improved Random Forest Regression with a combination of random Features. *Appl. Water Sci.* **2018**, *8*, 125. [CrossRef]
65. Yin, J.; Pham, H.V.; Tsai, F.T.-C. Multiobjective Spatial Pumping Optimization for Groundwater Management in a Multiaquifer System. *J. Water Resour. Plan. Manag.* **2020**, *146*, 04020013. [CrossRef]
66. Yoon, H.; Jun, S.-C.; Hyun, Y.; Bae, G.-O.; Lee, K.-K. A Comparative Study of Artificial Neural Networks and Support Vector Machines for Predicting Groundwater Levels in a Coastal Aquifer. *J. Hydrol.* **2011**, *396*, 128–138. [CrossRef]
67. Huang, M.; Tian, Y. Prediction of Groundwater Level for Sustainable Water Management in an Arid Basin using Data-driven Models. In Proceedings of the International Conference on Sustainable Energy and Environmental Engineering (SEEE), Bangkok, Thailand, 18–19 December 2015.

68. Li, R.; Ou, G.; Pun, M.; Larson, L. Evaluation of Groundwater Resources in Response to Agricultural Management Scenarios in the Central Valley, California. *J. Water Resour. Plan. Manag.* **2018**, *144*, 04018078. [CrossRef]
69. Bierkens, M.F.; Wada, Y. Non-Renewable Groundwater Use and Groundwater Depletion: A Review. *Environ. Res. Lett.* **2019**, *14*, 063002. [CrossRef]
70. Wada, Y.; Lo, M.-H.; Yeh, P.J.-F.; Reager, J.T.; Famiglietti, J.; Wu, M.-H.L.R.-J.; Tseng, Y.-H. Fate of Water Pumped from Underground and Contributions to Sea-Level Rise. *Nat. Clim. Chang.* **2016**, *6*, 777–780. [CrossRef]
71. Wada, Y. Modeling Groundwater Depletion at Regional and Global Scales: Present State and Future Prospects. *Surv. Geophys.* **2016**, *37*, 419–451. [CrossRef]
72. Wada, Y.; Bierkens, M.F. Sustainability of Global Water Use: Past Reconstruction and Future Projections. *Environ. Res. Lett.* **2014**, *9*, 104003. [CrossRef]
73. Wada, Y.; Wisser, D.; Bierkens, M.F. Global Modeling of Withdrawal, Allocation and Consumptive Use of Surface Water and Groundwater Resources. *Earth Syst. Dyn. Discuss.* **2014**, *5*, 15–40. [CrossRef]
74. Richey, A.S.; Thomas, B.F.; Lo, M.H.; Famiglietti, J.S.; Swenson, S.; Rodell, M. Uncertainty in Global Groundwater Storage Estimates in a Total Groundwater Stress Framework. *Water Resour. Res.* **2015**, *51*, 5198–5216. [CrossRef] [PubMed]
75. Rodell, M.; Famiglietti, J.S.; Wiese, D.N.; Reager, J.T.; Beaudoing, H.K.; Landerer, F.W.; Lo, M.-H. Emerging Trends in Global Freshwater Availability. *Nature* **2018**, *557*, 651, Erratum in *Nature* **2019**, *565*, E7. [CrossRef]
76. Zhao, W.G.; Wang, H.; Wang, Z.J. Groundwater Level Forecasting Based on Support Vector Machine. In *Applied Mechanics and Materials*; Trans Tech Publications: Freienbach, Switzerland, 2011; Volume 44, pp. 1365–1369.
77. Sahoo, S.; Russo, T.; Elliott, J.; Foster, I. Machine Learning Algorithms for Modeling Groundwater Level Changes in Agricultural Regions of the US. *Water Resour. Res.* **2017**, *53*, 3878–3895. [CrossRef]
78. Lueth, K.L. Why the Internet of Things is called the Internet of Things: Definition, History, Disambiguation. *IoT Anal.* **2014**, *19*. Available online: https://iot-analytics.com/internet-of-things-definition/ (accessed on 30 June 2021).
79. Romkey, J. Toast of the IoT: The 1990 Interop Internet Toaster. *IEEE Consum. Electron. Mag.* **2016**, *6*, 116–119. [CrossRef]
80. Maayan, G.D. The IoT Rundown for 2020: Stats, Risks, and Solutions. *Secur. Today* **2020**, *13*. Available online: https://securitytoday.com/articles/2020/01/13/the-iot-rundown-for-2020.aspx (accessed on 1 July 2021).
81. Premsankar, G.; di Francesco, M.; Taleb, T. Edge Computing for the Internet of Things: A Case Study. *IEEE Internet Things J.* **2018**, *5*, 1275–1284. [CrossRef]
82. Li, H.; Ota, K.; Dong, M. Learning IoT in Edge: Deep Learning for the Internet of Things with Edge Computing. *IEEE Netw.* **2018**, *32*, 96–101. [CrossRef]
83. Sun, X.; Ansari, N. EdgeIoT: Mobile Edge Computing for the Internet of Things. *IEEE Commun. Mag.* **2016**, *54*, 22–29. [CrossRef]
84. Kaur, N.; Sood, S.K. An Energy-Efficient Architecture for the Internet of Things (IoT). *IEEE Syst. J.* **2015**, *11*, 796–805. [CrossRef]
85. Corcoran, P.; Datta, S.K. Mobile-Edge Computing and the Internet of Things for Consumers: Extending Cloud Computing and Services to the Edge of the Network. *IEEE Consum. Electron. Mag.* **2016**, *5*, 73–74. [CrossRef]
86. Wu, W.-Y.; Lo, M.-H.; Wada, Y.; Famiglietti, J.S.; Reager, J.T.; Yeh, P.J.-F.; Ducharne, A.; Yang, Z.-L. Divergent Effects of Climate Change on Future Groundwater Availability in Key Mid-Latitude Aquifers. *Nat. Commun.* **2020**, *11*, 3710. [CrossRef]
87. Portmann, F.T.; Döll, P.; Eisner, S.; Flörke, M. Impact of Climate Change on Renewable Groundwater Resources: Assessing the Benefits of Avoided Greenhouse Gas Emissions Using Selected CMIP5 Climate Projections. *Environ. Res. Lett.* **2013**, *8*, 024023. [CrossRef]
88. Konikow, L.; Bredehoeft, J. *Groundwater Resource Development Effects and Sustainability*; The Groundwater Project: Guelph, ON, Canada, 2020.
89. Seyler, H.; Witthüser, K.; Holland, M. *The Capture Principle Approach to Sustainable Groundwater Use Incorporating Sustainability Indicators and Decision Framework for Sustainable Groundwater Use*; Water Research Commission: Pretoria, South Africa, 2016.
90. Van der Gun, J.; Lipponen, A. Reconciling Groundwater Storage Depletion due to Pumping with Sustainability. *Sustainability* **2010**, *2*, 3418–3435. [CrossRef]
91. Eltarabily, M.G.; Negm, A.M.; Yoshimura, C.; Saavedra, O.C. Modeling the Impact of Nitrate Fertilizers on Groundwater Quality in the Southern Part of the Nile Delta, Egypt. *Water Supply* **2017**, *17*, 561–570. [CrossRef]
92. Gorelick, S.M.; Voss, C.I.; Gill, P.E.; Murray, W.; Saunders, M.A.; Wright, M.H. Aquifer Reclamation Design: The Use of Contaminant Transport Simulation Combined with Nonlinear Programing. *Water Resour. Res.* **1984**, *20*, 415–427. [CrossRef]
93. Gordon, E.; Shamir, U.; Bensabat, J. Optimal Management of a Regional Aquifer under Salinization Conditions. *Water Resour. Res.* **2000**, *36*, 3193–3203. [CrossRef]
94. Heydari, F.; Saghafian, B.; Delavar, M. Coupled Quantity-Quality Simulation-Optimization Model for Conjunctive Surface-Groundwater Use. *Water Resour. Manag.* **2016**, *30*, 4381–4397. [CrossRef]
95. Izady, A.; Davary, K.; Alizadeh, A.; Ziaei, A.N.; Akhavan, S.; Alipoor, A.; Joodavi, A.; Brusseau, M.L. Groundwater Conceptualization and Modeling Using Distributed SWAT-Based Recharge for the Semi-Arid Agricultural Neishaboor Plain, Iran. *Hydrogeol. J.* **2015**, *23*, 47–68. [CrossRef]
96. Ayvaz, M.T. Application of Harmony Search Algorithm to the Solution of Groundwater Management Models. *Adv. Water Resour.* **2009**, *32*, 916–924. [CrossRef]
97. Ayvaz, M.T.; Karahan, H. A Simulation/Optimization Model for the Identification of Unknown Groundwater Well Locations and Pumping Rates. *J. Hydrol.* **2008**, *357*, 76–92. [CrossRef]

98. Coello, C.A.C.; Pulido, G.T.; Lechuga, M.S. Handling Multiple Objectives with Particle Swarm Optimization. *IEEE Trans. Evol. Comput.* **2004**, *8*, 256–279. [CrossRef]
99. El-Ghandour, H.A.; Elbeltagi, E. Optimal Groundwater Management Using Multiobjective Particle Swarm with a New Evolution Strategy. *J. Hydrol. Eng.* **2014**, *19*, 1141–1149. [CrossRef]
100. Gorelick, S.M. A Review of Distributed Parameter Groundwater Management Modeling Methods. *Water Resour. Res.* **1983**, *19*, 305–319. [CrossRef]
101. Bredehoeft, J.D.; Reichard, E.G.; Gorelick, S.M. If It Works, Don't Fix It: Benefits from Regional Ground-Water Management. In *Groundwater Models for Resources Analysis and Management*; CRC Press: Boca Raton, FL, USA, 1995; pp. 103–124.
102. Wagner, B.J. Recent Advances in Simulation-Optimization Groundwater Management Modeling. *Rev. Geophys.* **1995**, *33*, 1021–1028. [CrossRef]
103. Singh, A. Simulation and Optimization Modeling for the Management of Groundwater Resources. I: Distinct Applications. *J. Irrig. Drain. Eng.* **2014**, *140*, 04013021. [CrossRef]
104. Singh, A. Groundwater Resources Management through the Applications of Simulation Modeling: A Review. *Sci. Total Environ.* **2014**, *499*, 414–423. [CrossRef] [PubMed]
105. Gorelick, S.M. Large Scale Nonlinear Deterministic and Stochastic Optimization: Formulations Involving Simulation of Subsurface Contamination. *Math. Program.* **1990**, *48*, 19–39. [CrossRef]
106. Velasco-Levy, A.; Gómez, S. Sequential Gradient-Restoration Algorithm for the Optimization of a Nonlinear Constrained Function. *J. Astronaut. Sci.* **1982**, *30*, 131–142.
107. Dawoud, M.A.; Darwish, M.M.; El-Kady, M.M. GIS-Based Groundwater Management Model for Western Nile Delta. *Water Resour. Manag.* **2005**, *19*, 585–604. [CrossRef]
108. Mylopoulos, N.; Mylopoulos, Y.; Tolikas, D.; Veranis, N. Groundwater Modeling and Management in a Complex Lake-Aquifer System. *Water Resour. Manag.* **2007**, *21*, 469–494. [CrossRef]
109. Xu, X.; Huang, G.; Qu, Z.; Pereira, L.S. Using MODFLOW and GIS to Assess Changes in Groundwater Dynamics in Response to Water Saving Measures in Irrigation Districts of the Upper Yellow River Basin. *Water Resour. Manag.* **2011**, *25*, 2035–2059. [CrossRef]
110. Rajabi, M.M.; Ketabchi, H. Uncertainty-Based Simulation-Optimization Using Gaussian Process Emulation: Application to Coastal Groundwater Management. *J. Hydrol.* **2017**, *555*, 518–534. [CrossRef]
111. Cooper, H.H., Jr. The Equation of Groundwater Flow in Fixed and Deforming Coordinates. *J. Geophys. Res.* **1966**, *71*, 4785–4790. [CrossRef]
112. Remson, I.; Hornberger, G.M.; Molz, F.J. *Numerical Methods in Subsurface Hydrology*; Wiley: Hoboken, NJ, USA, 1971.
113. Pinder, G.F.; Bredehoeft, J. Application of the Digital Computer for Aquifer Evaluation. *Water Resour. Res.* **1968**, *4*, 1069–1093. [CrossRef]
114. Rahmati, O.; Choubin, B.; Fathabadi, A.; Coulon, F.; Soltani, E.; Shahabi, H.; Mollaefar, E.; Tiefenbacher, J.; Cipullo, S.; Bin Ahmad, B.; et al. Predicting Uncertainty of Machine Learning Models for Modelling Nitrate Pollution of Groundwater Using Quantile Regression and UNEEC Methods. *Sci. Total Environ.* **2019**, *688*, 855–866. [CrossRef] [PubMed]
115. Barzegar, R.; Moghaddam, A.A.; Deo, R.; Fijani, E.; Tziritis, E. Mapping Groundwater Contamination Risk of Multiple Aquifers Using Multi-Model Ensemble of Machine Learning Algorithms. *Sci. Total Environ.* **2018**, *621*, 697–712. [CrossRef]
116. Venkataraman, P. *Applied Optimization with MATLAB Programming*; John Wiley & Sons: Hoboken, NJ, USA, 2009.
117. Rosen, J. The Gradient Projection Method for Nonlinear Programming. Part II. Nonlinear Constraints. *J. Soc. Ind. Appl. Math.* **1961**, *9*, 514–532. [CrossRef]
118. Hastie, T.; Tibshirani, R.; Friedman, J. *The Elements of Statistical Learning: Data Mining, Inference, and Prediction*, 2nd ed.; Springer Science & Business Media: New York, NY, USA, 2017.
119. John Lu, Z.Q. The Elements of Statistical Learning: Data Mining, Inference, and Prediction. *J. R. Stat. Soc. Ser. A* **2010**, *173*, 693–694. [CrossRef]
120. Breiman, L. *Random Forests*; TR567; UC Berkeley: Berkeley, CA, USA, 1999.
121. Cutler, A.; Cutler, D.R.; Stevens, J.R. Random Forests. In *Ensemble Machine Learning*; Springer: Berlin/Heidelberg, Germany, 2012; pp. 157–175.
122. Zhang, T. Sequential Greedy Approximation for Certain Convex Optimization Problems. *IEEE Trans. Inf. Theory* **2003**, *49*, 682–691. [CrossRef]
123. Friedman, J.H. Greedy Function Approximation: A Gradient Boosting Machine. *Ann. Stat.* **2001**, *29*, 1189–1232. [CrossRef]
124. Haykin, S. *Neural Networks, a Comprehensive Foundation*; Prentice-Hall: Upper Saddle River, NJ, USA, 1999; pp. 161–175.
125. Samarasinghe, S. *Neural Networks for Applied Sciences and Engineering: From Fundamentals to Complex Pattern Recognition*; Auerbach Publications: Boca Raton, FL, USA, 2006.
126. ASCE. Artificial Neural Networks in Hydrology. I: Preliminary Concepts. *J. Hydrol. Eng.* **2000**, *5*, 115–123. [CrossRef]
127. ASCE. Artificial Neural Networks in Hydrology. II: Hydrologic Applications. *J. Hydrol. Eng.* **2000**, *5*, 124–137. [CrossRef]
128. Cho, S.; Lim, B.; Jung, J.; Kim, S.; Chae, H.; Park, J.; Park, S.; Park, J.K. Factors Affecting Algal Blooms in a Man-made Lake and Prediction using an Artificial Neural Network. *Measurement* **2014**, *53*, 224–233. [CrossRef]

129. Latt, Z.Z.; Wittenberg, H.; Urban, B. Clustering Hydrological Homogeneous Regions and Neural Network Based Index Flood Estimation for Ungauged Catchments: An Example of the Chindwin River in Myanmar. *Water Resour. Manag.* **2015**, *29*, 913–928. [CrossRef]
130. Nastos, P.; Paliatsos, A.; Koukouletsos, K.; Larissi, I.; Moustris, K. Artificial Neural Networks Modeling for Forecasting the Maximum Daily Total Precipitation at Athens, Greece. *Atmos. Res.* **2014**, *144*, 141–150. [CrossRef]
131. Afan, H.A.; El-Shafie, A.; Yaseen, Z.M.; Hameed, M.M.; Mohtar, W.H.M.W.; Hussain, A. ANN Based Sediment Prediction Model Utilizing Different Input Scenarios. *Water Resour. Manag.* **2015**, *29*, 1231–1245. [CrossRef]
132. He, Z.; Wen, X.; Liu, H.; Du, J. A Comparative Study of Artificial Neural Network, Adaptive Neuro-Fuzzy Inference System and Support Vector Machine for Forecasting River Flow in the Semiarid Mountain Region. *J. Hydrol.* **2014**, *509*, 379–386. [CrossRef]
133. Jang, J.-S. ANFIS: Adaptive-Network-Based Fuzzy Inference System. *IEEE Trans. Syst. Man Cybern.* **1993**, *23*, 665–685. [CrossRef]
134. Jang, J.-S.R.; Sun, C.-T.; Mizutani, E. Neuro-Fuzzy and Soft Computing—A Computational Approach to Learning and Machine Intelligence. In *IEEE Transactions on Automatic Control*; Prentice-Hall: Hoboken, NJ, USA, 1997; Volume 42, pp. 1482–1484.
135. Goyal, M.K.; Bharti, B.; Quilty, J.; Adamowski, J.; Pandey, A. Modeling of Daily Pan Evaporation in Sub Tropical Climates Using ANN, LS-SVR, Fuzzy Logic, and ANFIS. *Expert Syst. Appl.* **2014**, *41*, 5267–5276. [CrossRef]
136. Awan, J.A.; Bae, D.-H. Improving ANFIS Based Model for Long-Term Dam Inflow Prediction by Incorporating Monthly Rainfall Forecasts. *Water Resour. Manag.* **2014**, *28*, 1185–1199. [CrossRef]
137. Hipni, A.; El-shafie, A.; Najah, A.; Karim, O.A.; Hussain, A.; Mukhlisin, M. Daily Forecasting of Dam Water Levels: Comparing A Support Vector Machine (SVM) Model with Adaptive Neuro-Fuzzy Inference System (ANFIS). *Water Resour. Manag.* **2013**, *27*, 3803–3823. [CrossRef]
138. Vernieuwe, H.; Georgieva, O.; de Baets, B.; Pauwels, V.R.; Verhoest, N.E.; de Troch, F.P. Comparison of Data-Driven Takagi–Sugeno Models of Rainfall–Discharge Dynamics. *J. Hydrol.* **2005**, *302*, 173–186. [CrossRef]
139. Moghaddamnia, A.; Gousheh, M.G.; Piri, J.; Amin, S.; Han, D. Evaporation Estimation Using Artificial Neural Networks and Adaptive Neuro-Fuzzy Inference System Techniques. *Adv. Water Resour.* **2009**, *32*, 88–97. [CrossRef]
140. Shiri, J.; Kişi, Ö. Comparison of Genetic Programming with Neuro-Fuzzy Systems for Predicting Short-Term Water Table Depth Fluctuations. *Comput. Geosci.* **2011**, *37*, 1692–1701. [CrossRef]
141. Vapnik, V.N. *The Nature of Statistical Learning (Theory)*; Springer: New York, NY, USA, 1995.
142. Vapnik, V. *Statistical Learning Theory Biology*, 2nd ed.; Wiley: New York, NY, USA, 1998; p. 624.
143. Wen, X.; Si, J.; He, Z.; Wu, J.; Shao, H.; Yu, H. Support-Vector-Machine-Based Models for Modeling Daily Reference Evapotranspiration with Limited Climatic Data in Extreme Arid Regions. *Water Resour. Manag.* **2015**, *29*, 3195–3209. [CrossRef]
144. Ch, S.; Anand, N.; Panigrahi, B.K.; Mathur, S. Streamflow Forecasting by SVM with Quantum Behaved Particle Swarm Optimization. *Neurocomputing* **2013**, *101*, 18–23. [CrossRef]
145. Çimen, M.; Kisi, O. Comparison of Two Different Data-Driven Techniques in Modeling Lake Level Fluctuations in Turkey. *J. Hydrol.* **2009**, *378*, 253–262. [CrossRef]
146. Noori, R.; Karbassi, A.R.; Moghaddamnia, A.; Han, D.; Zokaei-Ashtiani, M.H.; Farokhnia, A.; Gousheh, M.G. Assessment of Input Variables Determination on the SVM Model Performance Using PCA, Gamma Test, and Forward Selection Techniques for Monthly Stream Flow Prediction. *J. Hydrol.* **2011**, *401*, 177–189. [CrossRef]
147. Tabari, H.; Kisi, O.; Ezani, A.; Talaee, P.H. SVM, ANFIS, Regression and Climate Based Models for Reference Evapotranspiration Modeling Using Limited Climatic Data in a Semi-Arid Highland Environment. *J. Hydrol.* **2012**, *444*, 78–89. [CrossRef]
148. Wang, Y.-H.; Yeh, C.-H.; Young, H.-W.V.; Hu, K.; Lo, M.-T. On the Computational Complexity of the Empirical Mode Decomposition Algorithm. *Phys. A Stat. Mech. Appl.* **2014**, *400*, 159–167. [CrossRef]
149. Humeau-Heurtier, A.; Abraham, P.; Mahé, G. Analysis of Laser Speckle Contrast Images Variability Using a Novel Empirical Mode Decomposition: Comparison of Results with Laser Doppler Flowmetry Signals Variability. *IEEE Trans. Med. Imaging* **2015**, *34*, 618–627. [CrossRef]
150. Raghavendra, N.; Deka, P. Support Vector Machine Applications in the Field of Hydrology: A Review. *Appl. Soft Comput.* **2014**, *19*, 372–386. [CrossRef]
151. Yoon, H.; Hyun, Y.; Ha, K.; Lee, K.-K.; Kim, G.-B. A Method to Improve the Stability and Accuracy of ANN and SVM Based Time Series Models for Long-Term Groundwater Level Predictions. *Comput. Geosci.* **2016**, *90*, 144–155. [CrossRef]
152. Wu, Z.; Huang, N.E. Ensemble Empirical Mode Decomposition: A Noise-Assisted Data Analysis Method. *Adv. Adapt. Data Anal.* **2009**, *1*, 1–41. [CrossRef]
153. Lee, C.H.; Cheng, R.T.S. On Seawater Encroachment in Coastal Aquifers. *Water Resour. Res.* **1974**, *10*, 1039–1043. [CrossRef]
154. Tyson, H.N.; Weber, E.M. Use of Electronic Computer in the Simulation of Dynamic Behaviour of Groundwater Basin. In Proceedings of the ASCE Water Resources Engineering Conference, Milwaukee, WI, USA, 15 May 1963.
155. D'Urso, G.; Menenti, M.; Santini, A. Regional Application of One-Dimensional Water Flow Models for Irrigation Management. *Agric. Water Manag.* **1999**, *40*, 291–302. [CrossRef]
156. Kite, G.W.; Droogers, P. *Comparing Estimates of Actual Evapotranspiration from Satellites, Hydrological Models, and Field Data: A Case Study from Western Turkey*; IWMI: Anand, India, 2000.
157. Van Dam, J.; Singh, R.; Bessembinder, J.J.E.; Leffelaar, P.A.; Bastiaanssen, W.G.M.; Jhorar, R.K.; Kroes, J.G.; Droogers, P. Assessing Options to Increase Water Productivity in Irrigated River Basins Using Remote Sensing and Modelling Tools. *Water Resour. Dev.* **2006**, *22*, 115–133. [CrossRef]

158. Hassan, A.E. Validation of Numerical Ground Water Models Used To Guide Decision Making. *Groundwater* **2004**, *42*, 277–290. [CrossRef] [PubMed]
159. Pint, C.D.; Hunt, R.J.; Anderson, M.P. Flowpath Delineation and Groundwater Age, Allequash Basin, Wisconsin. *Groundwater* **2003**, *41*, 895–902. [CrossRef]
160. Budge, T.J.; Sharp, J.M., Jr. Modeling the Usefulness of Spatial Correlation Analysis on Karst Systems. *Groundwater* **2009**, *47*, 427–437. [CrossRef] [PubMed]
161. Zhu, Y.; Shi, L.; Yang, J.; Wu, J.; Mao, D. Coupling Methodology and Application of a Fully Integrated Model for Contaminant Transport in the Subsurface System. *J. Hydrol.* **2013**, *501*, 56–72. [CrossRef]
162. Van Walsum, P.; Veldhuizen, A. Integration of Models Using Shared State Variables: Implementation in the Regional Hydrologic Modelling System SIMGRO. *J. Hydrol.* **2011**, *409*, 363–370. [CrossRef]
163. Yang, Z.; Lu, W.; Long, Y.; Li, P. Application and Comparison of Two Prediction Models for Groundwater Levels: A Case Study in Western Jilin Province, China. *J. Arid Environ.* **2009**, *73*, 487–492. [CrossRef]
164. Anderson, M.P.; Woessner, W.W.; Hunt, R.J. *Applied Groundwater Modeling: Simulation of Flow and Advective Transport*; Academic Press: Cambridge, MA, USA, 2015.
165. Singh, A.; Panda, S.N. Optimization and Simulation Modelling for Managing the Problems of Water Resources. *Water Resour. Manag.* **2013**, *27*, 3421–3431. [CrossRef]
166. Poeter, E.P.; Hill, M.C. Inverse Models: A Necessary Next Step in Groundwater Modeling. *Groundwater* **1997**, *35*, 250–260. [CrossRef]
167. Poeter, E. All Models Are Wrong, How Do We Know Which Are Useful. *Groundwater* **2007**, *45*, 390–391.
168. Michael, H.A.; Voss, C.I. Evaluation of the Sustainability of Deep Groundwater as an Arsenic-Safe Resource in the Bengal Basin. *Proc. Natl. Acad. Sci. USA* **2008**, *105*, 8531–8536. [CrossRef] [PubMed]
169. Zhang, L.; Potter, N.; Hickel, K.; Zhang, Y.; Shao, Q. Water Balance Modeling over Variable Time Scales Based on the Budyko Framework—Model Development and Testing. *J. Hydrol.* **2008**, *360*, 117–131. [CrossRef]
170. Harou, J.J.; Pulido-Velazquez, M.; Rosenberg, D.E.; Medellín-Azuara, J.; Lund, J.R.; Howitt, R.E. Hydro-Economic Models: Concepts, Design, Applications, and Future Prospects. *J. Hydrol.* **2009**, *375*, 627–643. [CrossRef]
171. Raes, D. *UPFLOW Water Movement in a Soil Profile from a Shallow Water Table to the Topsoil (Capillary Rise)*; Reference Manual; Department of Land Management: Leuven, Belgium, 2009.
172. Wondzell, S.M.; LaNier, J.; Haggerty, R. Evaluation of Alternative Groundwater Flow Models for Simulating Hyporheic Exchange in a Small Mountain Stream. *J. Hydrol.* **2009**, *364*, 142–151. [CrossRef]
173. Wondzell, S.M.; Gooseff, M.N.; McGlynn, B.L. An Analysis of Alternative Conceptual Models Relating Hyporheic Exchange Flow to Diel Fluctuations in Discharge During Baseflow Recession. *Hydrol. Process.* **2010**, *24*, 686–694. [CrossRef]
174. Yang, C.-C.; Chang, L.-C.; Chen, C.-S.; Yeh, M.-S. Multi-Objective Planning for Conjunctive Use of Surface and Subsurface Water Using Genetic Algorithm and Dynamics Programming. *Water Resour. Manag.* **2009**, *23*, 417–437. [CrossRef]
175. Sanford, W.E.; Pope, J.P. Current Challenges Using Models to Forecast Seawater Intrusion: Lessons from the Eastern Shore of Virginia, USA. *Hydrogeol. J.* **2010**, *18*, 73–93. [CrossRef]
176. Sherif, M.; Kacimov, A.; Javadi, A.; Ebraheem, A.A. Modeling Groundwater Flow and Seawater Intrusion in the Coastal Aquifer of Wadi Ham, UAE. *Water Resour. Manag.* **2012**, *26*, 751–774. [CrossRef]
177. Su, Y.-S.; Ni, C.-F.; Li, W.-C.; Lee, I.-H.; Lin, C.-P. Applying Deep Learning Algorithms to Enhance Simulations of Large-Scale Groundwater Flow in IoTs. *Appl. Soft Comput.* **2020**, *92*, 106298. [CrossRef]
178. Narendran, S.; Pradeep, P.; Ramesh, M.V. An Internet of Things (IoT) based Sustainable Water Management. In Proceedings of the 2017 IEEE Global Humanitarian Technology Conference (GHTC), San Jose, CA, USA, 19–22 October 2017; pp. 1–6.
179. Manguinhas, H.; Martins, B.; Borbinha, J.; Siabato, W. A Geo-Temporal Web Gazetteer Service Integrating Data from Multiple Sources. In Proceedings of the 3rd IEEE International Conference on Digital Information Management, London, UK, 13–16 November 2008.
180. Talari, S.; Shafie-Khah, M.; Siano, P.; Loia, V.; Tommasetti, A.; Catalão, J.P. A Review of Smart Cities Based on the Internet of Things Concept. *Energies* **2017**, *10*, 421. [CrossRef]
181. Jones, W.R.; Spence, M.J.; Bowman, A.W.; Evers, L.; Molinari, D.A. GWSDAT—GroundWater Spatiotemporal Data Analysis Tool. *arXiv* **2014**, arXiv:1310.8158.
182. Laraichi, S.; Hammani, A.; Bouignane, A. Data Integration as the Key to Building a Decision Support System for Groundwater Management: Case of Saiss Aquifers, Morocco. *Groundw. Sustain. Dev.* **2016**, *2*, 7–15. [CrossRef]
183. Van der Gun, J. Data, Information, Knowledge and Diagnostics on Groundwater. In *Advances in Groundwater Governance*; CRC Press: Boca Raton, FL, USA, 2018; pp. 193–213.
184. IGRAC. *Groundwater in a Changing World*; IGRAC Strategy 2019–23; IGRAC: Delft, The Netherlands, 2019; Available online: https://www.un-igrac.org/resource/igrac-strategy-2019-23-groundwater-changing-world (accessed on 7 July 2020).
185. Famiglietti, J.S. The Global Groundwater Crisis. *Nat. Clim. Change* **2014**, *4*, 945–948. [CrossRef]
186. Famiglietti, J.; Cazenave, A.; Eicker, A.; Reager, J.; Rodell, M.; Velicogna, I. Satellites Provide the Big Picture. *Science* **2015**, *349*, 684–685. [CrossRef] [PubMed]
187. Kombo, O.H.; Kumaran, S.; Bovim, A. Design and Application of a Low-cost, Low-Power, LoRa-GSM, IoT Enabled System for Monitoring of Groundwater Resources with Energy Harvesting Integration. *IEEE Access* **2021**, *9*, 128417–128433. [CrossRef]

188. Robles, T.; Alcarria, R.; de Andrés, D.M.; de la Cruz, M.N.; Calero, R.; Iglesias, S.; Lopez, M. An IoT Based Reference Architecture for Smart Water Management Processes. *J. Wirel. Mob. Netw. Ubiquitous Comput. Dependable Appl.* **2015**, *6*, 4–23.
189. Mao, F.; Khamis, K.; Krause, S.; Clark, J.; Hannah, D.M. Low-Cost Environmental Sensor Networks: Recent Advances and Future Directions. *Front. Earth Sci.* **2019**, *7*, 221. (In English) [CrossRef]
190. Chan, K.; Schillereff, D.N.; Baas, A.C.; A Chadwick, M.; Main, B.; Mulligan, M.; O'Shea, F.T.; Pearce, R.; EL Smith, T.; van Soesbergen, A.; et al. Low-Cost Electronic Sensors for Environmental Research: Pitfalls and Opportunities. *Prog. Phys. Geogr. Earth Environ.* **2021**, *45*, 305–338. [CrossRef]
191. Vijayakumar, N.; Ramya, R. The Real-Time Monitoring of Water Quality in IoT Environment. In Proceedings of the 2015 International Conference on Innovations in Information, Embedded and Communication Systems (ICIIECS), Coimbatore, India, 19–20 March 2015; pp. 1–5.
192. Brunner, P.; Franssen, H.-J.H.; Kgotlhang, L.; Bauer-Gottwein, P.; Kinzelbach, W. How Can Remote Sensing Contribute in Groundwater Modeling? *Hydrogeol. J.* **2007**, *15*, 5–18. [CrossRef]
193. Kemna, A.; Vanderborght, J.; Kulessa, B.; Vereecken, H. Imaging and Characterisation of Subsurface Solute Transport Using Electrical Resistivity Tomography (Ert) and Equivalent Transport Models. *J. Hydrol.* **2002**, *267*, 125–146. [CrossRef]
194. Dams, J.; Dujardin, J.; Reggers, R.; Bashir, I.; Canters, F.; Batelaan, O. Mapping Impervious Surface Change from Remote Sensing for Hydrological Modeling. *J. Hydrol.* **2013**, *485*, 84–95. [CrossRef]
195. Brunner, P.; Li, H.; Kinzelbach, W.; Li, W.; Dong, X. Extracting Phreatic Evaporation from Remotely Sensed Maps of Evapotranspiration. *Water Resour. Res.* **2008**, *44*, W08428. [CrossRef]
196. Brunner, P.; Bauer, P.; Eugster, M.; Kinzelbach, W. Using Remote Sensing to Regionalize Local Precipitation Recharge Rates Obtained from the Chloride Method. *J. Hydrol.* **2004**, *294*, 241–250. [CrossRef]
197. Choudhury, B.J. Synergism of Multispectral Satellite Observations for Estimating Regional Land Surface Evaporation. *Remote Sens. Environ.* **1994**, *49*, 264–274. [CrossRef]
198. Zhou, Y.; Dong, D.; Liu, J.; Li, W. Upgrading a Regional Groundwater Level Monitoring Network for Beijing Plain, China. *Geosci. Front.* **2013**, *4*, 127–138. [CrossRef]
199. McGuire, V.L. *Water-Level and Recoverable Water in Storage Changes, High Plains Aquifer, Predevelopment to 2015 And 2013–15*; 2328-0328; US Geological Survey: Reston, VA, USA, 2017.
200. Xiao, M.; Koppa, A.; Mekonnen, Z.; Pagán, B.R.; Zhan, S.; Cao, Q.; Aierken, A.; Lee, H.; Lettenmaier, D.P. How Much Groundwater Did California's Central Valley Lose during the 2012–2016 Drought? *Geophys. Res. Lett.* **2017**, *44*, 4872–4879. [CrossRef]
201. Brush, C.F.; Dogrul, E.C.; Kadir, T.N. *Development and Calibration of the California Central Valley Groundwater-Surface Water Simulation Model (C2vsim), Version 3.02-Cg*; Bay-Delta Office, California Department of Water Resources: Sacramento, CA, USA, 2013.
202. Faunt, C.C.; Belitz, K.; Hanson, R.T. Development of a Three-Dimensional Model of Sedimentary Texture in Valley-Fill Deposits of Central Valley, California, USA. *Hydrogeol. J.* **2010**, *18*, 625–649. [CrossRef]
203. Reddy, N.S.; Saketh, M.S.; Dhar, S. Review of Sensor Technology for Mine Safety Monitoring Systems: A Holistic Approach. In Proceedings of the 2016 IEEE First International Conference on Control, Measurement and Instrumentation (CMI), Kolkata, India, 8–10 January 2016; pp. 429–434.
204. Neyens, D.; Baïsset, M.; Lovighi, H. Monitoring the Groundwater Quality/Quantity from Your Desktop—Application To Salt Water Intrusion Monitoring EMI: Environmental data Management Interface. *E3S Web Conf.* **2018**, *54*, 00021. [CrossRef]
205. Drage, J.; Kennedy, G. Building a Low-Cost, Internet-of-Things, Real-Time Groundwater Level Monitoring Network. *Groundw. Monit. Remediat.* **2020**, *40*, 67–73. [CrossRef]
206. Alessio, B.; Walter, D.; Valerio, P.; Antonio, P. Integration of Cloud computing and Internet of Things: A survey. *Future Gener. Comput. Syst.* **2016**, *56*, 684–700. [CrossRef]
207. Fu, P.; Sun, J. *Web GIS: Principles and Applications*; ESRI Press: Redlands, CA, USA, 2010.
208. Li, S.; Dragicevic, S.; Veenendaal, B. *Advances in Web-Based GIS, Mapping Services and Applications*; CRC Press: Boca Raton, FL, USA, 2011.
209. Heywood, I.; Cornelius, S.; Carver, S. *An Introduction to Geographical Information Systems*; Pearson Education Limited: Harlow, UK, 2011.
210. Peng, Z.-R.; Tsou, M.-H. *Internet GIS: Distributed Geographic Information Services for the Internet and Wireless Networks*; John Wiley & Sons: Hoboken, NJ, USA, 2003.
211. Chenini, I.; Mammou, A.B. Groundwater Recharge Study in Arid Region: An Approach Using GIS Techniques and Numerical Modeling. *Comput. Geosci.* **2010**, *36*, 801–817. [CrossRef]
212. Stefan, C.; Ansems, N. Web-Based Global Inventory of Managed Aquifer Recharge Applications. *Sustain. Water Resour. Manag.* **2018**, *4*, 153–162. [CrossRef]
213. Maidment, D.R.; Morehouse, S. *Arc Hydro: GIS for Water Resources*; ESRI Inc.: Redlands, CA, USA, 2002.
214. Jones, D.; Jones, N.; Greer, J.; Nelson, J. A Cloud-Based MODFLOW Service for Aquifer Management Decision Support. *Comput. Geosci.* **2015**, *78*, 81–87. [CrossRef]
215. Balram, S.; Dragicevic, S. Collaborative Geographic Information Systems: Origins, Boundaries, And Structures. In *Collaborative Geographic Information Systems*; IGI Global: Hershey, PA, USA, 2006; pp. 1–23.
216. Werner, M.; Schellekens, J.; Gijsbers, P.; van Dijk, M.; van den Akker, O.; Heynert, K. The Delft-FEWS Flow Forecasting System. *Environ. Model. Softw.* **2013**, *40*, 65–77. [CrossRef]

217. Hsu, Y.-C.; Chang, Y.-L.; Chang, C.-H.; Yang, J.-C.; Tung, Y.-K. Physical-Based Rainfall-Triggered Shallow Landslide Forecasting. *Smart Water* **2018**, *3*, 3. [CrossRef]
218. Foglia, L.; Borsi, I.; Mehl, S.; De Filippis, G.; Cannata, M.; Vasquez-Suñe, E.; Criollo, R.; Rossetto, R. FREEWAT, A Free And Open Source, GIS-Integrated, Hydrological Modeling Platform. *Groundwater* **2018**, *56*, 521–523. [CrossRef]
219. Rossetto, R.; De Filippis, G.; Borsi, I.; Foglia, L.; Cannata, M.; Criollo, R.; Vázquez-Suñé, E. Integrating Free and Open Source Tools and Distributed Modelling Codes in GIS Environment for Data-Based Groundwater Management. *Environ. Model. Softw.* **2018**, *107*, 210–230. [CrossRef]
220. De Filippis, G.; Pouliaris, C.; Kahuda, D.; Vasile, T.A.; Manea, V.A.; Zaun, F.; Panteleit, B.; Dadaser-Celik, F.; Positano, P.; Nannucci, M.S.; et al. Spatial Data Management and Numerical Modelling: Demonstrating the Application of the QGIS-Integrated FREEWAT Platform at 13 Case Studies for Tackling Groundwater Resource Management. *Water* **2020**, *12*, 41. [CrossRef]
221. Snow, A.D.; Christensen, S.D.; Swain, N.R.; Nelson, E.J.; Ames, D.P.; Jones, N.L.; Ding, D.; Noman, N.S.; David, C.H.; Pappenberger, F.; et al. A High-Resolution National-Scale Hydrologic Forecast System from a Global Ensemble Land Surface Model. *J. Am. Water Resour. Assoc.* **2016**, *52*, 950–964. [CrossRef] [PubMed]
222. Morsy, M.M.; Goodall, J.L.; Castronova, A.; Dash, P.; Merwade, V.; Sadler, J.M.; Rajib, M.A.; Horsburgh, J.S.; Tarboton, D. Design of a Metadata Framework for Environmental Models with an Example Hydrologic Application in HydroShare. *Environ. Model. Softw.* **2017**, *93*, 13–28. [CrossRef]
223. Swain, N.R.; Christensen, S.D.; Snow, A.; Dolder, H.; Espinoza-Dávalos, G.; Goharian, E.; Jones, N.L.; Nelson, E.J.; Ames, D.P.; Burian, S. A New Open Source Platform for Lowering the Barrier for Environmental Web App Development. *Environ. Model. Softw.* **2016**, *85*, 11–26. [CrossRef]
224. MAGNET. MAGNET4water. Available online: https://www.magnet4water.net/ (accessed on 31 July 2021).
225. Malche, T.; Maheshwary, P. Internet of Things (IoT) Based Water Level Monitoring System for Smart Village. In Proceedings of the International Conference on Communication and Networks, Vancouver, BC, Canada, 31 July–3 August 2017; Springer: Berlin/Heidelberg, Germany, 2017; pp. 305–312.
226. Horsburgh, J.S.; Morsy, M.M.; Castronova, A.; Goodall, J.; Gan, T.; Yi, H.; Stealey, M.J.; Tarboton, D.G. Hydroshare: Sharing Diverse Environmental Data Types and Models as Social Objects with Application to the Hydrology Domain. *J. Am. Water Resour. Assoc.* **2016**, *52*, 873–889. [CrossRef]
227. Brewer, E.A. Towards Robust Distributed Systems. In Proceedings of the Nineteenth Annual ACM Symposium on Principles of Distributed Computing, Portland, OR, USA, 16–19 July 2000; Volume 7, pp. 343477–343502.
228. Hung, H.-C.; Liu, I.-F.; Liang, C.-T.; Su, Y.-S. Applying Educational Data Mining to Explore Students' Learning Patterns in the Flipped Learning Approach for Coding Education. *Symmetry* **2020**, *12*, 213. [CrossRef]
229. Lai, Y.-H.; Chen, S.-Y.; Lai, C.-F.; Chang, Y.-C.; Su, Y.-S. Study on Enhancing AIoT Computational Thinking Skills by Plot Image-Based VR. *Interact. Learn. Environ.* **2019**, *29*, 482–495. [CrossRef]
230. Anumalla, S.; Ramamurthy, B.; Gosselin, D.C.; Burbach, M. Groundwater Monitoring Using Smart Sensors. In Proceedings of the 2005 IEEE International Conference on Electro Information Technology, Lincoln, NE, USA, 22–25 May 2005; p. 6.
231. Afifi, M.; Abdelkader, M.F.; Ghoneim, A. An IoT System for Continuous Monitoring and Burst Detection in Intermittent Water Distribution Networks. In Proceedings of the 2018 International Conference on Innovative Trends in Computer Engineering (ITCE), Aswan, Egypt, 19–21 February 2018; pp. 240–247.
232. Kamaruidzaman, N.; Rahmat, S.N. Water Monitoring System Embedded with the Internet of Things (IoT) Device: A Review. *IOP Conf. Ser. Earth Environ. Sci.* **2020**, *498*, 012068. [CrossRef]
233. Tauro, F.; Selker, J.; van de Giesen, N.; Abrate, T.; Uijlenhoet, R.; Porfiri, M.; Manfreda, S.; Caylor, K.; Moramarco, T.; Benveniste, J.; et al. Measurements and Observations in the XXI century (MOXXI): Innovation and Multi-disciplinarity to Sense the Hydrological Cycle. *Hydrol. Sci. J.* **2018**, *63*, 169–196. [CrossRef]
234. Mesquita, E.; Paixão, T.; Antunes, P.; Coelho, F.; Ferreira, P.; Andre, P.; Varum, H. Groundwater Level Monitoring Using a Plastic Optical Fiber. *Sens. Actuators A Phys.* **2016**, *240*, 138–144. [CrossRef]
235. Verma, P.; Kumar, A.; Rathod, N.; Jain, P.; Mallikarjun, S.; Subramanian, R.; Amrutur, B.; Kumar, M.M.; Sundaresan, R. Towards an IoT Based Water Management System for a Campus. In Proceedings of the 2015 IEEE First International Smart Cities Conference (ISC2), Guadalajara, Mexico, 25–28 October 2015; pp. 1–6.
236. Odli, Z.S.M.; Izhar, T.N.T.; Razak, A.R.A.; Yusuf, S.Y.; Zakarya, I.A.; Saad, F.N.M.; Nor, M.Z.M. Development of Portable Water Level Sensor for Flood Management System. *ARPN J. Eng. Appl. Sci.* **2016**, *11*, 5352–5357.

*sustainability*

MDPI

*Article*

# Construction Sector Contribution to Economic Stability: Malaysian GDP Distribution

**Wesam Salah Alaloul [1], Muhammad Ali Musarat [1,*], Muhammad Babar Ali Rabbani [2], Qaiser Iqbal [2], Ahsen Maqsoom [3] and Waqas Farooq [4]**

1 Department of Civil and Environmental Engineering, Universiti Teknologi PETRONAS, Bandar Seri Iskandar, Tronoh 32610, Perak, Malaysia; wesam.alaloul@utp.edu.my
2 Department of Civil Engineering, Sarhad University of Science & Information Technology, Peshawar 25000, Pakistan; babaralirabbani@yahoo.com (M.B.A.R.); qi.civil@suit.edu.pk (Q.I.)
3 Department of Civil Engineering, COMSATS University Islamabad Wah Campus, Wah Cantt 47040, Pakistan; ahsen.maqsoom@ciitwah.edu.pk
4 Department of Electrical Engineering, Sarhad University of Science & Information Technology, Peshawar 25000, Pakistan; waqasfarooq.ee@gmail.com
* Correspondence: muhammad_19000316@utp.edu.my

**Abstract:** The construction sector exerts an exceptional impact on economic development all over the world. Adequate buildings and infrastructures made by the construction sector ensure that a country reaches certain targets like social development, industrialization, freight transportation, sustainable development, and urbanization. This study aims to determine the construction sector's connectivity with other sectors through complex linkages that contribute immensely to the economy and gross domestic product (GDP). The data were collected from the Department of Statistics Malaysia and the World Bank from the year 1970 to 2019, and the Pearson correlation test, the cointegration test, and the Granger causality test were conducted. The vector error correction model (VECM) was created for short-term and long-term equilibrium analysis and impulse response function (IRF) was performed to study construction industry behavior. Afterwards, the forecasting was done for the year 2020 to 2050 of the Malaysian economy and GDP for the required sectors. It was revealed that some sectors, such as agriculture and services, have forward linkages while other sectors, such as manufacturing and mining, are independent of construction sector causality, which signifies the behavior of the contributing sectors when a recession occurs, hence generating significant revenue. The Malaysian economy is moving towards sustainable production with more emphasis on the construction sector. The outcome can be used as a benchmark by other countries to achieve sustainable development. The significance of this study is its usefulness for experts all over the world in terms of allocating resources to make the construction sector a sustainable sector after receiving a shock. A sustainable conceptual framework has been suggested for global application that shows the factors involved in the growth of the construction industry to ensure its sustainable development with time.

**Keywords:** construction sector; economy; intersectoral linkages; VECM; forecasting; sustainable development

**Citation:** Alaloul, W.S.; Musarat, M.A.; Rabbani, M.B.A.; Iqbal, Q.; Maqsoom, A.; Farooq, W. Construction Sector Contribution to Economic Stability: Malaysian GDP Distribution. *Sustainability* **2021**, *13*, 5012. https://doi.org/10.3390/su13095012

Academic Editors: Marc A. Rosen and Shervin Hashemi

Received: 24 March 2021
Accepted: 27 April 2021
Published: 29 April 2021



## 1. Introduction

A country's prosperity is linked to its economic growth where all the sectors, such as primary, secondary, tertiary, and quaternary sectors, contribute to stabilizing the economy [1,2]. The construction sector has prime importance that signifies the prosperity, health, and quality of life for the country's citizens [3,4]. The construction sector acts as a backbone of the economic growth of any country, therefore, it influences every sector's role on all levels in an economy [5,6]. Developing countries are mostly dependent on the construction sector to implement their sustainable development [7,8]. The construction sector exerts a direct influence on social and economic development in money circulation [9]. Therefore, the absence of adequate building infrastructure results in undergrowth of a country's sectors, an underdeveloped economy, a substandard level of living, and imbalanced income distribution, which are the contributing factors towards a country's economic failure [10,11]. The effective management of the construction sector results in improved life quality, including boosted tourism, sustainable environment, money circulation, and job creation throughout the country [12,13]. In the economy, the gross domestic product (GDP) plays a vital role in balancing various sectors. Overall, the construction sector accounts for USD 1.7 trillion worldwide, and in most countries, it impacts 5–7% of the total GDP [14]. The construction sector contributes significantly to the national GDP, thus the need for development in this sector is important [15].

It has been reported that most countries are investing in research and development (R&D) to develop sustainable techniques for the construction sector [16]. Hence, revolutionizing the construction sector will ensure sustainable development [17]. Any variation in the construction sector will have a direct influence on all the related sectors [12,18,19]. The construction sector has a significant role in the economy, therefore, its influence in a country's economy is associated with forward and backward linkages with other sectors [17]. These linkages with other sectors are dependent on performance, therefore, a change in any of the sectors will have significant repercussions on the country's economy [20]. Thus, the construction sector has a considerable influence on the socio-economic aspects of any nation [21].

It is estimated that one-third of countries depend on resources which are necessary for social and economic development [22,23]. However, establishing sustainable goals does not require a significant workforce or abundance of resources, rather, it requires technological innovation, resourceful use of the materials, and smart industrial processes in developing and developed countries [24]. In developing countries, the construction sector is greatly influenced by globalization, as most of the complex, innovative techniques, and skilled manpower required for sustainable development are not available in the local market, therefore, they have to depend on the inescapable fact of globalization [25]. Most developed countries have norms that are no different from those of others, for example, in the construction sector, most practices in developing countries are influenced by the practices followed in developed countries [26]. Therefore, the conventional method of construction is still in effect in commonwealth countries [27,28].

The Malaysian construction sector (MCS) is facing various problems, like shortages of manpower, environment solutions, quality of the work, and a dearth of productivity, which have raised many challenges [29,30]. One of the contributing factors that hinders the performance of the construction sector in Malaysia is the conventional construction approach. The country must shift to modern construction technology like Hong Kong, Singapore, and the United Kingdom (UK) [31]. There is an absence of a framework to enhance the competitiveness, resource allocation, and funding laws, hence the quality level decreases in the national and international market [32]. The new challenges that the MCS is facing are: slow progress in terms of sustainability, excessive and unregularized use of resources, the absence of construction techniques causing less pollution, and energy efficiency [30]. The R&D for the construction sector in Malaysia is mostly carried out by educational institutes, which is not scalable to practical uses as per construction requirements [33]. The lack of a fair bidding process and transparent tendering [34] and the failure of timely release of funding [35] have contributed considerably to the underperformance of the MCS. The minimum growth of overall GDP was recorded in quarter (Q)1'2019 at just RM 341.6 billion as compared to the maximum of RM 3701 billion in Q4'2019. The percentage change shows a maximum growth of the Malaysian economy in Q3'2017, while from Q1'2016 to Q3'2019, there was not such a significant change in the percentage of overall GDP. However, a considerable fall in GDP production in Q1'2020 and Q2'2020 occurred because of the lockdown situation due to COVID-19, with a reduction of 17.1% in Q2'2020, whereas in Q3'2020, a reduction percentage was recorded as being 2.7% [36].

However, it should be noted that unseen events like the COVID-19 outbreak can significantly affect the performance of the overall GDP as well as the contributing sectors.

The overall GDP began to crumble as COVID-19 started to show its effects. For comparison purposes, it is evident that the Malaysian economy grew 3.6% in Q4′2019. With the effect of lockdown in Malaysia after the first case of COVID-19 was reported, there was a fall in the major sector of construction (CONST) and other sectors like services (SERV), manufacturing (MANU), agriculture (AGRI), and mining (MINI) in Q1′2020, but the effects were not quick as there was a surplus of material available in all sectors. The situation got worse when the COVID-19 outbreak was prolonged and people were forced to stay in their homes, which showed its repercussions in Q2′2020. With the prolonged lockdown, there was a major decrease in the Malaysian economy, as it contracted to 17.1% in Q2′2020 due to the fall in construction GDP to 44.5%. Hence, it is clear that the suspension of the construction industry is a major threat to the stability of the economy. At the end of Q3′2020, the economy saw a smaller decline of 2.7% in the overall GDP of all the sectors. In general, the effects of COVID-19 were not present in Q4′2019. The values are shown in Table 1 [37–39].

**Table 1.** Sector-wise quarterly gross domestic product (GDP) of Malaysia.

| Sector | GDP in Q4′2019 | GDP in Q1′2020 | GDP in Q2′2020 | GDP in Q3′2020 |
|---|---|---|---|---|
| Construction | 1.0% | −7.9% | −44.5% | 12.4% |
| Services | 6.1% | 3.1% | −16.2% | −4.0% |
| Manufacturing | 3.0% | 1.5% | −18.3% | 3.3% |
| Agriculture | −5.7% | −8.7% | 1.0% | −0.7% |
| Mining | −2.5% | −2.0% | −20.0% | −6.8% |

The above table shows the construction sector in comparison with the other sectors because the construction sector is considered as the major sector of Malaysian GDP and it has considerable volatility in the economy, therefore, MCS analysis was essential.

In Malaysia, the highest value of construction work was recorded in 2019, at approximately RM 146.37 billion. The value of construction work increased annually by approximately RM 10 billion from 2012 to 2018 [40]. There is not much difference between the value of construction work in 2018 and 2019 because the Malaysian government suspended construction activity to decrease debt. Since the classification of COVID-19 as a pandemic by the World Health Organization (WHO), most countries prepared for nationwide lockdown, which disrupted the economy, social life, and caused the shutting down of businesses. Among other sectors, the construction sector was also affected, which resulted in financial implications, overrunning of costs, halted projects, delays in projects, and loss of jobs [41]. Therefore, this research focused on studying the effects of shocks on the construction industry and what short- and long-term policies have to be applied to optimize the social and economic sustainability in the region. Moreover, this study also provides an essential sustainable framework as a guideline for international applications in order to establish a sustainable construction industry. Due to such importance, the following research questions were established:

1. In which direction does the construction industry move to recover from external shocks and the time taken to regain its original position?
2. What short- and long-term policies and decisions will affect the behavior of construction to achieve sustainability?
3. What sustainable framework should be followed to achieve sustainability in the construction industry that has global application?

The Granger causality test was used to assess the direction of the shock, which was produced using the impulse response function (IRF). The short- and long-term policies were assessed after creating a system of equations using the vector error correction model (VECM). A sustainable framework was proposed based on the results of the IRF and VECM along with its international applicability. The fundamental objective of conducting this study was to assess the role of the MCS in the prosperity of the Malaysian economy and how any change in the growth of the construction sector would impact the behavior of other sectors. This study addressed the issue of how much would the MCS be impacted when there is a unit investment in other sectors and vice versa, and what long-term actions the government should take to make the industry sustainable. The answer to this question was found by using the identification of the linkage direction of the MCS and use of the VECM equation by introducing a shock to study MCS behavior and assess the time required to fully recover from any shock, like a recession or lack of investment. Additionally, a conceptual framework was proposed to guide the governments of their respective countries about the necessary steps that they should take to make the construction sector shockproof or sustainable. Forecasting was performed to predict the sustainability of the construction sector if the present condition prevailed in the MCS. The application of unit shocks has never been tested in the construction sector which could move it one step closer towards its goal to attain sustainability in the region. This study is beneficial in terms of defining the socio-economical guidelines for the MCS to estimate the linkage model with other sectors. The financial planners and policymakers of Malaysia can use the results of this study to improve the construction sector by using advanced strategies, thus enabling Malaysia to be a progressive and prosperous country by the year 2050.

## 2. Literature Review

Numerous studies have been conducted to study and highlight the issues that are faced by the construction sector. VECM, regression, and Box–Jenkins methods were used for analyzing and predicting the construction tender prices index (TPI) in the Hong Kong market. With the value of mean absolute percentage error (MAPE) as 2.9%, it was found that the VECM outperforms the other two methods in prediction in terms of short- and long-term price movements [42]. The data from Melbourne and Sydney were used to perform an econometric approach to study the short- and long-term construction prices and their direction in Australia using the VECM and Granger causality test, respectively. The linkages and diffusion patterns were found among six states and two territories. It was concluded that there exists one returning diffusion path for every territorial market, hence, the shares go back to the funding of the federal state [43]. The VECM, along with the Box–Jenkins method and regression analysis, were put into use to predict the construction prices. The models were evaluated using the predictive ability of Theil coefficients and MAPE with values of 0 and 10%, respectively. It was found that the VECM with the inclusion of dummy variables predicts the prices better than the other two models [44]. The data of construction demand were obtained from the Australian Bureau of Statistics for the period of Q3'1998 to Q4'2013. The VECM indicated the links between the construction sector and state prices, population, and unemployment. The MAPE value of the VECM was lowest among all the other models, which was less than 10%, indicating robust reliable forecasting [45]. The forecasting of the cost of highway construction was performed based on the data collected from the State Department of Transportation. The VECM illustrated that raw oil and construction costs fit the diagnostics tests better. The presence of multivariate variables significantly increased the accuracy of the results as compared to univariate forecasting using the VECM [46]. The prices of construction materials like cement, steel, and asphalt were forecasted based on the data of the United States of America (USA). The methodology of the VECM was employed for examining short- and long-term effects on the mentioned prices. Based on the lowest values of MAPE and Theil coefficients, the VECM with asphalt prices outperformed the other variables [47]. The demand for construction workers was forecasted based on the data of Q1'1983-Q3'2005 in Hong Kong. After using the VECM as the method of analysis, it was revealed that labor skill and construction output were the major factors that influence the employment opportunities in this sector [48]. The construction cost index (CCI) short- and long-term effects on the consumer price index were tested and forecasting was performed using the unrestricted vector autoregression (VAR) model. The model was found to accurately forecast the cost trends of construction based on the lowest values of the mean squared errors (MSEs) and the mean absolute errors (MAEs) [49]. The construction material prices were generated for long-term effects using regression methods and VAR. The method of forecasting indicated that the presence of autocorrelation and higher values of $R^2$ greatly affected the performance of VAR processes like the VECM [50]. A study was conducted based on data from 2001 to 2006 of China, and the real estate demand and construction demand were forecasted. Regression analysis and cointegrating vectors were used for the purpose of forecasting It was found that economic planning is the reason for the change in investments in the Chinese market [51]. Economic development in developing countries saw a considerable rise without paying any heed to sustainability, which resulted in global warming [52], increased $CO_2$ emissions [53], and the generation of hazardous waste emanating from construction machinery [52] and steel [54] and cement manufacturing [55]. Based on the literature, there is an absence of work proposing a sustainable framework for the construction sector. The current practices of the construction sector of Malaysia were assessed for their sustainability based on questionnaires and interviews. To enable the construction sector in Malaysia to be fully sustainable, three major principles, namely environmental, social, and economic development, are needed [56]. Sustainability in the Malaysian construction sector is not new. Every skilled worker of construction sector is well aware of the sustainable work protocols but they turn a blind eye due to its need of bigger workforce, initial high costs and little acceptability in the traditional construction methods and absence of strict legislation [57]. Currently, Malaysia is facing major problems regarding $CO_2$ emissions and sustainability, therefore, the inflation effects on $CO_2$ emission were assessed. It was revealed that due to the decrease in the rate of inflation, the construction sector flourished, however, this poses a threat to the environment and has negative consequences for sustainability. By following the proposed inflation rate and $CO_2$ framework, Malaysia can overcome this problem [58]. The need for sustainable construction in Malaysia will ensure less usage of resources, reduced production costs, and holistic management of wastes. The lack of government enforcement, tax exemptions, and low level of interest of stakeholders in investing in sustainable processes and proper frameworks of sustainability are the reasons that the Malaysian construction sector is lagging in sustainable construction processes [59].

## 3. Methodology

To summarize the methodological procedure, the acquired data were shaped into the same base price index of constant year 2015. Later, the Pearson correlation test, unit root test, optimal lag length, cointegration test, causality test, and vector error correction model (VECM) steps were conducted. Furthermore, the forecast for the MCS was performed from the year 2020 to 2050. This study adopts a quantitative approach where the forecasting was performed on collected data of the Malaysian economy. Econometric and descriptive analyses were performed simultaneously, which led to the conclusion.

### 3.1. Data Collection

The data collection of the contributions of the construction sector (CONST), agriculture sector (AGRI), manufacturing sector (MANU), service sector (SERVC), and mining sector (MINING) to GDP was done with the Department of Statistics Malaysia [60] and World Bank Statistics from the year 1970 to 2019 [61]. The reason for selecting these variables is that they are the major sectors of the Malaysian economy. Any change in these sectors will show its effect on the national GDP of Malaysia. The data are in local currency units (LCUs), i.e., in the constant year 2015, as presented in Appendix A.

### 3.2. Data Analysis

After the data collection, the Pearson correlation test was performed to find if any correlation exists among the sectors. Then, the augmented Dickey–Fuller (ADF) test was conducted to find the presence of the unit root. As the order of integration was the same, the VECM was applied. The Granger causality test was conducted to find the direction of the dependence of each sector. The VECM equation was obtained along with coefficient values. The IRF was used by giving a shock to the construction sector to find how other sectors behave. Finally, forecasting from the year 2020 to 2050 was performed.

#### 3.2.1. Initial Tests for Conducting Analysis

To assess the strength of any two variables, the Pearson correlation test was performed by using Equation (1) between the MCS and other concerned sectors:

$$\mathrm{r} = \frac{n\,(\sum XY) - (\sum X)(\sum Y)}{\sqrt{\left[n\,(\sum X^2) - (\sum X)^2\right][n(\,(\sum Y^2) - (\sum Y)^2)]}} \quad (1)$$

where $X$ and $Y$ are the variables, $n$ shows the number of observations, and r is the correlation coefficient.

For the determination of the presence of the unit root, the ADF test was used. The added advantage of the ADF test is that if a correct lag order is identified, the accuracy of the ADF test could be increased [62]. The mathematical form of the ADF test is given in Equation (2):

$$\Delta Y_t = \alpha_1 + \alpha_2\, T + \tau\, Y_{t-1} + \gamma_i \sum_{t=1}^{n} \Delta Y\, t\_1 \; + \mu_t \quad (2)$$

where $Y_t$ is the time series, $\Delta Yt = Y_t - Y_1$, $T$ is the trend with time, $\alpha$ is the drift term, $\mu_t$ is the error term, and n is the number of lags to capture the white noise [63].

#### 3.2.2. Cointegration Tests

The linear relation of stationary time series data is called a cointegration model equation. This study used the Johansen cointegration test [64], which was developed by Johansen and Juselius [65]. This method was used to determine the relationship between the multidirectional time series. The null hypothesis was tested for a $p$-value of 0.05.

#### 3.2.3. Pairwise Granger Causality

This test was first developed by Granger [66] and its underlying principle is that if a time series "$X$" can be used to estimate the time series "$Y$" past data, then the previous values of "$X$" have significant power to estimate the present "$Y$", where "$X$" might be causing "$Y$" [67]. The mathematical form is shown in Equations (3) and (4):

$$Y_t = \beta_0 + \sum_{j=1}^{J} \beta_j Y_{t-j} + \sum_{k=1}^{K} \gamma_k X_{t-k} + \mu_t \quad (3)$$

$$X_t = \beta_0 + \sum_{j=1}^{J} \beta_j X_{t-j} + \sum_{k=1}^{K} \gamma_k Y_{t-k} + \mu_t \quad (4)$$

where $\mu_t$ is uncorrelated white noise, $\gamma_k$ is a measure of the influence of $X_{t-k}$ on $Y_t$. If $\gamma_k$ is statistically significant for both the equations, then causality is bidirectional.

#### 3.2.4. Vector Error Correction Model (VECM)

The error correction model (ECM) is recommended when there is the presence of the unit root and the variables are cointegrated. The ECM is used to introduce short-term and long-term equilibrium adjustment mechanisms when the variables diverge from the equilibrium. When the direction of the variable is required, the ECM is changed into the VECM. The mathematical form was given by Gujarati [68] and Granger [69], which is provided in Equation (5):

$$\Delta Y_t = \prod Y_{t-1} + \sum_{i=1}^{m-1} \Phi_i \Delta Y_{t-i}\, D_i + \mu_t \quad (5)$$

where $\Delta Y_t$ is the independent variable, $\prod$ is the matrix of cointegrating vectors, $\Phi$ represents a matrix of independent variables [68]. VECM captured short-term and long-term effects among the variable of stock prices [70]. The reason this study used VECM as a method of analysis is because of its applicability in multivariate time series, secondly, more precise short-term and long-term interdependencies are obtained if the equations are cointegrated [71].

### 3.2.5. Cumulative Sum (CUSUM) Tests

This test was developed by Brown et al. [72] for structural stability, and the mathematical form is shown in Equation (6). The principle behind this test is that the model is tested for a structural break because the presence of a structural break in the unit root will lead to inaccurate results of the model [73].

$$w_m = \frac{1}{\hat{\sigma}} \sum\nolimits_{t=k+1}^{T} w_t,\ m = k+1, \ldots, t \quad (6)$$

where $w_t$ the recursive residual, m is the sample number. The analysis is rejected if the plot deviates from the suggested boundary by the test level of confidence of 95%.

### 3.2.6. Impulse Response Functions (IRFs)

The IRF usage in this study signifies how a dependent variable responds to the vector autoregressive (VAR) model when a standard deviation shock is introduced to the error parameter. This study uses the Cholesky degree of freedom (DOF) test [74].

### 3.2.7. Forecasting

The final step was forecasting, which was performed to check the future behavior based on the past values of the construction sector. The reason for conducting forecasting was to guide statisticians and the government in terms of the forthcoming rise or fall in the construction GDP. Based on the proposed framework and forecasting, the government can take preventive measures to naturalize the sector shock in less time. The model validation is based on the Theil coefficient, RMSE, and MAPE that ensure the goodness-of-fit of the VECM forecasted values.

## 4. Results and Discussion

### *4.1. Relationship between Sectors and Stationarity Determination*

The relationship between the selected sectors of the Malaysian economy was analyzed based on the results of the Pearson correlation test, as shown in Table 2. The reason to include the GDP in the Pearson correlation test was to observe its fluctuation with the shock in the construction sector. The result shows that all the sectors are highly correlated with each other.

After the Pearson test, the unit root (ADF) test was used. As per the results, the unit root exists at the level intercept, as shown in Table 3. On the basis of ADF results, the first difference with a p-value of 0.05 shows that the data is stationary and a single order of integration exists in the data. Therefore, the data are suitable for the analysis of causality, the VECM, the IRF, and forecasting. The value shows that there exists a long-term relationship between the variables.

**Table 2.** Pearson correlation test.

| | CONST | AGRI | GDP | MANU | MINING | SERVC |
|---|---|---|---|---|---|---|
| CONST | 1 | - | - | - | - | - |
| AGRI | 0.957372 | 1 | - | - | - | - |
| GDP | 0.979313 | 0.978232 | 1 | - | - | - |
| MANU | 0.968410 | 0.971241 | 0.99501 | 1 | - | - |
| MINING | 0.967215 | 0.968167 | 0.994983 | 0.998431 | 1 | - |
| SERVC | 0.980794 | 0.980149 | 0.99728 | 0.993659 | 0.990488 | 1 |

**Table 3.** Unit roots test (ADF) results.

| Variable | Lag Order | t* | ADF Results | Order of Integration | Decision at Significance Level |
|---|---|---|---|---|---|
| LAGRI | 4 | 2.922 | 6.40 | I (1) | non-stationary |
| LMINING | 4 | 2.923 | 6.22 | I (1) | non-stationary |
| LMANU | 4 | 2.923 | 6.95 | I (1) | non-stationary |
| LGDP | 4 | 2.923 | 5.76 | I (1) | non-stationary |
| LSERVC | 4 | 2.923 | 4.23 | I (1) | non-stationary |
| LCONST | 4 | 2.923 | 4.78 | I (1) | non-stationary |

t* = t-statistic critical value at various significance levels.

If t* < ADF critical value, the null hypothesis is rejected, i.e., the unit root does not exist. The data are non-stationary at this level. After one difference, the data become stationary with t* < ADF, therefore, the order of integration was selected as I (1), which shows that the values are a good fit for analysis.

### 4.2. Optimal Lag Length

The lag length analysis was performed where a lag length of 4 was selected based on the lowest value of the Akaike information criterion (AIC), as shown in Table 4. Based on the results, a lag order of 2 was selected, which was used for building the VECM.

**Table 4.** Lag length selection.

| Lag | LR | FPE | AIC | SC | HQ |
|---|---|---|---|---|---|
| 0 | NA | $4.73 \times 10^{-12}$ | −9.049402 | −8.815502 | −8.961011 |
| 1 | 648.5405 | $2.90 \times 10^{-18}$ * | −23.36746 | −21.73016 * | −22.74872 * |
| 2 | 54.57443 * | $2.93 \times 10^{-18}$ | −23.42673 * | −20.38603 | −22.27765 |

* Indicates lag order selected by the criterion, LR = Sequential modified LR test statistic (each test at 5% level), FPE = Final prediction error, AIC = Akaike information criterion, SC = Schwarz information criterion and HQ = Hannan-Quinn information criterion.

### 4.3. Pairwise Granger Causality Analysis

Granger pairwise causality tests the null hypothesis that ARGI does not Granger cause CONST. If the value of Prob. is less than 0.05 ($p$-value < 0.05), the null hypothesis is rejected. The results show that AGRI Granger causes CONST. In the case of AGRI–CONST and SERVC–CONST, the null hypothesis was rejected as the change in CONST does have causal effects on AGRI. The same is the case with SERVC and CONST because the $p$-value is less than 0.05, as shown in Table 5.

**Table 5.** Empirical results of Granger causality.

| Null Hypothesis | Lag | Alternate Hypothesis | F-Stat | $p$-Value | Null Hypo Result |
|---|---|---|---|---|---|
| AGRI does not Granger Cause CONST | 2 | AGRI Granger causes CONS | 4.775 | 0.013 | Reject |
| CONST does not Granger Cause AGRI | 2 | - | 0.610 | 0.547 | Accept |
| GDP does not Granger Cause CONST | 2 | - | 2.851 | 0.068 | Accept |
| CONST does not Granger Cause GDP | 2 | - | 1.853 | 0.169 | Accept |
| MANU does not Granger Cause CONS | 2 | - | 1.778 | 0.181 | Accept |
| CONST does not Granger Cause MANU | 2 | - | 0.474 | 0.625 | Accept |
| MINI does not Granger Cause CONS | 2 | - | 1.135 | 0.330 | Accept |
| CONST does not Granger Cause MINI | 2 | - | 0.859 | 0.430 | Accept |
| SERV does not Granger Cause CONS | 2 | SERVC Granger causes CONST | 4.159 | 0.022 | Reject |
| CONST does not Granger Cause SERV | 2 | - | 1.485 | 0.237 | Accept |

### 4.4. Linkage Direction

It is evident that unidirectional causality exists between AGRI and CONST, and SERV and CONST. There are three independent variables, namely GDP, MANU, and MINI. There are no bidirectional linkages between any other sectors, as shown in Table 6.

**Table 6.** Direction of linkage.

| Description | Direction of Linkages |
|---|---|
| Agriculture and Construction | Unidirectional (Agriculture to Construction) |
| GDP and Construction | Independent |
| Manufacturing and Construction | Independent |
| Mining and Construction | Independent |
| Services and Construction | Unidirectional (Services to Construction) |

### 4.5. Johansen–Jusilus Cointegration Examination

The null hypothesis was tested, which states that no cointegrating equations existed, which was rejected for the first two equations based on the $p$-values being less than 0.05, as shown in Table 7. The null hypothesis was rejected at a $p$-value of 0.05 and the number of equations selected for analysis was two.

**Table 7.** Unrestricted cointegration rank test (maximum eigenvalue).

| Hypothesized No. of Cointegrating Eqn(s) | Eigenvalue | Max-Eigen Statistic | 0.05 Critical Value | $p$-Value |
|---|---|---|---|---|
| None * | 0.730538 | 61.63243 | 40.07757 | 0.0001 |
| At most 1 * | 0.527047 | 35.19165 | 33.87687 | 0.0347 |
| At most 2 | 0.420552 | 25.64697 | 27.58434 | 0.0867 |
| At most 3 | 0.317081 | 17.92481 | 21.13162 | 0.1327 |

* Max-eigenvalue test indicates 2 cointegrating eqn(s) at the 0.05 level.

If the variables are not cointegrated (have a long-term relationship), then unrestricted VAR is used for analysis. If the variables are cointegrated, then the usage of restricted VAR is preferred, i.e., the VECM.

### 4.6. VECM for Construction Sector

The VECM performs better while observing the short- and long-term causality through the estimated variables. The proposed relationship of this study is given in Equation (7):

$$\text{CONST} = f\,(\text{AGRI, CONST, GDP, MANU, MINING, SERVC}) \tag{7}$$

Equation (7) was generated by the VECM for short- and long-run equilibria. The model equation consists of two major parts, namely cointegrating equations and error correction mechanisms, which show the speed of adjustment that the sector will undergo to come back to its original state. The latter indicates the short-run coefficients' causality among the variables, which is represented by Equation (8):

$$D(LCONST) = C(1) \times (LCONST(-1) + 1.67180917629 \times LMINING(-1) - 2.68403475444 \times LSERVC(-1) + 0.941037693325 \times LGDP(-1) - 3.78428477371) + C(2) \times (LAGRI(-1) - 1.86519424503 \times LMINING(-1) + 9.43226731209 \times LSERVC(-1) - 10.9476629704 \times LGDP(-1) + 24.4605216091) + C(3) \times (LMANU-1) - 0.470043904124 \times LMINING(-1) - 3.81093319687 \times LSERVC(-1) + 4.20254250326 \times LGDP(-1) - 8.35346448657) + C(4) \times D(LCONST(-1)) + C(5) \times D(LCONST(-2)) + C(6) \times D(LAGRI(-1)) + C(7) \times D(LAGRI(-2)) + C(8) \times D(LMANU(-1)) + C(9) \times D(LMANU(-2)) + C(10) \times D(LMINING(-1)) + C(11) \times D(LMINING(-2)) + C(12) \times D(LSERVC(-1)) + C(13) \times D(LSERVC(-2)) + C(14) \times D(LGDP(-1)) + C(15) \times D(LGDP(-2)) + C(16) \quad (8)$$

*4.7. Short- and Long-Run Causality Coefficients*

To validate the produced model using the VECM equation, the first coefficient must always be negative and significant. In this case, the negative sign and its probability value are less than 0.05, which indicates the absence of any problem with the data and their ability to bounce back to equilibrium, as shown in Table 8. The coefficient value of C(1) is the smallest, therefore, forecasting can be performed for this model. The *p*-value of 0.05 was used for rejecting the null hypothesis.

**Table 8.** Coefficients' value and probabilities of CONST.

| | Coefficient | Std. Error | *t*-Statistic | Prob. |
|---|---|---|---|---|
| C(1) | −0.306733 | 0.101781 | −3.013650 | 0.0029 |
| C(2) | 0.011071 | 0.163561 | 0.067690 | 0.9461 |
| C(3) | 0.706349 | 0.362084 | 1.950790 | 0.0526 |
| C(4) | 0.244731 | 0.319261 | 0.766553 | 0.4443 |
| C(5) | 0.179515 | 0.220620 | 0.813686 | 0.4169 |
| C(6) | −0.137599 | 0.442141 | −0.311211 | 0.7560 |
| C(7) | −0.280829 | 0.425516 | −0.659974 | 0.5101 |
| C(8) | −1.390151 | 1.479392 | −0.939677 | 0.3486 |
| C(9) | −0.576630 | 1.530035 | −0.376873 | 0.7067 |
| C(10) | 1.823168 | 0.845907 | 2.155283 | 0.0324 |
| C(11) | 0.407999 | 0.937179 | 0.435347 | 0.6638 |
| C(12) | −0.526471 | 0.539764 | −0.975374 | 0.3306 |
| C(13) | −0.003194 | 0.486360 | −0.006567 | 0.9948 |
| C(14) | 0.771292 | 0.455573 | 1.693015 | 0.0921 |
| C(15) | 0.063258 | 0.430021 | 0.147105 | 0.8832 |
| C(16) | 0.006754 | 0.060987 | 0.110740 | 0.9119 |

*4.8. Explanatory Power and Efficiency of Equation for CONST Model*

Three checks were performed on the model, namely coefficient of $R^2$, F-statistics, and a DW test. In this case, the values of $R^2$ were calculated as 52.7%, which shows the explanatory power of the model. The values of the DW test were recorded as 2, which shows no autocorrelation. Similarly, the probability value of the F-statistics is significant (less than 0.05), as shown in Table 9.

**Table 9.** Result of CONST equation.

| Parameters | Value |
|---|---|
| Coefficient of determination ($R^2$) | 0.527 |
| Adjusted $R^2$ | 0.456 |
| Probability of F-statistic | 0.000 |
| Durbin–Watson statistic | 2.053 |

If the R-squared value is $0.5 < r < 0.7$, then it is considered a moderate effect size, which is acceptable, hence, this model is fit for analysis [75].

## *4.9. Validation of the Estimated Equation for CONST Model*

The model should always be nonspurious and should be unbiased. To have a nonspurious econometric equation, two validation checks were performed: (1) $R^2$ should be less than a Durbin–Watson statistic and (2) the residual should be stationary and white noise.

The first condition is satisfied, which is that the values of $R^2$ are less than the DW value (0.527 < 2.053), indicating the nonspurious equation. The second condition is that the model should be free from serial correlation, autocorrelation, and heteroskedasticity.

### 4.9.1. Serial Correlation Test for Residual

The Breusch–Godfrey serial correlation LM test was performed. The results shown in Table 10 indicate the probability of the chi-square with a *p*-value of 5%, and the model was free from serial correlation, hence, it was suitable for forecasting.

**Table 10.** Breusch–Godfrey serial correlation LM test.

| | | | |
|---|---|---|---|
| F-statistic | 0.661 | Prob. F (2, 39) | 0.5216 |
| Observed R-squared | 1.575 | Prob. chi-square (2) | 0.4549 |

### 4.9.2. Residual Heteroskedasticity Test

The autoregressive conditional heteroskedasticity (ARCH) test was conducted with the $H_0$ (null hypothesis) as there was no heteroskedasticity in the residual of the model equation. Hence, the results, as shown in Table 11 indicated that the null hypothesis could not be rejected as the *p*-value of the chi-square test was greater than 5%.

**Table 11.** Heteroskedasticity test: Breusch–Pagan–Godfrey.

| | | | |
|---|---|---|---|
| F-statistic | 1.947642 | Prob. F (7, 40) | 0.0872 |
| Observed R-squared | 12.20148 | Prob. chi-square (7) | 0.0941 |

### 4.9.3. Residual Correlogram

To check for autocorrelation and partial autocorrelation, residual correlograms were made. Figure 1 indicates that the spikes are below the dotted line representing the 2% significance standard deviation. The *p*-value was greater than 0.05, implying that there is no autocorrelation in the series.

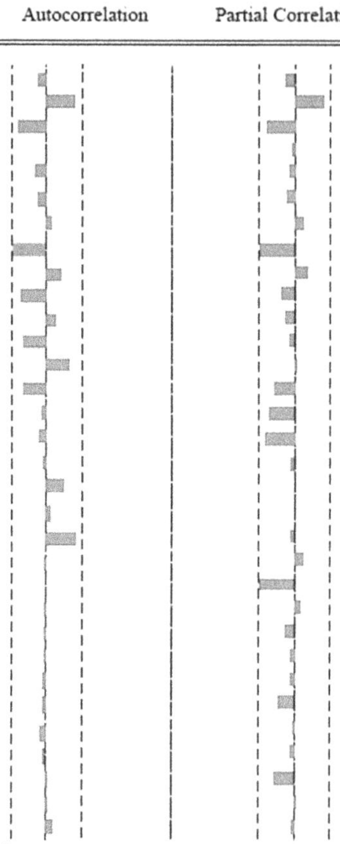


| | AC | PAC | Q-Stat | Prob |
|---|---|---|---|---|
| 1 | -0.057 | -0.057 | 0.1638 | 0.686 |
| 2 | 0.197 | 0.195 | 2.1600 | 0.340 |
| 3 | -0.120 | -0.198 | 4.4677 | 0.215 |
| 4 | 0.007 | -0.047 | 4.4700 | 0.346 |
| 5 | -0.134 | -0.061 | 5.4511 | 0.363 |
| 6 | -0.075 | -0.123 | 5.7688 | 0.450 |
| 7 | 0.066 | 0.098 | 6.0158 | 0.538 |
| 8 | -0.312 | -0.350 | 11.780 | 0.161 |
| 9 | 0.189 | 0.151 | 13.941 | 0.124 |
| 10 | -0.228 | 0.147 | 17.169 | 0.071 |
| 11 | 0.123 | -0.087 | 18.130 | 0.079 |
| 12 | -0.180 | -0.059 | 20.257 | 0.062 |
| 13 | 0.190 | 0.038 | 22.717 | 0.054 |
| 14 | -0.159 | -0.168 | 24.490 | 0.363 |
| 15 | -0.073 | -0.179 | 24.873 | 0.052 |
| 16 | -0.088 | -0.191 | 25.445 | 0.062 |
| 17 | -0.061 | -0.020 | 25.730 | 0.080 |
| 18 | 0.163 | 0.002 | 27.828 | 0.065 |
| 19 | 0.026 | -0.007 | 27.885 | 0.086 |
| 20 | 0.209 | -0.015 | 31.628 | 0.057 |
| 21 | -0.012 | 0.103 | 31.640 | 0.064 |
| 22 | 0.007 | -0.269 | 31.645 | 0.084 |
| 23 | -0.021 | 0.077 | 31.685 | 0.107 |
| 24 | -0.008 | -0.099 | 31.692 | 0.135 |
| 25 | -0.012 | -0.039 | 31.706 | 0.167 |
| 26 | -0.037 | -0.042 | 31.853 | 0.198 |
| 27 | -0.029 | -0.143 | 31.953 | 0.234 |
| 28 | -0.065 | -0.020 | 32.465 | 0.256 |
| 29 | -0.014 | -0.037 | 32.491 | 0.299 |
| 30 | 0.016 | -0.165 | 32.525 | 0.344 |
| 31 | 0.012 | 0.023 | 32.547 | 0.391 |
| 32 | 0.071 | -0.055 | 33.314 | 0.403 |

**Figure 1.** Residual correlation of CONST.

### 4.10. Structure Stability Test

To perform the impulse response function (IRF), the CUSUM test and CUSUM square tests were performed. The only difference between the CUSUM and CUSUM square tests is that the former shows the structural break in the model, while the latter shows the 5% significance of the residual data in the estimated model. In the case of the CUSUM test, the null hypothesis was "there is an absence of a structural break in the system". To reject the null hypothesis, the estimated model should be between the 5% significance lines of the upper and lower bounds. The CUSUM square test depicts the behavior of the residual of the estimated model. The null hypothesis was that the residual lies within the 5% significance level of the upper and lower bounds of the estimated model. The results of CUSUM square tests in Figure 2a,b indicate that it cannot reject the null hypothesis and there are no structural breaks in the model.

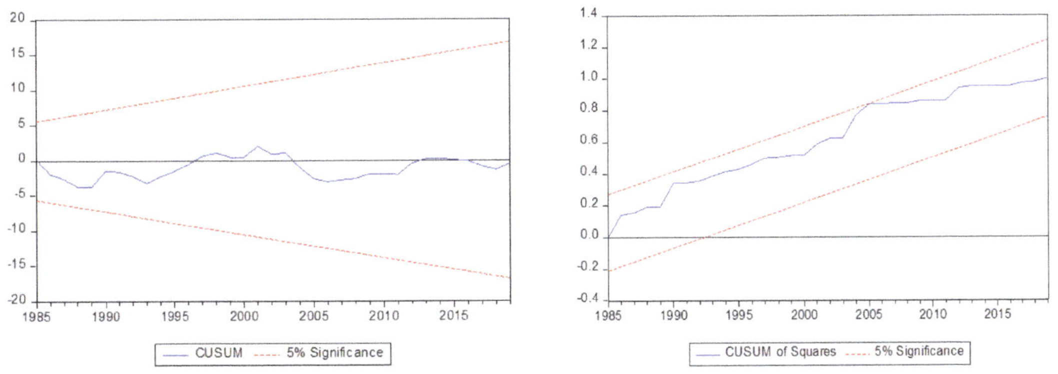


(**a**) (**b**)

**Figure 2.** Structural stability test. (**a**) CUSUM test. (**b**) CUSUM square test.

### *4.11. Impulse Response Functions (IRFs)*

The IRF shows how a dependent variable will react to a unit shock to the independent variables [76]. In other words, the IRF performs a shock to endogenous variables and checks their behavior with other variables, when one standard deviation is observed in CONST, and how other sectors correspond to this shock after coming back to their original position. In Figure 3 the red line refers to a 95% confidence interval, while the blue line refers to the impulse response function which should always be inside the red lines.

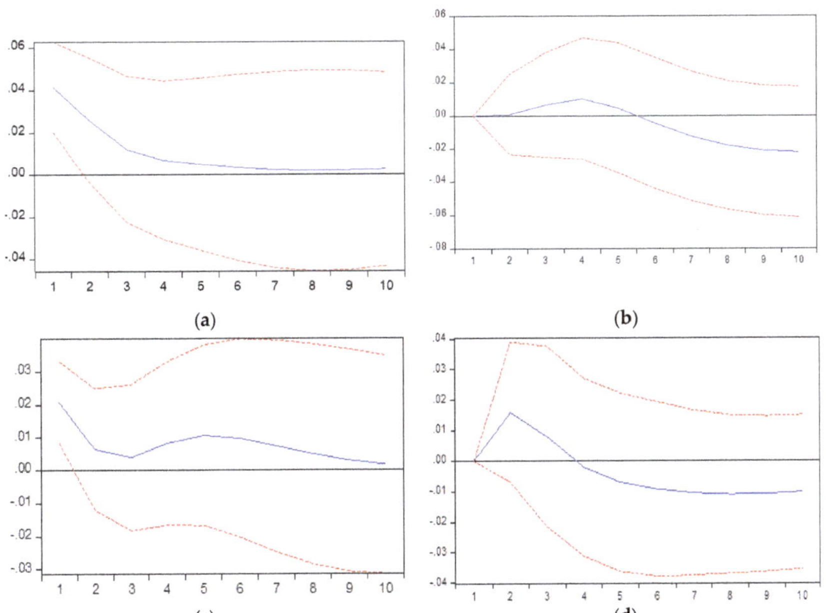

(**a**) (**b**) (**c**) (**d**)

**Figure 3.** *Cont.*

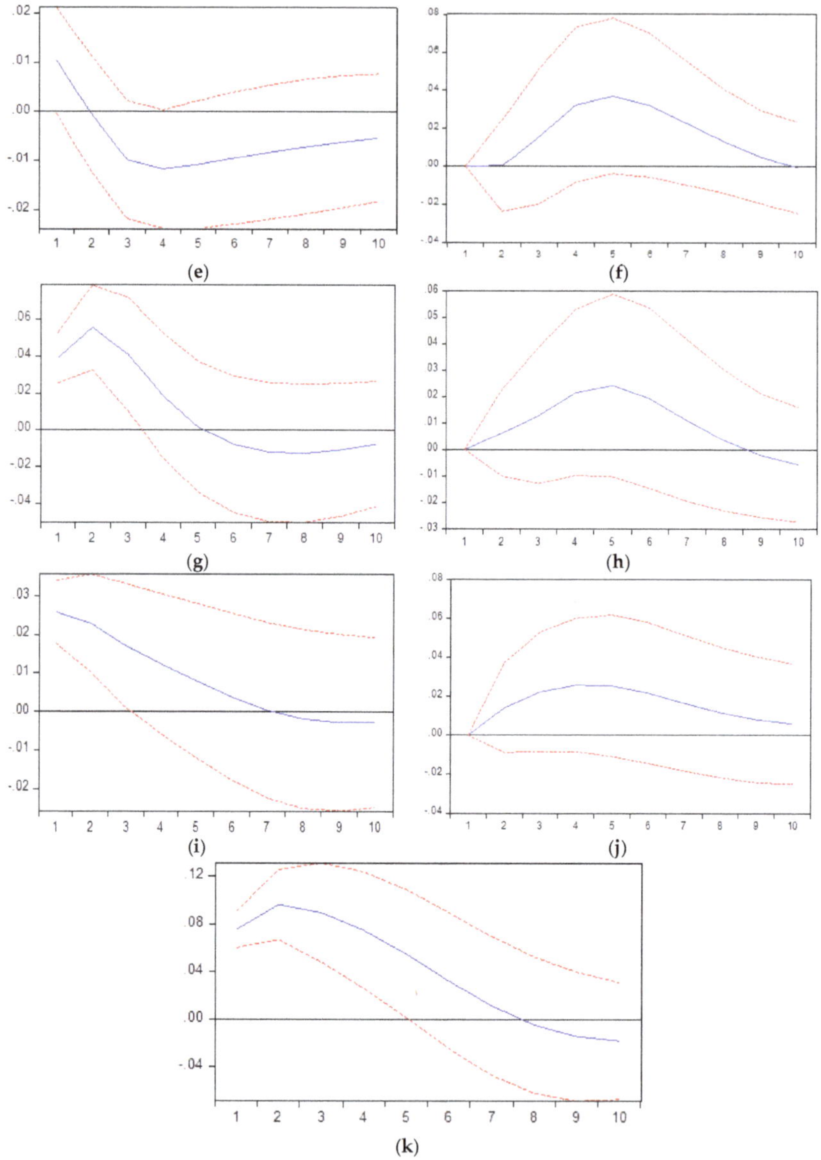

**Figure 3.** Various sector responses. (**a**) Response of MANU to CONST. (**b**) Response of CONST to MANU. (**c**) Response of MINING to CONST. (**d**) Response of CONST to MINING. (**e**) Response of AGRI to CONST. (**f**) Response of CONST to AGRI. (**g**) Response of SERVC to CONST. (**h**) Response of CONST to SERVC. (**i**) Response of GDP to CONST. (**j**) Response of CONST to GDP. (**k**) Response of CONST to its shock.

Figure 3a shows the behavior of MANU against the CONST shock, where it can be observed that MANU's response is positive for at least 5 or 6 years, and after that it becomes stable. This signifies that construction increases the output in the manufacturing sector. This makes sense as construction requires a lot of manufacturing products like tools, machinery, glass, steel, paint, and various resources. In Figure 3b, the behavior of CONST in reaction to the MANU sector is not that harmful in the first five years. After five years, there is a negative trend in the CONST sector from the shock produced by manufacturing because of the number of manufactured things already available in the market for at least five years. Figure 3c reveals the information about shocks between MINING and CONST. It is evident that the mining sector response was positive against the shock, which was similar to the manufacturing sector, and the mining sector will regain its original state after eight years. The overall effect will be positive in the mining sector from the shock produced by the construction sector. Figure 3d shows the behavior of the construction sector from the shock produced by the manufacturing sector. It conveys that there will be a positive reaction in the construction sector from shocks in the manufacturing sector for two years, and after that, there will be a decrease in aftershocks for ten years. Figure 3e shows the shock behavior between the agriculture and construction sectors. A positive response for up to two years followed by a negative trend for at least 8 years will occur in the agriculture sector after a shock produced in the construction sector. In Figure 3f, the construction sector suffers significantly from the shock produced in the agriculture sector. For the first two years, there was no response to the shock, but after two years and towards the end of the ninth year, there was a positive shock in the construction sector. Towards the end of the ninth period, the shock stabilizes after a considerable positive response in the construction sector from the shock produced by the agriculture sector. The shock produced by the construction sector creates a positive response in the service sector for at least five years. After five years, the service sector will overcome the shock from the construction sector to stabilize and come back to its former position before the shock. It is shown in Figure 3g,h that the response of the construction sector is positive for at least eight years, with a maximum point between the fourth and fifth years. After that, a negative response is shown in the construction sector. In Figure 3i, the shock and reaction of the GDP and the construction sector are shown. A shock to the construction sector will result in a positive response of the GDP for seven consecutive years, implying that GDP was more sensitive than the construction sector. After that period, the GDP will regain its initial state. Figure 3j depicts the shock in GDP and its reaction in the construction sector. Its behavior was different from other sectors, as there was no negative shock and the shock response was positive overall. After ten years, the response of GDP to the construction sector will fade away and it will regain its initial position before the shock. Figure 3k signifies the response of the construction sector to itself and how it will react for the next ten years. A positive shock can be seen for up to eight years. There will be a negative impact starting from the eighth year. This shows that the construction sector growth will increase significantly but the effect will be temporary.

### 4.12. Forecasting and Validation

The model was validated using the Theil U statistic, which shows the forecasting accuracy. A value of Theil of less than 1 indicates that the forecasting model is accurate. If the Theil value is greater than one, the forecasting method is poor. A value of Theil equal to 1 indicates that the forecasting model is the same as the analyzed model [77]. In Figure 4, the value of Theil U statistics (inequality coefficient) is 0, implying that the forecasted model has zero errors. If the Thiel value was 1, it would have indicated a poor forecast with the errors in the model.

Figure 5 shows the fitting of the forecasted line. The dotted line indicates the forecasting performed with the VECM equation ranging from 2020 to 2050. As the forecasted line (blue) lies within the red lines, with a 2% standard error line, there is no significant error in the forecasted model. Additionally, the value of the Theil coefficient is zero, implying that the predictive power of the model is the best, where the descriptive values of the CONST_FF are provided in Appendix B.

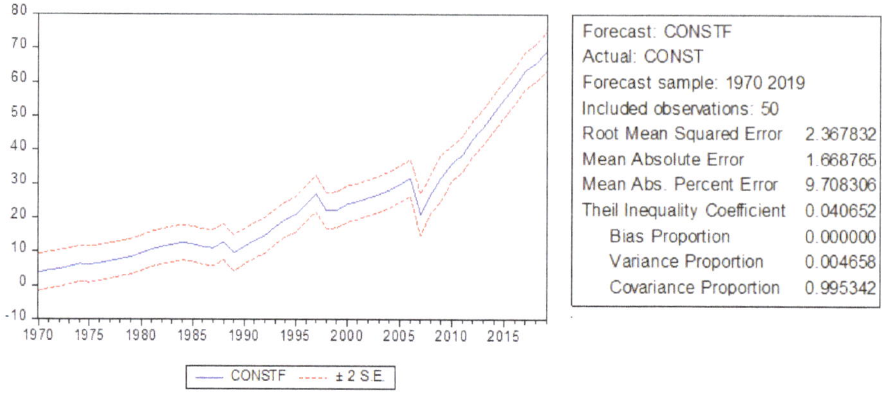


**Figure 4.** Fitted forecasted line.

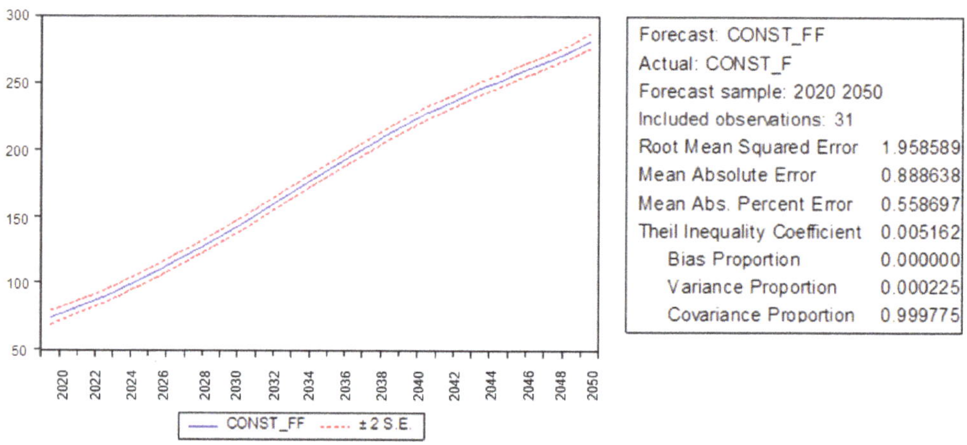


**Figure 5.** Fitted forecasted line for the years 2020–2050.

Based on the findings, it was revealed that the contribution of the construction sector to the Malaysian GDP will increase almost three times from 2020 to 2050. Figure 6 indicates the time frame of the predicted values from 2020–2029, 2030–2039, and 2040–2050. The blue line indicates that the construction sector will add RM 131.96 billion to the national GDP in the year 2029. By the year 2039, there will be an increase of RM 100 billion of the construction sector's contribution to the GDP in 2039 as compared to the year 2020. It was also predicted that RM 80 billion will be added from 2039 to 2050. The forecasted values are based on the current and past real-time values, provided the conditions remain the same. They indicate that the sector will continue to grow and it will generate a fair share of capital for the national economy. Moreover, the performance of the construction sector is greatly dependent on current and future laws of the government that might affect the productivity of the MCS in a positive and negative way. The failure to acknowledge the importance of the MCS by the government might seriously affect the growth of the construction sector. Similarly, subsidizing the construction sector could increase its performance, hence, the forecasted values would become different from real-time values.

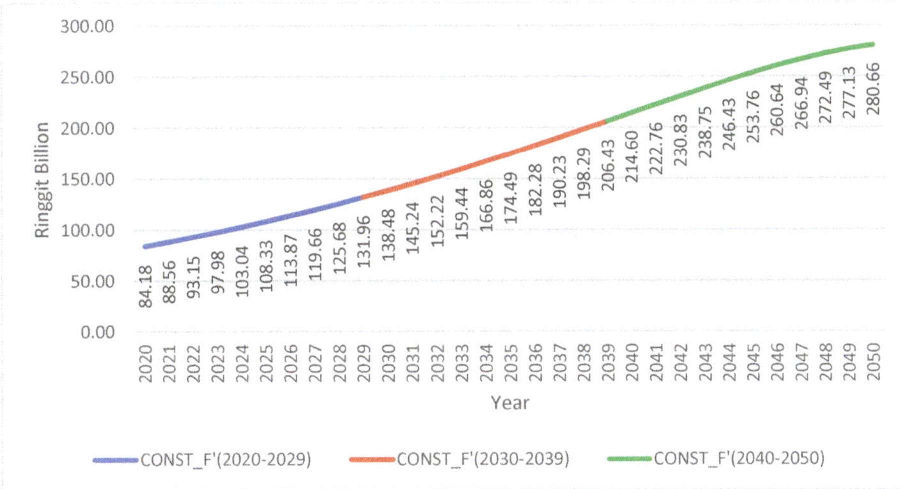


**Figure 6.** Graphical representation of original and forecasted values 2020–2050.

The significance of this study is that these results can be utilized to construct a detailed framework to bring sustainability to the construction sector in case of any inflation in the sector or GDP of the country. The application of this study will also help to make the construction sector a sustainable industry by studying the long-term effects of the shocks to prevent any major loss in shares that could plunge the sector into recession, which could affect millions of lives in the country. The applicability of this study can be extended to assess the global impact as well as the steps necessary to achieve sustainability in construction sectors all over the world.

This study made use of the time series from the years 1970 to 2019. The question arises as to why opt for pre-COVID-19 data in analysis? COVID-19 impacted the construction sector statistics and they could not be made a part of this study due to the unavailability of officially released data. Additionally, its impact is not easy to inspect as there are a lot of hidden factors. Moreover, still, the circumstances are changing rapidly, hence, it is difficult to make reliable forecasting based on the COVID-19 scenario. As the first case in Malaysia was reported on 25th January 2020 [78], therefore, the COVID-19 impacts on the construction industry took some time to show their effects on the overall economy, which delayed the timely publication of statistics. In this study, these COVID-19 effects can be regarded as one-time period (in this study, one year) shock to all industries and the behavior of this shock produced negative and positive variation in the individual sectors. The failure or poor contribution of the construction sector to the GDP has immensely dragged down the country's economic conditions. The suspension of all these sectors affected the overall performance of the country's GDP. Any future pandemic or airborne diseases, as per scientists' predictions, could threaten human life, as humans are greatly dependent on the growth of the construction sector. Specifically, the suspension of the construction sector exerted serious impacts on financial issues as a decrease in the sales and manufacturing industry, reduced household spending, a decrease in domestic spending due to high financial volatility, and limited operational movement, which is a fundamental need of the prosperity in the construction sector, were shown. The suspension of ongoing construction projects due to COVID-19 also had negative consequences on the timely completion of the projects, including regulation compliance, an increase in costs, limited resources, unavailability of materials, and contradiction in releasing of funds. Due to this, the Malaysian government released the Rakyat Economic Stimulus (PRIHATIN) package to cope with the loss in the construction sector. Therefore, the real values might deviate from the mentioned forecasted values in Figure 6.

### 4.13. Predicted Contribution of the Construction Sector to GDP

This study addresses the challenges and the obstacles faced by the MCS to enable competitiveness, self-sufficiency, and effectiveness. The MCS must concentrate on strengthening and upgrading sustainable methods, integrated solutions, timely financing, transparent bidding processes, timely payment and loan releases, resource allocation, ensuring adequate manpower, preference for local skilled workforce than foreign workforce, and R&D to develop innovative methods of construction. These are the factors that hinder the growth of the MCS as compared to neighboring countries.

The government of Malaysia introduced the Shared Prosperity Vision 2030 to provide a decent living standard to all Malaysians by 2030. Under this act, the goal is set to achieve RM 3.4 trillion GDP with a 4.7% increase in annum growth rate within the period of 2021 to 2030. Additionally, this includes increasing the productivity of its major sectors by introducing technology like manufacturing and services by up to 50% and 30%, respectively. The commitment was made to reduce the gap between different income classes, ethnicities, and regions along with the use of the full potential of manpower and resources to optimize the Malaysian economy to its highest potential. It is worth mentioning that developing countries like Malaysia have to face another factor, i.e., the development of the construction sector despite harming the environment [79]. Therefore, a blend of long-term policy and the ethical environmental plan was introduced in Malaysia, commonly referred to as National Transformation Plan 2050 (TN50), under which the country will introduce decisive strategies to cope with household sullage, industrial pollutants, shifting to renewable power sources, and increases in sustainable building practices by 2050. Thus far, the country is heading in the right direction as per TN50 but the effectiveness of this plan can only be assessed in the future [80].

It is evident from the IRF analysis that the Malaysian construction sector is sensitive to shocks in the agriculture and mining sector. That is, a shock in both agriculture and mining sectors will cause a fluctuation in construction sector performance, and the construction sector will take some time before coming back to its position as it was before the shock. In terms of generalization, this study was conducted on Malaysian statistics to assess the behavior of the construction sector in terms of other sectors of the economy, therefore, these results can only be applied to the Malaysian economy as each country has different mechanics of their respective sectors. However, the proposed framework and methodology can be followed in a wider aspect for international sustainable assessment of the construction sector all over the world.

Unlike the study conducted in Australia [45], this work used the IRF to study the behavior of the construction sector and proposed a framework that should be followed to bring the construction sector one step closer to achieving sustainability. When any major contributing sector reaches the maximum capacity of its growth, it can be easily impacted by various factors. This will result in the crash of the nationwide economy, therefore, the government of Malaysia must take into account the results produced by the shock and apply them to the sustainable planning of the sector. This study adds a theoretical contribution towards achieving a sustainable construction sector by studying the effects of short- and long-term initiatives introduced due to the effects of shocks in the construction sector all over the world. The VECM suggests that the construction sector will need both short- and long-term speed adjustments in terms of attaining a sustainable industry when a sustainable policy or greener technology incentives are introduced to the sector. The IRF results indicate that a shock to the construction sector will attract government support for the construction sector, which will incontrovertibly increase the output of the construction sector in the long run and will cause damage to the environment due to increased $CO_2$ levels unless green technology is made cheaper and affordable. From a global perspective, the VECM results predict that supporting the construction sector will boost the output in the long run, and based on the linkages, it is safe to say that if one sector underperforms, a huge industry like the construction sector will cope with it.

## 5. Conceptual Framework

The sectorial and sustainable framework consists of three essential steps, namely input, economy-contributing sector, and output. As shown in Figure 7, the input from the government, like budget allocation and resource allocation, is added to the economy-contributing sectors. Each sector has a unique working mechanism in terms of its influence on the country's economy. Concerning this study, only the construction sector mechanism is shown in detail. This framework suggests that the three inputs in each sector generate their respective outputs. As the scope of this study is the construction sector, it is further broken down into its subcategories. The construction sector is a single industry of collective assistances, namely: stakeholders, productivity measures, and collaborative procurements. These three assistances should be fueled by the latest use of technology, sustainable policies and techniques, and environmentally friendly procedures. Hence, the collaborative support of this framework helps in developing the construction sector. This framework is useful for all the countries that are struggling with embracing sustainability, in particular, in the construction sector. Therefore, the government needs to follow the sustainable construction framework introduced in this study to make the construction industry sustainable and enable this sector to be shockproof from unseen events like a pandemic.

The construction sector depends on stakeholder policies as they define the strategies and actions to improve the prevailing system of construction methods. Productivity measures in the sector ensure that the sector keeps up with the modern needs of the economy. Failure in this innovation and productive plans would lead to underdevelopment and slow projects, which would bring financial strain not only on clients but also on the national funds, therefore, construction method evaluation is necessary over time. The need for collaboration in this sector is of prime interest as it is a huge sector and, without collaborative work, it will collapse. Collapsing and underdevelopment in the construction sector mean less infrastructure, which will reduce the national GDP as the national economy needs adequate infrastructure. The construction sector mechanism revolves around three subcategories, namely administrative policies, technological advancements, and environmental allocation. These three subcategories are interlinked with each other. The administrative policies combined with the modern usage of technology in construction give rise to the sector's prosperity. Sustainable techniques are employed to give rise to development in the construction sector. The sustainable construction sector in turn ensures social prosperity and its fair share of the national GDP and also avoids environmental degradation, which is possible with resource allocation by the government. This relationship is beneficial from the economical perspective as it clearly states the impact of the construction sector on the economy as well as other sectors. When keeping an eye on the current scenario, it is important to understand how the construction sector contributes to the future. Moreover, this study can be beneficial for other developing countries with similar characteristics to the Malaysian economic sector.

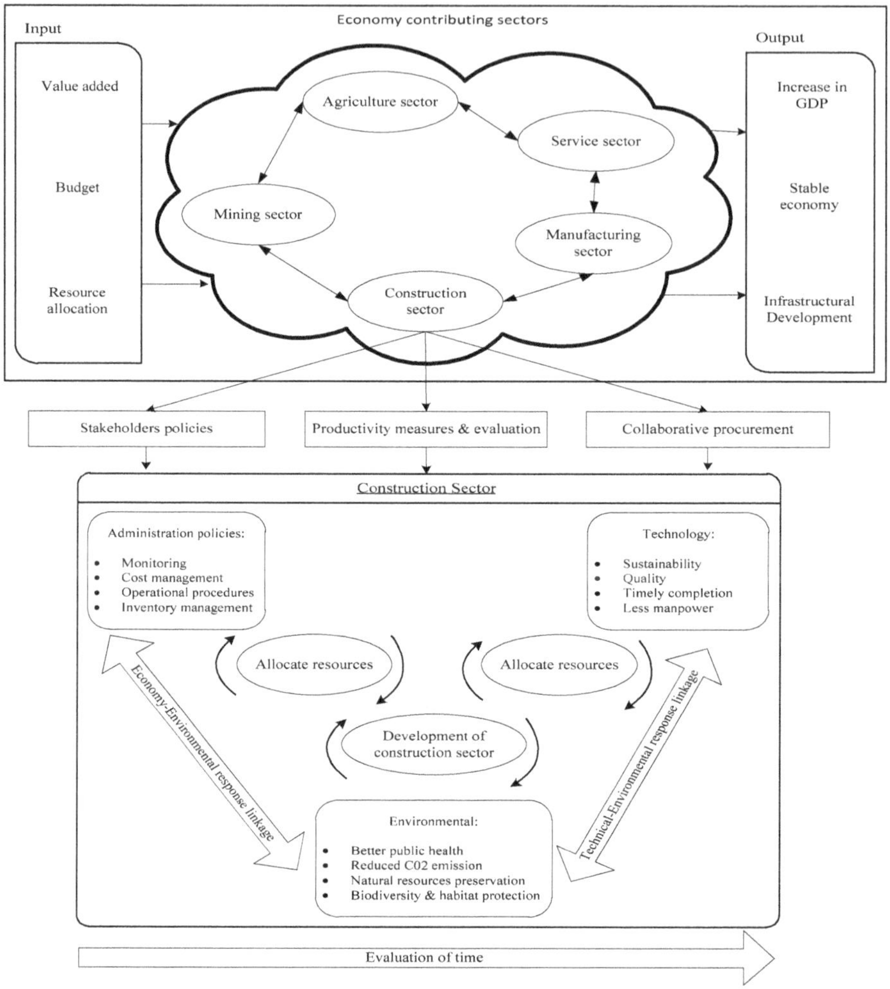


**Figure 7.** The construction sector and sustainable development as a conceptual framework.

## 6. Conclusions

Based on the obtained results, it is concluded that the construction sector of Malaysia is connected with other sectors and any direct or indirect change in its input will produce short-term as well as long-term effects on the sector itself and the cumulative GDP. The intersectoral links of SERV, MINI, AGRI, MANU, and CONST are more volatile as compared to the overall GDP, which shows that these sectors are influenced by investors' trust and other factors apart from the government input. Based on the lowest AIC value of −23.426, the number of lags was selected as two for Granger causality. The linkage direction was assessed based on a $p$-value less than 0.05. The number of the cointegrating equation was selected as two based on its value of less than 0.05. The selected value of cointegrating and lags was used in the VECM. The VECM with coefficient C(1) = −0.306 and a $p$-value less than 0.05 indicates the model's ability to bounce back, which shows the viability of the VECM equation. A DW test value of 2.0 indicated the absence of autocorrelation. The residual and heteroskedasticity with a $p$-value greater than 0.05 indicate the best forecasting model. Forecasting of the MCS output shows a rise to 131.96 RB, 100 RB, and 80 RB in the years 2020–2029, 2030–2039, and 2040–2050, respectively. The countercyclicality was revealed in this study, which signifies the presence of negative relationships between macroeconomic time series. The results of Granger causality models revealed that the construction sector is greatly influenced by the output of AGRI and SERV sectors and the growth of the CONST sector can be modeled based on the activity in these sectors. The long-term equilibrium of the model established the significance of the construction sector, which shows that overall GDP exhibits a large proportion of the effect on construction sector growth. Hence, this idea supports the fact that the construction industry has the largest contribution to the overall GDP of Malaysia.

The central theme of this study is based on the economic development of all the sectors with fundamental reference to the construction sector and how a change in one sector influences the behavior of other sectors. The empirical results have indicated that all sectors are prone to shocks and might have deteriorating effects on themselves in the short term and on the national economy in the long term. In other words, all sectors must have flexibility and adaptability in case of output shocks in the shortest time possible as failure to perform and generate revenue after the shock might collapse the sector completely, which would result in disastrous effects on the economy. This study has also confirmed the common belief that the construction sector is interdependent on other sectors, illustrating that the construction sector's behavior will fluctuate with the performance of other sectors in Malaysia.

The outcome of this analysis will be helpful for policymakers, industrialists, stakeholders, and investors, not only in Malaysia but everywhere in the world. Policymakers in Malaysia can understand more about the regulatory measures imposed on the construction sector to determine its role in the national GDP. Similarly, industrialists can determine the effects on the construction sector by examining causal effects in one sector. To protect themselves from atypical circumstances of the economy, like inflation or recession of a country's economy, investors and stakeholders can make strategies, monetize, and take decisive actions for improving their business which will lead to a sustainable sector. With respect to Malaysia, the joint venture programs between multinational firms and local authorities must be increased to grow sustainability in terms of human capital, increase skilled labor, executing required work in less time, and increasing workers skilled in modern techniques. The sharing of knowledge should be encouraged to increase the trust of local and overseas investors, which will raise foreign direct investments in the construction industry. For wider applicability of the policies, it is recommended that the governments of their respective countries must pay attention to the emphasis of the sustainable construction processes in the sector in terms of legislation, stakeholder trust, human resources, and capacity-building programs.

The VECM has its limitations as the coefficients in it are limited in their ability to forecast the behavior of exogenous and endogenous variables in case of any change in the subsectors of the construction industry, which could give rise to ambiguity in the analysis, therefore, constant updating of the parameters will be required in the VECM. Moreover, any external disruptive shift in technological assets, inflation, and higher market volatility will decrease the forecasting accuracy as the VECM is unable to predict sudden peak changes.

**Author Contributions:** Supervision, W.S.A., M.A.M.; Visualization, W.S.A., Q.I., A.M., W.F.; Project administration, W.S.A.; Writing—review & editing, W.S.A., M.A.M.; Formal analysis, M.A.M., M.B.A.R.; Investigation, M.A.M., M.B.A.R., Q.I., A.M.; Methodology, M.A.M., M.B.A.R.; Validation, M.A.M., M.B.A.R., W.F.; Writing—original draft, M.B.A.R., W.F.; Resources, Q.I., A.M. All authors have read and agreed to the published version of the manuscript.

**Funding:** This research received no external funding.

**Institutional Review Board Statement:** Not applicable.

**Informed Consent Statement:** Not applicable.

**Data Availability Statement:** All the data are available within this manuscript.

**Acknowledgments:** The authors would like to appreciate the YUTP-PRF project (cost center # 015LC0-088) awarded to Wesam Alaloul for the support.

**Conflicts of Interest:** The authors declare no conflict of interest.

## Appendix A

**Table A1.** Data description of the Malaysian sectors.

| Year | CONST (RB) | MANU (RB) | AGRI (RB) | SERVC (RB) | MINING (RB) | GDP (RB) |
|---|---|---|---|---|---|---|
| 1970 | 3.079 | 7.346 | 27.112 | 9.668 | 30.434 | 73.752 |
| 1971 | 4.217 | 8.271 | 27.504 | 10.312 | 35.006 | 81.152 |
| 1972 | 4.451 | 9.113 | 29.601 | 12.525 | 37.712 | 88.771 |
| 1973 | 5.075 | 11.165 | 33.087 | 17.341 | 38.856 | 99.159 |
| 1974 | 5.683 | 12.322 | 35.373 | 22.863 | 38.500 | 107.407 |
| 1975 | 5.098 | 12.686 | 34.300 | 24.935 | 38.876 | 108.268 |
| 1976 | 5.558 | 15.034 | 38.498 | 24.512 | 46.435 | 120.787 |
| 1977 | 6.236 | 16.626 | 39.407 | 29.844 | 48.642 | 130.152 |
| 1978 | 7.164 | 18.168 | 40.056 | 32.155 | 53.135 | 138.812 |
| 1979 | 8.026 | 20.227 | 42.359 | 33.455 | 61.024 | 151.790 |
| 1980 | 9.416 | 22.092 | 42.902 | 41.322 | 62.383 | 163.086 |
| 1981 | 10.778 | 23.136 | 44.986 | 49.372 | 62.042 | 174.408 |
| 1982 | 11.840 | 24.433 | 47.896 | 51.197 | 66.220 | 184.773 |
| 1983 | 13.066 | 26.356 | 47.589 | 53.142 | 74.465 | 196.325 |
| 1984 | 13.618 | 29.596 | 48.940 | 56.010 | 84.204 | 211.564 |
| 1985 | 12.479 | 28.464 | 49.913 | 55.440 | 82.519 | 209.395 |
| 1986 | 10.728 | 30.607 | 51.993 | 51.292 | 88.158 | 211.993 |
| 1987 | 9.446 | 34.708 | 55.648 | 54.629 | 93.157 | 222.999 |
| 1988 | 9.721 | 40.605 | 57.157 | 61.984 | 103.179 | 245.160 |
| 1989 | 10.847 | 48.857 | 59.877 | 77.719 | 114.517 | 267.371 |
| 1990 | 12.860 | 56.327 | 58.842 | 92.127 | 122.379 | 291.457 |
| 1991 | 14.859 | 64.212 | 59.454 | 105.857 | 133.093 | 319.278 |
| 1992 | 16.457 | 68.708 | 63.532 | 119.495 | 141.521 | 347.646 |
| 1993 | 18.234 | 78.727 | 61.538 | 143.397 | 150.759 | 382.046 |
| 1994 | 20.995 | 87.682 | 60.372 | 162.454 | 164.675 | 417.240 |
| 1995 | 25.415 | 97.642 | 58.842 | 174.221 | 191.903 | 458.251 |
| 1996 | 29.527 | 115.394 | 61.510 | 190.431 | 213.159 | 504.088 |
| 1997 | 32.655 | 127.068 | 61.923 | 209.897 | 225.727 | 541.001 |
| 1998 | 24.832 | 110.018 | 60.211 | 196.350 | 210.295 | 501.187 |
| 1999 | 23.752 | 122.859 | 60.499 | 207.194 | 229.778 | 531.948 |
| 2000 | 23.882 | 145.358 | 64.165 | 222.718 | 254.844 | 579.072 |
| 2001 | 24.662 | 139.151 | 64.054 | 232.493 | 247.705 | 582.070 |
| 2002 | 25.234 | 144.885 | 65.890 | 249.865 | 258.539 | 613.449 |
| 2003 | 25.694 | 158.162 | 69.864 | 259.763 | 278.441 | 648.959 |
| 2004 | 25.475 | 173.279 | 73.130 | 278.332 | 298.934 | 692.981 |
| 2005 | 25.103 | 182.283 | 75.027 | 302.180 | 308.681 | 729.932 |
| 2006 | 24.970 | 195.825 | 79.405 | 323.017 | 321.966 | 770.697 |
| 2007 | 27.104 | 213.471 | 86.414 | 383.277 | 331.237 | 819.242 |
| 2008 | 28.288 | 211.961 | 83.438 | 416.388 | 330.964 | 858.826 |
| 2009 | 30.032 | 164.335 | 74.864 | 426.613 | 306.427 | 845.827 |
| 2010 | 33.444 | 207.244 | 85.641 | 449.182 | 330.023 | 908.628 |
| 2011 | 34.998 | 218.513 | 91.503 | 481.438 | 337.544 | 956.703 |
| 2012 | 41.347 | 228.161 | 92.382 | 513.389 | 349.987 | 1009.097 |
| 2013 | 45.745 | 235.946 | 94.216 | 544.324 | 360.138 | 1056.461 |
| 2014 | 51.109 | 250.346 | 96.146 | 581.164 | 378.810 | 1119.920 |
| 2015 | 55.382 | 262.379 | 97.539 | 612.173 | 397.148 | 1176.941 |
| 2016 | 59.508 | 273.899 | 93.977 | 647.149 | 412.679 | 1229.312 |
| 2017 | 63.522 | 290.463 | 99.381 | 688.267 | 430.651 | 1299.897 |

**Table A1.** *Cont.*

| Year | CONST (RB) | MANU (RB) | AGRI (RB) | SERVC (RB) | MINING (RB) | GDP (RB) |
|---|---|---|---|---|---|---|
| 2018 | 66.218 | 304.847 | 99.470 | 735.834 | 444.009 | 1361.533 |
| 2019 | 71.850 | 316.355 | 101.287 | 781.024 | 453.070 | 1420.490 |

## Appendix B

**Table A2.** Forecast of CONST.

| Year | CONST_F (RB) |
|---|---|
| 2020 | 84.18 |
| 2021 | 88.56 |
| 2022 | 93.15 |
| 2023 | 97.98 |
| 2024 | 103.04 |
| 2025 | 108.33 |
| 2026 | 113.87 |
| 2027 | 119.66 |
| 2028 | 125.68 |
| 2029 | 131.96 |
| 2030 | 138.48 |
| 2031 | 145.24 |
| 2032 | 152.22 |
| 2033 | 159.44 |
| 2034 | 166.86 |
| 2035 | 174.49 |
| 2036 | 182.28 |
| 2037 | 190.23 |
| 2038 | 198.29 |
| 2039 | 206.43 |
| 2040 | 214.60 |
| 2041 | 222.76 |
| 2042 | 230.83 |
| 2043 | 238.75 |
| 2044 | 246.43 |
| 2045 | 253.76 |
| 2046 | 260.64 |
| 2047 | 266.94 |
| 2048 | 272.49 |
| 2049 | 277.13 |
| 2050 | 280.66 |

## References

1. Musarat, M.A.; Alaloul, W.S.; Liew, M. Impact of inflation rate on construction projects budget: A review. *Ain Shams Eng. J.* **2020**, *12*, 407–414. [CrossRef]
2. Müller, R.; Veser, M. The current state of nonfinancial reporting in Switzerland and beyond. *Die Unternehm.* **2020**, *74*, 296–311. [CrossRef]
3. Lean, C.S. Empirical tests to discern linkages between construction and other economic sectors in Singapore. *Constr. Manag. Econ.* **2001**, *19*, 355–363. [CrossRef]
4. Alaloul, W.; Altaf, M.; Musarat, M.; Javed, M.F.; Mosavi, A. Systematic Review of Life Cycle Assessment and Life Cycle Cost Analysis for Pavement and a Case Study. *Sustainability* **2021**, *13*, 4377. [CrossRef]
5. Hillebrandt, P.M. *Economic Theory and the Construction Industry*; Springer: Berlin/Heidelberg, Germany, 1985.
6. Alaloul, W.S.; Musarat, M.A.; Liew, M.; Qureshi, A.H.; Maqsoom, A. Investigating the impact of inflation on labour wages in Construction Industry of Malaysia. *Ain Shams Eng. J.* **2021**. [CrossRef]
7. Pradip Ajugiya, J.J.B.; Pitroda, J. Comparison of Modern and Conventional Construction Techniques for Next Generation: A Review. In Proceedings of the 21st ISTE State Annual Faculty Convention and National Conference on "Emerging Trends in Engineering", Tolani Foundation Gandhidham Polytechnic, Adipur, India, 27 December 2016.

8. Rafiq, W.; Musarat, M.; Altaf, M.; Napiah, M.; Sutanto, M.; Alaloul, W.; Javed, M.; Mosavi, A. Life Cycle Cost Analysis Comparison of Hot Mix Asphalt and Reclaimed Asphalt Pavement: A Case Study. *Sustainability* **2021**, *13*, 4411. [CrossRef]
9. Field, B.; Ofori, G. Construction and Economic Development: A Case Study. *Third World Plan. Rev.* **1988**, *10*, 41. [CrossRef]
10. Gaal, O.H.; Afrah, N. Lack of Infrastructure: The Impact on Economic Development as a case of Benadir region and Hir-shabelle, Somalia. *Dev. Ctry. Stud.* **2017**, *7*, 212581063.
11. Musarat, M.A.; Alaloul, W.S.; Liew, M.; Maqsoom, A.; Qureshi, A.H. Investigating the impact of inflation on building materials prices in construction industry. *J. Build. Eng.* **2020**, *32*, 101485. [CrossRef]
12. Khan, R.A. Role of construction sector in economic growth: Empirical evidence from Pakistan economy. In Proceedings of the First International Conference on Construction in Developing Countries (ICCIDC), Karachi, Pakistan, 4–5 August 2008.
13. Alaloul, S.W.; Musarat, M.A. Impact of Zero Energy Building: Sustainability Perspective. In *Sustainable Sewage Sludge Management and Resource Efficiency*; IntechOpen: London, UK, 2020.
14. Kenny, C. *Construction, Corruption, and Developing Countries*; The World Bank: Washington, DC, USA, 2007.
15. Estache, A. PPI partnerships vs. PPI divorces in LDCs. *Rev. Ind. Organ.* **2006**, *29*, 3–26. [CrossRef]
16. Lopes, J.P. Interdependence between the Construction Sector and the National Economy in Developing Countries: A Special Focus on Angola and Mozambique. Ph.D. Thesis, University of Salford, Salford, UK, 1997.
17. Geadah, A.N.K. Financing of Construction Investment in Developing Countries Through Capital Markets. Master Thesis, Massachusetts Institute of Technology, Department of Civil and Environmental, Cambridge, MA, USA, 2003.
18. Lopes, J. Investment in construction and economic growth. In *Economics for the Modern Built Environment*; Taylor & Francis: Oxford, UK, 2008; pp. 94–112.
19. Rameezdeen, R. *Study of Linkages between Construction Sector and Other Sectors of the Sri Lankan Economy*. Sri Lanka. 2005. Available online: https://www.iioa.org/conferences/15th/pdf/rameezdeen.pdf (accessed on 2 January 2021).
20. Ofori, G. Construction industry and economic growth in Singapore. *Constr. Manag. Econ.* **1988**, *6*, 57–70. [CrossRef]
21. Khan, R.A.; Liew, M.S.; Bin Ghazali, Z. Malaysian Construction Sector and Malaysia Vision 2020: Developed Nation Status. *Procedia Soc. Behav. Sci.* **2014**, *109*, 507–513. [CrossRef]
22. Yu-Chin, C.; Kenneth, R. Are the commodity currencies an exception to the rule? *Glob. J. Econ.* **2012**, *1*, 1250004. [CrossRef]
23. Cashin, P.; Céspedes, L.F.; Sahay, R. Commodity currencies and the real exchange rate. *J. Dev. Econ.* **2004**, *75*, 239–268. [CrossRef]
24. Olamade, O.O.; Oyebisi, T.O.; Olabode, S.O. Strategic ICT-use intensity of manufacturing companies in Nigeria. *J. Asian Bus. Strategy* **2014**, *4*, 1.
25. Drewer, S. Construction and development: A new perspective. *Habitat Int.* **1980**, *5*, 395–428. [CrossRef]
26. Ofori, G. Challenges of construction industries in developing countries: Lessons from various countries. In Proceedings of the 2nd International Conference on Construction in Developing Countries: Challenges Facing the Construction Industry in Developing Countries, Gaborone, Botswana, 15–17 November 2000.
27. Latham, M. *Constructing the Team: Final Report of the Government/Industry Review of Procurement and Contractual Arrangements in the UK Construction Industry*; Hmso: London, UK, 1994.
28. Oladinrin, T.; Ogunsemi, D.; Aje, I. Role of Construction Sector in Economic Growth: Empirical Evidence from Nigeria. *Futy J. Environ.* **2012**, *7*, 50–60. [CrossRef]
29. Alfan, E.; Zakaria, Z. Review of financial performance and distress: A case of Malaysian construction companies. *Br. J. Arts Soc. Sci.* **2013**, *12*, 147.
30. Kamar, A.M.; Hamid, Z.A.; Ghani, M.K.; Egbu, C.; Arif, M. Collaboration initiative on green construction and sustainability through Industrialized Buildings Systems (IBS) in the Malaysian construction industry. *Int. J. Sustain. Constr. Eng. Technol.* **2010**, *1*, 119–127.
31. Hamid, Z.; Kamar, K. Modernising the Malaysian construction industry. In Proceedings of the W089-Special Track 18th CIB World Building Congress, Salford, UK, 10–13 May 2010.
32. Malaysia, C. *Construction Industry Master Plan Malaysia 2006–2015*; Kuala Lumpur; Construction Industry Development Board: Negeri Sembilan, Malaysia, 2007.
33. Bin Ibrahim, A.R.; Roy, M.H.; Ahmed, Z.; Imtiaz, G. An investigation of the status of the Malaysian construction industry. *Benchmarking Int. J.* **2010**, *17*, 294–308. [CrossRef]
34. Kanan, P.D.P.; Bahaman, M.A.T.S. *The Key Issues in the Malaysian Construction Industry: Public and Private Sector Engagement*; Master Builders Association Malaysia: Kuala Lumpur, Malaysia, 2011.
35. Munaaim, C.M.; Dauurl, M.M.; Abdul-Rahman, H. Is late or non-payment a significant problem to Malaysian contractors. *J. Des. Built Environ.* **2007**, *3*, 1.
36. Department of Statistics Malaysia. Malaysia Economic Performance Fourth Quarter 2020. 2021. Available online: https://www.dosm.gov.my/v1/index.php?r=column/cthemeByCat&cat=100&bul_id=Y1MyV2tPOGNsVUtnRy9SZGdRQS84QT09&menu_id=TE5CRUZCblh4ZTZMODZIbmk2aWRRQT09 (accessed on 2 January 2021).
37. Department of Statistics Malaysia. Malaysia Economic Performance Third Quarter 2020. 2020. Available online: https://www.dosm.gov.my/v1/index.php?r=column/cthemeByCat&cat=100&bul_id=ZlRNZVRDUmNzRFFQQ29lZXJoV0UxQT09&menu_id=TE5CRUZCblh4ZTZMODZIbmk2aWRRQT09 (accessed on 2 January 2021).
38. Department of Statistics Malaysia. Malaysia Economic Performance Third Quarter 2019. 2020. Available online: https://www.dosm.gov.my/v1/index.php?r=column/pdfPrev&id=VDNYYXBkczhkUm9taU9pWUQvSEJBQT09 (accessed on 2 January 2021).

39. Department of Statistics Malaysia. Malaysia Economic Performance Fourth Quarter 2019. 2020. Available online: https://www.dosm.gov.my/v1/index.php?r=column/cthemeByCat&cat=100&bul_id=WWk2MDA3R1k1SlVsTjlzU3FZcjVlUT09&menu_id=TE5CRUZCblh4ZTZMODZIbmk2aWRRQT09 (accessed on 2 January 2021).
40. Hirschmann, R. Value of Construction Work in Malaysia from 2011 to 2019. 2020. Available online: https://www.statista.com/statistics/665028/value-of-construction-work-malaysia/#:~{}:text=Value%20of%20construction%20work%20in%20Malaysia%202011-2019&text=In%202019%2C%20the%20value%20of,a%20bid%20to%20decrease%20debt (accessed on 2 January 2021).
41. Gamil, Y.; Alhagar, A. The Impact of Pandemic Crisis on the Survival of Construction Industry: A Case of COVID-19. *Mediterr. J. Soc. Sci.* **2020**, *11*, 122. [CrossRef]
42. Wong, J.M.; Ng, S.T. Forecasting construction tender price index in Hong Kong using vector error correction model. *Constr. Manag. Econ.* **2010**, *28*, 1255–1268. [CrossRef]
43. Fu, C.; Liu, C. An economic analysis of construction industries producer prices in Australia. *Int. J. Econ. Financ.* **2010**, *2*, 3–14.
44. Jiang, H.; Xu, Y.; Liu, C. Construction price prediction using vector error correction models. *J. Constr. Eng. Manag.* **2013**, *139*, 04013022. [CrossRef]
45. Jiang, H.; Liu, C. A panel vector error correction approach to forecasting demand in regional construction markets. *Constr. Manag. Econ.* **2014**, *32*, 1205–1221. [CrossRef]
46. Shahandashti, S.M.; Ashuri, B. Highway Construction Cost Forecasting Using Vector Error Correction Models. *J. Manag. Eng.* **2016**, *32*, 04015040. [CrossRef]
47. Faghih, S.A.M.; Kashani, H. Forecasting Construction Material Prices Using Vector Error Correction Model. *J. Constr. Eng. Manag.* **2018**, *144*, 04018075. [CrossRef]
48. Wong, J.M.; Chan, A.P.; Chiang, Y. Forecasting construction manpower demand: A vector error correction model. *Build. Environ.* **2007**, *42*, 3030–3041. [CrossRef]
49. Xu, J.-W.; Moon, S. Stochastic forecast of construction cost index using a cointegrated vector autoregression model. *J. Manag. Eng.* **2013**, *29*, 10–18. [CrossRef]
50. Hyndman, J.R.; Athanasopoulos, G. *Forecasting: Principles and Practice*; OTexts: Melbourne, Australia, 2018.
51. Mak, S.; Choy, L.; Ho, W. Region-specific estimates of the determinants of real estate investment in China. *Urban Stud.* **2012**, *49*, 741–755. [CrossRef]
52. Hassani, M. *Construction Equipment Fuel Consumption During Idling: Characterization Using Multivariate Data Analysis at Volvo CE*; Digitala Vetenskapliga Arkivet: Vasteras, Sweden, 2020.
53. Zhang, X.; Chen, Y.; Jiang, P.; Liu, L.; Xu, X.; Xu, Y. Sectoral peak CO2 emission measurements and a long-term alternative CO2 mitigation roadmap: A case study of Yunnan, China. *J. Clean. Prod.* **2020**, *247*, 119171. [CrossRef]
54. Wang, Y.; Wen, Z.; Yao, J.; Dinga, C.D. Multi-objective optimization of synergic energy conservation and CO2 emission reduction in China's iron and steel industry under uncertainty. *Renew. Sustain. Energy Rev.* **2020**, *134*, 110128. [CrossRef]
55. Liu, Z.; Ciais, P.; Deng, Z.; Davis, S.J.; Zheng, B.; Wang, Y.; Cui, D.; Zhu, B.; Dou, X.; Ke, P.; et al. Carbon Monitor, a near-real-time daily dataset of global CO2 emission from fossil fuel and cement production. *Sci. Data* **2020**, *7*, 1–12. [CrossRef] [PubMed]
56. Momade, H.M.; Hainin, M.R. Review of sustainable construction practices in Malaysian construction industry. *Int. J. Eng. Technol.* **2018**, *7*, 5018–5021.
57. Abidin, N.Z. Sustainable construction in Malaysia–Developers' awareness. *World Acad. Sci. Eng. Technol.* **2009**, *53*, 807–814.
58. Musarat, M.; Alaloul, W.; Liew, M.; Maqsoom, A.; Qureshi, A. The Effect of Inflation Rate on $CO_2$ Emission: A Framework for Malaysian Construction Industry. *Sustainability* **2021**, *13*, 1562. [CrossRef]
59. Hock, I.L.C. Malaysia Needs Sustainable Construction. Malaysia. 2015. Available online: https://www.malaysiakini.com/letters/315928 (accessed on 2 January 2021).
60. Malaysia Department Of Statistics. Available online: https://www.dosm.gov.my/v1/index.php?r=column/cthemeByCat&cat=155&bul_id=aWJZRkJ4UEdKcUZpT2tVT090Snpydz09&menu_id=L0pheU43NWJwRWVSZklWdzQ4TlhUUT09 (accessed on 2 January 2021).
61. Bank, W. Malaysia. 2020. Available online: https://data.worldbank.org/country/MY (accessed on 2 January 2021).
62. Harris, R.I. Testing for unit roots using the augmented Dickey-Fuller test: Some issues relating to the size, power and the lag structure of the test. *Econ. Lett.* **1992**, *38*, 381–386. [CrossRef]
63. Gujarati, D.N. *Basic Econometrics*; Tata McGraw-Hill Education: New York, NY, USA, 2009.
64. Johansen, S. Statistical analysis of cointegration vectors. *J. Econ. Dyn. Control.* **1988**, *12*, 231–254. [CrossRef]
65. Johansen, J.S.; Juselius, K. Test. Struct. Hypothesis A Multivar. *Cointegration Anal. PPP UIP UK J. Econom.* **1992**, *53*, 1–244.
66. Granger, C.W. Investigating causal relations by econometric models and cross-spectral methods. *Econom. J. Econom. Soc.* **1969**, *37*, 424–438. [CrossRef]
67. Koop, G.; Quinlivan, R. *Analysis of Economic Data*; Wiley: Chichester, UK, 2000; Volume 2.
68. Gujarati, D.N. *Basic Econometrics*; MacGraw-Hill Inc.: New York, NY, USA, 1995; 838p.
69. Granger, C. Some properties of time series data and their use in econometric model specification. *J. Econ.* **1981**, *16*, 121–130. [CrossRef]
70. Suharsono, A.S. Forecasting Share Price Index In Indonesia and the World with the Univariate and Multivariate Time Series Model. *J. Sains Dan Seni Pomits* **2013**, *2*, 2.

71. Suharsono, A.; Aziza, A.; Pramesti, W. Comparison of Vector Autoregressive (VAR) and Vector Error Correction Models (VECM) for Index of ASEAN Stock Price. In *AIP Conference Proceedings*; AIP Publishing LLC: New York, NY, USA, 2017.
72. Brown, R.L.; Durbin, J.; Evans, J.M. Techniques for testing the constancy of regression relationships over time. *J. R. Stat. Soc. Ser. B Stat. Methodol.* **1975**, *37*, 149–163. [CrossRef]
73. Perron, P. The great crash, the oil price shock, and the unit root hypothesis. *Econom. J. Econom. Soc.* **1989**, *57*, 1361–1401. [CrossRef]
74. Lu, C.; Xin, Z. Impulse-Response Function Analysis: An Application to Macroeconomy Data of China. 2010. Available online: http://www.statistics.du.se/essays/D10_Xinzhou_lucao.pdf (accessed on 2 January 2021).
75. Zikmund, G.W.; Carr, G. *Business Research Methods*, 7th ed.; Cengage Learning, Dryden: New York, NY, USA, 2000.
76. James, H.A. *Time Series Analysis*; Princton University Press: Princton, NJ, USA; Oxford University: Oxford, UK, 1944.
77. Bliemel, F. *Theil's Forecast Accuracy Coefficient: A Clarification*; SAGE Publications: Los Angeles, CA, USA, 1973.
78. Elengoe, A. COVID-19 outbreak in Malaysia. *Osong Public Health Res. Perspect.* **2020**, *11*, 93–100. [CrossRef]
79. Government, Malaysia. *Shared Prosperity Vision 2030*; Percetakan Nasional Malaysia Berhad: Kuala Lumpur, Malaysia, 2019; pp. 1–30.
80. Nielsen, C.; Aalbers, R. *Optimal Discount Rates for Investments in Mitigation and Adaptation*; CPB Discussion Paper 257; CPB Netherlands Bureau for Economic Policy Analysis: The Hague, The Netherlands, 2013.

sustainability MDPI

Article

# Yield and Cost Effects of Plot-Level Wheat Seed Rates and Seed Recycling Practices in the East Gojam Zone, Amhara Region, Ethiopia: Application of the Dose–Response Model

Yirgalem Eshete [1,*], Bamlaku Alamirew [2] and Zewdie Bishaw [3]

1 College of Agriculture and Natural Resource, Debre Markos University, Debre Markos P.O. Box 269, Ethiopia
2 Yom Institute of Economic Development, Addis Ababa P.O. Box 62539, Ethiopia; bamlakalamirew@gmail.com
3 International Center for Agricultural Research in the Dry Areas (ICARDA), Addis Ababa P.O. Box 5689, Ethiopia; z.bishaw@cgiar.org
* Correspondence: etalem456@gmail.com; Tel.: +251-0913-655-606

**Abstract:** Previous studies investigated the effects of seed rates and seed recycling practices on the yield and yield-related variables. However, higher yield does not always guarantee cost-efficiency. This study aimed to investigate the yield effects of plot-level seed rate and the cost–benefit analysis of seed recycling practices. This study has introduced the dose–response model to the existing analytical methods used in analyzing the effect of different agrochemicals on crop yield. A multi-stage stratified sampling technique was used to select a total of 450 sample respondents. Data were gathered using a mix of data collection tools. Descriptive statistics along with the dose–response model were applied for data analysis. Farmers of the study were found to be dissimilar in terms of their seed rate application. A dose–response analysis indicated that the highest average wheat yield was associated with a seed rate of 50 kg ha$^{-1}$ above what is recommended. The yield effect of seed recycling was also assessed, and a one-time seed recycling has caused a yield decline of 665 kg ha$^{-1}$ compared to the non-recycled seeds. The cost reduced using recycled seed is by far lower than the economic gains associated with using unrecycled and fresh seeds. The cost–benefit analysis made clear that farmers can reduce their seed costs through seed recycling, but their yields and net income can be best improved by using unrecycled certified bread wheat seed (CBWS). Thus, farmers must be encouraged to use unrecycled seed by establishing agricultural credit schemes geared towards seed procurement and seed price subsidy as key strategies to reduce economically wasteful seed recycling practices.

**Keywords:** wheat; seed rate; yield effect; dose–response; seed recycling; cost–benefit analysis

**Citation:** Eshete, Y.; Alamirew, B.; Bishaw, Z. Yield and Cost Effects of Plot-Level Wheat Seed Rates and Seed Recycling Practices in the East Gojam Zone, Amhara Region, Ethiopia: Application of the Dose–Response Model. *Sustainability* **2021**, *13*, 3793. https://doi.org/10.3390/su13073793

Academic Editor: Shervin Hashemi

Received: 3 March 2021
Accepted: 24 March 2021
Published: 29 March 2021



## 1. Introduction

Wheat productivity is substantially dependent on the use of crop genetics and the application of improved agronomic management practices. The role of quality seeds cannot be overemphasized in enhancing agricultural productivity and improving the livelihood of smallholder farmers [1,2]. Among other agronomic management practices, the seeding rate can be considered as one of the major factors determining the ability of the crop to capture the available resources and hence increase yield. In most cropping systems, seed rate is a factor under farmer's control and has remained to be a vital crop production factor and one of the best decisions required to be made. However, the optimum seed rate for maximum bread wheat production greatly varies between regions according to climatic conditions, soil types, sowing time, cultivars, spacing, and other agronomic practices [3].

The majority of farmers in the developing countries are using either below or above the optimum recommended seed rates causing a wider yield gap between the potential and the actual production. In Ethiopia, despite the need for optimum wheat seed rates disaggregated on spatial, temporal, and varietal levels, seed rate recommendations are still based on blanket recommendations of a "one-size-fits-all" approach. Moreover, the contribution of certified seed in boosting productivity and enhancing the livelihood of farmers can be realized if seeds are not recycled or are replaced by new seeds based on expected utility. The expected productivity gains concomitant with the use of certified seeds of improved varieties are likely to be lost over time as the genetic merits conferred by breeding will break down and make resource-poor smallholder farmers vulnerable to risks associated with biotic and abiotic stresses. The use of improved certified seeds without a parallel adoption in other agronomic practices, such as seed rates, seed replacement, spacing, and tillage, did not bring a substantial increase in bread wheat productivity [4]. Yet, information on the yield effect of different agronomic practices is meager. Especially, there is inadequate information about the yield effect of farmer's plot level bread wheat seed rates and seed recycling practices in the study area. Thus, investigating wheat yield response to different plot level seed rates and seed recycling practices has been the major objective of this study. The output in this study will help farmers to decide whether they can reduce their costs of seed by correcting the seed rate. Moreover, the use of different seed rates and seed recycling practices are often evaluated in terms of one indicator, usually yield gains. However, the use of different seed rates and seed recycling practices should be evaluated from multiple dimensions. This study is thus meant to investigate the yield effects of plot-level seed rates along with the cost–benefit analysis of using recycled and unrecycled seeds. The cost–benefit analysis is expected to help farmers decide if they can reduce their seed cost through seed recycling and if their yields can be improved by using unrecycled certified bread wheat seed (CBWS) as compared to recycled seeds.

This research is expected to contribute to the literature in the following three ways. First, it will introduce a different analytical approach in its effort to investigate the yield effect of plot-level seed rates. Previous studies measured the yield effect of seed rate simply as the amount of crop yield per hectare in response to a specified seed rate, while others employ linear regression method and average treatment effect models [3,5,6]. In this study, the dose–response model is applied. It is a common analytical method in medical sciences [7,8]. To the researcher's knowledge, this paper is the first to use the generalized linear dose–response model to investigate the yield effect of plot-level seed rates in the study area. Second, it will add to the empirical literature on the yield effect of plot-level seed rates [5,9]. Third, this study analyzed the yield effect of seed rates of recycled and unrecycled CBWSs along with the cost–benefit analysis of seed recycling. Therefore, the output in this study will also help farmers to decide whether they can reduce their costs by recycling seeds or not.

## 2. Research Methods

### 2.1. Description of the Study Area

East Gojam zone has a total of 17 districts. Its headquarters, Debre Markos, is located 300 and 260 km, far from Addis Ababa (the capital city of the country) and Bahir Dar (the capital city of the region), respectively. It is located at 10°19′0.00″ North and 38°00′0.00″ East. The mixed farming system is the main occupation of farmers in the East Gojam zone. The area is dominated by mixed agricultural systems [10]. Figure 1 below indicates the map of the study areas.

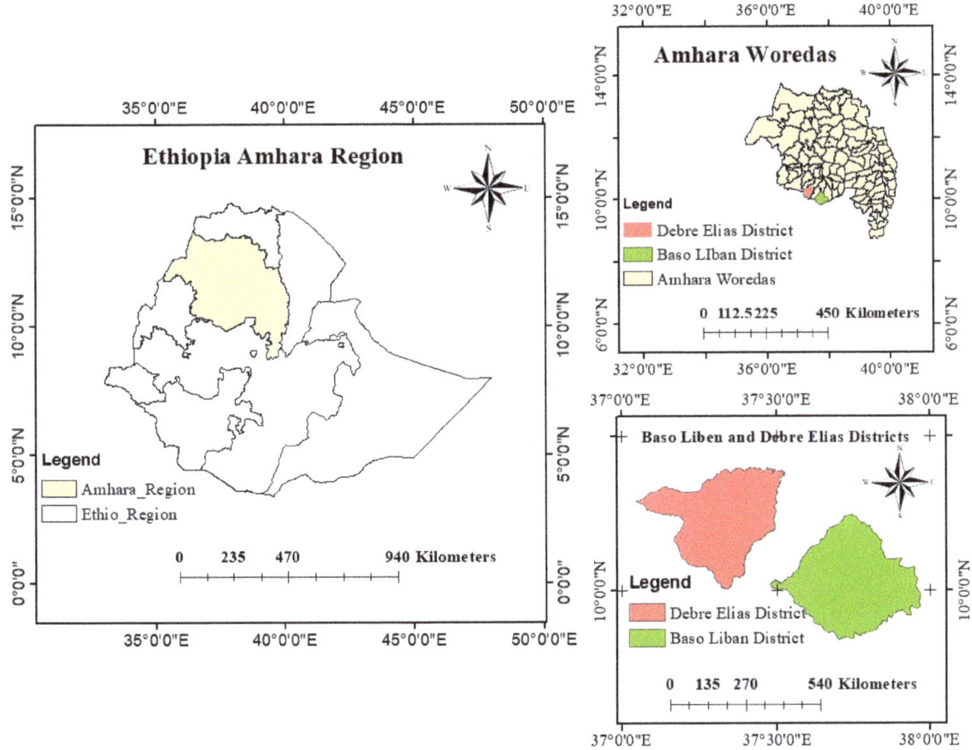


**Figure 1.** Map of the study area.

### 2.2. Sampling Technique and Sample Size Determination

Multi-stage purposive and random sampling methods together with probability proportional to size (PPS) were employed in the selection of sample study areas. In the first stage, the East Gojam Zone of the Amhara National Regional State (ANRS) was purposefully considered due to its high potential for wheat production. In the second stage, Baso Liban and Debre Elias districts were considered as specific study areas. The potential for wheat production, availability of sufficient representative sample respondents, and the long year's acquaintance of the study places by the researcher were a few of the reasons to consider these districts. Debre Elias and Baso Liban districts have 15 and 22 kebeles (Kebele is an Amharic term that refers to the lowest administrative region), respectively. Given the financial, time, and other resource constraints, this study has considered a total of 8 rural kebeles. These eight kebeles are thought to be sufficient and representative as there is homogeneity between most of the kebeles in terms of their agronomic practices in wheat production. Then 3 and 5 kebeles were considered as specific study areas by using the PPS sampling technique. A total of 450 farmers, determined by [11] sample size determination formula, was considered for the study.

### 2.3. Data Collection Methods

This study employs both primary and secondary data sources. However, it has heavily relied on primary data sources, including smallholder farmers involved in wheat production, development agents, crop experts, and key informants. Moreover, the information gathered from these primary sources was supported and triangulated from available secondary data sources, including regional and district annual reports, research findings, journals, publications, thesis, dissertations, books, proceedings, and so forth. Both qualitative and quantitative data types were gathered using structured interview schedules, questionnaires, FGDs (Focus Group Discussions), KIIs, (Key Informant Interviews), and personal observations.

### *2.4. Data Analysis Methods*

#### 2.4.1. Descriptive Data Analysis Tools

Both measures of central tendency (mean, median, and mode) and measures of central dispersion (range, variance, and standard deviation) were frequently used.

#### 2.4.2. Specification of Dose–Response Model

Dose–response models are regression models where the explanatory variable is usually referred to as the dose or concentration, while the dependent variable is usually referred to as response or effect. A dose was also known as "metameter", is defined as any pre-specified amount of biological, chemical, or radiation stress eliciting a certain, well-defined response. Other kinds of exposure or stress could also be imagined, e.g., the time elapsed in germination experiments. The dose is a non-negative quantity, and it is often, but not always, assumed to be measured without error, as is often the case in designed experiments [7]. In this study, the seed rate applied by the farmers will be considered as the dose or the concentration, while the amount of wheat yield is going to be considered as response or effect.

There are various types of dose–response models. Specifically, responses are defined to a given dose as the quantification of a biologically relevant effect, and as such, it is subject to random variation. The most common type is a continuous response, such as biomass, enzyme activity, or optical density. A binary or aggregated binary (binomial) response is also frequently used to describe results, such as dead/alive, immobile/mobile. The response may also be discrete as in the number of events observed in a specific time interval, such as the number of juveniles, offspring, or roots. Dose–response curves may also be used to summarize experiments where the response is an event time that is the time elapsed before some specific event is being observed absent [8]. This study will undertake a linear dose–response regression between wheat seed rates (dose/concentration) and wheat yield (response/effect).

The full specification of a statistical dose–response (regression) model involves specifying how the mean is described by a parametric function of dose as well as specifying assumptions about the distribution of the response. We will focus on ways to model the mean trends through mostly S-shaped or related biphasic functions because these functions have in common that they reflect a basic understanding of the causal relationship between the dose and the response, e.g., when a dose increases, the response monotonically decreases or increases one-way or another towards minimum or maximum response limits, respectively. Consequently, these functions have turned out to be extremely versatile for describing various biological mechanisms involving model parameters that allow the interpretation of observed effects within a biologically plausible framework. So we define dose–response models to be a collection of statistical models having a certain mean structure in common; this is not a strict mathematical definition, but rather a definition driven by applications. Consequently, dose–response models encompass a range of statistical models from nonlinear regression, generalized (non) linear regression, and parametric survival analysis. Let y denote an observed response value, possibly aggregated in some way, corresponding to a dose value $x \geq 0$. The values of y are often positive but may take arbitrary positive or negative values. Furthermore, we will assume that observation of y is subject to sampling variation, necessitating the specification of a statistical model describing the random variation. Specifically, we will focus on characterizing the mean of y (denoted E(y) below) in terms of a model function f that depends on the dose x:

$$E(y) = f(x, \beta_0) \quad (1)$$

So, for a given dose x, the corresponding observed response values will be distributed around f(x, β). The function f is completely known as it reflects the assumed relationship between x and y, except for the values of the model parameters β = (β1, βp), which will be estimated from the data to obtain the best-fitting function. The remaining distributional assumptions on y will depend on the type of response. For instance, for a continuous response, the normal distribution is commonly assumed, whereas, for a binary response, the binomial distribution is commonly assumed.

## 3. Result and Discussion

### 3.1. Bread Wheat Seed Sources

Several bread wheat varieties have been recommended for highland (Dega) and mid-highland (Weyina-Dega) agro-ecological zones of the country. Ogolcho (ETBW5520), Kubsa, Danda'a, Kakaba, Digalu, Tay, Katar, Abola, Dereselegn, K6290 Bulk, K6295-4A, ET-13 A2, Pavon 76, Dashen, Mitike, Galema (HAR-604), Tusa, Tura, Katar, Shina, Simba, Guna, and Densa have been introduced in both study districts [12].

However, almost all farmers have already abandoned using them except Ogolcho (ETBW5520), released in 2012, and it has become an exclusive wheat variety being grown in the study areas during the production season considered in this study. Several agricultural experiments and trials are carried out in the study area by different academic and research institutes and other governmental and non-governmental organizations. Despite these efforts, farmers in the study area are unable to access several alternative wheat varieties during the production season. Farmer's access to seed is quite supply-driven that the major formal source for CBWS (largely the Amhara Seed Enterprise) does not usually supply more than one variety during a specific production season. As a result, farmers are compelled to stick to a single variety supplied to them. Furthermore, [13] seemed to give a similar impression of low spatial diversity of wheat varieties where only a few dominant varieties appeared to occupy a large proportion of the wheat area. Although varietal diversity is low, the use of certified seed has been exceptionally high as compared to findings in several studies [13–15]. This may be because farmers are benefited from the spillover effects of research efforts being undertaken by several governmental and Non-Governmental Organizations. This may have encouraged farmers to allocate larger plots for wheat production using certified seeds. Given its huge potential for wheat production, farmers of the study area are often preferred targets for seed multiplication for the regional and national seed enterprises. This has allowed farmers to have easy and timely access to certified seeds, which could be one important reason for exceptionally high certified bread wheat seed use. Moreover, farmers' long years of experience and perceived benefits in using certified seeds could be another explanation for high rates of certified seed use. The spread effect of agricultural research to large-scale application of farm innovation is witnessed in other studies, too [16,17].

Farmers in the study area acquire seeds from four important sources (Appendix A Table A3). The Amhara Seed Enterprise has been a dominant seed supplier for almost 73% of the farmers, followed by farmer's cooperatives that supply seed for over 15% of the farmers. Farmers of the study area participate in seed multiplication. Yet, these farmers cannot sell seeds by themselves but through the cooperative to which they belong once the seed is approved and certified. Zonal and district experts working in the seed quality department will control the quality of seeds produced by the farmers, whereas the Ethiopian and the Amhara Seed Enterprises provide the technical support for seed production. Farmer's own saved seeds, along with research and academic institutions, have also been important seed sources for few sections of the farming community in the study area.

### 3.2. Yield Effects of Plot-Level Seed Rates

Among others, an appropriate plant population determined by seed rates is one of the most important agronomic practices for optimizing wheat yield by reducing competition for minerals [1,10,18]. Yet, wheat yield response to seed rate is inconsistent across spatial and temporal bases due to several biotic and abiotic factors. Ceteris paribus seed rate is an important yield determinant in bread wheat production. A better understanding of the relationship between seed rates and yield in wheat production could have paramount importance in improving seeding rate recommendations according to specific spatial, temporal, and bread wheat variety-related characteristics [5,9]. The application of the best quality seed may lead to lower gains if the recommended seed rates are not applied [14,19].

The Ministry of Agriculture generally recommends a seed rate of 80–100 kg of seed per hectare. However, based on location-specific wheat variety trials conducted in the East Gojam zone, a seed rate of 100–150 kg ha$^{-1}$ is recommended for Baso Liban and Debre Elias districts. However, farmers in the study area do not stick to the recommendation and use their own seed rates. The yield and yield effects of seed rates of the CBWS of the Ogolcho variety were evaluated when it was introduced in 2012 in the selected farmer training centers (FTCs) in Basoliban and Debre Elias districts. Since then, the yield effect of farmer's own plot level seed rates has never been empirically assessed. Therefore, exploring the effect of the plot level seed rates on wheat yield is an important objective of this study. Respondent farmers are classified into three groups for the study. The 1st category (below-recommended seed rate group) includes farmers with a seed rate ranging from 50–99 kg ha$^{-1}$ (50 kg ha$^{-1}$ being the lowest observed seeding rate). The 2nd group (recommended seed rate users group) includes respondents with a seed rate ranging between 100 and 150 kg ha$^{-1}$. The 3rd group ("above recommended") constitutes farmers with a seeding rate greater than 150 kg ha$^{-1}$.

The survey results showed that about 4.9%, 48% and 47.1% of the respondents use seed rates below, above, and within the recommended seed rates, respectively (Table 1). This means that above 95% of the farmers in the study area use seed rates either above the recommendation or within the recommended seed rate range. On average, farmers of the study area have used a seed rate of 194.4 kg ha$^{-1,}$ which is higher than what is recommended. A similar finding was reported by [12], where they had reported an average seed rate of 264.56 kg ha$^{-1}$, about 100 kg above what is recommended. Several studies also found the use of higher seed rates. Late seeding dates, planting methods, the need to compensate for possible damages due to ants, rats, and birds, intentions of grain yield maximization, and weed control are reported as some of the reasons for higher seed rates [3,20]. These explanations for higher seed could also be adopted as reasons for the higher seed rates witnessed in this study. The Pearson's correlation indicates a negative nexus between seed rate and hand weeding frequency (r = −0.5248, $p$ = 0.05). This could be interpreted as that farmers reduce the seed rate for making frequent hand weeding easy and convenient. Moreover, farmers in the study area have reportedly used lower seed rates due to perceived high seed prices and insufficient supply of certified seeds. Though it is not witnessed from the findings of this study, studies indicated several reasons that push farmers to use seed rates below what is recommended. For instance, superior agronomic practices, such as row planting and transplanting, reduce the seed rate required in crop production. These practices do not only reduce seed rates but also allow more spacing between seedlings, permit easy weeding, reduce competition between seedlings and allow for better branching out (tillering) of crops [6,21].

A good section of the respondents (48%) has used a seed rate above the recommendation. The Pearson's correlation indicates a positive nexus between the frequency of seed recycling and the amount of seed rate (r = 0.5534, $p$ = 0.05). This implies that higher seed rates are associated with the frequency of seed recycling. The recycled seeds are cheaper than fresh and unrecycled CBWS that may encourage farmers to purchase and use more amounts of bread wheat seed. Several studies provided explanations as to why farmers use more seed rates. For instance, quite often, late seeding dates cause higher seeding rates because a delay in sowing normally reduces individual plant growth and tiller production [20]. The result in this study does not portray a significant yield difference between farmers who use recommended and below recommended seed rates. The lowest average wheat productivity was associated with the farmers, who apply seed rates above what is recommended.

**Table 1.** Distribution of average productivity across varying seed rates.

| Seed Rate Group | Obs. | Seed Rate (kg ha$^{-1}$) | | Productivity (Qt ha$^{-1}$) | |
|---|---|---|---|---|---|
| | | Mean | Std. Dev | Mean | Std. Dev. |
| Below Recommended | 22 (4.9) | 67.0455 | 11.9183 | 40.3636 | 3.8365 |
| Recommended | 212 (47.1) | 133.9623 | 21.2714 | 40.3726 | 4.80499 |
| Above Recommended | 216 (48) | 266.6667 | 93.3859 | 36.7361 | 6.2424 |
| Total | 450 (100) | 194.3889 | 97.0926 | 38.62667 | 5.7883 |

Respondents of the study were asked about the reasons for the higher seed rate they are applying. Expected better yield gains associated with higher seed rates have been the most frequently reported reason for higher seed rates. The practice of using higher seed rates was reported in other studies as well [19,20,22,23]. These studies unfold that farmers usually use higher seed rates than what is recommended due to land scarcity, availability of credit markets for agricultural inputs, and access to markets. Though not implied from the data gathered here, these same reasons could also explain the findings in this study. Similarly, farmers in least developed countries (LDCs) prefer to use higher seed rates beyond what is recommended hoping that it will be a good strategy to control weeds and help them in coping with risks of crop production. High seed rate application beyond what is recommended is discouraged due to the negative consequences on seed quality, such as seed size and weight [19,24].

One cannot have a clear picture of the yield effects of seed rate from the descriptive statistics presented above. The best analytical model to understand the effect of a dose (concentration) on response is the dose–response function based on the generalized linear regression model. This regression model produces such estimates more quickly. However, the use of post estimation command, such as margins plot in STATA (StataCorp, TX, USA), enables to paint a clear picture of the effect of a dose (seed here) on the response (wheat yield per hectare). To do so, the data gathered was arranged in six seed rate groups to include all observations within ±50 kg ha$^{-1}$ of the recommended seed rates. All farmers that apply a seed rate of 50 kg ha$^{-1}$ or below are categorized as the 1st group, and the seed rate is calibrated as 50 kg ha$^{-1}$ (50 kg ha$^{-1}$ being the lowest seed rate). The seed rates above 50–100 are calibrated as 100, while seed rates above 100–150, above 150–200, above 200–250, above 250–300, and above 300 were calibrated as 150, 200, 250 and 300 kg ha$^{-1}$, respectively. The regression coefficient is calculated using the generalized linear model to investigate the effect of the seed rate applied by farmers on wheat productivity. The result indicated that a 1 kg increase in seed rate is associated with a probability of 0.054 increases in wheat productivity (See Appendix A Table A1). To sum up, Figure 2 below shows the relationship between plot level seed rate and yield in the study area.

Further analysis was carried out using predictive margins at the mean to investigate the magnitude of change in wheat productivity when the seed rate applied by the sample respondents increases from 50 kg ha$^{-1}$ (the minimum seed rate reported) to 300 kg ha$^{-1}$ (the maximum seed rate applied). Results of the analysis depicted that seed rates below what is recommended were associated with lower wheat productivity. Interestingly, a seed rate of 200 kg ha$^{-1}$ is associated with the highest average wheat production per hectare (4225 kg ha$^{-1}$). This means that an extra 50 kg ha$^{-1}$ application beyond the recommended seed rate is associated with higher productivity. This suggests that there might be a need to revise the seed rate recommendation for the study area. As presented in Table 2, an increase in the successive seed rate from 50–200 kg ha$^{-1}$ was associated with an increase in wheat productivity. However, additional seed rates being applied beyond 200 kg ha$^{-1}$ did not show an increasing effect on wheat productivity. Rather, productivity started to decline after the 200 kg ha$^{-1}$. Though the recommended seed rate in the study area was between

100 and 150 kg ha$^{-1}$, higher yield gains were reported for seed rates from 150–200 kg ha$^{-1}$. This is in agreement with other field-level experiments and surveys results by different authors that reported increasing the seed rate of wheat from 100–200 kg ha$^{-1}$ increased the grain yield and straw yields [21,24]. This study found that about 48% of the respondents reportedly used higher seed rates than what is recommended. Yet, the seed rate is best when it is associated with the maximum grain yield. The economic theory of diminishing return to input was proved from the findings of this study. The wheat yield hardly changes with additional units of seed rates once the maximum yield level was achieved. This could be witnessed from the flatten curve after 200 kg ha$^{-1}$ seed rates in Figure 1 above. Seed rates used above what is required to reach the flat part of the curve is money wasted that could have been used for other development expenditures.

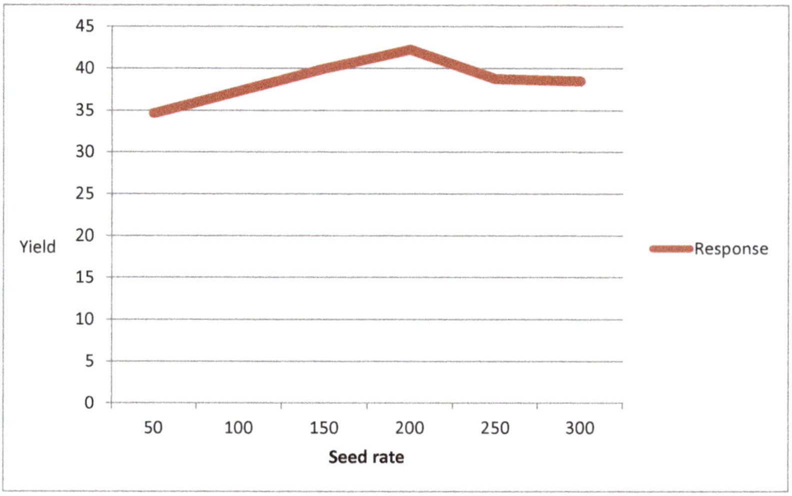


**Figure 2.** Seed rate–yield response curve.

**Table 2.** Marginal effects of different seed rates on wheat yield.

| Seed Rate (kg) | Delta-Method | | t | $p >$ \|t\| | 95% Confidence Interval | |
|---|---|---|---|---|---|---|
| | dydx | Std. Err | | | | |
| 50 | 34.635 | 0.37588 | 92.15 | 0.000 | 33.89965 | 35.37706 |
| 100 | 37.321 | 0.25025 | 149.13 | 0.000 | 36.82926 | 37.81289 |
| 150 | 40.004 | 0.25237 | 158.52 | 0.000 | 39.50783 | 40.49977 |
| 200 | 42.246 | 0.75394 | 56.03 | 0.000 | 40.76464 | 43.72809 |
| 250 | 38.733 | 0.37375 | 103.63 | 0.000 | 37.99811 | 39.46745 |
| 300 | 38.480 | 0.48983 | 78.56 | 0.000 | 37.51762 | 39.44294 |

### 3.3. Yield and Economic Effects of Seed Recycling

One important finding of this study is the practice of seed recycling. The study respondents were asked whether they do recycle wheat seeds or not and how often they do recycle seeds. A good section of the farmers (about 72.89%) considered in the study does not recycle seed beyond what is recommended (Table 1). Yet, it must be noted that about 27.11% of the study respondents have recycled seed. Studies in Ethiopia and Tanzania indicated that farmers often used obsolete or old varieties whose seeds are recycled. In Ethiopia, about 76% and 10% of the farmers grew modern bread wheat varieties and obsolete varieties, respectively. Several studies also indicated widespread seed recycling practices [19,25–27].

This study also solicited the reasons for the existing seed recycling practices. Appendix A Table A4 summarizes the distribution of farmers in terms of their seed recycling practice and reasons for seed recycling. Expensive prices, problems of timely seed supply, and supply shortage have been the most frequently cited reasons for seed recycling. Furthermore, during the FGD, participants mentioned that they see no noticeable difference in yield and other desirable qualities between non–recycled and recycled seeds if the recycling is made only once or twice. Not all studies indeed show a better yield advantage for certified seed over farmer saved seed. Proper seed production and seed saving practices, maintaining a seed free from contamination and varietal purity affect the yield differences significantly [3,28]. Farmers also indicated that the application of different coping mechanisms in times of seed supply shortage and higher seed prices. One important reported strategy was cleaning and saving seed from the existing production to be used in the coming production. Thus, seed recycling in the study area has been practiced as a seed shortage handling and cost reduction strategy. Farmers complain about the expensive price of CBWS and use it as an excuse for seed recycling. However, from findings of the cost–benefit analysis presented in Appendix A, Table A2, one can easily learn that these same farmers still incur costs well above the price of the certified seed. This is mainly because the magnitude of recycled seed rate has been so high, which is above equivalent to the costs incurred in using unrecycled CBWS. However, it must be noted that recycled seeds are cheaper and can be easily obtained informally from friends, relatives, neighbors, and consequently, it may encourage farmers to apply higher seed rates.

The cost–benefit analysis was used to assess the economic impact of seed recycling. The prevailing domestic market price was adopted to calculate the costs and benefits of seed recycling. The cost–benefit analysis is concerned with two important issues. One is costs reduced in the production of goods and services. The second is income increased as a result of the marketing of goods and services. In this study, the cost reduced as a result of seed recycling practice made at the expense of using fresh and unrecycled certified seed was calculated, followed by the income gained as a result of using fresh and unrecycled certified seed at the expense of recycled seed. During the 2019/20 cropping season, the unit cost of 100 kg of certified unrecycled and recycled bread wheat seeds were 2300 and 1800 Ethiopian Birr (ETB), respectively. Farmers that used unrecycled and fresh certified seeds had a mean seed rate of 169.2835 kg ha$^{-1}$ against a mean seed rate of 263.6364 kg ha$^{-1}$ in a one-time recycled seed (Appendix A Table A2). The cost reduction (CR) as a result of using a one-time recycled seed at the expense of unrecycled certified seed could be equated as follows:

$$CR = \frac{A1 \times B2}{100} - C1 \tag{2}$$

where CR = is cost reduction as a result of seed recycling, A1 = Mean seed rate of the unrecycled seed, B2 = cost of one-time recycled seeds, C1 = total seed cost of unrecycled seeds.

Unrecycled certified seed users, on average, used a seed rate of 169.2835 kg ha$^{-1}$ with a total seed cost of 3893.51 ETB. If unrecycled certified seed users have turned into use of the recycled seed, then the total seed cost would have been 3047.10 ETB. Following the formula above, it is clear that farmers can reduce their seed cost per hectare by 846.42 ETB by reverting from the use of unrecycled CBWS to a one-time recycled seed use. Thus, the economic rationale of farmers for seed recycling does hold water in terms of cost reduction. A study by [28] supports this finding. These authors found that farmers in Canada try to reduce production costs without incurring large decreases in yield from their own saved seed from a current crop to use for next year's planting, and it is common practice with most cereal and pulse crops. There could be several explanations behind costs reduced as a result of using recycled seed. The first explanation is lower transaction costs. Unrecycled seeds could be purchased from nearby informal sources at local markets. As a result, no extra transportation costs are incurred. In addition to transaction costs, costs pertinent to chemical treatment, labeling, and packaging do not apply to recycled seeds. However, cost reduction alone is not a guarantee to an economically profitable agricultural business. The economic gains as a result of using the other alternative (unrecycled certified seed here) or the economic loss as a result of sticking to an old practice (using recycled seed here) are also critical. This calls for the need to calculate the net income gain of a crop enterprise. The next equation in the cost–benefit analysis is to check whether the economic gains from seed recycling are better than unrecycled seed. The income gained (IG) as a result of using unrecycled certified seed at the expense of recycled seed is calculated using the following formula:

$$\mathrm{IG} = \frac{\mathrm{D2} \times \mathrm{F1}}{\mathrm{D1}} - \mathrm{F2} \tag{3}$$

where IG is income gained, D2 is the mean yield of recycled seed, F1 is mean income from unrecycled seed, D1 is mean yield of unrecycled seed, and F2 is mean income from the recycled seed.

Following the formula above and based on the numerical figures presented in Appendix A, Table A2, the use of fresh and unrecycled CBWS was associated with a higher yield (4044 kg ha$^{-1}$) and higher market prices (1950 Ethiopian Birr). Similarly, a one-time seed recycling is associated with a lower yield (3333 kg ha$^{-1}$) and market prices (1800 Ethiopian Birr). Thus, the average income per hectare given the price of a one-time recycled is 60,823.62 ETB. However, a farmer reverting to use fresh and unrecycled seed receive an annual income of 65,892.26 ETB per hectare. This means that the use of fresh and unrecycled CBWS leads to an economic gain of 5068.635 ETB than a one-time recycled seed. The net income per hectare associated with the use of unrecycled CBWS and a one-time recycled wheat seed was 74,968.58 ETB and 56,078.16 ETB, respectively (Appendix A Table A2). This implies that the use of fresh and unrecycled CBWS ends up in a net income gain of 18,890.42 ETB per hectare than a one-time recycled seed. To sum up, the results of the cost–benefit analysis made clear that farmers can reduce their seed costs through seed recycling, but their yields and net income can be best improved by using unrecycled CBWS.

Several authors highly suggested that farmers should use unrecycled certified seeds as continuous recycling reduces crop yield [15,19,29]. The finding in this study, too, is quite in agreement with these authors. On average, the mean productivity of unrecycled wheat seed users was the highest (Table 3). This highest wheat yield per hectare could be attributed to the yield superiority of the fresh and unrecycled wheat seed. As it can be learned from the above table, on average, a one-time and a twice seed recycling cause a 665 and 711 kg ha$^{-1}$ wheat yield decline as compared to the unrecycled certified seed. Here it is worthy of mentioning that the average yield difference between farmers, who recycle once and twice, was not conspicuous. This indicates the fact that the magnitude of yield loss as a result of recycling remains insignificant irrespective of recycling frequency. It signifies the need to rely heavily on fresh and unrecycled CBWS. Nonetheless, during the FGD, farmers opined that a one-time or twice recycling did not cause noticeable yield reduction. Yet, [14,25] reported that fresh certified seeds increase productivity and farmer's efficiency in comparison to frequently recycled seeds. The productivity and efficiency gains indicate that promoting fresh certified wheat seeds has the potential to raise production using the available improved seed technologies at the most efficiency. The finding in this study thus calls for an intensive extension campaign and field-based demonstration to convince farmers of the yield losses as a result of seed recycling. Such a significant yield difference is not expected between fresh certified seed and a one-time seed recycling for self-pollinated crops. The marked difference in yield could, therefore, be attributed to the quality of the seed as indicated by inappropriate seed rate, seed counterfeiting, and unwise planting method [30].

**Table 3.** Distribution of average productivity and seed recycling.

| Seed Recycling Frequency | Obs. | Productivity (Qt ha$^{-1}$) | | Seed Rate (kg ha$^{-1}$) | |
|---|---|---|---|---|---|
| | | Mean | Std. Dev | Mean | Std. Dev. |
| Unrecycled | 328 | 40.4421 | 4.9718 | 169.2835 | 66.96013 |
| Recycled once | 110 | 33.7909 | 4.6730 | 263.6364 | 129.0241 |
| Recycled twice and above | 12 | 33.3333 | 7.3278 | 245.8333 | 130.0495 |
| Total | 450 | 38.6267 | 5.7883 | 194.3889 | 97.09256 |

Source: own calculations (2020).

## 4. Conclusions and Recommendation

### 4.1. Conclusions

The proportion of farmers that uses a seed rate within the recommended seed rate, above the recommended and below the recommended seed rates were found to be 47.11%, 48%, and 4.89%, respectively. One important finding of this study was the practice of seed recycling. In this regard, one can deduce that about 72.89% of the farmers in the study area do not recycle seed. Unrecycled seeds were associated with mean productivity of 4044 kg ha$^{-1}$, while seeds that were recycled once and more than once were associated with mean productivity of 3379 and 3333 kg ha$^{-1}$. This implies that seed recycling, on average, causes 665 kg yields loss per hectare. A regression coefficient was calculated using the generalized linear model to investigate the effect of the seed rate applied by farmers on wheat productivity. Keeping other factors constant, the result indicated that a 1 kg increase in seed rate is associated with a 0.054 unit increase in wheat productivity (Appendix A Table A1). From the dose–response relationship, one could learn that the maximum yield was associated with a seed rate of 200 kg ha$^{-1,}$ which is higher than the recommended seed rate by 50 kg ha$^{-1}$. Thus, one can rightly conclude that there exists an opportunity of increasing wheat productivity by adding an extra seed rate of up to 50 kg ha$^{-1}$ exists. However, it also has to be noted that any increase in seed rate beyond the optimum found here has led farmers to diminishing productivity. The average wheat productivity associated with farmers, who applied seed rates as per the recommendation, below the recommendation, and above the recommendation were 4037, 4036 and 3674 kg ha$^{-1}$, respectively. The use of fresh and unrecycled CBWS ends up in a net income gain of 18,890.42 ETB per hectare than a one-time recycled seed. To sum up, the results of the cost–benefit analysis made clear that farmers can reduce their seed costs through seed recycling, but their yields and net income can be best improved by using unrecycled CBWS.

### 4.2. Recommendations

The researcher strongly believes that upon the implementation of the suggestions made here, proper seed rate use will be enhanced. Moreover, the researcher is strongly convinced that these recommendations can be used by farmers, extension agents, development practitioners, policymakers, and other stakeholders involved in wheat production in their efforts to improve wheat productivity. The following are the most important recommendations made based on the findings of the study discussed above.

- Though cost reduction and high seed prices are frequently reported as main factors for seed recycling, the cost–benefit analysis indicated that the use of unrecycled certified seed leads to better economic gains than recycled seeds. Therefore, this study recommends strategies that will encourage the use of fresh and unrecycled certified seeds. Moreover, agricultural credit schemes and price subsidies geared towards helping farmers use unrecycled certified are recommended as key strategies to reduce economically wasteful recycling practices. Implementation of these recommendations is expected to bring multiple benefits. First, it will raise farmer's productivity, income, and food security. Second, seed scientists, seed producers, and other individuals and institutions will benefit the most from their research and innovations. This profit could also be reinvested for more research and development efforts. Third, agro-processing could be enhanced that may increase government revenue and the creation of wider employment opportunities;
- Seed recycling practices have caused yield loss as compared to the non–recycled seeds. Therefore, ensuring the supply of fresh certified seeds is highly recommended. This could be achieved through encouraging seed parastatals, establishing new seed production and marketing cooperatives, and encouraging the existing ones;
- Wheat varietal diversity in the study area is found to be very low. This study, therefore, strongly recommends varietal diversification through the promotion of different varieties in the study areas instead of relying on one variety;

- Grain yield maximization was the most important reason to use seed rates beyond the recommendation. The evidence from the existing seed rate application of the farmers proved that a higher seed rate of up to 50 kg ha$^{-1}$ above the existing recommendation was associated with the highest yield per hectare. Thus, this study strongly suggests that the MoA, agricultural research institutes, and other stakeholders undertake location-specific seed rate trials and validations to keep or modify the existing seed rate application;
- This study has several significant implications for future research efforts that further contribute to the yield effects of plot-level seed rate and seed recycling practices. Thus, research efforts in the following thematic areas will enrich the existing discussions and knowledge as far as the yield effects of seed rate and seed recycling practices are concerned;
- The use of time-series data will provide clear evidence about the yield effects of plot-level seed rates and seed recycling practices of CBWS on a temporal basis;
- The yield effects of wheat seed rates and seed recycling practices for different wheat varieties in different socio-economic settings, weather conditions, and spatial arrangements should be investigated.

**Author Contributions:** Y.E. generated the idea and study design, collected data, carried out data analysis and writeup. B.A. and Z.B. provided constructive suggestions, statistical assistance, read, edit and revised and shape the manuscript. All authors have read and agreed to the published version of the manuscript.

**Funding:** This research received no external funding.

**Data Availability Statement:** The data used to support the findings of this study are available from the corresponding author upon request.

**Acknowledgments:** The authors of the study want to acknowledge farmers, who devoted their precious time during the interview process, the FGDs, and the KIIs. We also want to extend our gratitude to extension workers and district experts involved in data collection and supervision businesses. We also kindly appreciate the Ministry of Science and Higher Education of the Federal Democratic Republic of Ethiopia and Debre Markos University for their budget support.

**Conflicts of Interest:** The authors declare no conflict of interest.

## Appendix A

**Table A1.** Effect of seed dose on wheat productivity.

| Source | SS | df | MS | | | |
|---|---|---|---|---|---|---|
| Model | 4321.22144 | 1 | 4321.22144 | | | |
| Residual | 10,722.0586 | 448 | 23.9331664 | | | |
| Total | | 449 | 33.5039644 | | | |
| **Productivity** | **Coef.** | **Std. Err** | **t** | $p > \|t\|$ | **[95% Conf. Interval]** | |
| Seed_dose | 0.0536544 | 0.003993 | 13.44 | 0.000 | 0.045807 | 0.0615018 |
| _Cons | 31.95563 | 0.5474157 | 58.38 | 0.000 | 30.87981 | 33.03145 |

Scheme 2020. Number of obs = 450. Root MSE = 4.8922. F (1, 448) = 180.55. R-squared = 0.2873. Prob > F = 0.0000. Adjusted R-squared = 0.2857.

**Table A2.** Cost–benefit analysis of recycled and non-recycled seeds (n = 450).

| Seed Recycling Frequency | Obs. | Mean Seed Rate (kg ha$^{-1}$) (A) | Unit Seed Cost (Birr Qt$^{-1}$) (B) | Total Seed Cost (Birr Ha$^{-1}$) (C) | Yield (Qt ha$^{-1}$) (D) | Unit Selling Price (Birr Qt$^{-1}$) (E) | Mean Income (Birr Ha$^{-1}$) (F) | Net Income (Birr Ha$^{-1}$) (G) |
|---|---|---|---|---|---|---|---|---|
| 1. Unrecycled | 328 (72.89) | 169.2835 | 2300 | 3893.52 | 40.4421 | 1950 | 78,862.095 | 74,968.575 |
| 2. Recycled once | 110 (24.44) | 263.6364 | 1800 | 4745.46 | 33.7909 | 1800 | 60,823.62 | 56,078.16 |
| 3. Recycled twice | 12 (2.67) | 245.8333 | 1800 | 4424.999 | 33.3333 | 1800 | 59,999.94 | 55,574.991 |

Source: own calculations (2020).

**Table A3.** Frequency distribution of respondents by seed source (n = 450).

| Seed Sources | Frequency | Percent | Cumulative |
|---|---|---|---|
| Farmers saved seed (informal sources) | 42 | 9.33 | 9.33 |
| Amhara Seed Enterprise | 328 | 72.89 | 82.22 |
| Research and academic institutions | 11 | 2.44 | 84.67 |
| Cooperatives | 69 | 15.33 | 100.00 |
| Total | 450 | 100 | |

Source: own calculations (2020).

**Table A4.** Status and reasons for seed recycling (n = 450).

| Status and Reasons for Seed Recycling | Frequency | Percent | Cumulative |
|---|---|---|---|
| I do not recycle seed | 328 | 72.89 | 72.89 |
| Cost reduction | 61 | 13.56 | 86.44 |
| Expensive fresh seed | 27 | 6.00 | 92.44 |
| Timely availability and supply shortage | 19 | 4.22 | 96.67 |
| No yield difference | 15 | 3.33 | 100 |
| Total | 450 | 100 | |

Source: own calculations (2020).

## References


1. Mango, N.; Makate, C.; Tamene, L.; Mponela, P.; Ndengu, G. Impact of the adoption of conservation practices on cereal consumption in a maize-based farming system in the Chinyanja Triangle, Southern Africa. *Sustain. Future* **2020**, 2. [CrossRef]
2. Louwaars, N.P.L.; De Boef, W.S. Integrated seed sector development in Africa: A conceptual framework for creating coherence between practices, programs, and policies. *J. Crop Improv.* **2012**, *26*, 39–59. [CrossRef]
3. Zecevic, V.; Boskovic, J.; Knezevic, D.; Micanovic, D. Effect of seeding rate on grain quality of winter wheat. *Chil. J. Agric.* **2014**, *74*, 23–28. [CrossRef]
4. Krishna, V.V.; Spielman, D.J.; Veettil, P.C.; Ghimire, S. An Empirical Examination of the Dynamics of Varietal Turnover in Indian Wheat. 2014. IFPRI Discussion Paper 01336. Available online: https://ssrn.com/abstract=2417342 (accessed on 30 January 2020).
5. Lollato, R.P.; Ruiz Diaz, D.A.; De Wolf, E.; Knapp, M.; Peterson, D.E. Agronomic practices for reducing wheat yield gaps: A quantitative appraisal of progressive producers. *Crop Sci.* **2019**, *59*, 333. [CrossRef]
6. Vandercasteelen, J.; Dereje, M.; Minten, B.; Taffesse, A.S. Scaling-up adoption of improved technologies: The impact of the promotion of row planting on farmers' tef yields in Ethiopia. *LICOS-Discuss. Pap. Ser.* **2013**, *344*, 1–25. Available online: http://www.econ.kuleuven.be/licos (accessed on 1 July 2019).
7. Ritz, C.; Baty, F.; Streibig, J.C.; Gerhard, D. Dose-Response Analysis Using R. *PLoS ONE* **2015**, *10*, e0146021. [CrossRef]
8. Van der Vliet, L.; Ritz, C. Statistics for Analyzing Ecotoxicity Test Data. In *Encyclopedia of Aquatic Ecotoxicology*; Férard, J., Blaise, C., Eds.; Springer: New York, NY, USA, 2013; pp. 1081–1096.
9. Mehring, G.H. *Determining Optimum Seeding Rates for Diverse Hard Red Spring Wheat (Triticum aestivum L.) Cultivars*; North Dakota State University: Fargo, ND, USA, 2016.
10. MoA (Ministry of Agriculture of the Federal Democratic Republic of Ethiopia). *Technological Package for Wheat Production: A Training Manual for Development Practitioners*; MoA: Addis Ababa, Ethiopia, 2015.
11. Yamane, T. *Statistics: An Introductory Analysis*, 2nd ed.; Harper and Row: New York, NY, USA, 2015.
12. Teshome, A.; Abate, E. Wheat Technologies from Where to Where? The Case of East Gojam Zone of Amhara Region, Ethiopia. *J. Agric. Econ. Dev.* **2013**, *2*, 226–236. Available online: http://academeresearchjournals.org/journal/jaedISSN2327-3151 (accessed on 29 March 2021).
13. Bishaw, Z.; Struik, P.C.; Van Gastel, A.J.G. Assessment of on-farm diversity of wheat varieties and land-races: Evidence from farmer's field in Ethiopia. *Afr. J. Agric. Res.* **2014**, *9*, 2948–2963.
14. Furtas, R. Overview of Certified Seed and Farmer Saved Seed. In *An Updated Version of External Release by the Economic and Competitiveness Branch of the Alberta Government*; Economics and Competitiveness Branch, Economics Section, Alberta Government: Stettler, AB, Canada, 2016.
15. Bonny, S. Taking stock of the genetically modified seed sector worldwide: Market, stakeholders, and prices. *Food Secur.* **2014**, *6*, 525–540. [CrossRef]
16. Tiruneh, S.; Yigesu, A.Y.; Bishaw, Z. Measuring the effectiveness of extension innovations for out scaling agricultural technologies. *Afr. J. Agric. Sci. Technol.* **2015**, *3*, 316–326.
17. Bishaw, Z.; Atilaw, A. Enhancing Agricultural Sector Development in Ethiopia: The Role of Research and Seed Sector. *Ethiop. J. Agric. Sci.* **2016**, *27*, 101–130.

18. Gebre, G.G.; Isoda, H.; Rahut, D.B.; Amekawa, Y.; Nomura, H. Gender differences in the adoption of agricultural technology: The case of improved maize varieties in southern Ethiopia. *Women's Stud. Int. Forum* **2019**, *76*, 102264. [CrossRef]
19. Bezabih, T. Study on Intensity and Adoption of Improved Wheat Varieties and Associated Agronomic Practices in Kaffa Zone, the Case of Gesha Woreda. In *A Thesis Submitted in Partial Fulfillment of the Requirements of MA Degree in Rural Development*; Indira Gandhi National Open University (IGNOU): Delhi, India, 2012.
20. Abboye, A.D.; Teto, A.M. The response of Seed Rates and Row Spacing on Growth, Yield and Yield Components of Wheat (*Triticum aestivum* L.) Crop. *J. Nat. Sci. Res.* **2020**, *10*. [CrossRef]
21. Abate, G.T.; Bernard, T.; de Brauw, A.; Minot, N. The Impact of the Use of New Technologies on Farmers Wheat Yield in Ethiopia: Evidence from a Randomized Control Trial. *Agric. Econ.* **2018**, *49*, 409–421. [CrossRef]
22. Anteneh, A.; Asrat, D. Wheat Production and Marketing in Ethiopia: A Review Study. *Cogent Food Agric.* **2020**, *6*, 1778893. [CrossRef]
23. Okusaa, C. Seed Replacement Rates and Its Impact on Annual Yield Growth of the Kisumu District. In *A Thesis Submitted to Jomo Kenyatta University of Agriculture and Technology*; Kenyatta University of Agriculture and Technology: Nairobi, Kenya, 2017.
24. Tigabu, R.; Asfaw, F. Effects of Seed Rate and Row Spacing on Yield and Yield Components of Bread Wheat (*Triticum aestivum* L.) in Dalbo Awtaru Woreda, Wolaita Zone, Southern Ethiopia. *J. Biol. Agric. Healthc.* **2016**, *6*, 58–67.
25. Mideksa, T.; Letta, T.; Bayisa, T.; Abinasa, M.; Tilahun, A.; Hundie, B.; Alemu, W.; Abera, M. Bread Wheat Varietal Development and Release in Southeastern Highlands of Ethiopia. *Am. J. Biol. Environ. Stat.* **2018**, *4*, 15–19. [CrossRef]
26. Tesfaye, S.; Bedada, B.; Mesay, Y. Impact of improved wheat technology adoption on productivity and income in Ethiopia. *Afr. Crop Sci. J.* **2016**, *24*, 127. [CrossRef]
27. Bishaw, Z.; Struik, P.C.; van Gastel, A.J.G. Wheat and barley seed system in Syria: Farmer's varietal perceptions, seed sources, and seed management. *Int. J. Plant Prod.* **2011**, *5*, 323–347.
28. Clayton, G.W.; Brandt, S.; Johnson, E.N.; O'Donovan, J.T.; Harker, K.N.; Blackshaw, R.E.; Smith, E.G.; Kutcher, H.R.; Vera, C.; Hartman, M. Comparison of Certified and Farm-Saved Seed on Yield and Quality Characteristics of Canola. *Agron. J.* **2009**, *101*, 1581–1588. [CrossRef]
29. ATA (Agricultural Transformation Agency). *A Training Manual on Cereal Production*; ATA (Agricultural Transformation Agency): Addis Ababa, Ethiopia, 2015.
30. Tsegaye, D. Profitability of Contractual Bread Wheat Seed Production in Mecha District of Amhara Region, Ethiopia. *J. Central Eur. Agric.* **2012**, *13*, 142–149. [CrossRef]

sustainability

MDPI

Article

# Hazardous Solid Waste Confined in Closed Dump of Morelia: An Urgent Environmental Liability to Attend in Developing Countries

**M. Lourdes González-Arqueros [1], Gabriela Domínguez-Vázquez [2], Ruth Alfaro-Cuevas-Villanueva [3], Isabel Israde-Alcántara [3] and Otoniel Buenrostro-Delgado [3,*]**

1 CONACYT, Instituto de Investigaciones en Ciencias de la Tierra, Universidad Michoacana de San Nicolás de Hidalgo, Morelia 58000, Mexico; lourdes.gonar@gmail.com

2 Facultad de Biología, Universidad Michoacana de San Nicolás de Hidalgo, Morelia 58000, Mexico; gdoguez@yahoo.com.mx

3 Instituto de Investigaciones en Ciencias de la Tierra, Universidad Michoacana de San Nicolás de Hidalgo, Morelia 58000, Mexico; ruth.alfaro@umich.mx (R.A.-C.-V.); isabel.israde@umich.mx (I.I.-A.)

* Correspondence: otoniel.buenrostro@umich.mx; Tel.: +52-443-325-6301

**Abstract:** In developing countries, landfills of urban solid waste (USW) are a major source of contamination. One reason is the common practice of the illegal confinement of hazardous waste (HW). The contamination is mainly due to deficitary design location, operation and lack of liner, which enables the dispersion of pollutants. The aim of our work is to demonstrate the presence of heavy metals (HM) and arsenic (As) in USW of the closed dump of Morelia, which clandestinely confined HW for 20 years. Solid samples of USW were collected from eight opencast wells with different age of confinement. Composition, degradation status, physical-chemical characterization and analysis of HM and As were carried out. The results showed the presence of Pb, Cu, Ni, Zn, Cr, Fe and high concentrations of As. This study provides evidence about the usual and illegal practice of landfill HW together with USW; the hazard due to the presence of HM and As; the deficiency in the operation and closure; and, the lack of competent legislation on the subject. This information is essential to establish background information for improving laws and help decision makers in territorial planning to improve public and environment health.

**Keywords:** metals; arsenic; pollution; Mexico; developing countries; landfill; urban solid waste; disposal; waste management

**Citation:** González-Arqueros, M.L.; Domínguez-Vázquez, G.; Alfaro-Cuevas-Villanueva, R.; Israde-Alcántara, I.; Buenrostro-Delgado, O. Hazardous Solid Waste Confined in Closed Dump of Morelia: An Urgent Environmental Liability to Attend in Developing Countries. *Sustainability* **2021**, *13*, 2557. https://doi.org/10.3390/su13052557

Academic Editor: Shervin Hashemi

Received: 26 January 2021
Accepted: 21 February 2021
Published: 26 February 2021



## 1. Introduction

Environmental pollution is a consequence of the irrational use of natural resources. In developing countries, the sites for disposal of urban solid waste (USW) are a major source of contamination [1] where, through leachates, pollutants can be dispersed to soil and groundwater [2,3]. This contamination results from the deficiency in the design and operation of the landfills in which the USW are deposited, as well as the lack of monitoring of closed dumps and scarcity of environmental legislation [4].

Typically, closed dumps are considered environmental liabilities, as well as geographic sites polluted by the release of materials. Some of the most common landfills pollutants are heavy metals (HM). Very often, the effect of the pollutants is increased when the landfill is closed; since it is a common practice of the use of closed dumps as farmlands in urban and sub-urban centers in developing countries [5,6].

Accelerated urbanization in developing countries, which has affected the urban metabolism of human settlements, has generated a problem that is reflected in urban and environmental functionality [7]. In general, the so-called landfills of developing countries and which more closely resemble the category expressed in [4] as open dumps or controlled dumps, are located in areas of high marginalization, characterized by low per capita income [2,8] and high population density, in addition to not complying with environmental legislation operating as dump sites.

The northwest of Morelia city, where the closed landfill is located, is no longer a predominantly rural area and has become a densely populated human settlement with low economic income, mixed with agricultural and forested areas. The conurbation of this area with the closed dump and the landfill results in greater exposure and vulnerability to highly polluting and dangerous particles for the population health. In parallel, this site is located in a fracture zone, with highly porous soils and in a groundwater recharge area.

Previous work at the study site [9] reported high concentrations of cadmium, nickel, arsenic, lead, hexavalent chromium, and total chromium in leachates. The central problem is that there should not be heavy metals in USW. However, the confinement of hazardous waste (HW) in these landfills has become a common practice in many emergent nations. Therefore, the concentration of heavy metals in USW and the toxicity of them depend on the amount of hazardous waste it contains [10].

The aim of our work is to demonstrate the presence of heavy metals in the USW, clandestinely confined together with hazardous waste due to non-compliance with current environmental legislation. In this way, our study provides current data on the increasing environmental and public health risk represented by closed dumps or controlled dumps in developing countries.

## 2. Materials and Methods

### *2.1. Physical Description of the Study Area*

The study area (~128 km$^2$) (Figure 1) is located in Morelia, Michoacán, Mexico, in the Trans-Mexican Volcanic Belt, within the Michoacán-Guanajuato volcanic complex. The area lies in a recharge site of an overexploited aquifer that provides water to more than 120,000 people [11].

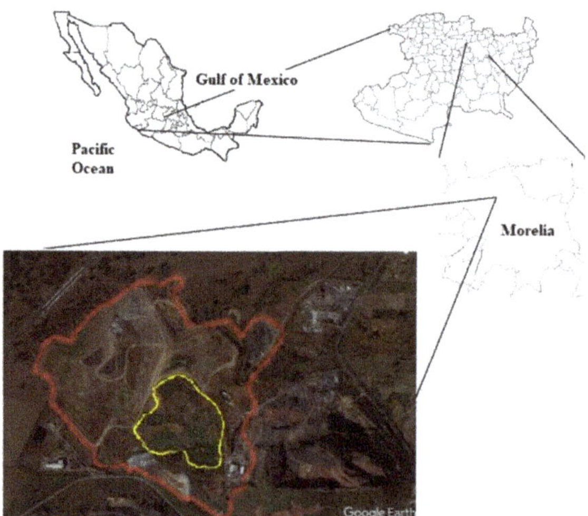


**Figure 1.** Location of the study area (Modified from Google Earth 2017®; INEGI). The red dotted line indicates the total perimeter dedicated to the landfill. The yellow dotted line indicates the closed dump area where the samples were collected.

The area presents two important hydrogeological conditions since it is located on permeable geological materials and is affected by a regional fault system. Two semi-shield volcanic bodies dominate the landscape in the northeast and in the southwest. The volcanism and the faults determine the relief, with dominant E–W structures, associated with the Morelia-Acambay fault system [12], in which two faults are notorious because of their size, "El Cerrito" and "Cointzio".

The closed dump of Morelia was built without liner or structure engineered and designed to control the percolation and infiltration of leachates and biogas generated in the site [9]. The site was in operation as an open-air dump since 1997 until 2007, in which an average of 900 tons of USW per day was deposited. The closure of the closed dump was carried out from the leveling and compaction of the solid waste, followed by a covering with a layer of approximately 10 cm of clay and tezontle (volcanic rock).

There is evidence that along with USW, hazardous waste was deposited illegally. Through visual identification of piles, open pits, and the type of waste at the time of truck discharge [13], 18 different sources of HW were recognized (Table 1); the varied list shows from inert (sharp objects) to organisms waste (animals and humans). The most frequent generators were medical offices 25%, automotive maintenance shops 21%, construction 14%, industries and garages 12%, and the remaining groups 28%. Surprisingly, waste from paper manufacturing and solids from wastewater treatment continued to be deposited even when the site was closed.

**Table 1.** Sources and type of waste identified in the illegal disposal of hazardous waste in the Morelia closed landfill. Source: [13].

| Source | Waste |
|---|---|
| Agricultural equipment stores | Plaguicide (pesticides, herbicides)<br>Fertilizers |
| Hospitals/clinics/doctor's offices/Clinical and radiological analysis laboratories | Sharp objects<br>Human waste<br>Healing material wastes<br>Syringes<br>Developing material/worn plates |
| Research laboratories | Reagent containers<br>Organisms remains |
| Veterinary clinics | Dead animals<br>Animals waste<br>Healing materials |
| Photographic developing workshops | Photographic and film wastes<br>Containers of developing products |
| Computer, photocopier and printer maintenance workshops | Ink cartridges<br>Accessories<br>Toner cartridges<br>Photocopier oil waste |
| Beauty salons<br>Slaughterhouses, butchers, chicken shops, guts dispensing | Beauty products<br>Dye residues<br>Animal waste |
| Housing | Household cleaners<br>Medicines and drugs<br>Cosmetics<br>Pests, garden waste and batteries |
| Sand mines/coal sale | Sand waste<br>Coal waste |
| Electrical workshops | Incandescent lamps<br>Light bulbs<br>Batteries |
| Paint stores | Paint containers<br>Paintbrushes<br>Containers of varnishes, thinner, turpentine, gasoline<br>Rag with solvents traces |
| Gas stations | Oil residues<br>Cleaning material residues |
| Hardware stores | Solvent containers |

**Table 1.** *Cont.*

| Source | Waste |
|---|---|
| Car parts stores | Oil containers<br>Used car parts<br>Antifreeze containers |
| Auto body shops | Paint containers<br>Car parts<br>Polisher containers |
| Garages | Oil containers<br>Tires<br>Used car parts<br>Inner tubes |
| Paper industry | Solids from wastewater treatment<br>Waste from paper manufacturing |
| Construction companies | Cement waste<br>Lime residues<br>Tile glue residues |

### 2.2. Sampling

The area of the closed dump was divided into four quadrants oriented from southwest to northeast (Figure 1). Eight sites were randomly selected, ensuring that half of the samples will have a confinement time of 5 years and the remaining half will have a confinement time of 10 years. Opencast wells were dug during the dry season to a depth of three meters with a backhoe loader with extension (Case 2002®). Subsequently, approximately three kilograms of solid samples of USW were taken. USW samples were placed in black polythene bags, labeled and placed in a cooler for their transfer to the laboratory. Inside the opencast well, the in-situ temperature was measured with a digital thermometer.

### 2.3. Sample Characterization

The solid samples of USW were characterized according to the Mexican Official Norm NMX-AA-022-1985 [14]. By-products were manually separated and grouped into two fractions: organic and inorganic. Then, the organic fraction was grouped into categories of degradability according to the classification proposed in [15].

Physicochemical analyses were performed according to the Mexican Official Norm NMX-AA-052-1985 [16]. The components of the samples were crushed with scissors and ground with an analytical mill (MF 10®) (with a one-millimeter sieve), deposited in plastic jars and frozen at −4 °C. Subsequently physical-chemical parameters as moisture (NOM NMX-AA-016-1984) [17], pH (NOM NMX-AA-25-1984) [18], total dissolved solids (TDS) NMX-AA-016-1984 [19], and volatile solids (VS) (2540G technique from Standard Methods) [20] were determined.

The concentration of heavy metals was determined from one gram of aliquot of each sample by acid digestion of sediments with a flame atomic absorption spectrophotometer (FLAA), according to the EPA method 3050B [21]. Arsenic was determined with the hydride generation method according to the Mexican Official Norm NMX-AA-051-SCFI-2016 [22]. Analyses were performed in duplicate.

In order to analyze the presence of significant differences among the metals concentrations and in the content of the organic fraction according to the confinement time of the USW, the results were captured in a database and processed with descriptive statistics and analysis of variance (ANOVA) using JMP 8 software.

## 3. Results

The organic fraction percentages per sample ranged from 48% to 67%. An average of 54% of the sample was the organic fraction and 46% the inorganic fraction. The categorization of the USW organic fraction samples (Table 2) showed that 82% of the by-products were of very rapid degradation since they mainly derived from food residues. Of note, 13%

of the samples showed moderately slow and slow degradation; this type of degradation is associated with the fraction of residues that, although organic, have a higher content of cellulose and lignin compared to those that come from food residues.

**Table 2.** Degradability of the organic fraction of the solid residues (% fresh weight).

| Degradability of the Organic Fraction | % |
|---|---|
| Very rapid | 82.0 |
| Moderately rapid | 4.0 |
| Moderately slow | 10.6 |
| Slow | 2.3 |

The physical-chemical characterization (Table 3) showed statistically significant differences between temperatures for both confinement periods. The predominantly basic pH values, together with the low moisture content, directly influenced the rate of degradation of the organic matter. Total solids (TS) values varied between 58% and 78%, with an average of ~68%. These values are considered high regardless of the USW confinement time. On the other hand, the VS values showed a wide variation, between 17% and 79%, and the averages of the samples according to confinement time did show significant differences. Likewise, the values of the ash, the values of VS, also showed a high variability and significant differences between averages. It reaffirms the previous results and corroborates a high variation in the degradation state of the organic fraction of solid waste within the study site.

**Table 3.** Average values of the physicochemical parameters of the urban solid waste (USW) with 5 and 10 years of confinement.

| Parameter | Average ± Standard Error | Years of Confinement | Significance |
|---|---|---|---|
| Temperature (°C) | 26.5 ± 0.44 | 5 | * |
| | 35.0 ± 0.86 | 10 | * |
| pH | 8.35 ± 0.03 | 5 | NS |
| | 8.14 ± 0.14 | 10 | NS |
| Moisture (%) | 31.7 ± 1.66 | 5 | NS |
| | 31.8 ± 1.24 | 10 | NS |
| Total solids (TS) (%) | 68.2 ± 1.66 | 5 | NS |
| | 68.1 ± 1.24 | 10 | NS |
| Volatile solids (VS) (%) | 73.1 ± 2.86 | 5 | * |
| | 52.1 ± 3.65 | 10 | * |
| Ash (%) | 26.8 ± 2.86 | 5 | * |
| | 49.4 ± 3.41 | 10 | * |

* = Significant. NS = Non-significant.

The heavy metals in the solid residues of the closed dump were lead (Pb), copper (Cu), nickel (Ni), zinc (Zn), chromium (Cr), iron (Fe), and the arsenic metalloid (As) (Table 4). The Pb values for samples 3, 6 and 7 were substantially higher compared to the rest (89, 149.67 and 108.67 mg/kg, respectively). The Cu values for sample 3 also showed an exceptionally higher value (744.17 mg/kg) compared to the rest of the samples. This sample also presented high values for Ni (217.33 mg/kg) and for As (60.56 mg/kg). The Zn values were homogeneous, except for sample 8, which showed a value considerably lower than the rest. On the contrary, sample 8 presented the highest value of Fe ($3.11 \times 10^4$ mg/kg). The Cr values for samples 1 and 5 were much higher than for the rest (383.00 and 127.50 mg/kg, respectively). Notwithstanding the disparity in the resulting values, no significant differences ($p = 0.8427$) were found regarding the heavy metal content among wells, despite the different confinement ages of the USW.

**Table 4.** Heavy metals and arsenic present in the USW from the closed dump (mg/kg).

| Well | Pb | Cu | Ni | Zn | Cr | Fe | As |
|---|---|---|---|---|---|---|---|
| 1 | 47.17 | 4.17 | 42.00 | 209.33 | 383.00 | $2.47 \times 10^4$ | 8.83 |
| 2 | 42.67 | 7.67 | 54.75 | 365.17 | 0.00 | $2.33 \times 10^4$ | 20.00 |
| 3 | 89.00 | 744.17 | 79.00 | 217.33 | 13.67 | $1.83 \times 10^4$ | 60.56 |
| 4 | 55.33 | 0.00 | 45.00 | 116.17 | 0.00 | $1.50 \times 10^4$ | 18.55 |
| 5 | 56.00 | 92.17 | 53.50 | 173.17 | 127.50 | $1.85 \times 10^4$ | 65.04 |
| 6 | 149.67 | 8.83 | 53.00 | 408.17 | 3.67 | $2.58 \times 10^4$ | 22.76 |
| 7 | 108.67 | 141.00 | 48.00 | 165.00 | 0.00 | $2.84 \times 10^4$ | 95.85 |
| 8 | 47.83 | 21.50 | 57.00 | 54.33 | 0.00 | $3.11 \times 10^4$ | 20.75 |
| Average | 74.54 | 127.44 | 54.03 | 213.58 | 57.19 | $2.31 \times 10^4$ | 39.04 |

## 4. Discussion

The degradation rate of the wastes and the physical-chemical characterization allowed the evaluation of handling of the dump during its operation stage, as well as the degradation behavior of the confined USW, and the effectiveness of the site closure measures.

The results showed a low moisture content of ~30%, still above that reported by [23] (15%) or by [24] (10%) as minimum values to favor USW optimal degradation. The degradation capacity is directly related to the moisture content of the samples. Therefore, these values of moisture are sufficient for 82% of the samples to show very fast decomposition because they are mainly composed of food wastes. However, higher moisture in the samples allows a better degradation of the residues since the optimal conditions for the establishment of the microbial consortia that degrade the wastes exist.

The failure to achieve optimal conditions for the establishment of organic matter degrading microorganisms delays the stabilization of the USW confined in the closed dump. In addition, the high holocellulose/lignin ratio of lignocellulosic compounds found in the 13% of the samples is also a delaying factor in stabilization of organic matter [25]. Still, we are aware that a larger number of samples should be analyzed to relate these results to the influence of the USW stabilization and confinement time.

Likewise, basic pH values indicate that acetogenic anaerobic bacteria would not be in their optimal environment since, according to [26], they develop optimally at a pH close to neutrality and are sensitive to pH variations. Despite this, the bacteria of these USW samples are active since they are totally inhibited at a pH below 6.0, which would be reflected in an accumulation of organic acids.

The deficiencies in waste stabilization are also directly related to the closure of the dump, whose work basically consisted of stabilizing the slopes and covering the USW with soil. Therefore, these deficiencies are probably not allowing the achievement of optimal conditions for the establishment of organic matter degrading microorganisms; which delays the stabilization of the USW confined in the closed dump.

Besides, moisture would also be affected by the closure measures. Previous studies in the study area [9] reported differences in the results obtained during the rainy and the dry season. Consequently, these measures also do not control the entry and dispersion of rainwater. In Morelia, precipitation increases 10 times during the rainy season, implying a considerable increase of the moisture values, as therefore in the amount of produced leachate. An increase in the amount of leachate generated would involve a greater dispersion of the pollutants they contain to surrounding soils and groundwater [27].

The results showed that five samples exceeded the maximum permissible level (MPL) for arsenic according to the national standard [28] about contaminated soils (Table 5). Although, these results should be reviewed with caution, because there is no other comparison parameter in Mexico for solid samples besides the MPL for sludge and biosolids, which were not exceeded. It highlights the urgency of legislation to regulate heavy metals and arsenic in USW, in order to control the disposal of HW in landfills that are not designed and prepared for this purpose.

**Table 5.** Maximum permissible limits for metals and arsenic established in the Official Mexican Norms.

| Chemical Constituent | Sludge and Biosolids (mg/kg) | Contaminated Soils ** (mg/kg) |
| --- | --- | --- |
| Pb | 300 | 400 |
| Cu | 1500 | NI * |
| Ni | 420 | 1600 |
| Zn | 2800 | NI * |
| Cr | 1200 | NI * |
| Fe | NI * | NI * |
| As | 41 | 22 |

** Total Reference Pollutant Concentrations ($PR_T$) by type of land use. NI *: Not included in the Mexican law. Modified from: NOM-004-SEMARNAT-2002 [29] anNOM-147-SEMARNAT/SSA1-2004 [28].

Besides the arsenic, we found that a worse scenario stands for copper, zinc and iron. For these pollutants, there are no reference values with which to establish whether the reported concentrations represent a danger to human and environmental health.

However, the heterogeneity in the values of the heavy metals and arsenic is due to the variability of the type of waste and the difference in the time of waste confinement; for which the Kruskal–Wallis test indicated a significant difference ($p = 0.01$), confirming the differences in the degradation state of the wastes of the different opencast wells.

Despite the fact that our findings are based on solid samples, this is in good agreement with the results of [9] in the same study area. Their geochemical analysis of groundwater from Morelia's municipal aquifer showed high concentration of heavy metals, exceeding the standards for drinking water of the World Health Organization. Subsequently, they analyzed samples of leachates during the rainy and dry seasons. Their results in leachates showed concentrations of Pb 1102 mg/L, Cu 2403 mg/L, Ni 10,678 mg/L, Cr 47,731 mg/L, and As 0.302 mg/L.

The results suggest that the possible percolation of leachates from the dump appears to be the most reasonable source of metal pollutants in the groundwater. The authors also explained that one of the factors in the production of leachate, and therefore dispersion of pollutants, was the joint deposition of organic and inorganic waste. In this regard, our study suggests that another major factor is the disposal of HW together with USW.

Many researchers, such as [30], have found high levels of heavy metals in the leachate as Fe, Pb, and Cd and a relatively small proportion of Zn, Cr, and Cu. Hence, our research stands out in that water pollution is a crucial problem that closed USW sites face due to dispersion of pollutants from leachates. The leachates would favor the solubility of the toxic components of USW due to its role as a catalyst for the degradation processes of hydrolysis, and dissolution of toxic components of organic and inorganic matter [31].

The aforementioned highlights the need for the implementation of monitoring to ensure the control of leachates produced at these sites during their operation and after closure. The situation is even more worrying in developing countries where research efforts towards monitoring the environment have not received the desired attention by stakeholders [1,2]. Furthermore, the retention of the pollutants in solution in the USW matrix is not assured due to the poor biodegradation of the organic matter in the waste, the lack of a liner, the absence of a leachates collection system and the poor coverage of the site [32].

## 5. Conclusions

The findings of this study indicate the presence of heavy metals and arsenic in the USW. These pollutants confirm that the solid wastes confined at the study site are a potential source of contamination.

Our research stressed that the HW confined clandestinely together with USW during 20 years reveal the illegality of the operation of the closed dump in Morelia, as well as the deficiencies in management during its operation and its later closure.

This work demonstrates the non-compliance with the Environmental Legislation regarding the USW disposal, the safe and adequate confinement. This study also underlines the importance of including the implementation of operational practices to control and/or avoid escape of those pollutants, likely from leachates which could reach the streams and/or water channels, affecting the quality of the water for human consumption.

Although our work includes only the analysis of one site, we provided further evidence that makes it urgent to review the methodologies for the USW disposal and the operation of landfills in developing countries. Further work needs to increase the number of solid samples and analyses of leachates and groundwater samples.

Despite the great environmental and health impact due to the poor management of landfills, in many developing countries, they still continue to be the main option for the treatment of USW. Studies of this type are therefore crucial for decision-makers and territorial planning.

**Author Contributions:** Conceptualization, O.B.-D. and M.L.G.-A.; methodology, O.B.-D.; validation, G.D.-V., R.A.-C.-V. and I.I.-A.; formal analysis, O.B.-D.; investigation, O.B.-D.; resources, O.B.-D.; data curation, G.D.-V., R.A.-C.-V. and I.I.-A.; writing—original draft preparation, O.B.-D. and M.L.G.-A.; writing—review and editing, O.B.-D. and M.L.G.-A.; visualization, G.D.-V., R.A.-C.-V. and I.I.-A.; supervision, O.B.-D. and M.L.G.-A.; project administration, O.B.-D.; funding acquisition, O.B.-D. All authors have read and agreed to the published version of the manuscript.

**Funding:** This work was funded by the National Council for Science and Technology of Mexico through Project Grant No. 62100 and the Co-ordination of Scientific Research of the Universidad Michoacana de San Nicolás de Hidalgo through Project Grant No. 5.9.

**Data Availability Statement:** The data presented in this study are available on request from the corresponding author.

**Acknowledgments:** The authors would like to thank Y.P. Meza-Cisneros for the laboratory analyses.

**Conflicts of Interest:** The authors declare no conflict of interest.

## References

1. Ferronato, N.; Torretta, V. Waste mismanagement in developing countries: A review of global issues. *Int. J. Environ. Res. Public Health* **2019**, *16*, 1060. [CrossRef] [PubMed]
2. Olayiwola, H.A.; Abudulawal, L.; Adewuyi, G.K.; Azeez, M.O. Heavy Metal Contents in Soil and Plants at Dumpsites: A Case Study of Awotan and Ajakanga Dumpsite Ibadan, Oyo State, Nigeria. *J. Environ. Earth Sci.* **2017**, *7*, 11–24.
3. Syeda, M.A.; Pervaiz, A.; Afzal, B.; Hamid, N.; Yasmin, A. Open dumping of municipal solid waste and its hazardous impacts on soil and vegetation diversity at waste dumping sites of Islamabad city. *J. King Saud Univ. Sci.* **2014**, *26*, 59–65. [CrossRef]
4. Mavropoulos, A. *Wasted Health: The Tragic Case of Dumpsites*; ISWA: Vienna, Austria, 2015.
5. Ogunyemi, S.; Awodoyin, R.O. Heavy metal contamination of some leafy vegetables growing within Ibadan metropolis, Southwestern Nigeria. *Trop. Agric. Res. Ext.* **2003**, *6*, 71–76.
6. Sola, O.; Awodoyin, R.O.; Opadeji, T. Urban agricultural production: Heavy metal contamination of amaranthus cruentus L. grown on domestic refuse landfill soils in Ibadan, Nigeria. *Emir. J. Food Agric.* **2003**, *15*, 87–94. [CrossRef]
7. Díaz-Alvárez, C.J. Metabolismo urbano: Herramienta para la sustentabilidad de las ciudades. *INTERdisciplina* **2014**, *2*, 51–70. [CrossRef]
8. Gutberlet, J.; Uddin, S.M.N. Household waste and health risks affecting waste pickers and the environment in low- and middle-income countries. *Int. J. Occup. Environ. Health* **2017**, *23*, 299–310. [CrossRef] [PubMed]
9. Israde-Alcántara, I.; Buenrostro Delgado, O.; Chavez, A.C. Geological characterization and environmental implications of the placement of the Morelia dump, Michoacán, central Mexico. *J. Air Waste Manag. Assoc.* **2005**, *55*, 755–764. [CrossRef] [PubMed]
10. Mackenzie, L.; Cornwell Davis, D.A. *Introduction to Environmental Engineering*; McGraw Hill: Boston, MA, USA, 2008.
11. Garduño-Monroy, V.H.; Ávila-Olivera, J.A.; Hernández-Madrigal, V.M.; Sámano, N.A.; Díaz, J.E. Estudio hidrogeológico del sistema acuífero de Morelia, Michoacán, para una correcta planificación del territorio. In *Urbanización, Sociedad y Medio Ambiente: Experiencias en Ciudades Medias*; Vieyra, A., Larrazábal, A., Eds.; UNAM: Mexico City, Mexico, 2014; p. 293.

12. Garduño-Monroy, V.H.; Medina-Vega, V.H.; Israde-Alcántara, I.; Hernández-Madrigal, V.M.; Ávila-Olivera, J.A. Unidades Geohidrológicas de la Región de Morelia-Cuitzeo. In *Atlas de la cuenca del lago de Cuitzeo: Análisis de su geografía y entorno socio ambiental. In Atlas de la Cuenca del Lago Cuitzeo: Análisis de su Geografía y Entorno Socioambiental*; UNAM: Mexico City, Mexico, 2010.
13. Carlón, A.T. *Generación y Disposición Final de los Residuos Sólidos Peligrosos en Morelia, Michoacán*; Universidad Michoacana de San Nicolás de Hidalgo: Morelia, México, 2004.
14. SECOFI NMX-AA-022-1985. Relación de Normas Oficiales Mexicanas Aprobadas por el Comité de Protección Al Ambiente. Contaminación Del Suelo. In *Residuos Sólidos Municipales. Selección y Cuantificación de Subproductos*; Secretaría de Comercio y Fomento Industrial: Mexico City, Mexico, 1985.
15. SCS Engineers Modelo Mexicano de Biogas—Versión 2; Guadalajara, Jalisco, 2009. Available online: http://www.epa.gov/lmop/documents/pdfs/manual_del_usuario_modelo_mexicano_de_biogas_v2_2009.pdf (accessed on 1 August 2020).
16. SECOFI NMX-AA-052-1985. Relación de Normas Oficiales Mexicanas Aprobadas por el Comité de Protección Al Ambiente. Contaminación Del Suelo. In *Residuos Sólidos Municipales Preparación de Muestras en el Laboratorio Para su Análisis*; Secretaría de Comercio y Fomento Industrial: Mexico City, Mexico, 1985.
17. SECOFI NMX-AA-016-1984. Relación de Normas Oficiales Mexicanas Aprobadas por el Comité de Protección al Ambiente. Contaminación Del Suelo. In *Determinación de Humedad*; Secretaría de Comercio y Fomento Industrial: Mexico City, Mexico, 1984.
18. SECOFI NMX-AA-25-1984. Relación de Normas Oficiales Mexicanas Aprobadas por el Comité de Protección al Ambiente. Contaminación Del Suelo. Residuos Sólidos. In *Determinación del pH. Método Potenciométrico*; Secretaría de Comercio y Fomento Industrial: Mexico City, Mexico, 1984.
19. SECOFI NMX-AA-016-1984. *Protección al Ambiente-Contaminación del Suelo-Residuos Sólidos Municipales-Determinación de Humedad*; Secretaría de Comercio y Fomento Industrial: Mexico City, Mexico, 1984.
20. *APHA Standard Methods for the Examination of Water and Wastewater*, 23nd ed.; Baird, R.B.; Eaton, A.D.; Rice, E.W. (Eds.) American Public Health Association: New York, NY, USA; Washington, DC, USA, 2005; ISBN 551979.
21. *EPA Method 3050: Acid Digestion of Sediments, Sludges, and Soils*; Environmental Protection Agency: Washington, DC, USA, 2002.
22. SE NMX-AA-051-SECFI-2016. Análisis de agua. In *Determinación de Metales por Absorción Atómica en Aguas Naturales, Potables, Residuales y Residuales Tratadas-Método de Prueba*; Secretaria de Economía: Mexico City, Mexico, 2016.
23. EPA EPA-600/SR-97/071. *Biodegradative Analysis of Municipal Solid Waste in Landfills*; Environmental Protection Agency: Washington, DC, USA, 2003.
24. Hartz, K.; Ham, R. Moisture level and movement effects on methane production rates in landfill samples. *Waste Manag. Res.* **1983**, *1*, 139–145. [CrossRef]
25. Hernández-Berriel, M.D.C.; Hernández-Paniagua, I.Y.; Clemitshaw, K.C.; Nila-Cuevas, J.A.; Buenrostro-Delgado, O. Evaluation of confinement conditions and content of lignocellulosic compounds on urban solid waste biodegradation rates. *Rev. Int. Contam. Ambient.* **2019**, *35*, 91–100. [CrossRef]
26. Robles Martínez, F. *Generación de Biogás y Lixiviados en los Rellenos Sanitarios*, 2nd ed.; Dirección de Publicaciones del IPN: Mexico City, Mexico, 2008; ISBN 970-36-0214-2.
27. Machado, S.L.; Carvalho, M.F.; Gourc, J.P.; Vilar, O.M.; do Nascimento, J.C.F. Methane generation in tropical landfills: Simplified methods and field results. *Waste Manag.* **2009**, *29*, 153–161. [CrossRef] [PubMed]
28. DOF Norma Oficial Mexicana NOM-147-SEMARNAT/SSA1-2004. *Criterios para Determinar las Concentraciones de Remediación de Suelos Contaminados por Arsénico, Bario, Berilio, Cadmio, Cromo Hexavalente, Mercurio, Níquel, Plata, Plomo, Selenio, Talio y/o Vanadio*; Diario Oficial de la Federación: Mexico City, Mexico, 2007.
29. DOF Norma Oficial Mexicana NOM-004-SEMARNAT-2002. Protección ambiental. Lodos y biosólidos. In *Especificaciones y Límites Máximos Permisibles de Contaminantes Para su Aprovechamiento y Disposición Final*; Diario Oficial de la Federación: Mexico City, Mexico, 2003.
30. Boateng, T.K.; Opoku, F.; Akoto, O. Heavy metal contamination assessment of groundwater quality: A case study of Oti landfill site, Kumasi. *Appl. Water Sci.* **2019**, *9*, 33. [CrossRef]
31. Mai, S.; Barampouti, E.M.; Koumalas, A.; Dounavis, A. Leachates from landfill sites in Thessaloniki, Greece: Effect of aging. *Environ. Res. Eng. Manag.* **2019**, *75*, 30–39. [CrossRef]
32. Liao, P.; Yuan, S.; Wang, D. Impact of Redox Reactions on Colloid Transport in Saturated Porous Media: An Example of Ferrihydrite Colloids Transport in the Presence of Sulfide. *Environ. Sci. Technol.* **2016**, *50*, 10968–10977. [CrossRef] [PubMed]

*sustainability*

MDPI

*Article*

# Modeling of 3R (Reduce, Reuse and Recycle) for Sustainable Construction Waste Reduction: A Partial Least Squares Structural Equation Modeling (PLS-SEM)

**Musa Mohammed [1,2,*], Nasir Shafiq [1,*], Ali Elmansoury [3,4], Al-Baraa Abdulrahman Al-Mekhlafi [5], Ehab Farouk Rached [4], Noor Amila Zawawi [1], Abdulrahman Haruna [2], Aminu Darda'u Rafindadi [6] and Muhammad Bello Ibrahim [7]**

1 Department of Civil & Environmental Engineering, University Technology PETRONAS, Seri Iskandar 32610, Malaysia; amilawa@utp.edu.my
2 Department of Building Technology, Abubakar Tafawa Balewa University (ATBU), Bauchi P.M.B 0248, Bauchi State, Nigeria; abkabo360@gmail.com
3 Architectural Department, College of Engineering, Al-Azhar University, Cairo 11311, Egypt; aamansoury74@gmail.com
4 Islamic Architecture Department, College of Islamic Architecture and Engineering, Um Al-Qura University, Makkah 24211, Saudi Arabia; efrached@uqu.edu.sa
5 Department of Management & Humanities, University Technology PETRONAS, Seri Iskandar 32610, Malaysia; albaraa901@gmail.com
6 Department of Civil Engineering, Bayero University, Kano P.M.B 3011, Kano State, Nigeria; aminuneti2013@gmail.com
7 Department of Civil Engineering, Hussaini Adamu Federal Polytechnic, Kazaure P.M.B 5004, Jigawa State, Nigeria; mbhafedpoly@gmail.com
* Correspondence: musaimohammed@gmail.com (M.M.); nasirshafiq@utp.edu.my (N.S.)

**Citation:** Mohammed, M.; Shafiq, N.; Elmansoury, A.; Al-Mekhlafi, A.-B.A.; Rached, E.F.; Zawawi, N.A.; Haruna, A.; Rafindadi, A.D.; Ibrahim, M.B. Modeling of 3R (Reduce, Reuse and Recycle) for Sustainable Construction Waste Reduction: A Partial Least Squares Structural Equation Modeling (PLS-SEM). *Sustainability* **2021**, *13*, 10660. https://doi.org/10.3390/su131910660

Academic Editor: Shervin Hashemi

Received: 15 August 2021
Accepted: 15 September 2021
Published: 25 September 2021



**Abstract:** There is a lack of awareness and knowledge among the Malaysian construction industry about waste management reduction. Numerous nations worldwide have understood and have incorporated the concept of the 3R (reduce, reuse, and recycle) in waste management, and it has worked out well. This study investigated construction waste issues and developed a model for sustainable reduction by applying 3R using a partial least squares structural equation modeling PLS-SEM in Malaysia. The research methodology adopted the quantitative and qualitative approaches by sending a survey questionnaire to the relevant stakeholders to obtain their views or perceptions and interviewing an expert in the related field about waste reduction in the Malaysian construction industry. Three hundred thirty questionnaires were collected within six months of submission. The significant factors are determined using mean ranking for the reduce, reuse and recycle elements. Based on the results, the exploratory power of the study model is considered sustainable with $R^2$ values of 0.83%. At the same time, the results of relationships between improving factors, policy-related factors, construction waste generated, and sustainable construction waste reduction were significant. Also, the findings revealed that the top factors for waste generation on reducing, reusing, and recycling are lack of design and documentation, and lack of guidance for effective construction waste-collecting. The paper will explore different and dynamic practices, such as recycling, reuse of construction waste management cost reduction, enabling stakeholders and managers to estimate and quantify the actual size of CWM costs and benefits for sustainable development goals.

**Keywords:** sustainable construction; waste management; construction waste reduction; modelling of waste (reduce, reuse and recycle); PLS-SEM

## 1. Introduction

The development of the economy in every nation depends on construction projects undertaken through several types of contract. The economy is mainly based on resources, which results in a significant amount of Waste that may violate the sustainable development goals set out in Paris in 2015 [1]. Over 23,000 tonnes of waste are created each day in Malaysia [2]. The problems that construction and demolition waste encounter can largely depend on the project lifecycle. Meadows and Randers [3] reported that construction sites has inappropriate raw material use, poor waste management practices, and a lack of awareness of the importance of waste minimization in the local area. It also contributes to excessive waste production and becomes an increasing problem for clients due to increased costs of this waste. Still, the issue of waste management continues to persist. More than 6 million tonnes of waste were generated from 2018, of which a quarter was produced in the Klang Valley alone [4]. It is estimated that 33% of construction waste directly resulted from the designer's inability to use waste minimization measures during the design phase [5]. However, waste management costs are often relatively higher than the benefits that the organization would gain more attention being given to it [6]. There is a lack of awareness and knowledge among the Malaysian construction industry about waste management issues [7]. Waste reduction by 3R is one of the most efficient construction industry steps towards sustainable waste management [7]. When reuse and recycling practices are applied to the large amount of construction waste generated at the sites, such practices lead to achieving sustainable development goals and, often, new product development [8].

Construction materials and building operations account for roughly half of overall carbon dioxide ($CO_2$) emissions [9]. According to a recent study, the construction industry exploited 30–40% of natural resources in developing countries [9]. As a result, successfully managing waste by incorporating environmental, social, and economic issues into long-term waste management strategies has become a significant challenge worldwide [10]. Furthermore, The rapid rate of urbanization is responsible not only for the increased use of non-renewable resources but also for the development of significant amounts of C&D waste, as well as the associated environmental concern [11]. More than 6 million tonnes of waste were created in Malaysia, with a quarter of that made in the country's most prosperous region, the Klang Valley [4]. It is estimated that about 33% of construction waste results from designers' failure to use waste management measures (WMM) throughout the design process [12]. Only 10% of construction waste is recycled. As a result, in Malaysia, disposing of construction waste remains a major challenge.

The 3R is considered a promising approach that would extend the life of landfills while reducing natural resource exploration. Recycling building wastes can help to recycle natural resources and minimize the cost of waste treatment before disposal, in addition to the environmental benefits of reducing the demand on land for waste disposal [13]. As a result, the general public, companies, developers, contractors, architects, and engineers must be enlightened on waste reduction and the value of recycling construction waste. To guide the construction industry in the right direction in a sustainable way, all relevant parties should ensure that waste is reduced by completing waste material reduction, reuse, and recycling. In other aspects, sustainability consists of a particular investment of advantages and costs of implementation planning, plus a cost-effective redistribution of capital available over time throughout waste management planning [14]. Esa [15] stated that implementing a construction and demolition waste management program begins with the planning and designing stage. The main purpose is to keep costs down in new construction projects, especially when they suffer from ineffective coordination, design errors, lack of activity planning, and bad procurement practices [16]. In due course, optimal planning of construction waste management policy demands deliberation over several issues, ranging from economic developments on emissions, health, and promotion of waste recycling to planning matters such as waste facilities and waste distribution networks to services. The previous study indicates inadequate methods and applications for minimizing construction waste during the construction and procurement phase [17,18]. Nevertheless, it is noticeable that there is a lack of focus on managing waste at the design phase based on an assessment carried out by [18]. The effectiveness of enhancement opportunities in construction can be tackled by embracing waste identification or reduction methods in the flow operations in parallel with value-adding techniques by developing new management tools and appropriate training programs [19].

The generation of waste from construction projects has had a significant impact on the environment, particularly when it comes to illegal dumping. The rising frequency of illegal dumping activities from Malaysian construction projects indicates that the country's construction waste management has to be addressed. Nonetheless, there is no clear definition of a comprehensive criterion for construction waste management, particularly in developing nations. As a result, more research on Malaysia's construction waste management is required [20]. The construction industry is highly unfriendly to the environment [21]. In terms of natural resource exploration, irreversible environmental transformation, and the accumulation of pollutants in the atmosphere, this industry contributes significantly to the environmental crisis [22]. During construction, waste is generated due to site clearing, material damage, material use, material non-use, excess procurement, and human errors. Moreover, statistics show that building and demolition projects account for 10–30% of overall trash [11]. Consequently, the primary objective was to develop a model for achieving sustainable construction waste reduction using PLS-SEM. By doing so, it could minimize the amount of waste generated during the construction processes.

## 2. Background of the Study

### 2.1. Overview of Construction and Demolition Waste Management in Malaysia and throughout the World

Based on the research goal, a categorization of C&DW management literature should be constructed to answer two key questions: the elements that affect C&DW management and the offered model for effective C&DW management to achieve sustainability. This study responds to C&DW definitions and generation questions, followed by a review of the research subjects of earlier studies in C&DW management, which paints a clear picture of previous C&DW management research techniques. Based on this study's research gap, the research questions and objectives are stated in depth at this phase. During the last few decades, industry and scholars worldwide have been increasingly aware of C&DW problems. Despite various efforts over the previous few years, it is estimated that the construction industry is still in its infancy and requires maturation to contribute to reducing environmental burdens [23] effectively. Many countries throughout the world are grappling with how to efficiently and effectively mitigate C&DW generation. Since the 1980s, significant study has been devoted to C&DW minimization to reduce the negative impacts of C&DW on building structures [23].

Researchers must have a thorough understanding of previous C&DW management studies and the C&DW management frameworks produced by other researchers. For example, Kabirifar [24] created a framework to improve C&DW management in Portugal, considering the project life cycle and stakeholders. Lu and Yuan [25] presented another framework in which research boundaries for C&DW management were described; this framework covers the amount, origins, and impacts of C&DW management, as well as C&DW reduction, reuse, and recycle strategies, C&DW tools, humans' role, performance, and regulatory environment.

Furthermore, Kabirifar [26] created a framework for C&DW environmental management practices in Europe. The author conducted a study to evaluate the effectiveness of C&DW management at three levels: national, regional, and project. At the national level, the effectiveness of techniques in the Malaysian construction industry, for example, was investigated [27]. Villoria Sáez [28] offered a framework for evaluating C&DW efficacy in another study. C&DW management has been studied from a hierarchical perspective, including C&DW reduction strategy; for example, the waste reduction strategy was deemed the most effective waste minimization strategy [25]. However, since some C&DW will be generated in the future, C&DW reuse and recycling should be implemented as feasible waste management methods to reduce waste to landfills. The term "C&DW reuse" refers to the practice of reusing the same building material for several purposes, even in various functions. If the generated waste cannot be reused, the materials should be transformed into new materials through recycling [24]. In addition, several scholars have investigated other elements that affect C&DW management. Identifying building activities that can accommodate reused construction materials [29], having waste targets for all linked trades [30], and having recycling targets for each construction project are just a few examples [31]. Specific places for cutting and storage of material [32], reuse material scraps [33], waste sorting, reusing, and recycling [34], and making sub-contractors responsible for waste disposal [35]. Although the variables listed above for C&DW management have addressed key essential themes related to C&DW management from both a hierarchical and influencing perspective, previous research has uncovered and categorized these elements in a structured manner. Then, to manage C&DW successfully, practical C&DW contributing elements should be found and classified, and these aspects should be linked with contributing factors to the C&DW management hierarchy.

### 2.2. Waste Reduction Hierarchy on 3R (Reduce, Reuse and Recycle)

The 3Rs principle of reducing, reusing, and recycling in a hierarchical order, classifying WM strategies as desirable. The 3Rs are intended to be a hierarchy, arranged from low to high in ascending order of their adverse environmental impacts [36]. This section discusses reducing, recycling, and reusing based on their hierarchical order and can be considered.

The summary of 3R factors used from previous research in construction waste management is shown in Table 1. We sought an idea from a professional with at least 20 years of working experience in construction management and who has participated in many construction projects in which he experienced construction first-hand information.

**Table 1.** Causes of 3R (reduce, reuse and recycle) application in building construction projects.

| S/N0 | Reduce Factors | References | Reused Factors | References | Reused Factors | References |
|---|---|---|---|---|---|---|
| 1 | Lack of design and documentation | [37,38] | Time pressure | [39,40] | Difficulties for delivery vehicles accessing construction sites | [41,42] |
| 2 | Transportation problem | [43,44] | Lack of supervision | [45] | Inefficient method of unloading | [46,47] |
| 3 | Incorrect Procurement | [48,49] | Lack of on-site waste management plans | [50,51] | Insufficient protections during unloading | [50,52] |
| 4 | Lack of design standards for reducing construction waste | [53,54] | Use of incorrect material requiring replacement | [54,55] | Waste resulting from cutting uneconomical shapes | [56,57] |
| 5 | Low cost for construction waste disposal | [58,59] | Poor craftsmanship | [60,61] | Damage to materials on-site | [62] |
| 6 | Poor Site operation | [63] | Improper planning for required quantities | [63] | Poor method of storage on-site | [64] |
| 7 | On-site management and planning | [62] | Delays in passing information on types and sizes of materials | [64] | Unnecessary inventories on site leading to waste | [65] |
| 8 | Poor Material storage and handling | [66] | Waste resulting from cutting uneconomical shapes | [67] | Materials supplied in loose form | [68] |
| 9 | Lack of Information about the 3R | [69] | Damage to materials on-site | [62] | Under-developed market for recycled construction waste products | [70] |
| 10 | Inappropriate urban planning | [54] | Poor method of storage on-site | [71] | Immature recycling market operation | [72] |

## 3. Methodology

A conceptual model is the first step in developing a research strategy. In addition, the conceptual model can be a supportive method by identifying a relevant literature review used to form intermediate theories (hypotheses) that can be checked through the empirical evidence [73]. The conceptual modeling method is divided into three stages: (1) identifying the model's constructs, (2) identifying these constructs, and (3) describing the relationships between these constructs [74–76].

This study carried on an in-depth evaluation of reports based on Scopus, Science Direct, Sage, Wiley Library, Emerald-based journal articles, and conference proceedings published on the subject matter to identify the model's constructs [77,78]. Previous research has only used measurement tools that can only measure a 3R error [79,80]. However, in this study, there is a greater awareness of the need to improve construction waste management 3R, and Malaysia's current CWM treatment condition is dire. Therefore, a survey was used to discover a new set of sustainable construction waste reduction factors, specifically in Malaysia [81]. The first part of the questionnaire gives a brief introduction to the survey—part two of the survey used a questionnaire to gather demographic information. The third part of the assessment included a semi-structured interview with an expert to obtain his opinion about the research's identified factors. The questionnaire utilized in this study was developed through focus groups with residential construction professionals, consultants, and other construction experts. Sampling is a process or technique for drawing a characteristic group of individuals or cases from a specific population [82]. For this study, a stratified random sampling technique and simple random sampling technique have been adopted. The sample size for this study has been determined based on Krejcie [83] table. The table indicates that for the population size of 950, the sample size should be 274 as the minimum. Following that, a validation process was carried out on four residential projects in Kuala Lumpur and the Perak Table 2 was used to establish the 3R of waste generated on-site, which was over 60% for high-rise construction and less than 25% for housing project sites. Table 2 depicts a high-level summary of the project under investigation.

**Table 2.** Overview of the project investigated.

| Project Name | Location | Project Types | Build-Up ($M^2$) |
|---|---|---|---|
| Green residential | Kuala Lumpur | High-rise | 73,271 |
| D'eco lake housing | Perak | Bungalow/terrace | 45,213 |
| Ridgewood | Perak | Bungalow/terrace | 24,992 |

### 3.1. Pilot Study

To ensure the validity of the data, we conducted a pilot study and evaluated the reliability of the data before distributing the questionnaire survey. Respondents' clarity and relevance were reviewed by an industry-experienced professor and some academics lectures of Malaysia's public and private university's construction industry. Following the recommendation of [84,85], a minimum Cronbach's alpha achieved acceptable values of 0.882 and 0.815, respectively. The value of the variables under the construction waste generation factors was 0.889. The Cronbach Alpha values obtained indicate an excellent internal consistency for the scale. A study sample should be between 150 and 300 for the study's EFA research [86,87].

Nonetheless [88], as previously reported, the researchers disagree about the size of a factor analysis sample, but the assumption is that a larger sample should be used. When it comes to the number of variables [89–91], statistical theory holds that working with 20–50 variables is more effective than individual factors because various factors fail to return the correct results variables exceeds that range. According to a study, reducing the number of variables is possible if the sample size is large enough [92]. For this study, the sample size was 330, which was considered a valid representation of the population as a whole within appropriate ranges [92–94].

### 3.2. Main Survey Design

The questionnaire survey was developed to collect data about Malaysia's sustainable construction waste reduction. First, a conceptual model is used to lay the groundwork for a research strategy. Then, a conceptual framework is used to investigate the topic literature to produce intermediate theories (hypotheses) that can be subjected to scientific investigation [95,96]. A questionnaire was created to gather cross-sections of all existing literature to undergo a comprehensive review. For the questionnaire, three main categories were chosen: contractors, consultants, and clients. Architects, electrical engineers, quantity surveyors, structural and mechanical engineers fall under these categories. The majority (32.0%) and 26.2% of survey respondents, respectively, have between 10 and 15 years and 15–20 years of industry experience. Just 7.5% of respondents have less than five years of experience [97].

However, this segment gives the results of the characteristics of respondents of this study. These findings show that the respondents possessed the required expertise and information to comment on the 3Rs (reduce, reuse, and recycle) for sustainable construction waste reduction, which led to them obtaining a significant amount of trust in their input and the validity of the findings. In addition, the results indicate that site engineers accounted for the majority (43.5%) of the participants' employment, followed by construction engineers (24.3%). For Bachelors (BSc), Masters (MS), and PhD degrees, the respondents' academic qualifications were 81%, 53%, and 35% [97], respectively. As shown in Table 3.

**Table 3.** Survey responds.

| S/N | Professional Background | Years of Experience | Qualification | Percent |
|---|---|---|---|---|
| 1 | Site engineers | 10 and 15 years | Bachelors (BSc) 18% | 43.5% |
| 2 | Construction engineers | 15–20 years | Masters (MS) 53% | 24.35 |
| 3 | Project managers | Less than 5 years | PhD degrees 35% | 7.5% |

*3.3. Assessing Goodness-of-Fit (GoF)*

The goodness of fit provides a global criterion for assessing the overall quality of the (PLS-SEM) model. However, unlike the covariance-based structural equation modeling (SEM), there is no generally accepted goodness of fit measure to evaluate PLS models. However, ref. [98] proposed a global criterion of the goodness of fit called the "GoF" index.

It can be determined using the formula below:

$$\text{GoF} = \sqrt{\overline{R^2} \times \overline{\text{AVE}}} \quad (1)$$

Choshin and Ghaffari [99] succumbed to a law of thumb for examining GoF. It is considered small, medium, and large if the values are 0.1, 0.25, and 0.36. Accordingly, the GoF of the structural research model is calculated as shown.

$$\text{GoF} = \sqrt{0.5226 \times 0.283}$$

$$\text{GoF} = \sqrt{0.148}$$

$$\text{GoF} = 0.39$$

Based on Behl [100] the model GoF of 0.39 is considered significant. Therefore, the research model fitted very well. The structural model is, consequently, good.

## 4. Results and Discussion

The survey data were analyzed using a combination of SPSS, Microsoft Excel, and partial structural equation modeling (PLS-SEM) software to determine the causal relationship between the exogenous (independent) constructs and the endogenous (dependent) construct. This study included measurement of individual item reliability, and these models have Cronbach's alpha, convergence, discriminant validity, path coefficients, coefficient of determination ($R^2$), effect size (f2), system predictive relevance ($Q^2$), and goodness-of-fit (GoF), PLS-SEM moderation results, and correlation. Out of 481 questionnaires, 330 were returned, resulting in an overall response rate of 69% of the sample size within 6 months of submission. This is deemed to be satisfactory for questionnaires compared to other related surveys in engineering and construction management. The findings of the Cronbach alpha for the main constructs are presented as follows. The field data indicated that (construction waste generation factors), (improving factors) achieved acceptable values of 0.882 and 0.815, respectively; the variables under the (sustainable construction waste generation) have a value of 0.889. The Cronbach's alpha values indicate an excellent internal consistency for the scale with the sample values above 0.7 considered acceptable [101]. In this study, hypotheses would contribute immensely to the established knowledge of the study's subject matter for kinds of material and what amount are waste, and how much is produced and classified as waste are explained in the following section.

### 4.1. *The Amount Types of Construction Waste Generated*

Many scholars lament the difficulties in obtaining reliable data on the amounts of construction waste generated, owing to the difficulty in determining the waste's precise quantity and composition. However, studies disclose some statistics and facts, such as that many contractors have waste percentage guidelines that they use when estimating bills of quantities. These proportions mainly depend on their labor force experience [102]. However, the contractor will likely engage with several subcontractors on various projects, and the subcontractors' labor will be motivated to complete the work as quickly as possible. As a result, waste levels will most likely differ from those anticipated [103]. Saunders and Wynn in the United Kingdom found that a waste level of roughly 10% is acceptable [104]. However, according to Harper, the typical waste allowance is 5% of the net measurements [105]. As a result, after a careful examination of different estimation approaches, estimates for construction waste for a Malaysian residential project have been completed.

#### 4.1.1. Built-Up Projects Assessment

This study is the first step in estimating the waste generated by residential construction projects in Malaysia. According to the overall findings, the built-up area of the entire project (179 projects) under evaluation was around 7,895,147.461m$^2$ throughout all states and across the three project categories based on the findings: high-rise, terrace, and bungalow. The high-rise project accounts for 41% of the total project, while the terrace and bungalow projects account for 51% and 8%, respectively. Furthermore, it reveals that most of these projects were located in Selangor (25%) and Johor Bahru (15%), respectively, and were located across the 12 states. Figure 1 depicts the summary of all the project locations and build-up areas under investigation.

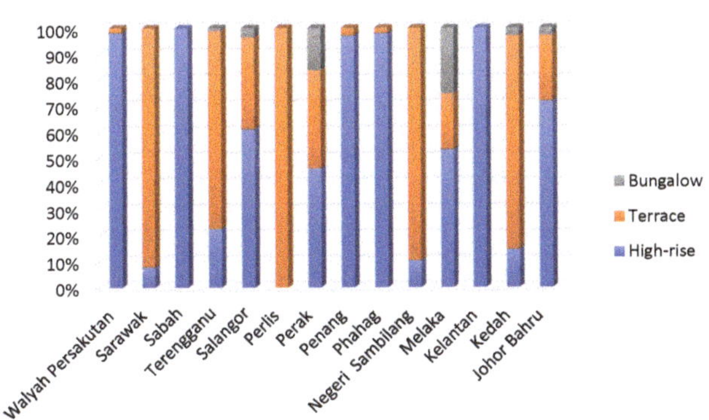


**Figure 1.** Build-up area (m$^2$).

#### 4.1.2. Estimated Waste for a Residential Project

The waste indicator approach was introduced to estimate the construction waste generated, defined as the quantity of construction waste generated in a unit of volume or weight per m$^2$ of gross floor area or area of activity [106]. Moreover, the estimated waste generated rate from Malaysian residential projects was determined by calculating the total gross floor area of the project's WGR of Singapore (30.2 kg/m$^2$ [64,65] and Thailand (21.38 kg/m$^2$) for residential construction.

$$\text{WGR} = \frac{30.2\ \text{kg/m}^2 + 21.38\ \text{kg/m}^2}{2} = 25.79\ \text{kg/m}^2 \tag{2}$$

Moreover, the approximate waste generated rate from Malaysia residential project was determined by calculating the total GFA of the project (7,895,147.461 m$^2$) as adopted by numerous researchers [107–109]. This can be calculated as Equation (3).

$$\text{EWG} = \text{GFA} \times \text{WGR} \tag{3}$$

EWG is the total expected waste generated by the project (m$^2$), GFA denotes the gross floor area per m$^2$ (based on CIDB data), and WGR indicates the waste rate generated per m$^2$.

$$\text{EWG} = 25.79\ \text{kg/m}^2 \times 7{,}895{,}147.461\ \text{m}^2 = 203{,}616\ \text{tons}$$

Consequently, based on the findings throughout the research period, from March to May 2018, the total estimated wastes accumulated by Malaysian residential buildings were 203,615 tonnes, as illustrated in Figure 2.

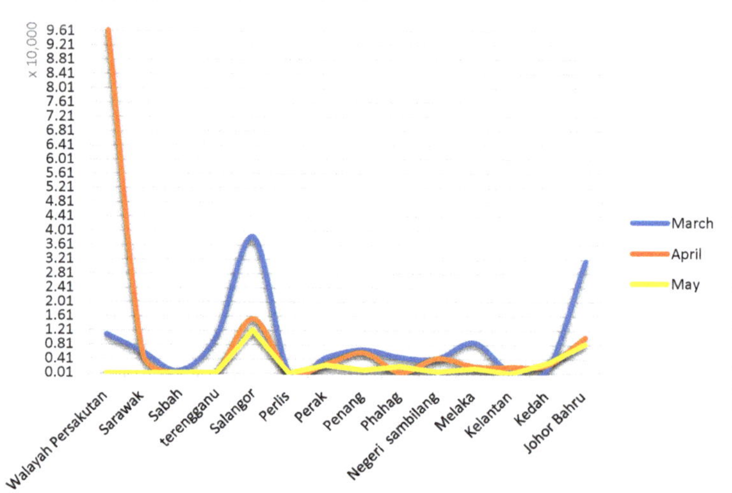


**Figure 2.** Waste generated in tonnes.

### 4.1.3. Mean Ranking for Waste Minimization Strategies

Following this finding, the respondent ranked 10 waste minimization strategies, with the reuse, reduction, and recycling of waste material on-site receiving the highest mean of 0.908, consistent with numerous writers [110]. Suppose improved material handling management is adopted on construction sites. In that case, the amount of waste that needs to be disposed of should be considerably decreased [111]. As a result, the respondent agrees that proper storage and material handling at a building site with a mean score of 0.874 will significantly reduce waste. A mean of 0.857 was followed by a list of recovered, reused, reduced, and recycled waste products. Poon, et al. [53], discovered that many materials are lost on construction sites due to poor material control. Due to improper storage and handling, this material wastage is not isolated on building sites since many resources can be spared. The respondent ranked the following factors as the least important: issuing waste segregation guidelines (mean = 0.829); and minimizing design changes (mean = 0.752), which is not surprising given that this factor was identified by Faniran and Cabanas as one of the major sources of waste generation [112]. Mean = 0.749 for analyzing site waste to be generated, 0.728 for minimizing waste at the source of origin, and 0.723 for organizing waste management meetings and training construction personnel; and designating waste disposal ranks last in the waste minimization strategies, with a mean score of 0.689, as shown in Table 4.

**Table 4.** Ranking for waste minimization strategies.

| Waste Minimization Strategies | Mean | Rank |
|---|---|---|
| Reusing or recycling some of the waste materials on site | 0.908 | 1 |
| Proper storage and handling of materials on-site | 0.874 | 2 |
| Prepare a list of each waste material to be salvaged, reduce, reused, or recycled | 0.857 | 3 |
| Issuing guidelines for waste segregation | 0.829 | 4 |
| Minimizing design changes | 0.752 | 5 |
| Analyzing site waste to be generated | 0.749 | 6 |
| Minimizing Waste at the source of origin | 0.728 | 7 |
| Organizing waste management meetings | 0.723 | 8 |
| Training of construction personnel | 0.723 | 9 |
| Designating waste disposal operators | 0.689 | 10 |

According to the findings of this study, approximately 4–6% of the concrete used will be lost to waste. Dry concrete was separated from other waste in some projects and repurposed to provide aggregate for road sub-bases. Wooden boards are used for falsework and formwork for concrete constructions and the erection of site boundaries and bamboo scaffolding. Some wooden boards were reused on the work site multiple times until the boards' quality degenerated, so they could no longer be used. Wood boards should be reused at least five times, resulting in a waste rate of about 20% [113].

### 4.2. Structure Equation Modeling PLS-SEM

The two-stage evaluation criteria in PLS-SEM modeling include assessing the structural model after evaluation of the measurement model. The measuring model must meet all the quality evaluation requirements before running the structural model [114]. Accordingly, the structural model is run after confirming the reliability and validity of the measurement models. The structural model established the causal relationships between the measurement models in the structural model [70]. The interrelationships identified are meant to address the research questions and test the research hypotheses. Results showed that all outer loads were related to the information phase of 3R (reduce, reuse, and recycle) for sustainable construction waste reduction from the initial measurement model because the loading factor was less than 0.5. The majority of these loads did not influence the related construct. All the models were examined using the model report with cr = 0.70 and, as a result, were proven to be accurate and valid [115,116].

#### 4.2.1. Measurement Model

The factor loadings, cross-loadings, Alpha, Rho_A C.R value in structural equation model (PLS-SEM), and the average variance extracted (AVE) were used to discover any problem with an item and evaluate a framework for achieving sustainable construction waste reduction the application of 3R. All items loaded effectively on their related constructs in the analysis were above 0.5 as the recommended threshold measures [117]. As displayed in Tables 4–6, all the items packed on factor loadings, cross-loadings, Alpha, Rho_A C.R, and AVE are in a lower range of 0.5278 upper range 0.9244. According to Bhandari and Naudeer [118], the AVE must be at least 0.5 for convergence constructs. However, both AVE interpretations are 0.5, with is the lowest acceptable value. This gave a relatively satisfactory result with adequate item loadings, composite reliability, higher-order constructs (HOC), as well as lower-order constructs (LOCs) such as Alpha Rho A, C.R, and AVE. As far as the individual items go, there are sufficient grounds to prove the indicators. The entire modified final model AVEs are above the recommended minimum of 0.5. The table shows that the minimum factor loading is above 0.6 and is significant ($t$-stat > 1.96; $p$-value < 0.005). The factors for improving waste management for sustainable development with applying 3R practical are 0.579, 0.761, 0.508, 0.736, 0.627, 0.564, 0.522, 0.717, 0.503, 0.591, 0.320, 0.551 and 0.563, respectively. Table 5 illustrates the final model for significant variables for sustainable construction waste reduction. The study used PLS-SEM, including various new methodological innovations to the current body. Previous literature mainly centered on the conceptualization and assessment of CWM in general.

**Table 5.** Measurement model.

| Higher-Order Constructs (HOC) | Lower Order Constructs (LOCs) | Items | Factor Loading | Alpha | Rho_A | C.R | AVE |
|---|---|---|---|---|---|---|---|
| Policy Related Factors (P) | | C | 0.733 | | | | 0.537 |
| | Waste Management Policy | PWM | 0.808 | 0.887 | 0.892 | 0.911 | 0.562 |
| | | WM1 | 0.772 | | | | |
| | | WM12 | 0.570 | | | | |
| | | WM13 | 0.777 | | | | |
| | | WM3 | 0.778 | | | | |
| | | WM4 | 0.760 | | | | |
| | | WM5 | 0.766 | | | | |
| | | WM7 | 0.769 | | | | |
| | | WM9 | 0.780 | | | | |
| | Industry Policy | PPF | 0.832 | 0.867 | 0.885 | 0.898 | 0.504 |
| | | PF1 | 0.855 | | | | |
| | | PF2 | 0.851 | | | | |
| | | PF3 | 0.837 | | | | |
| | | PF4 | 0.849 | | | | |
| | | PF5 | 0.530 | | | | |
| | | PF6 | 0.616 | | | | |
| Sustainable Construction Waste Management (S) | | S | 0.766 | | | | 0.587 |
| | Environmental Factor | SEE | 0.811 | 0.778 | 0.778 | 0.833 | 0.500 |
| | | EE1 | 0.719 | | | | |
| | | EE2 | 0.684 | | | | |
| | | EE3 | 0.736 | | | | |
| | | EE4 | 0.653 | | | | |
| | | EE9 | 0.741 | | | | |
| | Economic Factor | SEC | 0.647 | 0.888 | 0.933 | 0.915 | 0.551 |
| | | EC1 | 0.833 | | | | |
| | | EC10 | 0.063 | | | | |
| | | EC2 | 0.893 | | | | |
| | | EC3 | 0.652 | | | | |
| | | EC4 | 0.821 | | | | |
| | | EC5 | 0.833 | | | | |
| | | EC6 | 0.063 | | | | |
| | | EC7 | 0.893 | | | | |
| | Social Factors | SSC | 0.840 | 0.874 | 0.931 | 0.909 | 0.563 |
| | | SC1 | 0.885 | | | | |
| | | SC2 | 0.893 | | | | |
| | | SC3 | 0.833 | | | | |
| | | SC4 | 0.063 | | | | |
| | | SC5 | 0.893 | | | | |
| | | SC6 | 0.652 | | | | |

**Table 6.** Discriminant validity using Fornell.

| | C | CD | CESM | CM | CW | I | IDM | ILC | IME | ISCF | P | PPF | PWM | S | SEC | SEE | SSC |
|---|---|---|---|---|---|---|---|---|---|---|---|---|---|---|---|---|---|
| **C** | 0.616 | | | | | | | | | | | | | | | | |
| **CD** | 0.965 | 0.761 | | | | | | | | | | | | | | | |
| **CESM** | 0.187 | 0.145 | 0.734 | | | | | | | | | | | | | | |
| **CM** | 0.932 | 0.754 | 0.121 | 0.729 | | | | | | | | | | | | | |
| **CW** | 0.847 | 0.745 | 0.077 | 0.716 | 0.733 | | | | | | | | | | | | |
| **I** | 0.136 | 0.115 | −0.003 | 0.162 | 0.106 | 0.542 | | | | | | | | | | | |
| **IDM** | 0.054 | 0.025 | −0.013 | 0.065 | 0.092 | 0.722 | 0.707 | | | | | | | | | | |
| **ILC** | 0.174 | 0.153 | 0.052 | 0.207 | 0.114 | 0.630 | 0.206 | 0.792 | | | | | | | | | |
| **IME** | 0.077 | 0.075 | −0.054 | 0.105 | 0.031 | 0.847 | 0.630 | 0.228 | 0.722 | | | | | | | | |
| **ISCF** | 0.094 | 0.069 | 0.009 | 0.102 | 0.101 | 0.803 | 0.467 | 0.477 | 0.572 | 0.757 | | | | | | | |
| **P** | 0.152 | 0.142 | 0.019 | 0.140 | 0.143 | 0.409 | 0.325 | 0.179 | 0.415 | 0.242 | 0.595 | | | | | | |
| **PPF** | 0.207 | 0.191 | 0.093 | 0.179 | 0.206 | 0.285 | 0.214 | 0.097 | 0.309 | 0.182 | 0.832 | 0.710 | | | | | |
| **PWM** | 0.035 | 0.037 | −0.065 | 0.046 | 0.022 | 0.377 | 0.315 | 0.192 | 0.365 | 0.206 | 0.808 | 0.345 | 0.750 | | | | |
| **S** | 0.409 | 0.389 | 0.142 | 0.313 | 0.422 | 0.362 | 0.363 | 0.178 | 0.303 | 0.262 | 0.567 | 0.555 | 0.367 | 0.552 | | | |
| **SEC** | 0.143 | 0.130 | 0.081 | 0.042 | 0.245 | 0.128 | 0.232 | 0.085 | 0.052 | 0.047 | 0.275 | 0.150 | 0.306 | 0.647 | 0.743 | | |
| **SEE** | 0.689 | 0.656 | 0.158 | 0.613 | 0.609 | 0.364 | 0.267 | 0.263 | 0.249 | 0.362 | 0.329 | 0.411 | 0.117 | 0.811 | 0.361 | 0.707 | |
| **SSC** | 0.207 | 0.202 | 0.100 | 0.155 | 0.205 | 0.342 | 0.319 | 0.104 | 0.354 | 0.238 | 0.601 | 0.640 | 0.335 | 0.840 | 0.202 | 0.614 | 0.751 |

Assessment of Discriminant Validity

Discriminant validity is the degree of measurement of the model compared with other model constructs of the study. Voorhees, et al. [77,78] introduced criteria for evaluating validity. The idea that the variance items share with their measurement model is more significant than they associate in the research with other measurement models. Therefore, the AVE square root, which is the average correlation between the measurement models' indicator variables, should be greater than the correlation between the construct and any other structural model constructs. Since the square root of the correlation coefficient of each outer model is significantly higher than the square root of the AVE of each outer model [119,120]. Table 6 shows discriminant validity using heterotrait-monotrait (HTMT).

The HTMT ratio, Fornell used to assess the measurement models' discriminant validity as presented in Table 6, respectively. The assessment of discriminant validity using the HTMT ratio shown in the result indicates that the highest HTMT rate among the measurement models is 0.965 between C and CD. The highest value is below the recommended maximum conservative value of 0.85 [79]. This outcome shows that discriminant validity was achieved using the HTMT ratio criterion.

### 4.2.2. Assessment of Structural Model (Path Analysis)

The second-order constructs for their lower-order constructs are often recognized as the relative importance of construction waste generation. The factor with high path coefficients displays the relative strength of each element on the specific construct base on the developed model for achieving sustainability. Table 7 below demonstrates the necessary items on second-order constructs for their lower-order constructs for the relative strength for their particular construct.

**Table 7.** Second-order constructs for their lower order constructs.

| | Path Coefficients | T Statistics | *P* Values | Remark |
|---|---|---|---|---|
| C -> CD | 0.965 | 219.955 | 0.000 | Significant |
| C -> CESM | 0.187 | 2.677 | 0.008 | Significant |
| C -> CM | 0.932 | 109.719 | 0.000 | Significant |
| I -> IDM | 0.722 | 16.569 | 0.000 | Significant |
| I -> ILC | 0.630 | 9.570 | 0.000 | Significant |
| I -> IME | 0.847 | 31.893 | 0.000 | Significant |
| P -> PPF | 0.831 | 30.291 | 0.000 | Significant |
| P -> PWM | 0.809 | 25.539 | 0.000 | Significant |
| S -> SEC | 0.658 | 10.573 | 0.000 | Significant |
| S -> SEE | 0.808 | 40.817 | 0.000 | Significant |
| S -> SSC | 0.835 | 28.652 | 0.000 | Significant |

This study is the first stage of estimating the second-order constructs on their lower-order constructs for relative strength for their particular construct by anticipated waste generated base on 3R in Malaysia. The estimated wastes generated from Malaysian projects under the constructs of C -> CD with high path coefficients of 0.965 and T statistics of 219.955 have *p*-values of 0.000, which is very significant as shown in Table 7. Thus, the result revealed that I -> ISCF is the second factor leading to tremendous growth. They improve for waste generation with path coefficients of 0.803, 30.982 as *t* statistics and *p*-values of 0.000, as illustrated in Table 7.

The PLS-SEM research model on achieving sustainable construction waste reduction was used to investigate the impact of 3R (reduce, reuse, and recycle). The model is presented in (Figures 3 and 4). After reviewing the Bootstrapping approach, it was discovered that the hypothesis was of significance to the model. Bootstrapping is a random re-sampling of the original dataset used, in addition to the re-sampling process to generate new samples of the same size as the original dataset. The first approach tests the data set to determine correct path coefficients' statistical significance. It serves to determine whether the calculated path coefficients are accurate [121,122]. As seen in Figure 3, the pathway significance, pathway coefficients (β), and *p*-values (the *P*-value for the standardized pathway coefficient), as well as $R^2$ for each endogenous construct, were all tested. The result of the bootstrapping method showcases the *p*-values for every possible path. According to these results, the effect of 3R for sustainable construction waste reduction was positive and significant ($\beta = 0.743$, $p < 0.001$).

Impact of CDW Generation, Improving Factors, and Policy-Related Factors on Sustainable Construction Waste

A group of constructs agreed by respondents achieves the objectives/hypothesis of this study in a model using Smart-PLS 20 software [123]. The constructs presented in Table 8 identify the significant relationship of each group in improving design-out waste for sustainable development goals. Present causes for waste generation and policy-related factors significantly affect sustainable construction waste, as illustrated in Table 8.

**Table 8.** Effects of C, I, and P on S.

| Relationships | Path Coefficients | T Statistics | *P* Values | Remark |
|---|---|---|---|---|
| Construction waste generated -> S | 0.316 | 5.374 | 0.000 | Significant |
| Improving factors -> S | 0.124 | 2.065 | 0.039 | Significant |
| Policy related factors -> S | 0.466 | 8.191 | 0.000 | Significant |

Assessment of $R^2$ (Coefficient of Determination)

The determination coefficient ($R^2$) measures how the exogenous constructs describe or predict the endogenous construct. It measures the inner model predictive accuracy by stating the model percentage of construct variance [124]. Although there is no generally acceptable level of $R^2$ value, an $R^2$ value of 0.2 is considered high in the construction waste management area. The $R^2$ of the endogenous construct of the study is shown in Table 9. The $R^2$ level of the endogenous constructs.

**Table 9.** $R^2$ assessment.

| S/N | R Square | R Square Adjusted |
|---|---|---|
| Sustainable construction waste Reduction | 0.841 | 0.835 |

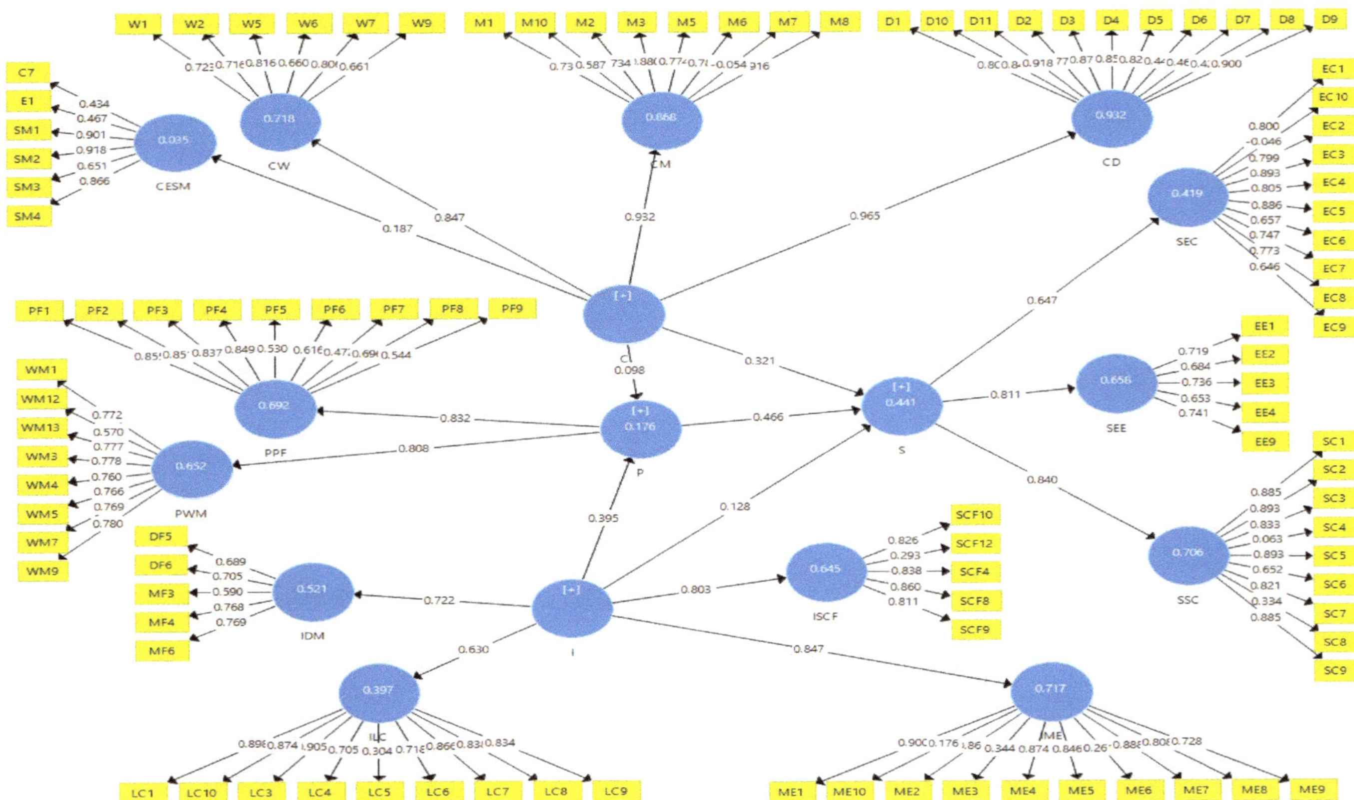


**Figure 3.** Study model.

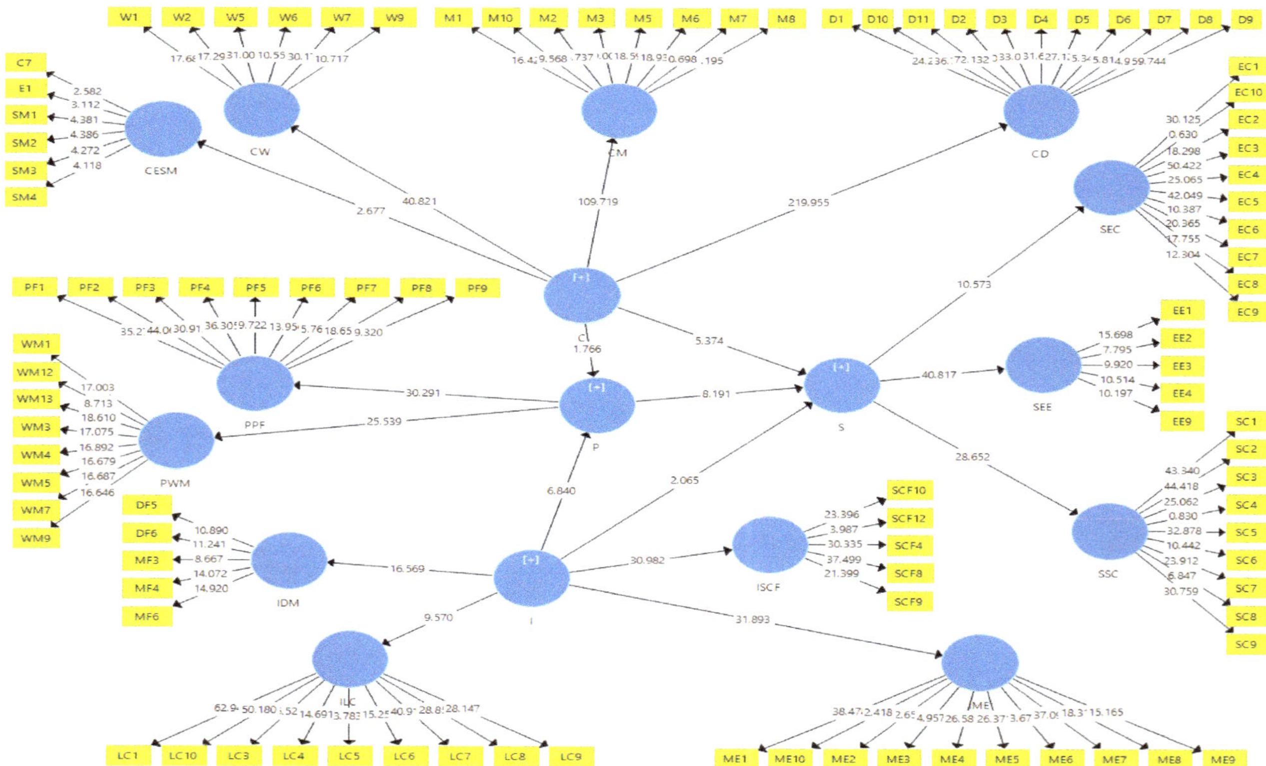


**Figure 4.** Model $t$-statistics and bootstrapping.

The results show that sustainable construction waste reduction $R^2$ value is 0.841 and adjusted R square 0.835, respectively, with the main endogenous construct, CD having the highest value. Thus, based on the evaluation criteria, the $R^2$ values in this study are within the acceptable moderate levels. However, the study is exploratory; its $R^2$ values can be considered high [125].

Predictive Relevance ($Q^2$) of the Structural Model

$Q^2$ assessed using the predictive relevance of Stone-Geisser ($Q^2$), which tested that all indicators' data points are predicted accurately in the outer model of endogenous constructs [126]. This quality evaluation criterion required the cross-validated redundancy ($Q^2$) value to be a positive integer above 0 to have sufficient predictive relevance [127]. According to the above submission, the study's final model was evaluated to determine the cross-validated redundancy ($Q^2$) using the blindfolding procedure with the aid of Smart-PLS 3 software [128]. This indicates the cross-validated redundancy of the models of the current waste practices and the factors for improving waste management. All the endogenous constructs have $Q^2$ values greater than 0. The $Q^2$ indicated that all the research models have good predictive relevance [129]. Therefore, the model is good.

### *4.3. Discussion*

A comparative study was very useful because each instance served as a guideline or structure for understanding the others. The waste-generated rates from the three projects under evaluation are explained. The construction waste generated from various building materials is shown in Figures 1 and 2. Furthermore, with an overall mean of 0.908, the data analysis suggests that the participants have a high level of agreement. The results indicate that a greater percentage of respondents believed that their criteria were a significant sign for efficient waste reduction, reuse, and recycling. The 5% trimmed-mean is tested once more to see if the data set's extreme values have changed the mean [130].

In this study, the mean (6.6714) and the 5% trimmed mean (7.9603) are quite close, showing that extreme values did not affect the data's analysis. Because of the requirement to complete construction work in Malaysia, the construction industry produces much waste. As a result, mitigating the environmental impact is a major concern in a construction project's implementation. It is also an essential aspect of the commitment to addressing global sustainability issues. Construction waste like demolition waste, which is mostly combined, can cause sorting, transforming, and disposal obstacles.

Furthermore, since these activities require many personnel, sorting and crushing might demand a high price in the recycling strategy. The predicted weight of waste generated per day for construction on the site was modified according to the data collection and analysis. For example, less than 1 tonne produces 43%, 1–5 tonnes produce 55%, and more than 5 tonnes produce only 3%. This signifies that a lot of waste is produced without regard for the project's environmental impact. In contrast to the high-rise building, which generated a lot of construction waste but had a very high rate of reducing, reusable, and recyclability, the mass housing (semi-detached and bungalow) had a level of reduction, reuse, and recycling of about 16% to 32% from projects three and four, respectively. It was found that concrete and aggregate generated the most waste in projects one and two, accounting for approximately 71% and 72% of total waste generated, respectively. This was because the two projects created less wood waste, as illustrated in Figure 1. However, the most significant proportion of waste came from wood in projects 2 and 3, accounting for around 87% and 82%, respectively. Moreover, all the three constructs that have a significant impact on sustainable construction waste are discussed in the following sections: identify the impact of construction waste generation on sustainability; identify the impact of improving factors on sustainable construction waste management; and identify the impact of policy-related factors on sustainability.

## 5. Conclusions and Implications

Based on the results and discussion above, this study investigated the modeling of 3R (reduce, reuse and recycle) for sustainable construction waste reduction. A questionnaire was used to analyze the perspective and comprehension of Malaysian construction industry experts regarding construction waste management. The majority of respondents believed that 20% of the construction trash generated at their site might be repurposed in their project while the minority of respondents reported that more than 50% of the waste produced on-site could be repurposed. Furthermore, the results revealed that the amount of waste generated at a construction site in Malaysia is an estimated average of 5 tonnes per day. The construction waste generated from high-rise building construction was calculated to be 4.4 tons/day. However, it was estimated to be 2.47 tons/day for housing projects, which was in line with the predicted amount.

On the other hand, the results of SEM-PLS indicated that the exploratory power of the study model is considered sustainable with $R^2$ values 0.83%. At the same time, the results of relationships between improving factors, policy-related factors, construction waste generated, and sustainable construction waste reduction were significant. Therefore, the results of this study demonstrated that all hypotheses were supported.

Regarding its theoretical contribution, this study will extend the body of knowledge regarding construction waste management strategies. This paper contributes to the literature to be a foundation for future researchers interested in construction waste management strategies. The results may also be helpful to many construction companies, particularly those in developing countries where there is a lot of construction waste with low awareness. Practically, the results will be beneficial to many construction companies, particularly those in developing countries where construction waste awareness is low. Furthermore, they can assist small and medium construction companies to use technologies for practical and efficient training toward sustainable development goals. Finally, this study has established a basis for improved specifications that could be critical for evaluating and removing waste. Construction waste prevention is significant, leading to the avoidance of design errors contributing to waste generation. The construction waste is identified chiefly through processes that involve conventional construction.

A study limitation is insufficient data on the amount of C&D waste reuse and recycling in Malaysia. However, the results revealed numerous concerns and problems contributing to the low percentage of reuse and recycling waste. Contamination, waste quality, collecting and transportation challenges, and difficulty sorting, converting, and disposing waste are the most pressing concerns. The model predicts waste generation, the 3R-produced waste, CO2, and waste assessment model for residential buildings. These elements of the CDW management model may guide how to handle construction waste more sustainably. Future researchers need to address the international standards in the construction industry, waste management, and construction demolition waste.

**Author Contributions:** Conceptualization and Investigation, M.M.; writing—original draft preparation M.M.; supervision and quality control, project leader, N.S. and N.A.Z.; Data collection, M.M. and N.S. Analysis of the literature, M.M.; N.S. and N.A.Z. software and data analysis, M.M.; A.-B.A.A.-M., visualization, M.M.; N.S. and N.A.Z. writing—review and editing, N.S., A.E., E.F.R., N.A.Z., A.H., A.D.R., M.B.I. and A.-B.A.A.-M. All authors have read and agreed to the published version of the manuscript.

**Funding:** This research received no external funding.

**Institutional Review Board Statement:** Not applicable.

**Informed Consent Statement:** Informed consent was obtained from all subjects involved in the study.

**Data Availability Statement:** The data supporting and used in this study findings are available from the corresponding author upon reasonable request.

**Acknowledgments:** The authors would like to express their gratitude to the postgraduate studies UTP for providing financial support for the Graduate Assistantship Scheme. The use of research facilities provided by Universiti Teknologi PETRONAS UTP Malaysia is acknowledged.

**Conflicts of Interest:** The authors declare that they have no established conflicting financial interests or personal relationships that may seem to have influenced the research presented in this paper.

## Abbreviations

| | |
|---|---|
| average variance extracted | AVE |
| coefficient | Rho_A C.R |
| Improving factors | I |
| effect size | (f2) |
| Policy related factors (P) | P |
| carbon dioxide | ($CO_2$) |
| construction and demolition | C&D |
| waste management measures | (WMM) |
| construction and demolition waste | C&DW |
| waste management | WM |
| Construction waste management | CWM |
| Specifications for building works | (SBW) |
| Construction industry development board | (CIDB). |
| Exploratory Factor Analysis | (EFA) |
| Bachelors (BSc), Masters (MS), and Doctorate Degree (PhD) | (BSc), (MS) and (PhD) |
| coefficient of determination | ($R^2$) |
| system predictive relevance | ($Q^2$) |
| goodness-of-fit | (GoF) |
| lower-order constructs | (LOCs) |
| Heterotrait-Monotrait ratio | (HTMT) |
| Construction waste generated | (C) |
| Sustainable construction waste Reduction | (S) |
| A partial least squares structural equation modelling | (PLS-SEM) |
| Reduce, reuse and recycle | 3R |

## References


1. Hickel, J. The contradiction of the sustainable development goals: Growth versus ecology on a finite planet. *Sustain. Dev.* **2019**, *27*, 873–884. [CrossRef]
2. Johari, A.; Alkali, H.; Hashim, H.; Ahmed, S.I.; Mat, R. Municipal solid waste management and potential revenue from recycling in Malaysia. *Mod. Appl. Sci.* **2014**, *8*, 37. [CrossRef]
3. Meadows, D.; Randers, J. *The Limits to Growth: The 30-Year Update*; Routledge: Milton Park, Miltonm, UK, 2012.
4. Cheam, W.-Y.; Lau, L.-S.; Wei, C.-Y. Factors influencing the residence's intention to adopt solar photovoltaic technology: A case study from Klang Valley, Malaysia. *Clean Energy* **2021**, *5*, 464–473. [CrossRef]
5. Arslan, H.; Coşgun, N.; Salgin, B. Construction and Demolition Waste Management in Turkey. In *Waste Management: An Integrated Vision*; IntechOpen: London, UK, 2012; pp. 313–332.
6. Starovoytova, D.; Namango, S. Solid waste management at university campus (Part 4/10): Perceptions, attitudes, and practices of students and vendors. *J. Environ. Earth Sci.* **2018**, *8*, 2224–3216.
7. Mitra, S.; Datta, P.P. Adoption of green supply chain management practices and their impact on performance: An exploratory study of Indian manufacturing firms. *Int. J. Prod. Res.* **2014**, *52*, 2085–2107. [CrossRef]
8. Schroeder, P.; Anggraeni, K.; Weber, U. The Relevance of Circular Economy Practices to the Sustainable Development Goals. *J. Ind. Ecol.* **2018**, *23*, 77–95. [CrossRef]
9. Neo, S.M.; Choong, W.W.; Bin Ahamad, R. Differential environmental psychological factors in determining low carbon behaviour among urban and suburban residents through responsible environmental behaviour model. *Sustain. Cities Soc.* **2017**, *31*, 225–233. [CrossRef]
10. Badgie, D.; Samah, M.A.A.; Manaf, L.A.; Muda, A.B. Assessment of Municipal Solid Waste Composition in Malaysia: Management, Practice, and Challenges. *Pol. J. Environ. Stud.* **2012**, *21*, 539–547.
11. Islam, R.; Nazifa, T.H.; Yuniarto, A.; Uddin, A.S.; Salmiati, S.; Shahid, S. An empirical study of construction and demolition waste generation and implication of recycling. *Waste Manag.* **2019**, *95*, 10–21. [CrossRef] [PubMed]
12. Ikechukwu, A.F.; Shabangu, C. Strength and durability performance of masonry bricks produced with crushed glass and melted PET plastics. *Case Stud. Constr. Mater.* **2021**, *14*, e00542.

13. Sarkodie, S.A.; Owusu, P.A. Global assessment of environment, health and economic impact of the novel coronavirus (COVID-19). *Environ. Dev. Sustain.* **2020**, *23*, 5005–5015. [CrossRef] [PubMed]
14. Babashamsi, P.; Yusoff, N.I.M.; Ceylan, H.; Nor, N.G.M.; Jenatabadi, H.S. Sustainable Development Factors in Pavement Life-Cycle: Highway/Airport Review. *Sustainability* **2016**, *8*, 248. [CrossRef]
15. Esa, M.R.; Halog, A.; Rigamonti, L. Strategies for minimizing construction and demolition wastes in Malaysia. *Resour. Conserv. Recycl.* **2017**, *120*, 219–229. [CrossRef]
16. Arayici, Y.; Egbu, C.O.; Coates, S.P. Building information modelling (BIM) implementation and remote construction projects: Issues, challenges, and critiques. *J. Inf. Technol. Constr.* **2012**, *17*, 75–92.
17. Kabirifar, K.; Mojtahedi, M.; Wang, C.C.; Tam, V.W. A conceptual foundation for effective construction and demolition waste management. *Clean. Eng. Technol.* **2020**, *1*, 100019. [CrossRef]
18. Won, J.; Cheng, J.C.; Lee, G. Quantification of construction waste prevented by BIM-based design validation: Case studies in South Korea. *Waste Manag.* **2016**, *49*, 170–180. [CrossRef]
19. Fewings, P.; Henjewele, C. *Construction Project Management: An Integrated Approach*; Routledge: Milton Park, Milton, UK, 2019.
20. Mah, C.M.; Fujiwara, T.; Ho, C.S. Life cycle assessment and life cycle costing toward eco-efficiency concrete waste management in Malaysia. *J. Clean. Prod.* **2018**, *172*, 3415–3427. [CrossRef]
21. Ametepey, S.; Aigbavboa, C.; Ansah, K. Barriers to Successful Implementation of Sustainable Construction in the Ghanaian Construction Industry. *Procedia Manuf.* **2015**, *3*, 1682–1689. [CrossRef]
22. Hessing, M.; Summerville, T. *Canadian Natural Resource and Environmental Policy: Political Economy and Public Policy*; UBC Press: Vancouver, BC, Canada, 2014.
23. Boyd, C.E.; D'Abramo, L.R.; Glencross, B.D.; Huyben, D.C.; Juarez, L.M.; Lockwood, G.S.; McNevin, A.A.; Tacon, A.G.J.; Teletchea, F.; Tomasso, J.R., Jr.; et al. Achieving sustainable aquaculture: Historical and current perspectives and future needs and challenges. *J. World Aquac. Soc.* **2020**, *51*, 578–633. [CrossRef]
24. Kabirifar, K.; Mojtahedi, M.; Wang, C.; Tam, V.W. Construction and demolition waste management contributing factors coupled with reduce, reuse, and recycle strategies for effective waste management: A review. *J. Clean. Prod.* **2020**, *263*, 121265. [CrossRef]
25. Lu, W.; Yuan, H. A framework for understanding waste management studies in construction. *Waste Manag.* **2011**, *31*, 1252–1260. [CrossRef] [PubMed]
26. Kabirifar, K.; Mojtahedi, M.; Wang, C. A Systematic Review of Construction and Demolition Waste Management in Australia: Current Practices and Challenges. *Recycling* **2021**, *6*, 34. [CrossRef]
27. Khalfan, M.M.; Maqsood, T. Current state of off-site manufacturing in Australian and Chinese residential construction. *J. Constr. Eng.* **2014**, *2014*, 164863. [CrossRef]
28. Sáez, P.V.; Merino, M.D.R.; Porras-Amores, C.; Astorqui, J.S.C.; Pericot, N.G. Analysis of Best Practices to Prevent and Manage the Waste Generated in Building Rehabilitation Works. *Sustainability* **2019**, *11*, 2796. [CrossRef]
29. Danso, H. Engineering, Identification of key indicators for sustainable construction materials. *Adv. Mater. Sci. Eng.* **2018**, *2018*, 6916258. [CrossRef]
30. Tlusty, M.F.; Rhyne, A.L.; Kaufman, L.; Hutchins, M.; Reid, G.M.; Andrews, C.; Boyle, P.; Hemdal, J.; McGilvray, F.; Dowd, S. Opportunities for Public Aquariums to Increase the Sustainability of the Aquatic Animal Trade. *Zoo Biol.* **2012**, *32*, 1–12. [CrossRef]
31. Karlsson, I.; Rootzén, J.; Johnsson, F. Reaching net-zero carbon emissions in construction supply chains—Analysis of a Swedish road construction project. *Renew. Sustain. Energy Rev.* **2019**, *120*, 109651. [CrossRef]
32. Wang, Y.; Zeng, H.C.; Lee, J.Y. Highly Reversible Lithium Storage in Porous SnO2 Nanotubes with Coaxially Grown Carbon Nanotube Overlayers. *Adv. Mater.* **2006**, *18*, 645–649. [CrossRef]
33. Klemeš, J.J.; Fan, Y.V.; Jiang, P. Plastics: Friends or foes? The circularity and plastic waste footprint. *Energy Sources Part A Recovery Util. Environ. Eff.* **2021**, *43*, 1549–1565. [CrossRef]
34. Adeniran, A.; Nubi, A.; Adelopo, A. Solid waste generation and characterization in the University of Lagos for a sustainable waste management. *Waste Manag.* **2017**, *67*, 3–10. [CrossRef] [PubMed]
35. Ogunmakinde, O.E.; Sher, W.; Maund, K. An Assessment of Material Waste Disposal Methods in the Nigerian Construction Industry. *Recycling* **2019**, *4*, 13. [CrossRef]
36. Nduneseokwu, C.K.; Qu, Y.; Appolloni, A. Factors Influencing Consumers' Intentions to Participate in a Formal E-Waste Collection System: A Case Study of Onitsha, Nigeria. *Sustainability* **2017**, *9*, 881. [CrossRef]
37. Dalsgaard, P.; Halskov, K. Reflective design documentation. In Proceedings of the Designing Interactive Systems Conference, Newcastle, UK, 11–15 June 2012.
38. Chybiński, M.; Kurzawa, Z.; Polus, Ł. Problems with Buildings Lacking Basic Design Documentation. *Procedia Eng.* **2017**, *195*, 24–31. [CrossRef]
39. Silver, D. Haste or Waste? Peer pressure and productivity in the emergency department. *Rev. Econ. Stud.* **2021**, *88*, 1385–1417. [CrossRef]
40. Granado, M.P.P.; Suhogusoff, Y.V.M.; Santos, L.R.O.; Yamaji, F.M.; De Conti, A.C. Effects of pressure densification on strength and properties of cassava waste briquettes. *Renew. Energy* **2020**, *167*, 306–312. [CrossRef]
41. Rajendran, P.; Gomez, C.P. Implementing BIM for waste minimisation in the construction industry: A literature review. In Proceedings of the 2nd International Conference on Management, Kuala Lumpur, Malaysia, 11–12 June 2012.

42. Lundesjö, G. Consolidation centres construction logistics. In *Urban Logistics: Management, Policy and Innovation in a Rapidly Changing Environment*; Kogan Page, Limited: London, UK, 2018; p. 210.
43. Sahoo, P.; Jana, D.K.; Pramanik, S.; Panigrahi, G. A novel reduction method for type-2 uncertain normal critical values and its applications on 4D profit transportation problem involving damageable and substitute items. *Int. J. Appl. Comput. Math.* **2021**, *7*, 123. [CrossRef]
44. Singh, G.; Singh, A. Solving fixed-charge transportation problem using a modified particle swarm optimization algorithm. *Int. J. Syst. Assur. Eng. Manag.* **2021**, 1–14. [CrossRef]
45. Negash, Y.T.; Hassan, A.M.; Tseng, M.-L.; Wu, K.-J.; Ali, M.H. Sustainable construction and demolition waste management in Somaliland: Regulatory barriers lead to technical and environmental barriers. *J. Clean. Prod.* **2021**, *297*, 126717. [CrossRef]
46. Vignesh, K.; Rajadesingu, S.; Arunachalam, K.D. Challenges, issues, and problems with zero-waste tools. In *Concepts of Advanced Zero Waste Tools*; Elsevier: Amsterdam, The Netherlands, 2021; pp. 69–90.
47. Rubinov, V.; Fetisov, V. Problems of a Modern Approach to the Technological Process of Waste Management. In Proceedings of the 2021 Wave Electronics and Its Application in Information and Telecommunication Systems (WECONF), St. Petersburg, Russia, May 31–June 4 2021.
48. Daoud, A.O.; Othman, A.A.; Robinson, H.; Bayati, A. Exploring the relationship between materials procurement and waste minimization in the construction industry: The case of Egypt. In Proceedings of the International Conference on Sustainability, Green Buildings, Environmental Engineering & Renewable Energy (SGER 2018), Kuala Lumpur, Malaysia, 27 January 2018.
49. Nagapan, S.; Rahman, I.A.; Asmi, A. A review of construction waste cause factors. In Proceedings of the Asian Conference on Real Estate: Sustainable Growth Managing Challenges (ACRE), Johor Bahru, Malaysia, 3–5 October 2011.
50. Osmani, M.; Villoria-Sáez, P. Current and Emerging Construction Waste Management Status, Trends and Approaches. In *Waste*; Elsevier: Amsterdam, The Netherlands, 2019.
51. Crawford, R.H.; Mathur, D.; Gerritsen, R. Barriers to Improving the Environmental Performance of Construction Waste Management in Remote Communities. *Procedia Eng.* **2017**, *196*, 830–837. [CrossRef]
52. Aldana, J.C.; Serpell, A. Methodology for the preparation of construction project waste management plans based on innovation and productive thinking processes: A case study in Chile. *J. Constr.* **2016**, *15*, 32–41. [CrossRef]
53. de Magalhães, R.F.; Danilevicz, Â.D.M.F.; Saurin, T.A. Reducing construction waste: A study of urban infrastructure projects. *Waste Manag.* **2017**, *67*, 265–277. [CrossRef] [PubMed]
54. Huang, B.; Wang, X.; Kua, H.; Geng, Y.; Bleischwitz, R.; Ren, J. Construction and demolition waste management in China through the 3R principle. *Resour. Conserv. Recycl.* **2018**, *129*, 36–44. [CrossRef]
55. Poon, C.S.; Yu, A.T.W.; Jaillon, L.C. Reducing building waste at construction sites in Hong Kong. *Constr. Manag. Econ.* **2004**, *22*, 461–470. [CrossRef]
56. Farrow, R.L. *ManTech Journal. Achieving A Breakthrough. Volume 7/Number 3. 1982*; Army Materials and Mechanics Research Center: Watertown, MA, USA, 1982.
57. Sundström, M. Connecting Social Science and Information Technology: Democratic Privacy in the Information Age. Lund University: Lund, Sweden, 2001.
58. Ferguson, J. *Managing and Minimizing Construction Waste: A Practical Guide*; Thomas Telford: Westerkirk, UK, 1995.
59. Dainty, A.R.J.; Brooke, R.J. Towards improved construction waste minimisation: A need for improved supply chain integration? *Struct. Surv.* **2004**, *22*, 20–29. [CrossRef]
60. Muthu, S.S. *Advances in Textile Waste Water Treatments*; Springer Nature: Basingstoke, UK, 2021.
61. König, H.; Kohler, N.; Kreißig, J.; Lützkendorf, T. *A Life Cycle Approach to Buildings*; Walter de Gruyter: Berlin, Germany, 2012.
62. Hasmori, M.F.; Zin, A.F.M.; Nagapan, S.; Deraman, R.; Abas, N.; Yunus, R.; Klufallah, M. The on-site waste minimization practices for construction waste. In Proceedings of the IOP Conference Series: Materials Science and Engineering, Chennai, India, 16–17 September 2020; IOP Publishing: Bristol, UK, 2020.
63. Ajayi, S.O.; Oyedele, L.O.; Bilal, M.; Akinade, O.; Alaka, H.A.; Owolabi, H.A. Critical management practices influencing on-site waste minimization in construction projects. *Waste Manag.* **2016**, *59*, 330–339. [CrossRef]
64. Al-Moghany, S.S. *Managing and Minimizing Construction Waste in Gaza Strip*; Islamic University of Gaza-Palestine: Gaza, Palestine, 2006.
65. Holdren, J.P.; Fetter, S. Contribution of Activation Products to Fusion Accident Risk: Part II. Effects Of Alternative Materials And Designs. *Nucl. Technol.-Fusion* **1983**, *4*, 599–619. [CrossRef]
66. Adewuyi, T.O.; Odesola, I.A. Factors affecting material waste on construction sites in Nigeria. *J. Eng. Technol.* **2015**, *6*, 82–99.
67. Wilson, R.E. Educational Specifications. In *AIA School Plant*; American Institute Of Architects: Washington, DC, USA, 1955; p. 18.
68. Polat, G.; Damci, A.; Turkoglu, H.; Gurgun, A.P. Identification of Root Causes of Construction and Demolition (C&D) Waste: The Case of Turkey. *Procedia Eng.* **2017**, *196*, 948–955. [CrossRef]
69. Aparcana Robles, S.R.; Hinostroza, M.L. Guidebook for the Development of Nationally Appropriate Mitigation Actions on Sustainable Municipal Waste Management. 2015. Available online: https://core.ac.uk/download/pdf/43252112.pdf (accessed on 1 September 2021).
70. Begum, R.A.; Siwar, C.; Pereira, J.J.; Jaafar, A.H. A benefit–cost analysis on the economic feasibility of construction waste minimisation: The case of Malaysia. *Resour. Conserv. Recycl.* **2006**, *48*, 86–98. [CrossRef]

71. Bao, Z.; Lee, W.M.; Lu, W. Implementing on-site construction waste recycling in Hong Kong: Barriers and facilitators. *Sci. Total Environ.* **2020**, *747*, 141091. [CrossRef] [PubMed]
72. Bishop, B.; Johnson, K.; King, E.; Lotts, C.; Miller, S.; Wurtz, M. *Environmental Assessment of Proposed Mixed-Use Business Park on an Enchanced Use Lease at Grand Forks Air Force Base, North Dakota*; Air Force Civil Engineer Center Joint Base: San Antonio, TX, USA, 2014.
73. LoBiondo-Wood, G.; Faan, P.R.; Haber, J.; Faan, P.R. *Nursing Research E-Book: Methods and Critical Appraisal for Evidence-Based Practice*; Elsevier: Amsterdam, The Netherlands, 2021.
74. Jarvis, C.B.; MacKenzie, S.B.; Podsakoff, P.M. A Critical Review of Construct Indicators and Measurement Model Misspecification in Marketing and Consumer Research. *J. Consum. Res.* **2003**, *30*, 199–218. [CrossRef]
75. Al-Mekhlafi, A.-B.A.; Isha, A.S.N.; Chileshe, N.; Abdulrab, M.; Kineber, A.F.; Ajmal, M. Impact of Safety Culture Implementation on Driving Performance among Oil and Gas Tanker Drivers: A Partial Least Squares Structural Equation Modelling (PLS-SEM) Approach. *Sustainability* **2021**, *13*, 8886. [CrossRef]
76. Al-Mekhlafi, A.-B.; Isha, A.; Chileshe, N.; Abdulrab, M.; Saeed, A.; Kineber, A. Modelling the Relationship between the Nature of Work Factors and Driving Performance Mediating by Role of Fatigue. *Int. J. Environ. Res. Public Health* **2021**, *18*, 6752. [CrossRef]
77. Al-Mekhlafi, A.B.A.; Isha, A.S.N.; Naji, G.M.A. The relationship between fatigue and driving performance: A review and directions for future research. *J. Crit. Rev.* **2020**, *7*, 134–141.
78. Al-Mekhlafi, A.B.A.; Isha, A.S.N.; Sabir, A.A.; Naji, G.M.A.; Ajmal, M.; Al-Harasi, A.H. Fatigue Assessment of Oil and Gas Tanker Drivers: Psychomotor Vigilance Test (PVT-192). *Solid State Technol.* **2020**, *63*, 4256–4262.
79. Yang, G.-L.; Fukuyama, H.; Song, Y.-Y. Measuring the inefficiency of Chinese research universities based on a two-stage network DEA model. *J. Inf.* **2018**, *12*, 10–30. [CrossRef]
80. Kineber, A.F.; Othman, I.; Oke, A.E.; Buniya, M.K. Value Management Activities in Building Projects in Developing Countries. in Advances in Civil Engineering Materials. In Proceedings of the International Conference on Architecture and Civil Engineering (ICACE2020), Kuala Lumpur, Malaysia, 4–5 January 2021; Springer: Singapore, 2021.
81. Siti–Nabiha, A.K.; George, R.A.; Wahid, N.A.; Amran, A.; Mahadi, R.; Abustan, I. The Development of a Green Practice Index for the Malaysian Hotel Industry. *Issues Soc. Environ. Account.* **2014**, *8*, 23. [CrossRef]
82. Ritchie, J.; Lewis, J.; Elam, R.G. Selecting Samples. In *Qualitative Research Practice: A Guide for Social Science Students and Researchers*; Sage: Los Angeles, CA, USA, 2013.
83. Krejcie, R.V.; Morgan, D.W. Determining Sample Size for Research Activities. *Educ. Psychol. Meas.* **1970**, *30*, 607–610. [CrossRef]
84. Jaso-Sánchez, M.A. *Evaluation of Research Collaboration in the Mexican Phyto-Pharmaceutical Sector*; The University of Manchester: Manchester, UK, 2007.
85. Kineber, A.; Othman, I.; Oke, A.; Chileshe, N.; Buniya, M. Identifying and Assessing Sustainable Value Management Implementation Activities in Developing Countries: The Case of Egypt. *Sustainability* **2020**, *12*, 9143. [CrossRef]
86. Kyriazos, T.A. Applied Psychometrics: Sample Size and Sample Power Considerations in Factor Analysis (EFA, CFA) and SEM in General. *Psychology* **2018**, *9*, 2207–2230. [CrossRef]
87. Kineber, A.F.; Othman, I.; Oke, A.E.; Chileshe, N.; Buniya, M.K. Impact of Value Management on Building Projects Success: Structural Equation Modeling Approach. *J. Constr. Eng. Manag.* **2021**, *147*, 04021011. [CrossRef]
88. Ferreira-Valente, A.; Costa, P.; Elorduy, M.; Virumbrales, M.; Costa, M.J.; Palés, J.; Ferreira-Valente, M.A. Psychometric properties of the Spanish version of the Jefferson Scale of Empathy: Making sense of the total score through a second order confirmatory factor analysis. *BMC Med. Educ.* **2016**, *16*, 242. [CrossRef]
89. Morgan, J.N.; Sonquist, J.A. Problems in the analysis of survey data, and a proposal. *J. Am. Stat. Assoc.* **1963**, *58*, 415–434. [CrossRef]
90. Ajmal, M.; Isha, A.S.N.; Nordin, S.M.; Kanwal, N.; Al-Mekhlafi, A.B.A.; Naji, G.M.A. A Conceptual Framework for the Determinants of Organizational Agility: Does Safety Commitment Matters? *Solid State Technol.* **2020**, *63*, 4112–4119.
91. Ajmal, M.; Isha, A.S.N.; Nordin, S.M.; Sabir, A.A.; Munir, A.; Al-Mekhlafi, A.B.A.; Naji, G.M.A. Safety Management Paradigms: COVID-19 Employee Well-Being Impact on Occupational Health and Safety Performance. *J. Hunan Univ. Nat. Sci.* **2021**, *48*, 128–142.
92. Mason, M. Sample size and saturation in PhD studies using qualitative interviews. In *Forum Qualitative Sozialforschung/Forum: Qualitative Social Research*; Institute for Qualitative Research and Center for Digital Systems (Free University of Berlin): Berlin, Germany, 2010; Volume 11.
93. Naji, G.M.A.; Isha, A.S.N.; Al-Mekhlafi, A.B.A.; Sharafaddin, O.; Ajmal, M. Implementation of leading and lagging indicators to improve safety performance in the upstream oil and gas industry. *J. Crit. Rev.* **2020**, *7*, 265–269.
94. Naji, G.M.A.; Isha, A.S.N.; Alzoraiki, M.; Al-Mekhlafi, A.B.A.; Sharafaddin, O.; Saleem, M.S. Impact Of Safety Culture and Psychosocial Hazard on Safety Performance Among Upstream Employees in Malaysia at Oil And Gas Industry. *Solid State Technol.* **2020**, *63*, 4120–4126.
95. Robinaugh, D.J.; Hoekstra, R.H.A.; Toner, E.; Borsboom, D. The network approach to psychopathology: A review of the literature 2008–2018 and an agenda for future research. *Psychol. Med.* **2019**, *50*, 353–366. [CrossRef] [PubMed]
96. Kineber, A.F.; Othman, I.; Oke, A.E.; Chileshe, N.; Zayed, T. Exploring the value management critical success factors for sustainable residential building—A structural equation modelling approach. *J. Clean. Prod.* **2021**, *293*, 126115. [CrossRef]

97. Reichmann, S.; Klebel, T.; Hasani-Mavriqi, I.; Ross-Hellauer, T.J.S.S. Between administration and research: Understanding data management practices in a mid-sized technical university. *SocArXiv* **2020**, *1*.
98. Wong, K.K.K. Partial least squares structural equation modeling (PLS-SEM) techniques using SmartPLS. *Mark. Bull.* **2013**, *24*, 1–32.
99. Choshin, M.; Ghaffari, A. An investigation of the impact of effective factors on the success of e-commerce in small-and medium-sized companies. *Comput. Hum. Behav.* **2017**, *66*, 67–74. [CrossRef]
100. Behl, A. Antecedents to firm performance and competitiveness using the lens of big data analytics: A cross-cultural study. *Manag. Decis.* **2020**. [CrossRef]
101. Malta, J.S.; Silveira, L.P.; Drummond, P.L.D.M.; Costa, N.L.; dos Santos, R.M.M.; Reis, I.A.; Reis, A.M.M.; de Pádua, C.A.M. Validity and reliability of the QLQ-MY20 module for assessing the health-related quality of life in Brazilians with multiple myeloma. *Curr. Med. Res. Opin.* **2021**, *37*, 1163–1169. [CrossRef] [PubMed]
102. Perez, M.V.; Navarro, P.X.S.; Morillas, A.V.; Valdemar, R.M.E.; Araiza, J.P.H.L. Waste management and environmental impact of absorbent hygiene products: A review. *Waste Manag. Res.* **2020**, *39*, 767–783. [CrossRef] [PubMed]
103. Sacks, R.; Eastman, C.; Lee, G.; Teicholz, P. *BIM handbook: A Guide to Building Information Modeling for Owners, Designers, Engineers, Contractors, and Facility Managers*; John Wiley & Sons: Hoboken, NJ, USA, 2018.
104. Kabirifar, K.; Mojtahedi, M.; Wang, C.C.; Tam, V.W. Effective construction and demolition waste management assessment through waste management hierarchy; a case of Australian large construction companies. *J. Clean. Prod.* **2021**, *312*, 127790. [CrossRef]
105. Slorach, P.C.; Jeswani, H.K.; Cuéllar-Franca, R.; Azapagic, A. Environmental and economic implications of recovering resources from food waste in a circular economy. *Sci. Total Environ.* **2019**, *693*, 133516. [CrossRef]
106. Wu, H.; Duan, H.; Zheng, L.; Wang, J.; Niu, Y.; Zhang, G. Demolition waste generation and recycling potentials in a rapidly developing flagship megacity of South China: Prospective scenarios and implications. *Constr. Build. Mater.* **2016**, *113*, 1007–1016. [CrossRef]
107. Subramaniam, S.; Nagapan, S.; Kupusamy, K.; Manian, H.; Daud, Z. Investigate How Construction Waste Generation Rate is Different for Every Types of Project in Peninsular Malaysia Using Site Visit Method. *Int. J. Integr. Eng.* **2018**, *10*, 150–156. [CrossRef]
108. Umar, U.A.; Shafiq, N.; Isa, M.H. Investigation of construction wastes generated in the Malaysian residential sector. *Waste Manag. Res.* **2018**, *36*, 1157–1165. [CrossRef]
109. Maniam, H.; Nagapan, S.; Abdullah, A.H.; Subramaniam, S.; Sohu, S. A Comparative Study of Construction Waste Generation Rate Based on Different Construction Methods on Construction Project in Malaysia. *Eng. Technol. Appl. Sci. Res.* **2018**, *8*, 3488–3491. [CrossRef]
110. Jin, R.; Li, B.; Zhou, T.; Wanatowski, D.; Piroozfar, P. An empirical study of perceptions towards construction and demolition waste recycling and reuse in China. *Resour. Conserv. Recycl.* **2017**, *126*, 86–98. [CrossRef]
111. Hiete, M.; Stengel, J.; Ludwig, J.; Schultmann, F. Matching construction and demolition waste supply to recycling demand: A regional management chain model. *Build. Res. Inf.* **2011**, *39*, 333–351. [CrossRef]
112. Ribić, B.; Voća, N.; Ilakovac, B. Concept of sustainable waste management in the city of Zagreb: Towards the implementation of circular economy approach. *J. Air Waste Manag. Assoc.* **2017**, *67*, 241–259. [CrossRef] [PubMed]
113. Kim, M.H.; Song, H.B. Analysis of the global warming potential for wood waste recycling systems. *J. Clean. Prod.* **2014**, *69*, 199–207. [CrossRef]
114. Henseler, J.; Ringle, C.M.; Sarstedt, M. A new criterion for assessing discriminant validity in variance-based structural equation modeling. *J. Acad. Mark. Sci.* **2014**, *43*, 115–135. [CrossRef]
115. Laato, S.; Islam, A.N.; Farooq, A.; Dhir, A. Unusual purchasing behavior during the early stages of the COVID-19 pandemic: The stimulus-organism-response approach. *J. Retail. Consum. Serv.* **2020**, *57*, 102224. [CrossRef]
116. Abu Aisheh, Y.I.; Tayeh, B.A.; Alaloul, W.S.; Almalki, A. Health and safety improvement in construction projects: A lean construction approach. *Int. J. Occup. Saf. Ergon.* **2021**, 1–26. [CrossRef] [PubMed]
117. Narayan, D.; Cassidy, M.F. A dimensional approach to measuring social capital: Development and validation of a social capital inventory. *Curr. Sociol.* **2001**, *49*, 59–102. [CrossRef]
118. Bhandari, S.; Naudeer, S. Improving efficiency and value in health care Intravenous iron management for anaemia associated with chronic kidney disease: Linking treatment to an outpatient clinic, optimizing service provision and patient choice. *J. Eval. Clin. Pract.* **2008**, *14*, 996–1001. [CrossRef] [PubMed]
119. Afthanorhan, W.M.A.B.W. A comparison of partial least square structural equation modeling (PLS-SEM) and covariance based structural equation modeling (CB-SEM) for confirmatory factor analysis. *Int. J. Eng. Sci. Innov. Technol.* **2013**, *2*, 198–205.
120. Purwanto, A. Partial Least Squares Structural Squation Modeling (PLS-SEM) Analysis for Social and Management Research: A Literature Review. *J. Ind. Eng. Manag. Res.* **2021**, *2*, 114–123.
121. Chytrý, M.; Tichý, L.; Holt, J.; Botta-Dukát, Z. Determination of diagnostic species with statistical fidelity measures. *J. Veg. Sci.* **2002**, *13*, 79–90. [CrossRef]
122. Dijkstra, T.K.; Henseler, J. Consistent partial least squares path modeling. *MIS Q.* **2015**, *39*, 297–316. [CrossRef]
123. Al-Maroof, R.S.; Salloum, S.A. An Integrated Model of Continuous Intention to Use of Google Classroom. In *Recent Advances in Intelligent Systems and Smart Applications*; Springer: Berlin/Heidelberg, Germany, 2021; pp. 311–335.

124. Hair, J.F., Jr.; Hult, G.T.M.; Ringle, C.M.; Sarstedt, M. *A Primer on Partial Least Squares Structural Equation Modeling (PLS-SEM)*; Sage Publications: Kowloon, Hong Kong, 2021.
125. Chakrabarti, S.; Makhija, M. Exploratory study on variables impacting display advertising spend of leading advertisers in the USA. *J. Mark. Commun.* **2021**, *27*, 176–206. [CrossRef]
126. Chin, W.; Cheah, J.-H.; Liu, Y.; Ting, H.; Lim, X.-J.; Cham, T.H. Demystifying the role of causal-predictive modeling using partial least squares structural equation modeling in information systems research. *Ind. Manag. Data Syst.* **2020**, *120*, 2161–2209. [CrossRef]
127. Roldán, J.L.; Sánchez-Franco, M.J. Variance-based structural equation modeling: Guidelines for using partial least squares in information systems research. In *Research Methodologies, Innovations and Philosophies in Software Systems Engineering and Information Systems*; IGI Global: Hershey, PA, USA, 2012; pp. 193–221.
128. Yahaya, M. Partial Least Square Structural Equation Modeling (PLS-SEM): A Note For Beginners; Sokoto, Nigeria. 2019. Available online: https://www.researchgate.net/publication/339433516_Partial_Least_Square_Structural_Equation_Modeling_PLS-SEM_A_NOTE_FOR_BEGINNERS (accessed on 1 September 2021).
129. Sharma, P.N.; Shmueli, G.; Sarstedt, M.; Danks, N.; Ray, S. Prediction-Oriented Model Selection in Partial Least Squares Path Modeling. *Decis. Sci.* **2018**, *52*, 567–607. [CrossRef]
130. Umar, U.A.; Shafiq, N.; Ahmad, F.A. A case study on the effective implementation of the reuse and recycling of construction & demolition waste management practices in Malaysia. *Ain Shams Eng. J.* **2020**, *12*, 283–291. [CrossRef]

sustainability MDPI

Article

# Fisheries in the Context of Attaining Sustainable Development Goals (SDGs) in Bangladesh: COVID-19 Impacts and Future Prospects

Atiqur Rahman Sunny [1,2,*], Mahmudul Hasan Mithun [3], Shamsul Haque Prodhan [1], Md. Ashrafuzzaman [1], Syed Mohammad Aminur Rahman [4], Md Masum Billah [5], Monayem Hussain [6], Khandaker Jafor Ahmed [4], Sharif Ahmed Sazzad [7], Md Tariqul Alam [8], Aminur Rashid [8] and Mohammad Mosarof Hossain [9]

1 Department of Genetic Engineering and Biotechnology, Shahjalal University of Science and Technology, Sylhet 3100, Bangladesh; shamsulhp@gmail.com (S.H.P.); azamanbt@gmail.com (M.A.)
2 Suchana Project, WorldFish, Bangladesh Office, Gulshan, Dhaka 1213, Bangladesh
3 Bangladesh Fisheries Research Institute, Mymensing 2201, Bangladesh; mithun622bsmrau@gmail.com
4 Department of Geography, Environment and Population, The University of Adelaide, Adelaide, SA 5005, Australia; syedamin22nd@gmail.com (S.M.A.R.); khandaker.ahmed@adelaide.edu.au (K.J.A.)
5 Ministry of Education, Bangladesh Secretariat, Dhaka 1000, Bangladesh; masumbahadur@gmail.com
6 Ecofish-ll Project, WorldFish, Bangladesh Office, Gulshan, Dhaka 1213, Bangladesh; hussain.monayem@gmail.com
7 Pathfinder Agro and Fisheries Consultation Center, Sylhet 3100, Bangladesh; sazzad139@gmail.com
8 Department of Aquaculture, Sylhet Agricultural University, Sylhet 3100, Bangladesh; talam.aq@sau.ac.bd (M.T.A.); aminur0579sau@gmail.com (A.R.)
9 Department of Coastal and Marine Fisheries, Sylhet Agricultural University, Sylhet 3100, Bangladesh; mosarofmh.cmf@sau.ac.bd
* Correspondence: atiksunny@yahoo.com

**Citation:** Sunny, A.R.; Mithun, M.H.; Prodhan, S.H.; Ashrafuzzaman, M.; Rahman, S.M.A.; Billah, M.M.; Hussain, M.; Ahmed, K.J.; Sazzad, S.A.; Alam, M.T.; et al. Fisheries in the Context of Attaining Sustainable Development Goals (SDGs) in Bangladesh: COVID-19 Impacts and Future Prospects. *Sustainability* **2021**, *13*, 9912. https://doi.org/10.3390/su13179912

Academic Editor: Shervin Hashemi

Received: 8 July 2021
Accepted: 30 August 2021
Published: 3 September 2021



**Abstract:** Fisheries and the aquaculture sector can play a significant role in the achievement of several of the goals of the 2030 Sustainable Development agenda. However, the current COVID-19 situation can negatively impact the fisheries sector, impeding the pace of the achievement of development goals. Therefore, this paper highlighted the performance and challenges of the fisheries sector in Bangladesh, emphasising the impact of COVID-19 and the significance of this sector for achieving the Sustainable Development Goals (SDGs), through primary fieldwork and secondary data. The total fish production in the country has increased more than six times over the last three decades (7.54 to 43.84 lakh MT) with improved culture techniques and extension services. Inland closed water contributions have increased to 16%, while inland open water has declined to 10%, and marine fisheries have dropped to 6% over the past 18 financial years (2000–2001 to 2018–2019). COVID-19, a significant health crisis, has also affected various issues associated with aquatic resources and communities. Transportation obstacles and complexity in the food supply, difficulty in starting production, labour crisis, sudden illness, insufficient consumer demand, commodity price hikes, creditor's pressure, and reduced income were identified as COVID-19 drivers affecting the fisheries sector. The combined effect of these drivers poses a significant threat to a number of the SDGs, such as income (SDG1), nutrition (SDG2), and food security (SDG3 and SDG12), which require immediate and comprehensive action. Several recommendations were discussed, the implementation of which are important to the achievement of the SDGs and the improved management of the aquatic sector (SDG14—life below, and SDG16—life above water).

**Keywords:** fisheries; aquaculture; food security; COVID-19; SDGs; Bangladesh

## 1. Introduction

Bangladesh, a nation blessed with diverse aquatic resources, is located in Southeast Asia, between 20°′34′ to 26°′38′ N latitude and 88°′01′ to 92°′42′ E longitude, and has an area of 147,570 km$^2$ [1]. It is one of the top fish-producing countries globally, with its vast inland, coastal, and marine water resources [2]. The fisheries sector in Bangladesh is one of the most productive and dynamic sectors, which has played an increasingly significant role in the economy for the last few decades [3,4]. Since its independence in 1971, Bangladesh has achieved tremendous progress in the fisheries sector, contributing significantly to the promotion of the dependent community's food security and socioeconomic status [5,6], which are crucially highlighted in the UN's Sustainable Development Goals (SDGs). There are different types of interactions between SDGs and aquatic food production systems and the well-being of the dependent community. Fish from capture fisheries and aquaculture help marginalised people maintain their food supply and livelihoods in terms of nutritional security, good health and well-being, poverty alleviation, and reduced inequalities. Around 12% of the total population of Bangladesh is involved in the fisheries sector directly and indirectly [7], contributing 3.50% to the national GDP, 25.72% to the agricultural GDP [7,8] and providing a significant share of animal proteins.

The diversified fisheries resources of Bangladesh are mainly divided into two groups: inland and marine fisheries [3,9]. Inland fisheries have two sub-sectors, inland capture and inland culture fisheries, covering 3.89 and 0.82 million ha, respectively (Table 1). The inland open water habitats include rivers and estuaries (853,863 ha), the Sundarbans (177,700 ha), beels (114,161 ha), Kaptai Lake (68,800 ha), and floodplains (2,675,758 ha) (Figure 1). On the other hand, inland closed water habitats include ponds (397,775 ha), seasonal cultured water bodies (144,217 ha), Baor (5671 ha), shrimp/prawn farms (258,553 ha), crab farms (9377 ha), pen culture (6330 ha), and 1.76 lakh m$^3$ for cage culture [7]. In 2018–2019, the inland capture, inland culture, and marine fisheries contributed approximately 28.19%, 56.76%, and 15.05% of the country's total fisheries production, respectively [7] (Table 1). Over the last three decades, the total fish production increased more than six times from 7.54 MT in 1983–1984 to BDT 43.84 lakh MT in 2018–2019. In 2018–2019, the country earned BDT 425,031.00 by exporting almost 73.17 thousand MT of fish and fishery products [7].

**Table 1.** Sector-wise fish production and water areas of Bangladesh in 2018–2019.

| Fisheries Types | Fisheries Sector | Water Area (ha) | Production (MT) | Production (%) |
|---|---|---|---|---|
| Inland Fisheries | Inland open water (capture) rivers, estuaries, Sundarbans, beels, Kaptai Lake, floodplains | 3,890,282 | 1,235,709 | 28.19 |
| | Inland closed water (culture) pond, seasonal cultured water body, Baor, shrimp/prawn farm, pen culture, cage culture | 821,923 | 2,488,601 | 56.76 |
| Total (Inland) | | 4,712,205 | 3,724,310 | 84.95 |
| Marine Fisheries | Industrial (trawl) fishing | | 107,236 | 2.45 |
| | Artisanal fishing | | 552,675 | 12.61 |
| Total (Marine) | | | 659,911 | 15.05 |
| Total production (Inland + Marine) | | | 4,384,221 | 100 |

Data source: Department of Fisheries (DoF) [7].

Although inland capture fisheries are an important source of total fish production, they have reduced remarkably due to several anthropogenic interventions and natural causes, e.g., pollution, over-exploitation, destructive fishing, and habitat degradation [10]. In 1983–1984, the total production of inland capture and culture fisheries was 62.59% and 15.53%, respectively, whereas, in 2018–2019, inland capture fisheries sharply dropped to 28.19%, and inland culture fisheries increased to 56.76% [7]. However, in recent years, several socio-eco-friendly programs were implemented to increase the productivity of inland open waters, such as a community-based fisheries management program, the establishment of fish sanctuaries, beel nursery management, the stocking of fingerlings in water bodies, the restoration of the aquatic habitats, and an increase in water area under cage and pen farming [11]. Bangladesh also has a huge potential for marine fisheries, comprising artisanal (fishing below 40 m depth) and industrial (trawl fishing above 40 m depth) fisheries and coastal sub-sectors. Despite having a long coastline (ca. 710 km) and a large marine water area, the marine fisheries sector is underdeveloped compared to other business sectors in Bangladesh [12]. Nonetheless, the recently settled maritime boundary with Myanmar and India, up to 200 nautical miles from the coastline, comprising 118,813 km$^2$ of maritime water, represent a huge blue economy development prospect. The potentiality and challenges of Bangladesh's fisheries and aquaculture sector were reviewed by Ghose (2014) up to 2012 and Shamsuzzaman et al. (2017) up to 2015. The impact of the COVID-19 pandemic on that sector is yet to be explored.

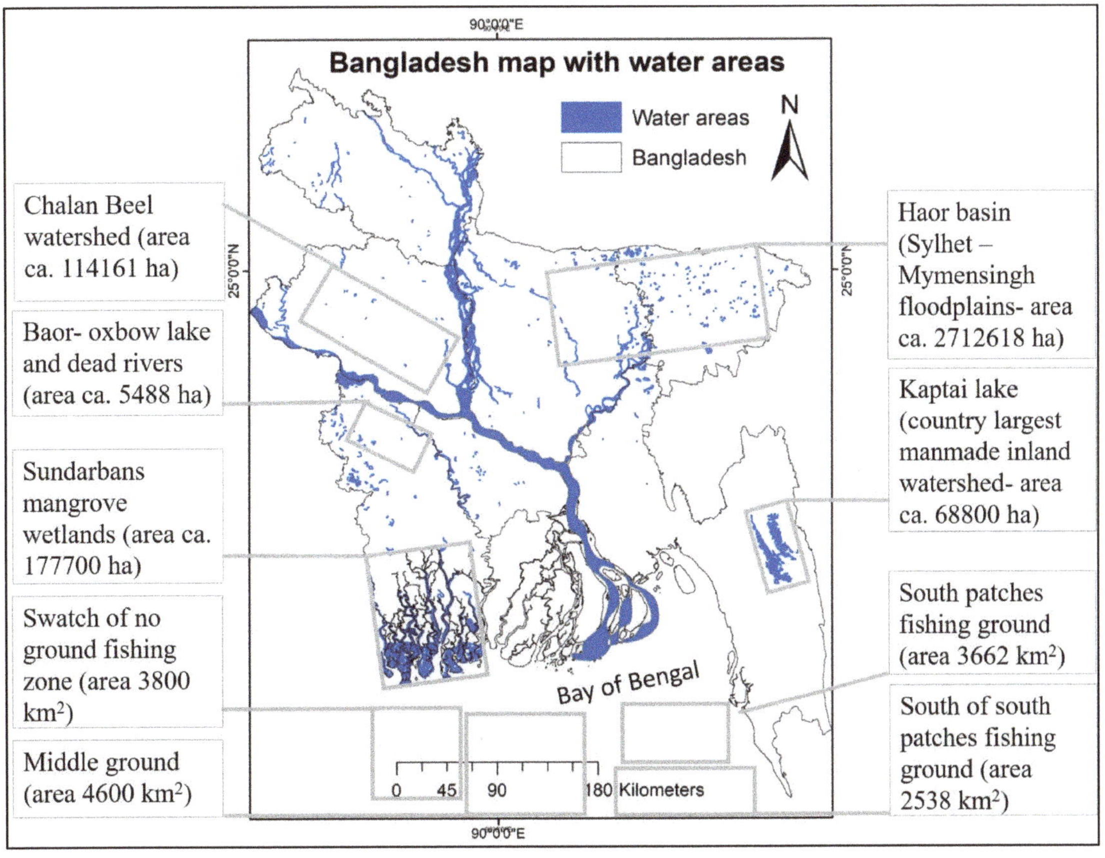


**Figure 1.** Aquatic resources and focal natural waters in Bangladesh, including major fishing grounds in the Bay of Bengal. This map illustrates the tentative location of major water areas of Bangladesh and does not necessarily reflect any precise administrative and or political boundary line with neighbouring countries.

This aquaculture and fisheries nexus requires observation through the lens of the COVID-19 pandemic because it is likely to impact the aquatic food system and the resilience of the dependent communities in Bangladesh. Bangladesh is struggling to cope with the adverse effects of COVID-19 due to its resource constraints. As of 1 August 2021, the country has confirmed 1,264,328 COVID-19 positive cases, 1,093,266 recovered cases, and the total death toll of 20,916. Bangladesh has adopted so-called preventive measures, such as social distancing, lockdowns, local and international travel restrictions, and work from home. These initiatives reduce the household income mainly for the wage earners and, as a result, people are facing difficulties managing their living expenses. Indirectly, COVID-19 has affected aquatic food production as well. The restrictions in movement, difficulty in fish and input transportation, low consumer demand, unsold mature fish, low market prices, disruption in the new farming cycle, the labour crisis, limited attendance of the service provider, debt cycle, ban period, and disease susceptibility have all impacted the livelihood of fish farmers, fishers, and associated stakeholders. There are limited studies on the impacts of the COVID-19 pandemic on small-scale fisheries and aquaculture. A more in-depth focus is needed to obtain a general overview of the impacts of this pandemic on the different sub-sectors of fisheries, dependent communities, and associated stakeholders. Aquaculture and fisheries sectors provide crucial provisions to the wider society and have solid linkages for achieving several SDGs within the context of Bangladesh. Nevertheless, literature addressing the linkage of fisheries and the aquaculture sector with relevance to development (e.g., the economic, social, and environmental dimensions of sustainability) and the achievement of the SDGs are also limited. Therefore, this research aimed to highlight the potential and challenges of the fisheries and aquaculture sector in Bangladesh by bridging the gap of the existing situation (e.g., provide an update on the status of the sector, the COVID-19 pandemic impact on the aquatic resource system, and the involved stakeholders and dependent community). This attempt also reflected this sector's potential linkage and significance for achieving several SDGs in Bangladesh because of the COVID-19 pandemic. Finally, this study recommends plans and policies to aid in the recovery of the fisheries sector following the COVID-19 pandemic and analogous inversion in the future.

## 2. Materials and Methods

### *2.1. Study Areas*

This cross-sectional study was purposively conducted in 15 communities of 10 districts under eight administrative divisions (Table 2).

**Table 2.** Selected study areas in eight administrative divisions in Bangladesh.

| Divisions (n = 8) | Districts (n = 10) | Communities (n = 15) |
|---|---|---|
| Khulna | Bagerhat | Chila and Joymonigul |
| | Chuadanga | Darsana |
| Rangpur | Gaibanda | Palashbari |
| Rajshahi | Pabna | Bera |
| Dhaka | Narsingdi | Polash and Narsingdi Sadar |
| Chattogram | Chandpur | Haimchar |
| | Laxmipur | Char Alexandar |
| Barishal | Barishal | Kawarchar and Laharhat |
| Sylhet | Sylhet | Osmaninagar and Balaganj |
| Mymensingh | Mymensingh | Phulpur and Mymensingh Sadar |

### *2.2. Empirical Primary Data Collection*

This study on Bangladesh's fisheries and aquaculture fishing communities emerged while the COVID-19 pandemic remained pervasive within the country. Hence, the study was based on field data administrating, such as key informant interviews (KIIs), focus group discussion (FGDs), in-depth interviews, and a primary survey questionnaire. In this case, appropriate ethical consideration was maintained, and the respondents' enthusiastic participation during the field visit, data collection, and preservation was ensured.

#### 2.2.1. In-Depth Tete-a-Tete Interviews

As the COVID-19 scenario of the country was gradually deteriorating and the government was enforcing various restrictions, it was difficult to move across the state for in-depth tete-a-tete interviews. In total, 50 in-depth tete-a-tete interviews were conducted with the aquaculture, fisheries, and coastal communities, hatcheries and fish production farmers to assess the impact of the pandemic on the fisheries sector. The COVID-19 health protocols and social distancing were strictly maintained during the field site visits.

#### 2.2.2. Key Informant Interviews

We conducted 50 key informant interviews (KIIs) with fisheries-dependent communities (fish, shrimp, crab, snail, and oyster harvesters) to determine the impacts of the COVID-19 pandemic. With strict lockdown and shutdowns, it was challenging to physically meet with eminent and noteworthy officials and various stakeholders for face-to-face interviews. In that case, around 30 telephonic stakeholder interviews were also conducted to cross-check the data accuracy.

#### 2.2.3. Focus Group Discussion

A total of 22 focus group discussions (FGDs) were conducted to understand the context of COVID-19 and the prospects of the fisheries sector tackling the health and food crisis. Respondents provided some exciting information during the FGDs. Participants from diverse departments of aquaculture and fisheries of the selected districts were considered for the FGDs. In every FGD, 6–8 respondents participated, ensuring the World Health Organization's COVID-19 health protocol was followed, face masks were worn, and social distance was maintained.

### *2.3. Secondary Data Gathering*

Secondary data on the COVID-19 pandemic was collected from the Directorate General of Health Services (DGHS), Institute of Epidemiology and Disease Control and Research (IEDCR). Relevant fisheries- and aquaculture-related secondary data, including, but not limited, to fish production, and food demand and supply, was collected from the Department of Fisheries (DoF). Other significant data sources were scientific articles and technical reports.

### *2.4. Data Analysis*

The recording of the KIIs and FGDs were transcribed and analysed. The other content and the summary of the data were analysed to develop a conceptual understanding of this study. MS Excel (version 2016) was used to formulate the quantitative data using descriptive statistics in frequencies and percentages. The data were presented in graphs, bars charts, and tables.

## 3. Present Status of Inland Fisheries Resources of Bangladesh

### *3.1. Inland Open Water Fish Production in Bangladesh*

Inland fisheries refer to the capturing and culturing of fishes in inland water. Bangladesh ranked third in inland captured fisheries and ranked fifth in world aquaculture production [7]. However, in production from single-species culture, explicitly the monoculturing of Tilapia (*Oreaochromis niloticus*), Bangladesh ranked fourth in the world and became third in Asia. The national fish, hilsa (*Tenualosa ilisha*), contributed around 2.15% to the total fish production in the country, with a total production of 5.33 lakh MT, and was awarded the national Geographical Indication Registration Certificate [7,8].

Inland fisheries are sub-divided into inland capture (fishing from inland open water) and inland culture fisheries (aquaculture). The inland capture fishery contains five habitats: the river estuary, the Sundarbans (the world's largest mangrove forest), beels, Kaptai Lake, and floodplains. The inland culture fishery contains seven types of habitats, such as ponds, seasonal culture water bodies, baors, shrimp and prawn farms, crab farms, pen culture, and cage culture [7]. In 2018–2019, the total fish production in Bangladesh was reported as 4.38 million MT, of which 1.23 million MT (28.19%) were from inland open waters, 2.48 million MT (56.76%) from inland closed waters, and 0.65 million MT (15.05%) from marine fisheries (Table 1). Inland open water fisheries are still a significant source of total fish production, but their share has declined from 38.68% in 2000–2001 [13] to only 28.19%

in 2018–2019 [7]. Conversely, inland closed water fisheries contributions have increased from 40.01% in 2000–2001 [13] to 56.76% in 2018–2019 [7]. In recent years, the average yield (annual fish harvest per ha in metric tons) in open inland waters improved sharply (Table 3). The beel fishery has a high productivity of 875 kg/ha compared to the other habitats of inland open water (Table 3). A total of 293 fish species noted by Hossain et al. 2012 [14], and 260 taxa with 12 exotic fish and 24 prawn species, were recorded in Bangladesh [15] and are known to inhabit the nation's freshwaters. Significant carp species, such as catla (*Catla catla)*, rui *(Labeo rohita)*, mrigal *(Cirrhinus cirrhosus)*, and kalbasu *(Labeo calbasu)*, along with some exotic carp species, such as silver carp (*Hypophthalmichthys molitrix*), grass carp (*Ctenopharyngodon idellus*), and common carp (*Cyprinus carpio*), are the most available carp species found in the market. About 40–50 small indigenous fish species grow to a maximum length of 25 cm [16]. Some more commonly found species of this variety include titputi (*Puntius ticto*), mola (*Amblypharyngodon mola*), kholisha (*Colisa lalius*), koi (*Anabas testudineus*), and baila (*Glossogobius giuris*). Several small indigenous species of fish are now endangered or critically endangered in Bangladesh [14], although the inland water resources of Bangladesh offer significant potential for the development of freshwater capture and culture fisheries [17,18]. The catch from the inland capture fishery is noteworthy, especially for fish supply and employment in rural areas.

**Table 3.** Inland open water fish production of Bangladesh in 2018–2019 [7].

| Inland Open Water (Capture Fisheries) | Water Area (ha) | Production (MT) | Productivity (Kg/ha) |
|---|---|---|---|
| Rivers and Estuaries | 853,863 | 320,598 | 381 |
| The Sundarbans | 177,700 | 18,225 | 103 |
| Beel | 114,161 | 99,197 | 875 |
| Kaptai Lake | 68,800 | 10,152 | 154 |
| Floodplains | 2,712,618 | 768,367 | 292 |

Data source: Department of Fisheries (DoF) [7].

### 3.2. *Inland Closed Water (Culture) Fish Production of Bangladesh*

The inland aquaculture has generally experienced the fastest growth in the country, with the adoption of advanced culture technologies, species intensification, and improvement of farming systems, particularly in pond aquaculture, among various segments of the fisheries sub-sector [19,20]. However, around half of the fish are provided by aquaculture for direct human consumption and is set to grow further. The aquaculture industry boosts the economy with increasing production capacity and high export opportunities. The average growth rate of the fisheries sector over the last 10 years was about 5.01%, while aquaculture grew by 8.59%. Generally, two types of aquaculture are practised in Bangladesh, namely, freshwater and coastal aquaculture. Marine aquaculture is growing in the country, and the farming of marine fish is significant. Freshwater aquaculture is mainly comprised of pond farming of carps (Indian major and Chinese carps), pangasius (*Pangasius pangasius*) and other catfishes, tilapia (*Oreochromis nilotics*), Mekong climbing perch (*Anabas testudineus*), and several other domesticated fish. Pen and cage culture are two new cultivation approaches in Bangladesh and contributed to 12,361 MT (0.28%) and 3802 MT (0.09%) to the total fish production in 2018–2019, respectively (Table 4). These new methods have the highest potential for increasing fish production in Bangladesh. Total fish production in pen culture declined in the previous years in 2018–2019, but cage culture is becoming more popular and the production rate has increased in recent years. Coastal aquaculture is mainly comprised of shrimp and prawn farming in *ghers* (a coastal pond or enclosures with dug-out soil used to create dykes) and, in 2018–2019, the total production of shrimp and prawn was about 258,039 MT. In Bangladesh, aquaculture production systems are mainly extensive and improved extensive, with some semi-intensive and intensive systems in very few cases [17]. The current aquaculture productions for the pond, seasonal water body, baor (oxbow lake) and shrimp *ghers* are 4.96, 1.50, 1.82, and 0.99 (MT/ha), respectively (Table 4). Inland pond culture represents the mainstay of aquaculture in Bangladesh, accounting for more than 80% of the total recorded aquaculture production. Pond aquaculture is the most practised type of closed water fishery in Bangladesh and contributed 45.04% (1,974,632 MT) to total fish production in 2018–2019 [7]. The main cultured species in the coastal areas of Bangladesh are giant tiger prawn (*Penaeus monodon*) and giant river prawn (*Macrobrachium rosenbergii*) [21].

**Table 4.** Inland closed water fish production in 2018–2019 in Bangladesh.

| Inland Closed Water | Water Area (ha) | Production (MT) | Productivity (Kg/ha) |
|---|---|---|---|
| Pond | 397,775 | 1,974,632 | 4964 |
| Seasonal cultured- water body | 144,217 | 217,340 | 1507 |
| Baor | 5671 | 10,343 | 1824 |
| Shrimp/Prawn farm | 258,553 | 258,039 | 998 |
| Crab farm (capture and fattening culture) | 9377 | 12,084 | 1289 |
| Pen culture | 6330 | 12,361 | 1953 |
| Cage culture | 176,000 $m^3$ | 3802 | 22 $kg/m^3$ |

Data source: Department of Fisheries (DoF) [7].

Shrimp and Prawn Production Trends in Bangladesh

Shrimp and prawn are major export items and are mainly produced in coastal districts, such as Khulna, Satkhira, Barishal, Patuakhali, Bagerhat, Bhola, Chattogram, and Cox's Bazar. As a result of increasing demand and prices, shrimp culture began expanding in the 1970s, aiming mainly at the export markets [22]. It was reported that the shrimp and prawn culture area had expanded from 217,177 ha in 2005–2006 to over 258,553 ha in 2018–2019. The production of shrimp has increased from 211,010 MT to 239,855 MT (Figure 2), and the average production of shrimp and prawn is 998 kg/ha [7]. In the Khulna region, shrimp is cultured in a modified rice field (also called *gher*), while in Cox's Bazar, shrimp and salt are produced alternately [23,24]. However, different programs and development projects are also being implemented for the increased production and promotion of shrimp aquaculture. The shrimp cluster farming approach was popularised for enhancing shrimp production and promoting a business-friendly supply chain by adopting good aquaculture practices. The government has maintained standards in all stages of fish and shrimp production, processing, and export. In Bangladesh, shrimp culture technology has intensified in recent years, but the production level has not increased satisfactorily. One of the main reasons for that is the unavailability of good quality post-larvae (PL) and feed. The specific, pathogen-free post-larvae should be supplied broadly to shrimp farms to reduce the different viral and bacterial disease related mortality.

### 3.3. *Marine Fisheries Production in Bangladesh*

The Bay of Bengal is blessed with rich coastal and marine ecosystems, hosting a wide range of biodiversity, including fish, shrimp, molluscs, crabs, mammals, reptiles, seaweeds, etc. (Figure 3) [25,26]. It has the potential to contribute to the economy of Bangladesh by the exploration, exploitation, and management of living and non-living resources [27,28]. Most specifically, after the recent decision of the Bangladesh-Myanmar maritime boundary in 2012 and the decision of the India-Bangladesh maritime boundary in 2014, the country established sovereign rights to more than 118,813 $km^2$ of territorial sea, 200 nautical miles of Exclusive Economic Zone (EEZ), and an assortment of living and non-living resources under the continental shelf up to 354 nautical miles from the Chittagong coasts.

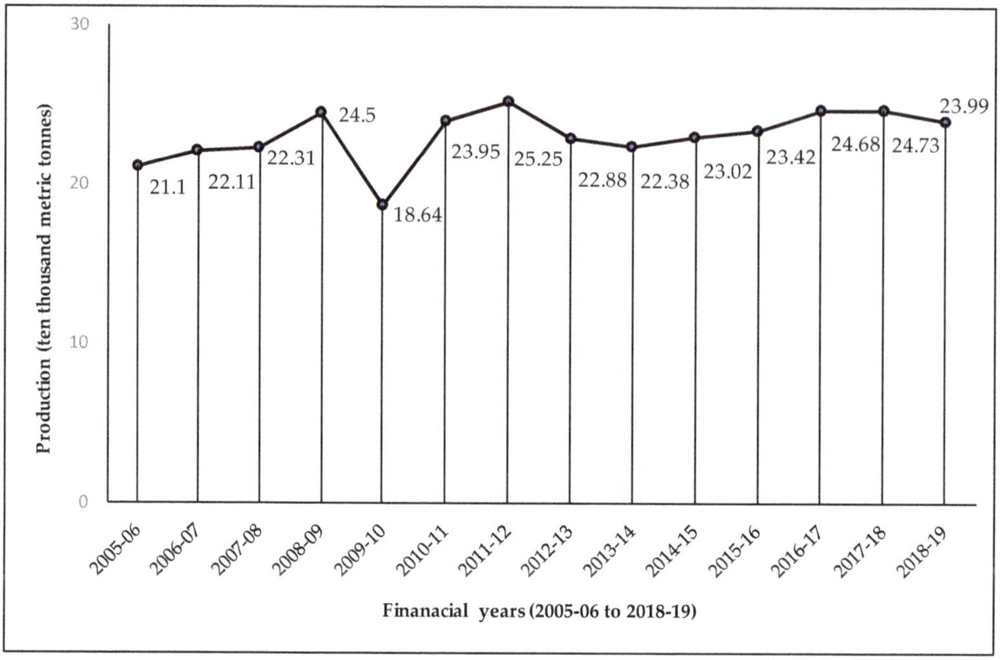


**Figure 2.** The trend of shrimp/prawn production from 2005–2006 to 2018–2019. Data source: Department of Fisheries (DoF) [7].

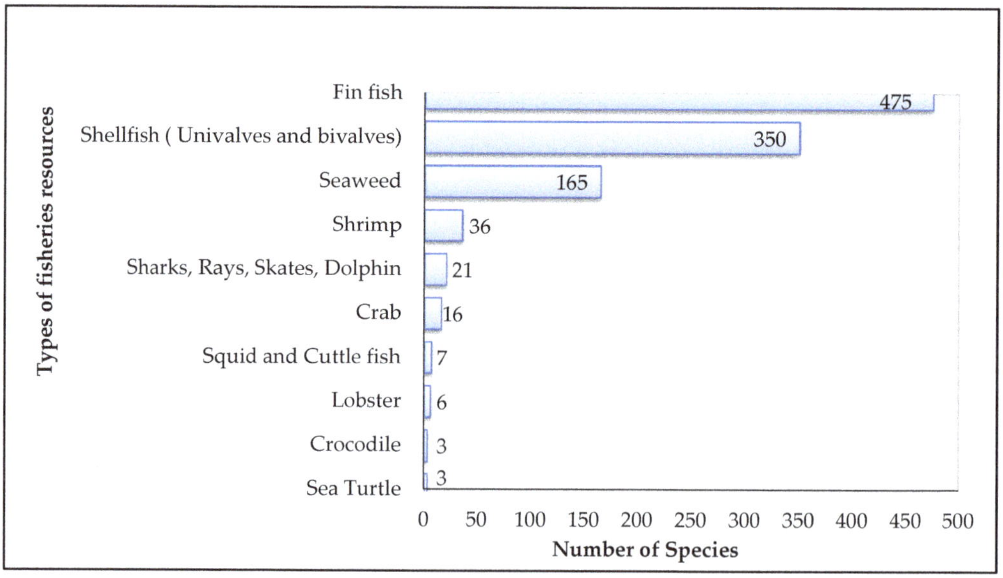


**Figure 3.** Coastal and marine fisheries resources in Bangladesh [29]. Data source: Food and Agricultural Organization of the United Nations (FAO) [29].

Marine fisheries production contributes only 15% to overall national fish production (Table 1) [29,30]. The coastal fisheries resources are over-exploited, and as a result, the fish stocks and other resources have declined. Numerous surveys examined the status of marine fisheries resources between the 1970s and 1980s, but no current or comprehensive knowledge is available on the fisheries stocks, nor the systematic, biological, and ecological aspects of Bangladesh's coastal and marine fisheries [31,32]. The main commercial fishing zones in the Bay of Bengal are Swatch of no Ground, Middle Ground, South Patches and South of South Patches [33,34]. Artisanal or subsistence fishing is conducted near the coastal areas of Bangladesh. Hilsa shad, sardine, Bombay duck, salmon, pomfret, jewfish, sharks, and rays, among others, are the dominant species captured from marine and coastal water areas [7].

Although the production growth rates fluctuate among the sub-sectors of inland fisheries, the total annual production had gradually increased over the past decade (financial years of 2008–2009 to 2018–2019) and developed a solid ground for stable aquaculture and fisheries production (Table 5).

**Table 5.** Sector-wise annual fish production from financial years of 2008–2009 to 2018–2019 in Bangladesh.

| A. Inland Fisheries | Year-Wise Fisheries Production (Metric Tonnes) | | | | | | | | | | |
|---|---|---|---|---|---|---|---|---|---|---|---|
| | 2008–2009 | 2009–2010 | 2010–2011 | 2011–2012 | 2012–2013 | 2013–2014 | 2014–2015 | 2015–2016 | 2016–2017 | 2017–2018 | 2018–2019 |
| (a) Inland (Capture) | | | | | | | | | | | |
| 1. River | 138,160 | 141,148 | 144,566 | 145,613 | 147,264 | 167,373 | 174,878 | 178,458 | 271,639 | 320,598 | 325,478 |
| 2. The Sundarbans | 18,462 | 20,437 | 22,451 | 21,610 | 15,945 | 18,366 | 17,580 | 16,870 | 18,086 | 18,225 | 18,282 |
| 3. Beel | 79,200 | 79,209 | 81,564 | 85,208 | 87,902 | 88,911 | 92,678 | 95,453 | 98,117 | 99,197 | 99,890 |
| 4. Kaptai Lake | 8590 | 7336 | 8980 | 8537 | 9017 | 8179 | 8645 | 9589 | 9982 | 10,152 | 10,578 |
| 5. Floodplains | 843,671 | 781,807 | 797,024 | 696,127 | 701,330 | 712,976 | 730,210 | 747,872 | 765,782 | 768,367 | 781,481 |
| (b) Inland (culture) | | | | | | | | | | | |
| 1. Pond | 912,178 | 1,140,484 | 1,219,736 | 1,392,412 | 1,446,594 | 1,526,160 | 1,613,240 | 1,719,783 | 1,833,118 | 1,900,298 | 1,974,632 |
| 2. Seasonal culture | 35,842 | 46,902 | 51,230 | 132,163 | 200,833 | 193,303 | 201,280 | 207,658 | 215,547 | 216,353 | 217,340 |
| 3. Baor | 5038 | 8727 | 4864 | 5186 | 6146 | 6514 | 7267 | 7729 | 8002 | 8072 | 10,343 |
| 4. Shrimp | 145,585 | 155,866 | 184,939 | 196,306 | 206,235 | 216,447 | 223,582 | 239,798 | 246,406 | 254,367 | 258,039 |
| 5. Crab | - | - | - | - | - | - | - | 13,160 | 14,421 | 11,787 | 12,084 |
| 6. Pen culture | - | - | - | - | - | 13,054 | 13,070 | 13,364 | 13,368 | 10,285 | 12,361 |
| 7. Cage culture | - | - | - | - | - | 1447 | 1969 | 2062 | 2490 | 4253 | 3802 |
| B.Marine Fisheries | | | | | | | | | | | |
| 1.Marine Industrial | 35,429 | 34,182 | 41,665 | 73,386 | 73,030 | 76,885 | 84,846 | 105,348 | 108,479 | 120,087 | 107,236 |
| 2. Marine Artisanal | 479,215 | 483,100 | 504,668 | 505,234 | 515,958 | 518,500 | 515,000 | 521,180 | 528,997 | 534,600 | 552,675 |
| Total (MT) | 2,701,370 | 2,899,198 | 3,061,687 | 3,261,782 | 3,410,254 | 3,548,115 | 3,684,245 | 3,878,324 | 4,134,434 | 4,276,641 | 4,384,221 |
| Growth rate (%) | 5.39 | 7.32 | 5.6 | 6.54 | 4.55 | 4.04 | 3.84 | 5.27 | 6.6 | 3.44 | 2.52 |

Data source: Department of Fisheries (DoF) [7].

## 4. Fish Production Trends in Bangladesh from 1983–1984 to 2017–2018

Bangladesh's total inland open water fish production in 2003–2004 was 732,067 MT, which slowly increased and reached 1,235,709 MT in 2018–2019 (Figure 4) [7]. Whereas inland closed water fish production in Bangladesh has doubled in the last decades, inland open water production has been slower to increase. It was reported that the country's total inland closed water fish production was 914,752 MT in 2003–2004 and had become 2,405,415 MT by 2017–2018. The country's total marine water fish production was 455,207 MT in 2003–2004 and had become 654,687 MT by 2017–2018. However, during the last three years, a slight increase in marine water fish production was observed, while it remained more or less constant in 2015–2018.

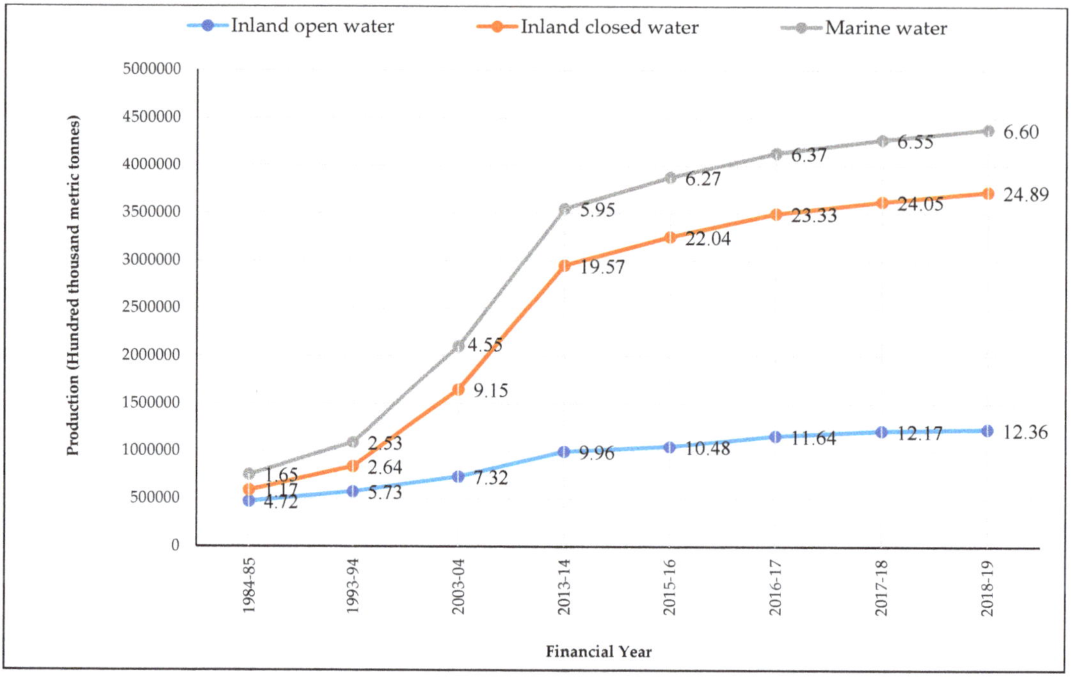


**Figure 4.** The trend of fish production in Bangladesh (1983–1984 to 2018–2019). Data source: Department of Fisheries (DoF) [7].

## 5. Export of Fish and Fishery Products

The fisheries sector of Bangladesh has emerged as the second most essential contributor to export earnings. Ten categories of fishery products (frozen freshwater fish, frozen marine water fish, frozen shrimp, chilled fish, live fish, dry fish, salted and dehydrated fish, kuchia (*Monopterus cuchia*) and crab (*Scylla* sp.) in live condition, and fish scales/shrimp skulls) are exported from Bangladesh to more than 55 countries. The shrimp export destinations are highly concentrated in the European Union (EU). More than 45% of fishery exports (by quantity) were exported to the EU in 2015. The remaining 55% were exported to the USA, Japan, Russia, China, Thailand, Vietnam, India, Malaysia, Philippines, and Saudi Arabia [11,35,36]. The year-wise annual export value was highest in 2013–2014, but during the last three years (2015–2019), the export value and quantity sharply decreased compared to previous years (Figure 5) [7].

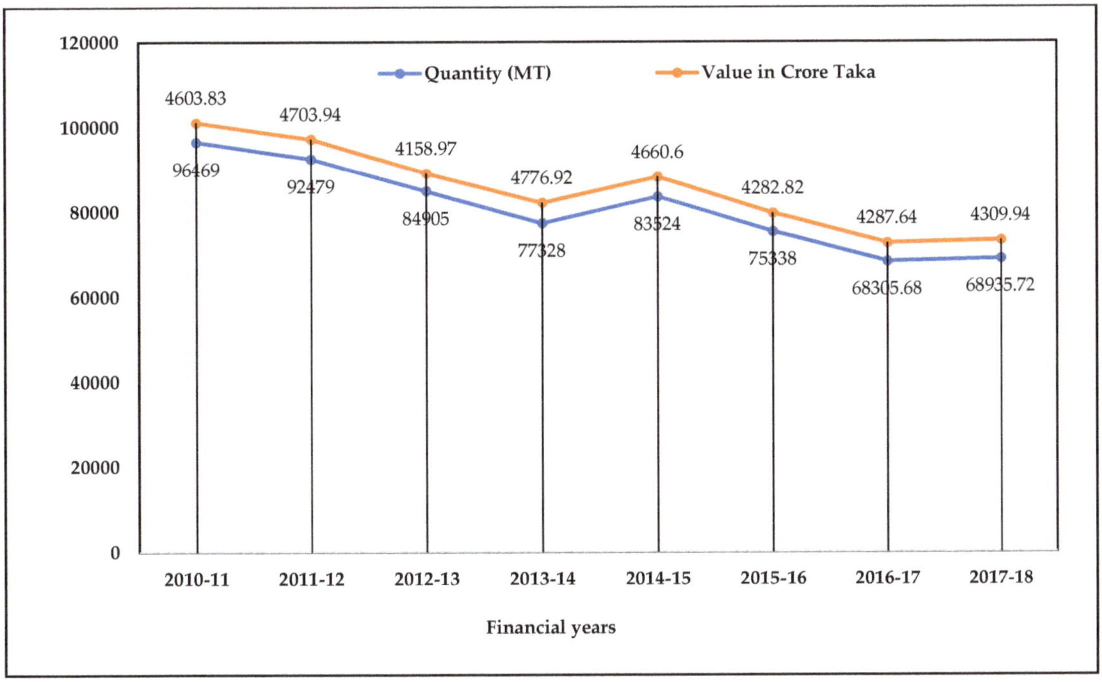


**Figure 5.** Year-wise annual export value of fish and fisheries product in BDT. Data source: Department of Fisheries (DoF) [7].

## 6. Fisher's Livelihood and Vulnerabilities: Observations from Fieldwork

The research findings showed that most fishers had inadequate income and poor housing facilities, which is concordant with other studies' findings in Bangladesh [37,38]. Fishers were identified as having not received proper institutional education and bearing the responsibilities of large family sizes [39,40]. They, in turn, cannot afford to send their children to school, which is relevant to SDG 4 which addresses access to quality education. However, most fishers reported having sanitary latrine facilities (related to SDG 3: good health and well-being) and using solar power as a primary source of electricity.

The fishers were not concerned about the nutritional value of foods and consumed fish as it was all they could afford fish. Fishers reported a reliance on loans to maintain their family during hard times and for the purchase of fishing equipment (e.g., net, fishing pot, boat, etc.). They faced several constraints, as outlined in Figure 6. The significant constraints included natural calamities, a reduction in fish catches, an unsatisfactory law and order situation, a lack of access to credit, ban periods, environmental degradation in coastal areas, a lack of alternative income-generating activities (IGAs), lack of community-based organisation, lack of infrastructure, the burden of *Dadon* (rent money), inadequate market facilities, and the loss of fishing equipment. Dependency on a single livelihood option, namely, fishing, increases their vulnerability. Conflicts between stakeholders, such as boat owners and moneylenders, also hampered financial stability and made illegal fishing an attractive option [41,42]. The livelihood status of the fishing communities was also affected by various types of seasonal stress. The ban period and the seasonal shifts in the fish availability make the fishers more vulnerable.

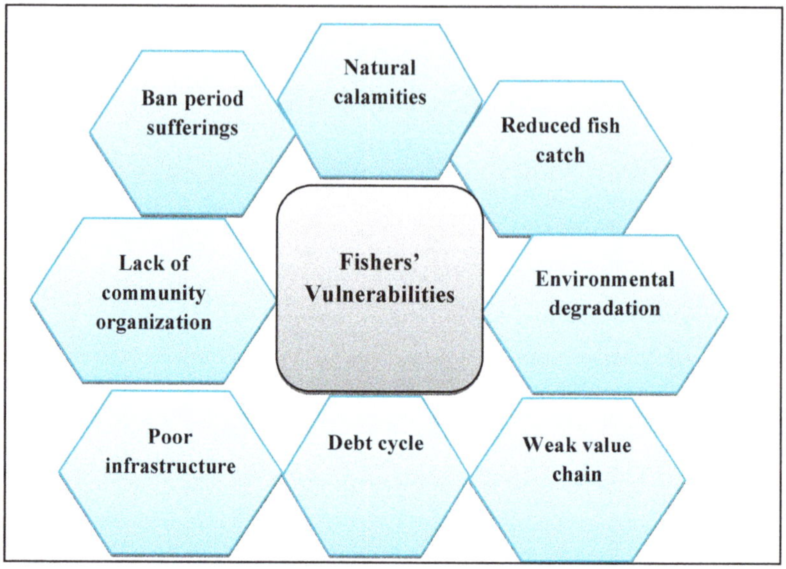


**Figure 6.** Fishers' vulnerabilities (contextually modified after Sunny et al., 2021 [8]).

Shocks, including natural disasters such as floods, droughts, cyclones, earthquakes, landslides, disease epidemics, and sudden economic changes, affect the livelihood of fishers. In the fishery context, cyclones and floods have a devastating effect on life and property (SDG 13: climate action). Loss of life (loss in human capital), physical infrastructure, and assets, such as the loss of fishing gear, roads, bridges and transport linkages, limit the access to health and education services and employment opportunities in other sectors. Compared with farming communities, coastal fishing communities are more susceptible to inclement weather conditions [43,44].

## 7. COVID-19 Scenario and Responses in Bangladesh

Bangladesh is facing momentous challenges in tackling the COVID-19 pandemic after confirming the first COVID-19 case within its borders on 8 March 2020 and the first death shortly thereafter on 18 March 2020. The media and some academics believe that COVID-19 may have been present in Bangladesh earlier, but not remained unidentified due to insufficient monitoring and inadequate testing [45]. While the number of infected cases remained low until the end of March 2020, in April, the suspected case rate showed a sharp rise with the new suspected rate in Bangladesh reaching 1155% in mid-April when 186% was the highest in Asia, placing the country ahead of India, Indonesia, Thailand, and Sri Lanka [8]. On 31 May 2020, a total of 312,061 suspected cases were tested, of which 47,153 cases were confirmed positive; there were also 9781 recovered cases, and the total number of deaths increased to 650. As time passed, the country experienced an increase of COVID-19 in all divisions. As of 1 August 2021, the highest increase in cases was observed in the Dhaka division (760,062 cases), followed by the Chattogram division (181,832 cases), Khulna division (93,644 cases), Rajshahi division (83,558 cases), Rangpur division (44,875 cases), Sylhet division (39,966 cases), Barishal division (33,870 cases), and Mymensingh division (26,521 cases). The number of recovered and death cases has also increased simultaneously (Figure 7).

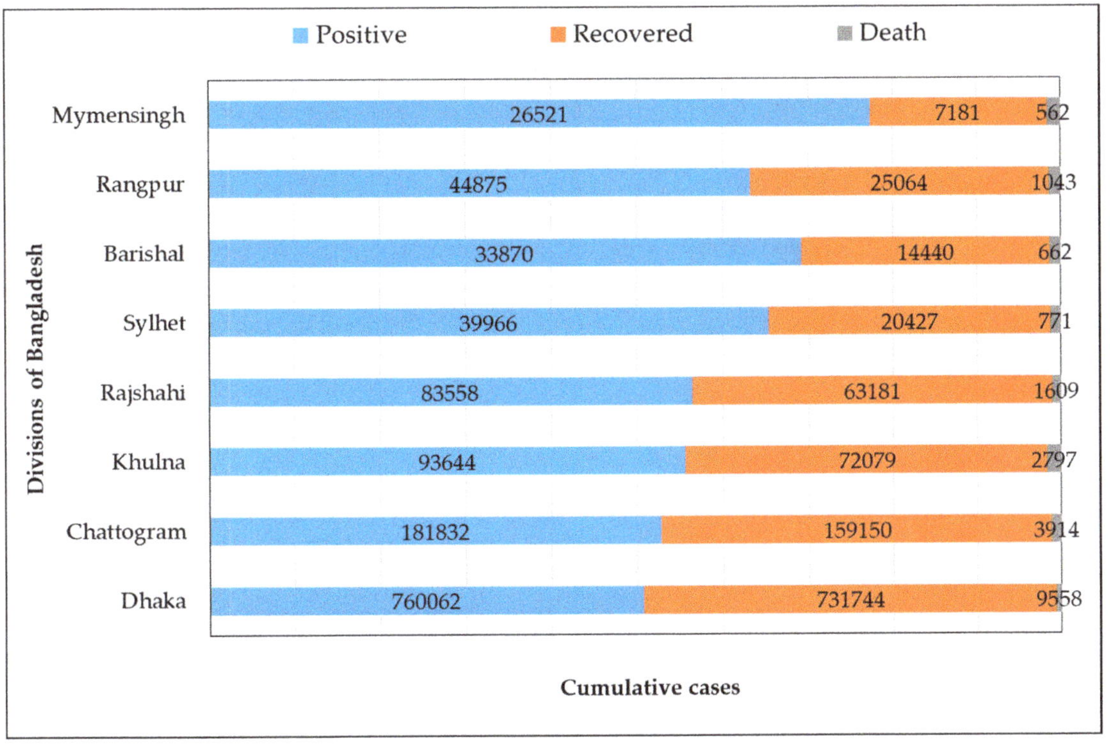


**Figure 7.** Positive, recovered, and death statistics of COVID-19 in Bangladesh.

In the early stages of COVID-19 in Bangladesh, there were no strict protocols nor lockdowns. To control the transmission of the virus, the Bangladesh government suspended on-arrival flights from all European countries on 14 March 2020, shut down all educational institutions on 17 March 2020, and deployed troops for quarantine supervision on 19 March 2020. The government of Bangladesh declared a ten-day nationwide lockdown for the first time on 23 March 2020, effective from 26 March to 04 April 2020. On 1 June 2020, the restrictions were eased despite the growing number of COVID-19 cases and associated deaths. All offices, shops, and public transportation were restricted, following the strict health guidelines. On 25 October 2020, the Bangladesh government implemented a new policy to wear face masks as "no mask, no service in government and private offices" (Table 6). In the second wave of the coronavirus outbreak in Bangladesh, seven days of controlled movement was imposed all over the country on 5 April 2021. On 14 April 2021, a strict lockdown was imposed until 21 April and later extended to May, closing all government and non-government organisations, except emergency services. The government imposed another lockdown, to control the spread, from 1 July until 07 July, but it was later extended until further notice.

**Table 6.** Snapshot of COVID-19 response in Bangladesh.

| | COVID-19 Response |
|---|---|
| 8 March 2020 | First COVID cases identified |
| 14 March 2020 | Shut down of all educational institution |
| 15 March 2020 | Banned flights from Europe |
| 18 March 2020 | First death and banning of all types of gatherings |
| 26 March 2020 | Nationwide lockdown declared |
| 1 June 2020 | Eased restrictions despite ongoing pandemic |
| 25 October 2020 | Implemented "no mask, no service" policy |
| 19 December 2020 | Closure of educational institutions extended |
| 5 April 2021 | Imposed seven days-controlled movement |
| 14 April 2021 | Strict lockdown imposed |
| 1 June 2021 | Eased restrictions despite ongoing pandemic |
| 1 July 2021 | Strict shutdown imposed |

## 8. Impact of COVID-19 on Bangladesh Fisheries

This section presents the quantitative and qualitative findings to identify the impact of the COVID-19 pandemic on the fisheries sector.

### *8.1. Perceived Impacts of COVID-19 on the Fisheries Sectors*

The COVID-19 pandemic has seriously hampered fishing, fish farming, and fish marketing activities and impacted the livelihood of fishery-depended communities. Respondent perceptions of the COVID-19 impact on the fisheries sector were explored through seven statements with five-point Likert scales (Table 7). Participants strongly believed that the pandemic negatively impacted fish production, fishing activities, household food consumption, and income. Some perceived the pandemic to be associated with increased conflicts and stress.

**Table 7.** Perceptions of the respondents on the influence of COVID-19 on the fisheries sector of Bangladesh.

| Questions Related to Impact COVID-19: COVID-19 Has | Responses (%) | | | | |
|---|---|---|---|---|---|
| | Strongly Agree | Agree | Neither Agree nor Disagree | Disagree | Strongly Disagree |
| Negatively influenced fish production | 59.00 | 22.00 | 19.00 | 0.00 | 0.00 |
| Negatively influenced fishing | 55.70 | 33.30 | 7.70 | 0.00 | 0.00 |
| Negatively impacted household food consumption | 27.90 | 47.20 | 15.40 | 5.20 | 4.30 |
| Negatively influenced income | 59.80 | 25.90 | 14.30 | 0.00 | 0.0 |
| Increased conflicts | 3.50 | 21.50 | 40.90 | 24.10 | 10.00 |
| Increased hygiene practice | 0.00 | 11.00 | 51.00 | 23.00 | 15.00 |
| Increased stress | 2.90 | 9.10 | 45.20 | 32.80 | 10.00 |

Data source: Field survey, 2021.

#### 8.1.1. Markets

COVID-19 has drastically transformed Bangladesh's fish markets, homestead, and commercial aquatic food production system (SDG 12: responsible production and consumption). Usually, marginalised fish farmers cultured fish to meet their nutritional demands and sold any remaining fish to supplement their family income. COVID-19 impacted the supply of farming inputs and restricted farmers to the home, which decreased business profitability. During the first weeks of lockdown, the fish prices fell in alignment with the reduced demand, despite fish being the most consumed animal-source food in Bangladesh (SDG 14: life below water). Fish farmers and vendors reported transportation of fish, fingerlings, feed, and other inputs as a main problem, and more than 58% of fish farmers indicated that they could not sell their marketable fish due to transportation complexity and low market demand. As a result, a good portion of mature fish remained unsold, with farmers spending extra money to feed them, which ultimately increased expenditures and reduced the family income (SDG 12: responsible production and consumption). Generally, low-income workers are the primary consumers of pangus (*Pangasianodon hypophthalmus*) and tilapia (*O. niloticus*), which are produced inexpensively all over the country. Nevertheless, at the start of COVID-19, millions of workers migrated back to their hometowns, thereby reducing demand, and the price of pangus in the retail markets decreased from BDT 120/kg to BDT 60/kg.

COVID-19 also affected the livelihood of crab farmers and traders as exports were indefinitely postponed. With China as the primary export market for crabs, and due to an import ban on 25 January, the export market of crabs remains closed. According to the Rampal-Bagerhat Crab Dealers Association, the main import target is the Chinese New Year festival, which was hindered due to the export ban with many crabs dying as a result of COVID-19 restrictions [45–48]. Though COVID-19 restrictions reduced the export opportunities, however, they opened the door to local markets and increased the access to fishery items. During the COVID-19 pandemic, local people were able to access fishery items that were previously relegated to the seafood processing industry and considered export items. The increased supply of fish in the local market reduced its price and the income of fishers.

#### 8.1.2. Fishers

In Bangladesh, April and May are the best time to release fry into farming ponds, and between March and April, fish farmers prepare the pond for the new season by selling the mature fish from the previous year. About 39% of the fish farmers and vendors claimed that they could not start a new farming cycle due to unsold mature fish (SDG 12: responsible production and consumption). It also affected the supply chain, as the local vehicles, such as trucks, pickups, and vans, were hesitant to transport fish fingerlings, feed, and other materials; because of the lockdown, transporters often face various obstacles and fines that adversely impact the fish landing centre, wholesale, and retail market. During the early phases of the pandemic, very little fishing is thought to have taken place even though no regulations were preventing it. Fishers were compelled to stay at home and were hesitant to fish when the demand for fish was decreasing [45]. Many fishers and entrepreneurs lost their income at the beginning of the crisis when they could not arrange transport for their most recent catch.

March to April was also the seasonal hilsa conservation ban period, which affects the fishing communities in the major rivers of Bangladesh. During the ban period, a restriction is imposed in all hilsa sanctuaries, and, with the seasonal ban period, the lack of alternative sources of income increased the vulnerability of these communities in addition to the COVID-19 crisis. The government-provided-subsidisation was far less than what was required for the fishers to maintain their wellbeing (SDG 1: no poverty). It was also reported that many fishers were unable to receive this support due to nepotism [8,49,50].

#### 8.1.3. Fish Farmers

In Bangladesh, March and April are the peak season for fish stocking, which coincided with the beginning of the COVID-19 crisis that began in March 2020, just before peak harvest season. As the prices and demand for fish were decreasing gradually and transportation systems broke down, most farmers decided to retain their fish in their ponds instead of harvesting them (SDG 12: responsible production and consumption). One interviewee predicted that the price of fish might remain low even after the pandemic when other farmers also begin harvesting fish, and that the surplus supply will depress prices further.

Traders usually pay fish farmers after the fish are sold. With COVID-19, farmers fear losing their income if the traders are unable to sell the fish, while unsold fish cannot be returned due to transport disruptions and a lack of storage options. Due to the delayed harvesting of mature fish, the fry cannot be stocked at the onset of the monsoon (the start of the primary growing season), which may reduce the supply of fish in the first half of 2021 [8].

8.1.4. Seed

In Bangladesh, the COVID-19 crisis started in March 2020; March is the peak season for hatchery owners to sell their fish seeds and ready themselves for a new production cycle. The demand for fish seed fell dramatically due to farmers' uncertainty and the inability to harvest (SDG 12: responsible production and consumption). In some hatcheries, due to the lack of separate nursing ponds, they were unable to store seed five to seven days more. One of the hatchery specialists from the Jessore district reported that, while the hatchery had remained closed, they had to pay the hatchery staff. Another hatchery manager from Mymensing informed us that, due to COVID-19, they had to sell their seed at a low price, and that most hatcheries and farmers doubted whether they would get their production cost or not. Hatchery owners from Northwest Bangladesh were reported to have reduced production by at least 30–40%, and four out of twenty-one have shut down completely. Credit facilities for supplying fry to farmers are no longer possible as nurseries face a liquidity crisis due to a lack of sales. Hatchery inputs, such as pituitary glands, hormones, and probiotics, are imported from outside Bangladesh. Due to lockdown, the inputs import has become more challenging. The price of the pituitary gland has tripled since the crisis started. Due to the COVID- 19 lockdown, transporting hatchery inputs and fry is a crucial challenge faced by fish hatcheries, nurseries, and fish farmers [51].

8.1.5. Feed

Most of the feed mills import their ingredients to produce feed. Due to the COVID-19 situation, importing goods has become more challenging as a result of the price hike and transportation complexity. Experts predicted a 42% reduction in feed supply due to input shortages and low demand. As the feed production costs increased, this led to speculation on whether millers will compromise the feed quality. Commercial small-scale producers are concerned about feed access.

## 9. Prospects, Development-Relevance and Linkage of Fisheries and Aquaculture Sectors in Achieving SDGs in Bangladesh

Bangladesh achieved a prestigious global position due to its outstanding production of inland open water capture fisheries (third) and culture fisheries (fifth) [7]. Globally, 60% of total hilsa (*Tenualosa ilisha)* production is from Bangladesh [8]. The fisheries and aquaculture sectors are inextricably linked to achieving several SDGs and have development relevance (e.g., economic, social, and environmental) in Bangladesh. Among the 17 SDGs, these sectors directly or indirectly contribute to achieving several SDGs, such as eliminating poverty (SDG 1), reducing hunger (SDG 2), promoting good health and well-being (SDG 3), establishing decent work and economic growth (SDG 8), influencing responsible consumption and production (SDG 12), organising climate action (SDG 13), developing life below water (SDG 14), and advancing life on land or life above water (SDG 15) (Table 8).

### *9.1. Prospects of Inland Closed Water*

Inland closed water fish production is almost saturated, but a five-fold increase in production will be needed within the next five decades to maintain current levels of aquatic food intake (SDG 12: responsible production and consumption). Bangladesh has advanced technology, such as RAS (re-circulatory aquaculture system), biofloc, and aquaponics [2]. These technologies are considered eco-friendly alternatives of fish culture, a less disease-prone system, which enhances productivity and, consequently, sustainable production.

**Table 8.** Prospective linkage of fisheries and aquaculture sectors with achieving the SDGs in the context of Bangladesh.

| The United Nations Sustainable Development Goals SDGs (1–17) | Direct or Indirect Linkage with Fisheries and Aquaculture Sub-Sectors | | | Linked with the Dimensions of Sustainable Development |
|---|---|---|---|---|
| | Inland Closed Water (Culture) | Inland Open Water (Capture) | Marine Water (Capture) | |
| Eliminate poverty | High | High | Medium | 1, 2 |
| Erase hunger | High | High | High | 1, 2 |
| Establish good health and promote well-being | High | Medium | Medium | 1, 2 |
| Provide inclusive and equitable quality education | Low | Low | Low | 1, 2 |
| Gender equality and women empowerment | Low | Low | Low | 2 |
| Ensure clean water and sanitation | Low | Low | Low | 3 |
| Affordable, reliable, and modern clean energy | Low | Low | Low | 3 |
| Create decent work and promote economic growth | Medium | Medium | Low | 1, 2 |
| Increase industry, innovation, and infrastructure | Medium | Medium | Medium | 1, 3 |
| Influence reduction of inequalities | Low | Low | Low | 2 |
| Mobilise sustainable cities and communities | Low | Low | Low | 2 |
| Ensure responsible consumption and production | High | High | High | 1, 2 |
| Climate action to combat climate change impacts | Low | Low | Low | 3 |
| Develop life below water or conserve and sustainable resource utilisation | Very high | Very high | Very high | 1, 2, 3 |
| Advance life above water or life on land | Low | Low | Low | 3 |
| Assurance peace, justice, and strong institutions | Low | Low | Low | 2 |
| Build and strengthen global partnerships for the sustainable development | Low | Low | Low | 1, 2 |

Criteria and scores are based on the authors' critical qualitative evaluations of field observations and experiences. Low (too little, scant or negligible linkage, score: 0–25); Medium (moderate or mild linkage exists but not well-defined, score: 26–50); High (good linkage, nevertheless, still needs to be defined clearly, score: 51–75); Very high (direct linkage with clear definition, score: 76–100) The idea of contextualising and generating this table benefited from Wang et al. (2020) and de Bisthoven et al. (2020) [52,53]). Potentially linked with dimensions of sustainable development indicates: (1) economic growth, (2) social inclusion, and (3) environmental protection.

### *9.2. Prospects of Inland Open Water*

Although Bangladesh is blessed with substantial open water resources (SDG 14: life below water), only 28% of fish production derives from closed inland waters. Increasing production requires the promotion of the pen and cage culture system. It is estimated that only 1.29 lakh $m^3$ and 5294 ha are under the cage and pen culture system. For example, the community-based fisheries management, i.e., the Daudkandi model of community floodplain aquaculture [54,55], is expected to improve fisher access, livelihoods, and the sustainability of the fisheries sector. Co-management of natural wetlands could be another option to protect and increase the biodiversity and livelihood of the dependent community [9].

### *9.3. Prospects of Marine Fisheries Resources*

The utilisation of marine resources by implementing proper management strategies is essential to meet SDG 14 (life below water). Following prospects will open the door for establishing a special economic zone to develop the southern part of Bangladesh.

#### 9.3.1. Blue Economy

The blue economy is an emerging concept in Bangladesh, which refers to the ocean, seas, and coast [56,57]. Ocean resources are a potential resource for increasing food security, improving nutrition and health, alleviating poverty, creating jobs, generating alternative energy, seaborne trade, and industrial profiles (SDG 1: no poverty; SDG 2: zero hunger; SDG 3: good health and well-being; SDG 10: reduced inequalities) [58,59]. According to expert opinion, Bangladesh would extract about USD 1.2 billion from its substantial marine resources [60,61]. It will be a stepping stone to meet the Vision-2041 of Bangladesh if the resources are sustainably extracted and adequately used. It is possible to achieve 5% of the GDP by 2030 through the utilisation of sea resources.

9.3.2. Biotechnology and Marine Genetic Resource

Marine biotechnology can address several global challenges, including, but not limited to, human health, sustainable food supplies, environmental remediation, and energy security [62]. Many marine resources are a significant source of potential drugs, particularly antibiotics. For example, in 2017, there were over 36 marine-derived drugs in clinical development, including 15 to treat cancer [63,64].

9.3.3. Research

Research works related to marine fisheries resources and their production performance are very limited in Bangladesh. Therefore, appropriate research works on marine fisheries are required. Several marine fishery academies and institutes are available in Bangladesh, which can significantly increase marine fisheries production.

## 10. Challenges of Fisheries and COVID-19 Nexus to Achieve SGDs and Recovery Plans

Short-term and long-term recovery plans can potentially tackle the COVID-19 crisis in terms of addressing the immediate need and durable requirements for the sustainable recovery of the fisheries sector. Some recovery plans elicited from community perception and the authors' suggestions are discussed below.

### *10.1. Need for Emergency Financial and Non-Financial Support*

The overall income of fish farmers, fishers, and relevant stakeholders was severely affected, influencing the violation of fisheries rules. Immediate financial support, such as a rationing system as well as psychological counselling, could be effective. An urgent management strategy needs to be developed and implemented to save the livelihoods of fishers, hatcheries, feed industries, and other stakeholders [53].

### *10.2. Building Resilient Fisheries and Aquaculture Sector*

During the COVID-19 lockdown and imposed restrictions, many fishers and fish labourers lost their income. In order to make the fisheries and aquaculture sector resilient, emphasis should be given to increase the capacity of the stakeholders by providing subsidies, incentives, interest-free loans, and alternating income-generating options.

### *10.3. Addressing Potential Threats in Aquatic Biodiversity*

The degradation of aquatic biodiversity is one of the primary concerns in Bangladesh. IUCN assessments show that 64 out of 253 fish species are threatened, 9 are critically endangered, 30 are endangered, and 25 are vulnerable [56,65]. Excessive use of surgical face masks, disinfectants, hand sanitiser, and other pharmaceutical chemicals may also induce threats and water pollution. The return of the unemployed in the fishing sectors may increase overfishing and impose threats to biodiversity. A comprehensive management strategy including, but are not limited to, an awareness program is required for biodiversity conservation and COVID-19 risk mitigation.

### *10.4. Integrating Climate Change Risks with the Responses of COVID-19*

Bangladesh is highly sensitive to climate change due to its geography, and experiences frequent floods and cyclones (SDG 13: climate action) [66,67]. The current pandemic affects the coping strategy of fishers and the domestic food supply [24,48,49]. Developing appropriate technology and management strategies for resilient inland and marine fisheries is a mandate for Bangladesh, integrating climate change and COVID-19.

### 10.5. Initiatives to Address the Risk of Orders Cancellation by Foreign Buyers

During the lockdown, foreign shellfish buyers cancelled their orders due to border travel restrictions, which was an alarming issue for the shrimp industry. Minimising the risk of such cancellations requires more initiatives by fisheries and transport sectors for the smooth and safe transport of fish products.

### 10.6. Developing Long-Term Strategies

A long-term management strategy is essential for the sustainable development of the fisheries sector and for tackling the COVID-19 pandemic threats. The human resources and budget allocation (0.56% of the national budget) for the fisheries sector should be increased. A package of development and extension projects need to be implemented at the marginal level to provide financial, technical, and moral support to the fish farmers and fishers.

## 11. Conclusions and Recommendations

Fisheries are crucial in the national macroeconomic, food and nutrition security perspectives (SDG 1: no poverty; SDG 2: zero hunger; SDG 14: life below water). The COVID-19 pandemic adversely affected the fisheries and aquaculture sectors in many ways. Therefore, efficient and sustainable management of aquatic resources is essential for the continued and significant contribution to the country's health and economic sector. Despite the enormous prospects and potential, several reasons, such as climate change, onshore coal-based power plant installation, a lack of scientific information, skilled human resources, and poor implementation of acts and rules related to marine fisheries, limit the fisheries resources, production, and performance. However, the supply of good quality seed, feed, and extension services could increase fisheries production in inland water bodies. To have a resilient coastal and marine aquaculture development, adopting appropriate technology is a prerequisite for Bangladesh. The development of communication and transportation systems for rapid access to information, coordination with the regional and international networks for updated technology, value chain, and proper utilisation of the marine resources is required to boost the total fisheries production of Bangladesh.

Policymakers should be more aware of their well-planned efforts to meet the SDGs that can also ensure precise and prompt functioning of the fisheries sector. Interdisciplinary coordination for enhancing investment, research infrastructure, environmental policies, and the introduction of modern storage and marketing facilities could rapidly improve the situation. Perceptions of small-scale fishers should be considered during the policymaking process. Particular strategies and emphasis are required to assess their vulnerability and sustainability. Attention should also be paid to the environmental abnormalities and constraints, such as frequent climatic disasters (SDG 13: climate action), a declining aquatic biodiversity trend (SDG 14: life below water), and increased water and air pollution trends, in connection with the success of the fisheries sector.

**Author Contributions:** A.R.S.: conceptualization, designed and performed research, methodology validation, field works, data curation, data analysis, writing the original draft, reviewing, and editing; M.H.M., S.H.P., and M.A.: designed research, methodology validation, formal analysis, data curation, data analysis, visualization, reviewing, and editing; S.M.A.R., M.M.B., M.H., K.J.A., S.A.S., M.T.A. and A.R.: methodology validation, formal analysis, investigation, visualization, reviewing, editing, and proof reading, M.M.H.: conceptualization, designed research, methodology validation, formal analysis, investigation, visualization, reviewing, and editing. All authors have read and agreed to the published version of the manuscript.

**Funding:** This research received no external funding.

**Institutional Review Board Statement:** Not applicable.

**Informed Consent Statement:** Written informed consent was obtained from the participants.

**Data Availability Statement:** Not applicable.

**Acknowledgments:** We would like to thank the editor of *Sustainability* and the anonymous reviewers for their comments and suggestions, which helped improve the paper. We would like to express our deepest gratitude to the respondents who provided valuable information. Thanks to Md. Abdul Baten for helping to prepare Figure 1. We acknowledge the logistic and technical support of the Department of Genetic Engineering and Biotechnology of Shahjalal University of Science and Technology, Sylhet, Bangladesh.

**Conflicts of Interest:** The authors declare no conflict of interest.

## References

1. Kuddus, M.A.; Datta, G.C.; Miah, M.A.; Sarker, A.K.; Hamid, S.M.A.; Sunny, A.R. Performance Study of selected Orange Fleshed Sweet Potato Varieties in North Eastern Bangladesh. *Int. J. Environ. Agric. Biotechnol.* **2020**, *5*, 673–682. [CrossRef]
2. Ghose, B. Fisheries and Aquaculture in Bangladesh: Challenges and Opportunities. *Ann. Aquac. Res.* **2014**, *1*, 1–5.
3. Shamsuzzaman, M.M.; Islam, M.M.; Tania, N.J.; Al-mamun, M.A.; Barman, P.P.; Xu, X. Fisheries resources of Bangladesh: Present status and future direction. *Aquac. Fish.* **2017**, *2*, 145–156. [CrossRef]
4. Sunny, A.R.; Masum, K.M.; Islam, N.; Rahman, M.; Rahman, A.; Islam, J.; Rahman, S.; Ahmed, K.J.; Prodhan, S.H. Analysing livelihood sustainability of climate-vulnerable fishers: Insight from Bangladesh. *J. Aquac. Res. Dev.* **2020**, *11*, 593. [CrossRef]
5. Islam, M.R.; Cansse, T.; Islam, M.S.; Sunny, A.R. Climate change and its impacts: The case of coastal fishing communities of the Meghna river in south central Bangladesh. *Int. J. Mar. Environ. Sci.* **2018**, *12*, 368–376. [CrossRef]
6. Islam, M.M.; Sunny, A.R.; Hossain, M.M.; Friess, D. Drivers of Mangrove Ecosystem Service Change in the Sundarbans of Bangladesh. *Singap. J. Trop. Geogr.* **2018**, *39*, 244–265. [CrossRef]
7. DoF. *Yearbook of Fisheries Statistics of Bangladesh, 2018–2019*; Fisheries Resources Survey System (FRSS), Department of Fisheries: Dhaka, Bangladesh, 2019; pp. 35–38.
8. Sunny, A.R.; Sazzad, S.A.; Prodhan, S.H.; Ashrafuzzaman, M.; Datta, G.C.; Sarker, A.K.; Rahman, M.; Mithun, M.H. Assessing impacts of COVID-19 on aquatic food system and small-scale fisheries in Bangladesh. *Mar. Policy* **2021**, *126*, 104422. [CrossRef] [PubMed]
9. Sunny, A.R.; Hassan, M.N.; Mahashin, M.; Nahiduzzaman, M. Present status of hilsa shad (*Tenualosa ilisha*) in Bangladesh: A review. *J. Entomol. Zool. Stud.* **2017**, *5*, 2099–2105.
10. Islam, M.M.; Islam, N.; Mostafiz, M.; Sunny, A.R.; Keus, H.J.; Karim, M.; Hossain MZ& Sarker, S. Balancing between livelihood and biodiversity conservation: A model study on gear selectivity for harvesting small indigenous fishes in southern Bangladesh. *Zool. Ecol.* **2018**, *28*, 86–93. [CrossRef]
11. DoF. *Yearbook of Fisheries Statistics of Bangladesh, 2017–2018*; Fisheries Resources Survey System (FRSS), Department of Fisheries: Dhaka, Bangladesh, 2018; Volume 35, p. 129.
12. Hussain, M.G.; Hoq, M.E. *Sustainable Management of Fisheries Resources of the Bay of Bengal*; Support to Sustainable Management of the BOBLME Project; Bangladesh Fisheries Research Institute: Mymensingh, Bangladesh, 2010.
13. DoF. *National Fish Week, Compendium*; Department of Fisheries, Ministry of Fisheries and Livestock, Government of Bangladesh: Dhaka, Bangladesh, 2001. (In Bengali)
14. Hossain, M.A.R.; Wahab, M.A.; Belton, B. *The Checklist of the Riverine Fishes of Bangladesh*; Fisheries and Aquaculture News (FAN Bangladesh): Dhaka, Bangladesh, 2012; pp. 18–27.
15. Rahman, A.K.A. *Freshwater Fishes of Bangladesh*; Zoological Society of Bangladesh: Dhaka, Bangladesh, 1989.
16. Felts, R.A.; Rajts, F.; Akhteruzzaman, M. *Small Indigenous Fish Species Culture in Bangladesh*; IFADEP Sub-Project-2, Development of Inland Fisheries, Integrated Food Assisted Development Project; BdFISH Feature: Dhaka, Bangladesh, 1996.
17. Hossain, M.A.R. An overview of fisheries sector of Bangladesh. *Res. Agric. Livest. Fish.* **2014**, *1*, 109–126. [CrossRef]
18. Sunny, A.R.; Reza, J.; Anas, M.; Hassan, M.N.; Baten, M.A.; Hasan, R.; Monwar, M.M.; Solaimoan, H.; Hossain, M.M. Biodiversity assemblages and conservation necessities of ecologically sensitive natural wetlands of north eastern Bangladesh. *Indian J. Geo-Mar. Sci.* **2020**, *49*, 135–148.
19. Beveridge, M.C.M.; Little, D.C. The history of aquaculture in traditional societies. In *Ecological Aquaculture: The Evolution of the Blue Revolution*; Costa-Pierce, B.A., Ed.; Blackwell Science Limited: Oxford, UK, 2002.
20. General Economics Division, Bangladesh Planning Commission. *Fisheries Sub-Sector, Seventh Five-Year Plan (FY 2016—FY 2020)*; Government of People's Republic of Bangladesh: Dhaka, Bangladesh, 2015; pp. 283–291.
21. Azim, M.E.; Wahab, M.A.; Verdegem, M.C.J. Status of aquaculture and fisheries in Bangladesh. *World Aquac.* **2002**, *67*, 37–41.
22. Mazid, M.A.; Gupta, M.V. Research and information needs for fisheries development and management. In Proceedings of the National Workshop on Fisheries Resources Development and Management in Bangladesh, Dhaka, Bangladesh, 29 October–1 November 1995.
23. Chakma, S.; Paul, A.K.; Rahman, M.A.; Mithun, M.H.; Sunny, A.R. Impact of Climate Change and Ongoing Adaptation Measures in the Bangladesh Sundarbans. *Preprints* **2021**. [CrossRef]
24. Sunny, A.R. Impact of oil Spill in the Bangladesh Sundarbans. *Int. J. Fish. Aquat. Stud.* **2017**, *5*, 365–368.
25. Hossain, M.S. Biological aspects of the coastal and marine environment of Bangladesh. *Ocean. Coast. Manag.* **2001**, *44*, 261–282. [CrossRef]

26. Islam, M.S. Perspectives of the coastal and marine fisheries of the Bay of Bengal, Bangladesh. *Ocean. Coast. Manag.* **2003**, *46*, 763–796. [CrossRef]
27. MoFA. Ministry of Foreign Affairs, Press Release: Press Statement of the Hon'ble Foreign Minister on the Verdict of the Arbitral Tribunal/PCA. Dhaka. 8 July 2014. Available online: http://www.mofa.gov.bd/PressRelease/PRDetails.php (accessed on 17 July 2016).
28. Islam, M.M.; Shamsuzzaman, M.M.; Mozumder, M.M.H.; Xiangmin, X.; Ming, Y.; Jewel, M.A.S. Exploitation and conservation of coastal and marine fisheries in Bangladesh: Do the fishery laws matter? *Marine Policy* **2017**, *76*, 143–151. [CrossRef]
29. Food and Agricultural Organization. *The State of World Fisheries and Aquaculture 2020. Sustainability in Action*; FAO: Rome, Italy, 2020. [CrossRef]
30. FRSS. *Fisheries Resources Survey System (FRSS), Fisheries Statistical Report of Bangladesh*; Department of Fisheries: Dhaka, Bangladesh, 2018; Volume 34, pp. 1–57.
31. Food and Agricultural Organization. *The State of World Fisheries and Aquaculture (Opportunities and Challenges)*; FAO: Rome, Italy, 2014.
32. Khan, M.I.; Islam, M.M.; Kundu, G.K.; Akter, M.S. Understanding the Livelihood Characteristics of the Migratory and Non-Migratory Fishers of the Padma River, Bangladesh. *J. Sci. Res.* **2018**, *10*, 261–273. [CrossRef]
33. Sunny, A.R.; Ahamed, G.S.; Mithun, M.H.; Islam, M.A.; Das, B.; Rahman, A.; Rahman, M.T.; Hasan, M.N.; Chowdhury, M.A. Livelihood Status of The Hilsa (*Tenualosa ilisha*) Fishers: The Case of Coastal Fishing Community of The Padma River, Bangladesh. *J. Coast. Zone Manag.* **2019**, *22*, 469.
34. Alok, K.P.; Shapon, K.B.; Mohammad, S.I.; Hussain, M.A. Comparative socioeconomic study with a review on fisherman's livelihood around Tulsiganga River, Joypurhat, Bangladesh. *J. Fish. Aquat. Sci.* **2018**, *13*, 29–38. [CrossRef]
35. Rana, M.E.U.; Salam, A.; Shahriar, N.K.M.; Hasan, M. Hilsa Fishers of Ramgati, Lakshmipur, Bangladesh: An Overview of Socio-Economic and Livelihood Context. *J. Aquac. Res. Dev.* **2018**, *9*, 541.
36. Islam, M.M.; Islam, N.; Sunny, A.R.; Jentoft, S.; Ullah, M.H.; Sharifuzzaman, S.M. Fishers' perceptions of the performance of hilsa shad (Tenualosailisha) sanctuaries in Bangladesh. *Ocean. Coast. Manag.* **2016**, *130*, 309–316. [CrossRef]
37. Islam, M.M.; Shamsuzzaman, M.M.; Sunny, A.R.; Islam, N. Understanding fishery conflicts in the hilsa sanctuaries of Bangladesh. In *Inter-Sectoral Governance of Inland Fisheries*; Song, A.M., Bower, S.D., Onyango, P., Cooke, S.J., Chuenpagdee, R., Eds.; Too Big to Ignore: St. John's, NL, Canada, 2017; pp. 18–31.
38. Haque, N.; Blowfield, M.E. *Socio-Economic Methodologies for Coastal Communities: The Example of Set Bagnet Communities in Bangladesh*; Information Bulletin 10, DFID Post-Harvest Fisheries Project; DFID: Chennai, India, 1997.
39. Sunny, A.R.; Prodhan, S.H.; Ashrafuzzaman Sazzad, S.A.; Rahman, M.A.; Billah, M.M.; Hussain, M.; Rahman, M.; Haider, K.M.N.; Alam, M.T. Livelihoods and vulnerabilities of small-scale fishers to the impacts of climate variability and change: Insights from the coastal areas of Bangladesh. *Egypt. J. Aquat. Biol. Fish.* **2021**, *25*, 549–571. [CrossRef]
40. Sunny, A.R.; Prodhan, S.H.; Ashrafuzzaman Ahamed, G.S.; Sazzad, S.A.; Mithun, M.H.; Haider, K.M.N.; Alam, M.T. Understanding Livelihood Characteristics and Vulnerabilities of Small-scale Fishers in Coastal Bangladesh. *J. Aquac. Res. Dev.* **2021**, *12*, 635.
41. Kuddus, M.A.; Alam, M.J.; Datta, G.C.; Miah, M.A.; Sarker, A.K.; Sunny, M.A.R. Climate resilience technology for year round vegetable production in northeastern Bangladesh. *Int. J. Agril. Res. Innov. Tech.* **2021**, *11*, 29–36. [CrossRef]
42. Sunny, A.R.; Alam, R.; Sadia, A.K.; Miah, Y.; Hossain, S.; Mofiz, S.B. Factors affecting the Biodiversity and Human Well-Being of an Ecologically Sensitive Wetland of North Eastern Bangladesh. *J. Coast. Zone Manag.* **2020**, *23*, 471.
43. Rana, M.S.; Uddin, M.M.; Uddin, M.S.; Alam, M.T. Status of Aquaculture and Livelihood of Fish Farmers in Golapgonj Upazila under Sylhet District, Bangladesh. *Int. J. Nat. Sci.* **2016**, *6*, 66–74.
44. Sunny, A.R.; Islam, M.M.; Rahman, M.; Miah, M.Y.; Mostafiz, M.; Islam, N.; Hossain, M.Z.; Chowdhury, M.A.; Islam, M.A.; Keus, J.H. Cost effective aquaponics for food security and income of farming households in coastal Bangladesh. *Egypt. J. Aquat. Res.* **2019**, *45*, 89–97. [CrossRef]
45. Islam, M.M.; Khan, M.I.; Barman, A. Impact of novel coronavirus pandemic on aquaculture and fisheries in developing countries and sustainable recovery plans: Case of Bangladesh. *Mar. Policy* **2021**, *131*, 104611. [CrossRef]
46. Rabby, A.F.; Hossain, M.A.; Dey, T.; Alam, M.T.; Uddin, M.S. Fish Marketing System and Socio-Economic Status of Arotdars (Commission Agents) in North-Eastern Part of Bangladesh. *Trends Fish. Res.* **2019**, *8*, 23–30.
47. Sunny, A.R.; Islam, M.M.; Nahiduzzaman, M.; Wahab, M.A. Coping with climate change impacts: The case of coastal fishing communities in upper Meghna hilsa sanctuary of Bangladesh. In *Water Security in Asia: Opportunities and Challenges in the Context of Climate Change*; Babel, M.S., Haarstrick, A., Ribbe, L., Shinde, V., Dichti, N., Eds.; Springer: Berlin/Heidelberg, Germany, 2018; ISBN 978-3-319-54612-4. Available online: http://www.springer.com/us/book/9783319546117 (accessed on 1 January 2018).
48. Alam, M.T.; Hussain, M.A.; Sultana, S.; Mazumder, S.K. Impact of Sanctuary on Fish Biodiversity and Production in Two Important Beels of Bangladesh. *Adv. Biol. Res.* **2017**, *11*, 348–356. [CrossRef]
49. Sunny, A.R. A review on the effect of global climate change on seaweed and seagrass. *Int. J. Fish. Aquat. Stud.* **2017**, *5*, 19–22.
50. Kawsar, M.A. Impact of Corona Virus on Fisheries: Bangladesh Perspective. Agrinews24.com. 2020. Available online: http://www.agrinews24.com/impact-of-corona-virus-on-fisheries-bangladesh-perspective/ (accessed on 24 April 2020).
51. Rashid, H. Problems Arising in Fish Farming Due to Corona Disaster. BonikBarta. 2020. Available online: https://bonikbarta.net/home/news_description/229391/ (accessed on 12 June 2020).
52. Wang, X.; Yuen, K.F.; Wong, Y.D.; Li, K.X. How can the maritime industry meet Sustainable Development Goals? An analysis of sustainability reports from the social entrepreneurship perspective. *Transp. Res. Part D Transp. Environ.* **2020**, *78*, 102173. [CrossRef]

53. De Bisthoven, L.J.; Vanhove, M.P.M.; Rochette, A.J.; Hugé, J.; Verbesselt, S.; Machunda, R.; Munishi, L.; Wynants, M.; Steensels, A.; Malan-Meerkotter, M.; et al. Social-ecological assessment of Lake Manyara basin, Tanzania: A mixed method approach. *J. Environ. Manag.* **2020**, *267*, 110594. [CrossRef]
54. Bayazid, Y. The Daudkandi model of community floodplain aquaculture in Bangladesh: A case for Ostrom's design principles. *Int. J. Commons* **2016**, *10*, 854–877. [CrossRef]
55. Hasan, M.M.; Hossain, B.S.; Alam, M.J.; Chowdhury, K.A.; Al Karim, A.; Chowdhury, N.M.K. The Prospects of Blue Economy to Promote Bangladesh into a Middle-Income Country. *Open J. Mar. Sci.* **2018**, *8*, 355. [CrossRef]
56. Hussain, M.G.; Failler, P.; Karim, A.A.; Alam, M.K. Major opportunities of blue economy development in Bangladesh. *J. Indian Ocean Reg.* **2018**, *14*, 88–99. [CrossRef]
57. MoFA. Blue Economy—Development of Sea Resources for Bangladesh. 2019. Available online: https://mofa.gov.bd/site/page/8c5b2a3f-9873-4f27-87612737db83c2ec/OCEAN/BLUE-ECONOMY--FOR-BANGLADESH (accessed on 31 October 2019).
58. Hunt, B.; Vincent, A.C. Scale and sustainability of marine bioprospecting for pharmaceuticals. *AMBIO: J. Hum. Environ.* **2006**, *35*, 57–64. [CrossRef]
59. World Bank. *Bangladesh: Climate Change and Sustainable Development. Report No. 21104-BD*; Rural Development Unit, South Asia Region, The World Bank (WB): Dhaka, Bangladesh, 2000; p. 95.
60. Chowdhury, M.T.H.; Sukhan, Z.P.; Hannan, M.A. Climate change and its impact on fisheries resource in Bangladesh. In Proceedings of the International Conference on Environmental Aspects of Bangladesh (ICEAB10), Kitakyushu, Japan, 4 September 2010.
61. Rahman, A.K.A. *Freshwater Fishes of Bangladesh*, 2nd ed.; Zoological Society of Bangladesh, Department of Zoology, University of Dhaka: Dhaka, Bangladesh, 2005; p. 394.
62. IUCN. *Red List of Bangladesh, A Brief on Assessment Result 2015*; International Union for Conservation of Nature: Dhaka, Bangladesh, 2015; 26p.
63. Haque, M.M.; Belton, B.; Alam, M.M.; Ahmed, A.G.; Alam, M.R. Reuse of fish pond sediments as fertiliser for fodder grass production in Bangladesh: Potential for sustainable intensification and improved nutrition. *Agric. Ecosyst. Environ.* **2016**, *216*, 226–236. [CrossRef]
64. Hossain, M.I. Climate Change: A Challenge to coastal agriculture in Bangladesh. In *Planned Decentralisation: Aspired Development*; World Bank: Dhaka, Bangladesh, 2013; pp. 60–65.
65. Chowdhury, A.H. Environmental impact of coal based power plant of Rampal on the Sundarbans (World Largest Mangrove Forest) and surrounding areas. *MOJ Ecol. Environ. Sci.* **2017**, *2*, 1–14. [CrossRef]
66. Chowdhury, A.R. Coast or Construction? The Daily Star. 2018. Available online: https://www.thedailystar.net/star-weekend/environment/coast-or-construction-1587958 (accessed on 8 June 2018).
67. Mazid, M.A. Development of fisheries in Bangladesh. *Policy* **2002**, *2*, 3.

sustainability

MDPI

Article

# Using a Product Life Cycle Cost Model to Solve Supplier Selection Problems in a Sustainable, Resilient Supply Chain

Yu-Jwo Tao [1,2,3], Yi-Shyuan Lin [1], Hsuan-Shih Lee [4,5], Guo-Ya Gan [6] and Chang-Shu Tu [7,*]

1 Graduate Institute of Technological & Vocational Education, National Taipei University of Technology, Taipei 10608, Taiwan; ta794260@gmail.com (Y.-J.T.); yishyuan@ntut.edu.tw (Y.-S.L.)
2 Department of Business, National Open University, Taipei 10607, Taiwan
3 Department of Business Administration, National Taiwan University of Science and Technology, Taipei 10607, Taiwan
4 Department of Shipping and Transportation Management, National Taiwan Ocean University, Keelung 20224, Taiwan; hslee@email.ntou.edu.tw
5 Department of Information Management, Ming Chuan University, Taipei 11103, Taiwan
6 College of Auditing and Evaluation, Nanjing Audit University, Nanjing 211815, China; ganguoya@foxmail.com
7 Department of Information Management, Chang Gung University, Taoyuan City 33302, Taiwan
* Correspondence: long.tree@msa.hinet.net; Tel.: +886-0927351833

**Abstract:** Supplier selection constitutes a crucial component of manufacturing procurement. We developed a product life cycle cost (PLCC) model to support Taiwanese light-emitting diode (LED) manufacturers in capacity planning for sustainable and resilient supply chain (SC) management. For firms, supply chain PLCC (SCPLCC) is a key consideration, but relevant evidence is scarce. We applied two types of goal programming, namely multiobjective linear programming and revised multichoice goal programming (RMCGP), to develop a PLCC-based model that minimizes net costs, rejections, and late deliveries. Moreover, we constructed a decision-making tool for application to a case of SC sustainable procurement management in a high-tech Taiwanese LED company. Managers can resolve relevant problems by employing the two approaches of the SCPLCC model with various parameters. The implementation of RMCGP with weighted linear goal programming sensitivity analysis produced sufficient findings, according to a study of five models for practical implications. The primary findings of the current model assist business decision-makers in minimizing PLCC, reducing PLCC cost, minimizing net cost, number of rejections, number of late deliveries, achieving PLCC goals, and selecting the best supplier in the context of sustainable SC development.

**Keywords:** supplier selection; product life cycle cost; geometric mean weighting; penalty weighting; multiobjective linear programming; revised multichoice goal programming

**Citation:** Tao, Y.-J.; Lin, Y.-S.; Lee, H.-S.; Gan, G.-Y.; Tu, C.-S. Using a Product Life Cycle Cost Model to Solve Supplier Selection Problems in a Sustainable, Resilient Supply Chain. *Sustainability* **2022**, *14*, 2423. https://doi.org/10.3390/su14042423

Academic Editor: Shervin Hashemi

Received: 31 December 2021
Accepted: 16 February 2022
Published: 20 February 2022



## 1. Introduction

Because of the globalization of commercial markets and improvements in information technology, a well-constructed supply chain management (SCM) system is an instrumental tool for gaining competitive advantage [1,2]. In higher SC levels, supplier selection constitutes an integral component of manufacturing procurement and a pivotal industrial activity [3–5]. With the growing globalization of industry, SCM has become an increasingly multifaceted undertaking. SCM is an essential element of most businesses and is critical to corporate success and customer satisfaction. SCs represent the pathway between suppliers and buyers, customers, or consumers [6].

The primary goal of a manufacturing company is to make the correct product in the correct amount for the correct customer in a timely manner. Purchasing companies demand advanced buyer interactions with their suppliers for a respectful relationship between the two parties.

Sustainability in industry has emerged as a crucial strategy that expands beyond organizational boundaries to include the entire SC. Concerns regarding SC resilience have also arisen. Research on the intersection between the sustainability and resilience of SCs is nascent and is a consequence of their mutual effects. However, confusion regarding concepts, measurements, and implementation methods of sustainable and resilient SCs remains.

Some life cycle studies have compared the environmental sustainability of organic and nonorganic pork SCs. The present study investigated how overall food SC performance depends on the performance of partners in a sustainable and energy-efficient SC [7–9].

A light-emitting diode (LED) is a semiconductor device that emits light through the recombination of electrons with electron holes within the apparatus. LEDs are a green technology with high potential for reducing global $CO_2$ emissions. McKinsey stated that the overall market share of LED manufacturers was likely to increase after 2012, with global annual revenues projected to reach €37 billion by 2016 and €64 billion by 2020. Fortune Business Insights reported that the value of the global lighting market reached an estimated US$118.33 billion in 2019 and is projected to reach US$163.72 billion by 2027 [10,11]. Using white LED lighting technology, according to the optoelectronics industry development organization (OIDA), may cut CO2 emissions by 2.5 billion tons per year globally.

The LED industry is a prominent example of sustainability and resilience concepts in SCM. Although the industry is highly competitive and dynamic, its products are resource and emission intensive in both the production and use phases. In the context of climate change, the growing focus placed on sustainability represents a response to demands for carbon emission reductions. Resilience constitutes a response to climate adaptation challenges. Sustainable SCM (SSCM) has been employed for objectives concerning climate-related adaptation. The basic concepts of resilience SCM (RSCM) are derived from the characteristic elements of resilience theory. Despite the increasing relevance of SSCM and RSCM to enterprises with respect to addressing climate change concerns, studies have largely neglected the need for the systematic integration of the two paradigms [12–16].

Resilience is key to the ability of organizations to effect positive change and overcome various challenges, even in crisis situations. The extent of an organization's ability to respond to various disturbances depends on that organization's objectives and level of maturity in crisis management. In human resource management, organizational resilience is acknowledged as a multidisciplinary, multidimensional quality. Suryaningtyas et al. [7] reported a positive association between organizational resilience and organizational performance. For general managers, this suggests that organizational resilience should be continually applied both operationally and strategically to maintain a company's sustainability.

To choose suppliers and distribute resources in the food industry, Kaviani et al. [17] develop a combined intuitionistic fuzzy analytic hierarchy process and fuzzy multi-objective optimization approach. Noppasorn et al. [18] provided a Fuzzy Programming approach for optimizing multi-objective Aggregate Production Planning problems in uncertain situations.

We developed a product life cycle cost (PLCC) model to support Taiwanese LED manufacturers in capacity planning for sustainable and resilient SCM. Taiwan has a unique political and economic environment. The present PLCC-based model integrates SSCM and RSCM, thus allowing for high degrees of flexibility and efficiency. In applying this model, LED firms can employ technologies that provide decision-makers with strategic guidance with regard to examining and comparing various alternative sustainability strategies.

Although SCPLCC is a vital corporate consideration, relevant evidence is limited. SCPLCC has become associated with deciding how one supplier should be selected from various alternatives.

The supplier base must be optimized to identify high-performing suppliers in SCs. Approaches for solving SCPLCC-related problems have been discussed. Details are provided as follows.

Few attempts have been made to incorporate information vagueness into the SCPLCC problem. The multiobjective linear programming (MOLP) model can account for the variation in or imprecision of a decision-maker's aspiration level (intermediate control variables), thus establishing a more certain key point for decision-making policy [17,19]. Herein, we implemented a fuzzy MOLP model and a revised multichoice goal programming (RMCGP) model. The application of weighted linear goal programming (WLGP) to the RMCGP model produced adequate results in a scenario involving the PLCC of a high-tech company.

We extended the work of Amid et al. [19] and referenced a study by Wang [5] to solve the PLCC problem. To solve the SCPLCC problem, we modified a sample data set of auto part manufacturers from a study by Kumar et al. [20]. For sustainable buying, we used real-world data from LED companies [21].

Third, to solve the SCPLCC problem, we incorporated geometric mean weighting into the MOLP and RMCGP models such that the decision-maker can understand differences between assigning and not assigning all objectives and constraints in an SC decision of equal importance [22]. Our SCPLCC model can help decision-makers optimize supplier selection and prioritization; moreover, it can enable purchasing managers to enhance SC performance through the fulfilment of PLCC goals and minimization of net costs, rejections, and late deliveries. Managers can resolve relevant problems by employing the two approaches of the SCPLCC model with various parameters [22].

To close the theory–practice gap, we integrated experiential information and the opinions of practitioners in considering industrial SCs. By proposing a novel PLCC framework encompassing aspects of both resilience and sustainability, we extended existing conceptual frameworks on SSCM and RSCM. We considered climate change the most critical environmental concern that warrants SCM responses in terms of both sustainability and resilience through climate change mitigation and adaptation measures, respectively. We selected the LED industry as the study subject because it contributes substantially to climate change and is vulnerable to its adverse effects (because of the globalized and highly complex SC).

The first goal programming was developed by Charnes and Cooper [1]. Most goal programming methods rely on mapping one goal to a single scalar aspiration level. However, one goal may map multiple choice aspiration levels [2]. Chang conceived multichoice goal programming (MCGP) to address such scenarios [2]. The two main characteristics of MCGP are as follows. First, the aspiration levels used in goal programming need not be scalars; they can be vectors. Second, in the problem formulation stage, the aspiration level should not be underestimated. Acknowledging the complexity of the multiplicative mixed binary terms employed in MCGP, Chang [4] devised an MCGP approach that obviates the need for such terms altogether. MCGP methods can be implemented to solve real-world decision-making problems.

Nasr et al. [23] demonstrated the applicability of a fuzzy goal programming technique in the garment manufacturing and distribution business through a case study. Beiki et al. [24] used a real-world case study of automotive manufacturing to apply the linguistic entropy weight approach and the multi-objective programming (MOP) method to the applicability of a sustainable supplier selection and order allocation issue.

The remainder of this paper is organized as follows: Section 2 presents a review of the literature on quantitative decision-making techniques concerning SCs. Section 3 introduces the explanation of the solution of the SCPLCC problem under the two present approaches. In Section 4, considering a Taiwanese LED firm, the problem-solving process of the two goal programming approaches based on a sample data set of auto part manufacturers from a study by Kumar et al. [20] is described. In Section 5, conclusions are drawn regarding the SCCM-related advantages provided by these two approaches.

## 2. Literature Review

Criteria concerning SC selection and supplier ratings have been a research focus since the 1960s. In a review article, Dickson [25] compiled a list of more than 50 distinct factors meaningful to SCM. Various methods have been used in SC research, including linear weighting methods and mathematical programming models. For example, linear programming and goal programming, as multicriteria decision-making techniques, aim to facilitate decision-making by minimizing undesirable deviations in goal-related values.

In the present study, we classified studies on SCs according to whether they employed the qualitative factor approach or the integrated factor approach. Table 1 provides information on SC studies in which mathematical programming models were applied. SC models involve the use of diverse approaches, such as simple weighted scoring methods and compound mathematical programming techniques. The incorporation of multiple elements in the selection criteria is desirable and is often associated with high subjectivity in the decision-making process [26].

**Table 1.** Summary of recent studies on supply chains (SCs) employing mathematical programming models.

| Author(s)/Year | Category | Methods | Cost | Rejection | Delivers | PLCC | Capacity | Budget |
|---|---|---|---|---|---|---|---|---|
| MO′ Ath et al. (2017) [27] | Integrated | WGP, LP | | | | ✓ | | ✓ |
| Umarusman (2018) [28] | Qualitative | criterion | ✓ | ✓ | ✓ | | | |
| Budzinski et al. (2019) [29] | Integrated | | | | | ✓ | | |
| Ojo et al. (2020) [30] | Qualitative | GP | | | | | | ✓ |
| Hocine et al. (2020) [31] | Integrated | WA-FMCGP | ✓ | | | | | ✓ |
| Hardy et al. (2020) [32] | Qualitative | | | | | | | |
| Al-Huaain et al. (2020) [33] | Qualitative | GP | ✓ | | | | ✓ | ✓ |
| Bibhas and Sushil (2020) [34] | Qualitative | Game Theoretic | ✓ | | | | | ✓ |
| Biswarup and Bibhas (2021) [35] | Qualitative | Game Theoretic | ✓ | ✓ | | | | ✓ |
| Bahareh et al. (2021) [36] | Qualitative | FGP | | ✓ | | | | |
| Tavan, et al. (2021) [37] | Qualitative | FGP | ✓ | | ✓ | | | |
| Mabrouk (2021) [38] | Qualitative | Fuzzy set | ✓ | | ✓ | | | |
| This study | Integrated | MOLP, RMCGP | ✓ | ✓ | ✓ | ✓ | ✓ | ✓ |

Few studies have considered fuzziness in the analysis of SCPLCC data. Kumar et al. [39] presented a fuzzy goal programming model that minimizes the total tolerable weight variations of variables. In the present study, we employed two fuzzy programming approaches.

## 3. Methods

Zimmermann [40] used the approach of Bellman and Zadeh [41] and reformulated a linear programming problem with a fuzzy goal and fuzzy restrictions such that it could be solved as a conventional linear programming problem. Indexes, decision variables, and parameters were accounted for in the construction of an enriched multiobjective linear programming model in compliance with a set of assumptions, which are listed as follows.

(i) Assumptions.
(ii) Only one item is purchased from one supplier.
(iii) Quantity discounts are not considered.
(iv) The suppliers have an adequate supply of the item.

The lead time and demand of the item are constant and known with confidence.

The indexes, parameters, and decision variables of the SCPLCC model are defined in Table 2.

**Table 2.** Definitions of indexes, parameters, and decision variables.

| Index | |
|---|---|
| i | key index for supplier, for all $i = 1, 2, \ldots, n$ |
| j | key index for objectives function, for all $j = 1, 2, \ldots, J$ |
| k | key index for constraints, for all $k = 1, 2, \ldots, K$ |
| Decision variable | |

**Table 2.** *Cont.*

| Index | |
|---|---|
| $x_i$ | Order quantity given to the supplier $i$ |
| Parameters | |
| $\widetilde{D}$ | Aggregate demand of the item over a fixed planning period |
| n | Number of suppliers competing for selection |
| $p_i$ | Price of a unit item of ordered quantity $xi$ to supplier $i$ |
| $R_i$ | Percentage of the rejected units delivered by supplier $i$ |
| $L_i$ | Percentage of the units delivered late by supplier $i$ |
| $C_i$ | PLCC cost for supplier $i$ |
| $\widetilde{U}_i$ | Upper limit of the quantity available for supplier $i$ |
| $r_i$ | Vendor rating value for supplier $i$ |
| P | Least total purchasing value that a vendor can have |
| $f_i$ | Supplier quota flexibility for supplier $i$ |
| F | Least value of flexibility in supply quota that a supplier should have |
| $B_i$ | Budget constraint allocated to each supplier |

Note: Fuzzy parameters are indicated by a tilde.

Model: SCPLCC problem.

We formulate the multiobjective SCPLCC problem with four fuzzy objectives and fuzzy crisp constraints as follows.

$$\text{Minimize } Z_1 \cong \sum_{i=1}^{n} P_i x_i \quad (1)$$

$$\text{Minimize } Z_2 \cong \sum_{i=1}^{n} R_i x_i \quad (2)$$

$$\text{Minimize } Z_3 \cong \sum_{i=1}^{n} L_i x_i \quad (3)$$

$$\text{Minimize } Z_4 \cong \sum_{i=1}^{n} C_i x_i \quad (4)$$

The problem is subject to the following constraints:

$$\sum_{i=1}^{n} x_i \geq \widetilde{D} \text{ (aggregate demand constraint)} \quad (5)$$

$$x_i \leq \widetilde{U}_i \; \forall \, i \,;\, i = 1, 2, \ldots, n \text{ (capacity constraint)} \quad (6)$$

$$\sum_{i=1}^{n} r_i(x_i) \geq P,\; i = 1, 2, \ldots, n \text{ (total items purchasing constraint for supplier } i) \quad (7)$$

$$\sum_{i=1}^{n} f_i(x_i) \leq F,\; i = 1, 2, \ldots, n \text{ (Supplier quota flexibility for supplier } i) \quad (8)$$

$$P_i x_i \leq B_i,\; i = 1, 2, \ldots, n \text{ (budget constraint)} \quad (9)$$

$$x_i \geq 0,\; i = 1, 2, \ldots, n \text{ (non} - \text{negativity constraint)} \quad (10)$$

### 3.1. Key Elements of Objectives or Constraints of the SCPLCC Model

This model has four main fuzzy objectives: minimizing the total net cost, minimizing rejected items, minimizing late deliveries, and realizing the supplier's goal regarding the PLCC. The aggregate demand constraint ensures that the item is available in the required quantity over a fixed planning period. The supplier production capacity constraint limits the supply on the basis of the uncertain aggregate demand being set at 10% of the deterministic model. The budget constraint means that no one supplier can exceed its budget. The nonnegativity constraint prohibits negative orders. In general, the tilde (~) indicates that a situation is fuzzy. In the present model, both the objective functions and demand constraints are fuzzy parameters [15,22].

Depending on the objectives and restrictions and on whether the weights are equal or unequal, the fuzzy decision is either symmetric or asymmetric, respectively [22,40]. This problem is defined as the following membership function:

$$\mu_d(x) = \min(\mu_g(x), \mu_c(x)).$$

A general linear MOLP model for supplier selection involving the minimization $(Z_1, Z_2, Z_3, Z_4)$ and maximization $(Z_2)$ of the fuzzy and crisp constraints of the objective function is expressed as follows [22]:

$Z_1 = Z_3 = Z_4 = \sum_{i=1}^{n} c_{ki}x_i$, $k = 1, 2, 3, 4$ $Z_2 = \sum_{i=1}^{n} c_{ti}x_i$, $t = 2$, for maximization of the fuzzy and crisp constraints of the objective function.

such that $\sum_{i=1}^{n} a_{ri}x_i \geq d_r$, $r = 1, 2, \ldots, m$ for fuzzy restrictions (capacity constraints 1–4),

$\sum_{i=1}^{n} b_{si}x_i \geq d_s$, $s = 1, 2, \ldots, m$ for crisp restrictions (budget constraints 1–4), and

$x_i \geq 0$, where $i = 1, 2, \ldots, n$.

In fuzzy linear programming, objectives and restrictions are treated the same because they are defined through an individual membership function. Figure 1 presents the fuzzy objective functions and constraints of the SCPLCC problem.

The membership function $(Z_k)$ and maximization goals $(Z_l)$ are expressed as

$$\mu_{zk}(x) = \begin{cases} 1 & \text{for} & Z_k \leq Z_k^- \\ f_{\mu_{zk}} = \frac{(Z_k^+ - Z_k(x))}{(Z_k^+ - Z_k^-)} & \text{for} & Z_k^- \leq Z_k(x) \leq Z_k^+ \\ 0 & \text{for} & Z_k \leq Z_k^+ \\ & (k = 1,\ 2,\ \ldots,\ p) & \end{cases} \quad (11)$$

$$\mu_{zl}(x) = \begin{cases} 1 & \text{for} & Z_l \geq Z_l^+ \\ f_{\mu_{zl}} = \frac{(Z_l(x) - Z_l^-)}{(Z_l^+ - Z_l^-)} & \text{for} & Z_l^- \leq Z_l(x) \leq Z_l^+ \\ 0 & \text{for} & \text{for} Z_l \leq Z_l^- \\ & (l = p+1,\ p+2, \ldots q) & \end{cases} \quad (12)$$

where $Z_k^-$ and $Z_l^+$ can be obtained by solving the multiobjective problem by considering only one objective at a time; $Z_k^+$ is the maximum value (i.e., the least optimal solution) of the negative objective $Z_k^-$, and $Z_l^-$ is the minimum value (i.e., the least optimal solution) of the positive objective function $Z_l^+$ [20,42,43].

### 3.2. Solving the SCPLCC Problem through the Weighted Additive Approach

The weighted additive (WA) model, which is convex, enables decision-makers to select different weights according to specific purposes.

When the SCPLCC problem is solved, the weights of the membership functions of the objectives and constraints are calculated according to a supertransitive approximation; thus, these weights are assigned separately. In Equations (13)–(19), $\alpha_j$ is the weighting coefficient that indicates the relative importance of the fuzzy goals and fuzzy restrictions.

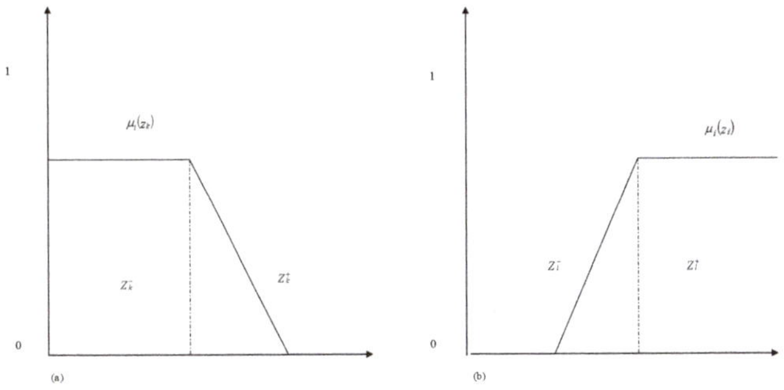


**Figure 1.** Presentation of the objective function as a fuzzy number: (**a**) minimum $Z_k^-$ and (**b**) maximum $Z_l^-$.

The following crisp simplex objective programming function is the same as that of the fuzzy model.

$$\text{Max} \sum_{j=1}^{S} \alpha_j \lambda^* \tag{13}$$

$$\lambda_j \leq \mu_{zj}(x),\ j = 1,\ 2,\ \ldots,\ q,\ \text{for all objective functions} \tag{14}$$

$$\gamma_r \leq \mu_{hr}(x),\ r = 1,\ 2,\ \ldots,\ h,\ \text{for all constraints} \tag{15}$$

$$g_m(x) \leq b_m,\ m = 1,\ \ldots,\ p, \tag{16}$$

$$\lambda \in [0.1] \tag{17}$$

$$\sum_{j=1}^{S} \alpha_j = 1,\ \alpha_j \geq 0, \tag{18}$$

$$x_i \geq 0,\ i = 1,\ 2,\ \ldots,\ n. \tag{19}$$

Refer to the study by Amid et al. [19] for a more detailed discussion of this model.

Weighted Max–Min Approach

Lin [44] devised and provided a proof for a weighted max–min approach that can capture an ideal resolution within a feasible area such that the ratio of the achievement level is as close to the ratio of the weight as possible. Lin noted that the WA model of Tiwari et al. [45] yields objectives involving heavy weights with relatively greater achievement values. However, the proportion of the achievement levels is not essentially the same as that of the objective weights [20,44,46]. To solve the present SCPLCC problem, the following equations corresponding to Lin's weighted max–min model were employed [20]:

$$\text{Max } \lambda \tag{20}$$

subject to

$$w_j \lambda \leq \mu_{zj}(x)\ j = 1,\ 2,\ \ldots, q,\ \text{for all objection functions} \tag{21}$$

$$\gamma_r \leq \mu_{hr}(x)\ r = 1,\ 2,\ \ldots, h\ \text{for all constraints} \tag{22}$$

$$g_m(x) \leq b_m,\ m = 1, \ldots,\ p, \tag{23}$$

$$\lambda \in [0.1] \tag{24}$$

$$\sum_{j=1}^{S} \alpha_j = 1,\ \alpha_j \geq 0, \tag{25}$$

$$x_i \geq 0,\ i = 1,\ 2,\ \ldots,\ n. \tag{26}$$

See Amid et al. [20] for a more detailed description of this model.

### 3.3. Revised MCGP Approach for Solving the SCPLCC Problem

We employed the RMCGP achievement function model developed by Chang [47]. Two RMCGP achievement function models were implemented. Details are presented as follows.

Specifically, the type I model is used for the scenario in which "more is better" (i.e., where achievement is the upper bound):

Minimize $\sum_{i=1}^{n} \left[w_i(d_i^+ + d_i^-) + \alpha_i(e_i^+ + e_i^-)\right]$

Subject to

$$f_i(X)b_i - d_i^+ + d_i^- = b_i y_i \; i = 1,\ 2,\ \ldots,\ n, \tag{27}$$

$$y_i - e_i^+ + e_i^- = g_{i,\max} \; i = 1,\ 2,\ \ldots,\ n, \tag{28}$$

$$g_{i,\min} \le y_i \le g_{i,\max} \tag{29}$$

$$d_i^+, d_i^-, e_i^+, e_i^- \ge 0,\ i = 1,\ 2,\ \ldots,\ n, \tag{30}$$

Here, $X \in F$, where $F$ is a feasible set and $X$ is unrestricted in sign.

Note that $b_i \in \{0, 1\}$ is a control variable attached to $|f_i(X) - y_i|$, which can be either achieved or released in Equation (27). In terms of real conditions, $b_i$ is subject to some constraints in guiding the relationships between the objectives of the SCPLCC model.

The type II model is used in the scenario in which "less is better" (i.e., where achievement is the lower bound):

Minimize $\sum_{i=1}^{n} \left[w_i(d_i^+ + d_i^-) + \alpha_i(e_i^+ + e_i^-)\right]$

Subject to

$$f_i(X)b_i - d_i^+ + d_i^- = b_i y_i \; i = 1,\ 2, \ldots,\ n, \tag{31}$$

$$y_i - e_i^+ + e_i^- = g_{i,\min} \; i = 1,\ 2,\ \ldots,\ n, \tag{32}$$

$$g_{i,\min} \le y_i \le g_{i,\max} \tag{33}$$

$$d_i^+, d_i^-, e_i^+, e_i^- \ge 0, i = 1,\ 2,\ \ldots,\ n, \tag{34}$$

Here, $X \in F$, where $F$ is a feasible set and $X$ is unrestricted in sign.

The definitions given in the definition of the type I model apply to the variables. All mixed-integer terms in Equations (27) and (29) can be easily linearized through the approach developed by Chang [47]. As presented in Equations (27), (29) and (30), no selection limitations were imposed on a specific objective, but some dependent relationships were observable between goals. Thus, we can add the auxiliary constraint $b_i \le b_{i+1} + b_{i+2}$ to the RMCGP achievement function model, where $b_i$, $b_{i+1}$, and $b_{i+2}$ are binary variables. Therefore, if $b_i = 1$, $b_{i+1}$ or $b_{i+2}$ must also equal 1. In other words, if goal 1 has been accomplished, then either goal 2 or goal 3 must have also been accomplished. Chang [47] presented a case regarding the managerial implications of these restrictions.

### 3.4. Process of SCPLCC Problem Resolution

A comprehensive solution to the SCPLCC problem was obtained through the following steps.

Step 1: Construct the model of the SCPLCC problem.

Step 2: Solve the MOLP problem through geometric mean weighting and obtain the lower and upper bounds (i.e., the max and min) of the ideal value of the four objectives, respectively.

Step 3: Replicate the procedure for each remaining objective. Define the lower and upper bounds of the optimal values for each objective according to the set of constraints.

Step 4: Process these values as the lower and upper bounds of the ideal values for the crisp construction of the SCPLCC problem.

Step 5: For each objective function, find a lower bound and an upper bound matching the established resolutions for each goal. Let $Z_j^-$ and $Z_j^+$ denote the lower and upper bounds of the $j$th objective goal $Z_j$, respectively [20].

Step 6: For each objective function, find the membership function in relation to Equations (11) and (12).

Step 7: Subject the criteria to geometric mean weighting.

Step 8: Establish and solve the corresponding WA model for the SCPLCC problem in relation to Equations (13)–(19).

Step 9: Establish and solve the corresponding max–min model for the SCPLCC problem in relation to Equations (20)–(25).

Step 10: Implement RMCGP through the geometric-mean-weighted and no-penalty-weighted construction of the fuzzy optimization problem, as expressed in Equations (27) and (30).

Step 11: Implement RMCGP through the geometric-mean-weighted and penalty-weighted construction of the fuzzy optimization problem, as presented in Equations (27) and (34).

Step 12: Solve the RMCGP model through the geometric-mean-weighted and penalty-weighted goal programming construction of the fuzzy optimization SCPLCC problem. Subsequently, compare the results obtained under the two goal programming approaches. Figure 2 shows a schematic representation of the entire methodology (MOLP and RMCGP) used to address the SCPLCC procurement challenge.

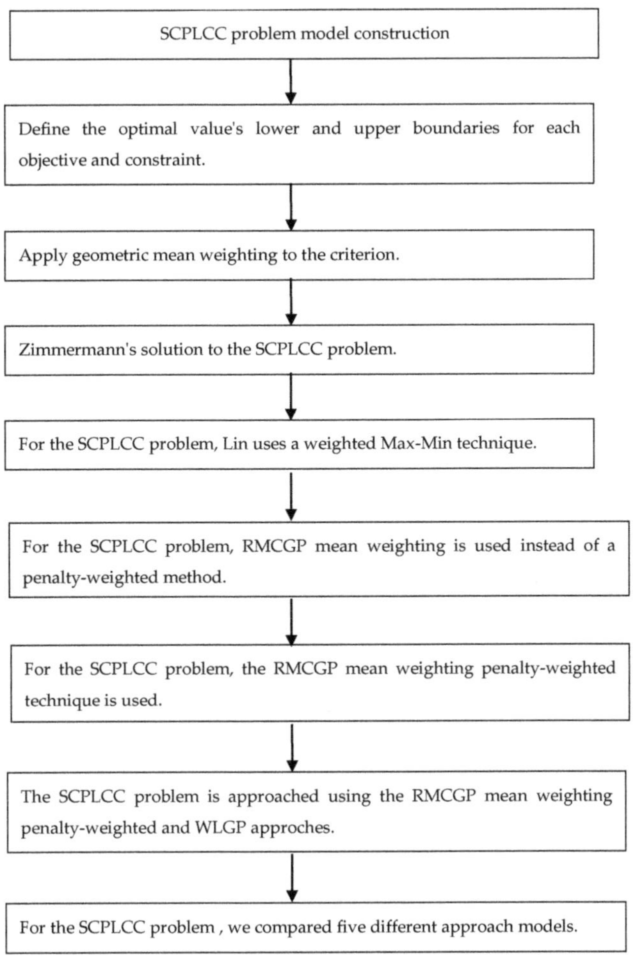


**Figure 2.** Graphic representation of the whole process.

## 4. Practical Example

The company on which the SCPLCC model was tested is part of a multinational group in the LED research and development sector. Specifically, the firm manufactures parts for made-to-order electric lights. External purchases accounted for more than 75% of the total annual costs of the company. The manager of the LED firm formed a team to propose three prospective providers. This group included managers from buying, marketing, quality control, manufacturing, engineering, and research and development.

Members of the team convened multiple sessions to develop profiles for competing providers and assembled a preliminary list of three companies for evaluation purposes [21]. The management is concerned with improving the efficiency of the purchasing process and optimizing the company's sourcing strategies. The managers believe that evaluating and certifying their vendors is essential for achieving inventory reduction and shortening the time to market. They had been instructed to develop longer-term trust-based relationships with a smaller group of vendors. In response, a team was formed to recommend three or four suitable vendors. The team consisted of managers from the purchasing, marketing, quality control, production, engineering, and research and development departments. The team convened several meetings to create profiles for the evaluation of three initial candidate vendors. An SCPLCC problem model was developed for vendor selection and quota allocation under the consideration of uncertain environmental conditions. Four main objective functions were investigated: (1) minimizing net cost, (2) minimizing net rejections, (3) minimizing late deliveries, and (4) minimizing PLCC. All four functions were under practical constraints regarding item demand, vendor capacity limitations, and vendor budgets, among other factors. The price quoted $P_i$, percentage of rejections $R_i$ on a scale from 0 to 1, percentage of late deliveries $L_i$, the PLCC (designated as $C_i$), suppliers' cost capacity $U_i$, suppliers' quota flexibility $F_i$ on a scale from 0 to 1, the vendors' ratings $r_i$, on a scale from 0 to 1, and suppliers' budget allocation $B_i$ were considered. Profiles of the three suppliers are presented in Table 3.

**Table 3.** Profiles of the three suppliers.

| Supplier No. | $P_i$ ($) | $R_i$ (%) | $L_i$ (%) | $C_i$ ($) | $U_i$ (Units) | $r_i$ | $F_i$ | $B_i$ ($) |
|---|---|---|---|---|---|---|---|---|
| 1 | 3 | 0.05 | 0.04 | 1.92 | 5000 | 0.88 | 0.02 | 25,000 |
| 2 | 2 | 0.03 | 0.02 | 1.04 | 15,000 | 0.91 | 0.01 | 100,000 |
| 3 | 6 | 0.02 | 0.08 | 3.94 | 6000 | 0.97 | 0.06 | 35,000 |

In this case demonstration, the linear membership function was employed to fuzzify the right-hand side of the restrictions in the preceding SCPLCC model. The net cost, rejections, late deliveries, and supplier capacity at the lowest and highest levels of the membership function must be defined according to Equations (11) and (12). The fuzzy parameters' uncertainty level was assumed to be 10% of that of the deterministic model. The values corresponding to the lowest and highest aspiration levels of the membership functions are listed in Table 4.

**Table 4.** Limiting values in the membership functions of net cost, rejections, late deliveries, PLCC, vendor capacities, and budget information.

| | (Min.) $\mu = 1$ | (Min.) $\mu = 0$ |
|---|---|---|
| Net cost objective goal | 57,000 | 71,833 |
| Rejection objective goal | 413 | 521 |
| Late deliveries objective goal | 604 | 816 |
| Product life cycle cost objective goal | 10,000 | 90,000 |
| Capacity constraints | | |
| Supplier 1 | 5000 | 5500 |
| Supplier 2 | 15,000 | 16,500 |
| Supplier 3 | 6000 | 6600 |
| Budget constraints | | |
| Supplier 1 | 25,000 | 27,500 |
| Supplier 2 | 100,000 | 110,000 |
| Supplier 3 | 35,000 | 38,500 |

*4.1. Application of the WA Approach to a Numerical Example*

We examined a numerical example through the WA approach employed by Tiwari et al. [45]. Before solving the problem, we first determined the weights of four goals and six restrictions through supertransitive approximation. The following binary comparison matrix was assumed to consist of the obtained net cost, rejections, late deliveries, PLCC, supplier 1 constraints, supplier 2 constraints, and supplier 3 constraints:

$$A = \begin{pmatrix} Net\ cost \\ Rejection \\ Late\ deliveries \\ Productlifecyclecost \\ Vendor1capacity \\ Vendor2capacity \\ Vendor3capacity \\ Vendor\ \ 1\ budget\ constraints \\ Vendor\ \ 2\ budget\ constraints \\ Vendor\ \ 3\ budget\ constraints \end{pmatrix} \Rightarrow \begin{pmatrix} 1 & 6 & 4 & 9 & 3 & 4 & 9 & 9 & 8 & 2 \\ 1/6 & 1 & 1/2 & 3 & 1/3 & 1/3 & 2 & 4 & 5 & 1/4 \\ 1/4 & 2 & 1 & 4 & 1/2 & 1/2 & 3 & 5 & 6 & 1/3 \\ 1/9 & 1/3 & 1/4 & 1 & 1/5 & 1/2 & 2 & 3 & 3 & 1/6 \\ 1/3 & 3 & 2 & 5 & 1 & 1 & 4 & 6 & 7 & 1/2 \\ 1/4 & 3 & 2 & 5 & 1 & 1 & 4 & 6 & 7 & 1/2 \\ 1/9 & 1/2 & 1/3 & 2 & 1/4 & 1/4 & 1 & 3 & 4 & 1/5 \\ 1/9 & 1/4 & 1/5 & 1/2 & 1/6 & 1/6 & 1/3 & 1 & 2 & 1/8 \\ 1/8 & 1/5 & 1/6 & 1/3 & 1/7 & 1/7 & 1/4 & 1/2 & 1 & 1/9 \\ 1/2 & 4 & 3 & 6 & 2 & 2 & 5 & 8 & 9 & 1 \end{pmatrix}$$

Supertransitive approximation [48] was conducted to generate the comparison matrix, and the following weights were obtained [9]: $w_1 = 0.2958$, $w_2 = 0.0579$, $w_3 = 0.0863$, $w_4 = 0.0365$, $w_5 = 0.1291$, $w_6 = 0.1254$, $w_7 = 0.0392$, $w_8 = 0.0199$, $w_9 = 0.0151$, and $w_{10} = 0.1949$.

4.1.1. Solving the SCPLCC Problem through the WA Approach

For this SCPLCC example, the optimal quota allocations (i.e., the purchasing orders), production capacity limitations, and budget restrictions were obtained through Zimmermann's WA method [49], as shown in Equations (13)–(19). This SCPLCC problem can now be reformulated as the following program:

Maximize $0.2958\lambda_1^* + 0.0579\lambda_2^* + 0.0863\lambda_3^* + 0.0365\lambda_4^* + 0.1291\lambda_5^* + 0.1254\lambda_6^* + 0.0392\lambda_7^* + 0.0199\lambda_8^* + 0.0151\lambda_9^* + 0.1949\lambda_{10}^*$

Subject to

$$\lambda_1^* \le (71833 - (3x_1 + 2x_2 + 6x_3))/(71833 - 57000)$$

$$\lambda_2^* \le (521 - (0.05x_1 + 0.03x_2 + 0.01x_3))/(521 - 413)$$

$$\lambda_3^* \le (816 - (0.04x_1 + 0.02x_2 + 0.08x_3)/(816 - 604)$$

$$\lambda_4^* \le (90000 - (1.92x_1 + 1.04x_2 + 3.94x_3))/(90000 - 10000)$$

$$\mu_1^* \le (71833 - (3x_1 + 2x_2 + 6x_3))/(71833 - 57000)$$

$$\mu_2^* \leq (521 - (0.05x_1 + 0.03x_2 + 0.01x_3))/(521 - 413)$$
$$\mu_3^* \leq (816 - (0.04x_1 + 0.02x_2 + 0.08x_3)/(816 - 604)$$
$$\mu_4^* \leq (90000 - (1.92x_1 + 1.04x_2 + 3.94x_3))/(90000 - 10000)$$
$$Z_1 = 3x_1 + 2x_2 + 6x_3$$
$$Z_2 = 0.05x_1 + 0.03x_2 + 0.01x_3$$
$$Z_3 = 0.04x_1 + 0.02x_2 + 0.08x_3$$
$$Z_4 = 1.92x_1 + 1.04x_2 + 3.94x_3$$
$$x_1 + x_2 + x_3 = 20000$$
$$\lambda_5^* \leq (5500 - x_1)/(5500 - 5000)$$
$$\lambda_6^* \leq (16500 - x_2)/(16500 - 15000)$$
$$\lambda_7^* \leq (6600 - x_3)/(6600 - 6000)$$
$$\mu_5^* \leq (5500 - x_1)/(5500 - 5000)$$
$$\mu_6^* \leq (16500 - x_2)/(16500 - 15000)$$
$$\mu_7^* \leq (6600 - x_3)/(6600 - 6000)$$
$$0.88x_1 + 0.91x_2 + 0.97x_3 \geq 18400$$
$$0.02x_1 + 0.01x_2 + 0.06x_3 \leq 600$$
$$\lambda_8^* \leq (27500 - 3x_1)/(27500 - 25000)$$
$$\lambda_9^* \leq (110000 - 2x_2)/(110000 - 100000)$$
$$\lambda_{10}^* \leq (38500 - 7x_3)/(38500 - 35000)$$
$$x_i \geq 0 \text{ and integer, } i = 1,\ 2,\ 3.$$

The numerical example of the SCPLCC problem was solved using LINGO software (2002) Chicago, IL, USA [50]:
$x_1 = 240$ $x_2 = 15{,}570$ $x_3 = 4190$ $\mu_1 = 0$ $\mu_2 = 0$ $\mu_3 = 0$
$\mu_4 = 0$ $\mu_5 = 0$ $\mu_6 = 0$ $\mu_7 = 0.3066$ $\mu_8 = 1$
$\mu_9 = 1$ $\mu_{10} = 0.0011$
$z_1 = 57{,}000$ $z_2 = 521$ $z_3 = 656.20$ $z_4 = 33{,}162.20$.

#### 4.1.2. Solving the SCPLCC Example through the Weighted Max–Min Approach

The optimal quota allocations (i.e., the purchasing orders), production capacity limitations, and budget constraints were obtained using the weighted max–min approach developed by Lin [44], as expressed in Equations (20)–(26). This SCPLCC problem can now be reformulated as the following program:

Maximize $\lambda$

Subject to

$$0.2958\lambda \leq (71833 - (3x_1 + 2x_2 + 6x_3))/(71833 - 57000)$$
$$0.0579\lambda \leq (521 - (0.05x_1 + 0.03x_2 + 0.01x_3))/(521 - 413)$$
$$0.0863\lambda \leq (816 - (0.04x_1 + 0.02x_2 + 0.08x_3))/(816 - 604)$$
$$0.0363\lambda \leq (90000 - (1.92x_1 + 1.04x_2 + 3.94x_3))/(90000 - 10000)$$
$$0.2958\mu \leq (71833 - (3x_1 + 2x_2 + 6x_3))/(71833 - 57000)$$
$$0.0579\mu \leq (521 - (0.05x_1 + 0.03x_2 + 0.01x_3))/(521 - 413)$$

$$0.0863\mu \leq (816 - (0.04x_1 + 0.02x_2 + 0.08x_3))/(816 - 604)$$

$$0.0363\mu \leq (90000 - (1.92x_1 + 1.04x_2 + 3.94x_3))/(90000 - 10000)$$

$$Z_1 = 3x_1 + 2x_2 + 6x_3$$

$$Z_2 = 0.05x_1 + 0.03x_2 + 0.01x_3$$

$$Z_3 = 0.04x_1 + 0.02x_2 + 0.08x_3$$

$$Z_4 = 1.92x_1 + 1.04x_2 + 3.94x_3$$

$$x_1 + x_2 + x_3 = 20000$$

$$0.1291\lambda \leq (5500 - x_1)/(5500 - 5000)$$

$$0.1254\lambda \leq (16500 - x_2)/(16500 - 15000)$$

$$0.0392\lambda \leq (6600 - x_3)/(6600 - 6000)$$

$$0.88x_1 + 0.91x_2 + 0.97x_3 \geq 18400$$

$$0.02x_1 + 0.01x_2 + 0.06x_3 \leq 600$$

$$0.0199\lambda \leq (27500 - 3x_1)/(27500 - 25000)$$

$$0.0151\lambda \leq (110000 - 2x_2)/(110000 - 100000)$$

$$0.1949\lambda \leq (38500 - 6x_3)/(38500 - 35000)$$

$$\text{where } x_i \geq 0 \text{ and } i = 1,\ 2, \text{ and } 3$$

The numerical example of the SCPLCC problem was solved using LINGO software (2002):
$x_1 = 0$ $x_2 = 15{,}750$ $x_3 = 4250$ $\lambda = 0.9595$ $\mu_1 = 0$
$\mu_2 = 0$ $\mu_3 = 0$ $\mu_4 = 0$ $\mu_5 = 0$ $\mu_6 = 0$
$\mu_7 = 0.3066$ $\mu_8 = 1$ $\mu_9 = 1$ $\mu_{10} = 0.0011$
$z_1 = 57{,}000$ $z_2 = 515$ $z_3 = 655$ $z_4 = 33{,}125$.

#### 4.1.3. Solving the SCPLCC Example by Using the RMCGP Approach

We now consider an RMCGP problem with goals and constraints that cannot be solved under current goal programming methods. The optimal quota allocations (i.e., the purchasing orders), supplier production capacity, and financial budget constraints were considered. This SCPLCC problem was then formulated as follows:

Main goals (with each goal designated as $G$):

$(G_1)$ $3x_1 + 2x_2 + 6x_3 = 57{,}000$ ($G_{1,\text{MIN.}}$) or 71,833 ($G_{1,\text{MAX.}}$)
$(G_2)$ $0.05x_1 + 0.03x_2 + 0.01x_3 = 413$ ($G_{2,\text{MIN.}}$) or 521 ($G_{2,\text{MAX.}}$)
$(G_3)$ $0.04x_1 + 0.02x_2 + 0.08x_3 = 604$ ($G_{3,\text{MIN.}}$) or 816 ($G_{3,\text{MAX.}}$)
$(G_4)$ $1.92x_1 + 1.04x_2 + 3.94x_3 = 10{,}000$ ($G_{4,\text{MIN.}}$) or 90,000 ($G_{4,\text{MAX.}}$)

Capacity constraints:

$(G_5)$ $x_1 = 5000$ ($G_{5,\text{MIN.}}$) or 5500 ($G_{5,\text{MAX.}}$) ($X_1$, production capacity of supplier 1)
$(G_6)$ $x_2 = 15{,}000$ ($G_{6,\text{MIN.}}$) or 165,000 ($G_{6,\text{MAX.}}$) ($X_2$, production capacity of supplier 2)
$(G_7)$ $x_3 = 6000$ ($G_{7,\text{MIN.}}$) or 165,000 ($G_{7,\text{MAX.}}$) ($X_3$, production capacity of supplier 3)
$x_1 + x_2 + x_3 = 20{,}000$ (Constraint of total demand)

Budget constraints:

$(G_8)$ $3x_1 = 25{,}000$ ($G_{8,\text{MIN.}}$) or 27,500 ($G_{8,\text{MAX.}}$) ($X_1$, budget constraint of vendor 1)
$(G_9)$ $2x_2 = 100{,}000$ ($G_{9,\text{MIN.}}$) or 110,000 ($G_{9,\text{MAX.}}$)($X_2$, budget constraint of vendor 2)
$(G_{10})$ $6x_3 = 35{,}000$ ($G_{10,\text{MIN.}}$) or 110,000 ($G_{10,\text{MAX.}}$) ($X_3$, budget constraint of vendor 3)

Through an RMCGP approach involving no penalty weighting [47] combined with geometric mean weighting [22,47], this SCPLCC problem was rewritten as the following program:

Min = $0.2958(d_1^- + e_1^-) + 0.0579(d_2^- + e_2^-) + 0.0863(d_3^- + e_3^-) + 0.0365(d_4^- + e_4^-)$
$+0.1291(d_5^- + e_5^-) + 0.1254(d_6^- + e_6^-) + 0.0392(d_7^- + e_7^-)$

$+0.0199(d_8^- + e_8^-) + 0.0151(d_9^- + e_9^-) + 0.1949(d_{10}^- + e_{10}^-)$
s.t.
$(3.1x_1 + 2x_2 + 7x_3)b_1 - d_1^+ + d_1^- = y_1 b_1$ for the net cost goal, with a lower value being more desirable
$y_1 - e_1^+ + e_1^- = 57000$ represents $|y_1 - g_{1,\min}|$,
where $57000 \leq y_1 \leq 71833$
$(0.05x_1 + 0.03x_2 + 0.01x_3)b_2 - d_2^+ + d_2^- = y_2 b_2$ represents the rejection goal, with a lower value being more desirable; $y_2 - e_2^+ + e_2^- = 413$ represents $|y_2 - g_{2,\min}|$, where

$$413 \leq y_2 \leq 521$$

$(0.04x_1 + 0.02x_2 + 0.08x_3)b_3 - d_3^+ + d_3^- = y_3 b_3$ corresponds to the goal of minimizing late deliveries goal, with a lower value being more desirable.
$y_3 - e_3^+ + e_3^- = 604$ represents $|y_3 - g_{3,\min}|$, where

$$604 \leq y_3 \leq 816$$

$(1.92x_1 + 1.04x_2 + 3.94x_3)b_4 - d_4^+ + d_4^- = y_4 b_4$ corresponds to the PLCC goal, with a lower value being more desirable.
$y_4 - e_4^+ + e_4^- = 10000$ represents $|y_4 - g_{4,\min}|$, where

$$1000 \leq y_4 \leq 90000$$

$x_1 b_5 - d_5^+ + d_5^- = y_5 b_5$ is the capacity constraint goal of supplier 1, with a lower value being more desirable.
$y_5 - e_5^+ + e_5^- = 5000$ represents $|y_5 - g_{5,\min}|$, where

$$5000 \leq y_5 \leq 5500$$

$x_2 b_6 - d_6^+ + d_6^- = y_6 b_6$ is the capacity constraint goal of supplier 2, with a lower value being more desirable.
$y_6 - e_6^+ + e_6^- = 15000$ represents $|y_6 - g_{6,\min}|$, where

$$15000 \leq y_6 \leq 16500$$

$x_3 b_7 - d_7^+ + d_7^- = y_7 b_7$ is the capacity constraint goal of supplier 3, with a lower value being more desirable.
$y_7 - e_7^+ + e_7^- = 6000$ represents $|y_7 - g_{7,\min}|$, where
$6000 \leq y_7 \leq 6600$
$3x_1 b_8 - d_8^+ + d_8^- = y_8 b_8$ is the budget constraint goal of supplier 1, with a lower value being more desirable.
$y_8 - e_8^+ + e_8^- = 25000$ represents $|y_8 - g_{8,\min}|$, where

$$25000 \leq y_8 \leq 27500$$

$2x_2 b_9 - d_9^+ + d_9^- = y_9 b_9$ is the budget constraint goal of supplier 1, with a lower value being more desirable.
$y_9 - e_9^+ + e_9^- = 100000$ represents $|y_9 - g_{9,\min}|$, where

$$100000 \leq y_9 \leq 110000$$

$6x_3 b_{10} - d_{10}^+ + d_{10}^- = y_{10} b_{10}$ is the budget constraint goal of supplier 1, with a lower value being more desirable.
$y_{10} - e_{10}^+ + e_{10}^- = 35000$ represents $|y_{10} - g_{10,\min}|$, where

$$35000 \leq y_9 \leq 38500$$

In this program, $b_1 = b_2 + b_3 + b_4 + b_5 + b_6 + b_7 + b_8 + b_9 + b_{10}$ so that the net cost, rejection, and late delivery goals, as well as zero, are achieved.

Moreover, $b_2 + b_3 + b_4 + b_5 + b_6 + b_7 + b_8 + b_9 + b_{10} = 1$; this addition of auxiliary constraints can force the net cost goal such that the lower-bound goal is achieved and either the rejection goal or the late delivery goal is achieved.

$Z_1 = 3x_1 + 2x_2 + 6x_3$
$Z_2 = 0.05x_1 + 0.03x_2 + 0.01x_3$
$Z_3 = 0.04x_1 + 0.02x_2 + 0.08x_3$
$Z_4 = 1.92x_1 + 1.04x_2 + 3.94x_3$
$x_1 + x_2 + x_3 = 20000$
$0.88x_1 + 0.91x_2 + 0.97x_3 \geq 18400$
$0.02x_1 + 0.01x_2 + 0.06x_3 \leq 600$
$x_i \geq 0$ and $I = 1, 2, 3$.
$d_i^+, d_i^-, e_i^+, e_i^- \geq 0$ ($i = 1, 2, \ldots, 10$)
$b_i \geq 0$ ($i = 1, 2, \ldots, 10$) are binary variable.

The numerical example of the SCPLCC problem was solved using LINGO software [50]:

$x_1 = 5000$ $x_2 = 9166.67$ $x_3 = 5833.33$
$y_1 = 57{,}000$ $y_2 = 410$ $y_3 = 604$ $y_4 = 10{,}000$ $y_5 = 5000$ $y_6 = 15{,}000$ $y_7 = 6000$ $y_8 = 25{,}000$ $y_9 = 100{,}000$ $y_{10} = 35{,}000$
$b_1 = 1$ $b_2 = 0$ $b_3 = 0$ $b_4 = 0$ $b_5 = 1$ $b_6 = 0$ $b_8 = 0$ $b_9 = 0$ $b_{10} = 0$
$z_1 = 68{,}383.33$ $z_2 = 583.33$ $z_3 = 850$ $z_4 = 42{,}116.67$

In sum, the RMCGP method involving geometric mean weighting and no penalty weighting solved the SCPLCC problem.

#### 4.1.4. Solving the SCPLCC Example through RMCGP, Geometric Mean Weighting, and Penalty Weighting

In this SCPLCC example, the optimal quota allocations (i.e., purchasing order), production capacity limitations, and budget constraints of the suppliers were determined through RMCGP with a penalty-weighted approach, as expressed in Equations (27) and (30). This SCPLCC problem can now be reformulated as the following program:

$$\begin{gathered}
\text{Min} = 0.2958(5d_1^+ + d_1^- + e_1^-) + 0.0579(4d_2^+ + d_2^- + e_2^-) + 0.0863(3d_3^+ + d_3^- + e_3^-) \\
+0.0365(2d_4^+ + d_4^- + e_4^-) + 0.1291(d_5^- + e_5^-) + 0.1254(d_6^- + e_6^-) \\
+0.0392(d_7^- + e_7^-) + 0.0199(d_8^- + e_8^-) + 0.0151(d_9^- + e_9^-) + 0.1949(d_{10}^- + e_{10}^-) \\
\text{s.t.} \\
(3.1x_1 + 2x_2 + 7x_3)b_1 - d_1^+ + d_1^- = y_1 b_1 \\
y_1 - e_1^+ + e_1^- = 57000 \\
57000 \leq y_1 \leq 71833 \\
(0.05x_1 + 0.03x_2 + 0.01x_3)b_2 - d_2^+ + d_2^- = y_2 b_2 \\
y_2 - e_2^+ + e_2^- = 413 \\
413 \leq y_2 \leq 521 \\
(0.04x_1 + 0.02x_2 + 0.08x_3)b_3 - d_3^+ + d_3^- = y_3 b_3 \\
y_3 - e_3^+ + e_3^- = 604 \\
604 \leq y_3 \leq 816 \\
(1.92x_1 + 1.04x_2 + 3.94x_3)b_4 - d_4^+ + d_4^- = y_4 b_4 \\
y_4 - e_4^+ + e_4^- = 10000 \\
10000 \leq y_4 \leq 90000 \\
y_5 - e_5^+ + e_5^- = 15000 \\
x_1 b_5 - d_5^+ + d_5^- = y_5 b_5 \\
y_6 - e_6^+ + e_6^- = 15000 \\
x_2 b_6 - d_6^+ + d_6^- = y_6 b_6 \\
15000 \leq y_6 \leq 16500
\end{gathered}$$

$$y_7 - e_7^+ + e_7^- = 6000$$
$$x_3b_7 - d_7^+ + d_7^- = y_7b_7$$
$$6000 \le y_7 \le 6600$$
$$y_8 - e_8^+ + e_8^- = 25000$$
$$3x_1b_8 - d_8^+ + d_8^- = y_8b_8$$
$$25000 \le y_8 \le 27500$$
$$y_9 - e_9^+ + e_9^- = 100000$$
$$2x_2b_9 - d_9^+ + d_9^- = y_9b_9$$
$$100000 \le y_9 \le 110000$$
$$y_{10} - e_{10}^+ + e_{10}^- = 35000$$
$$6x_3b_{10} - d_{10}^+ + d_{10}^- = y_{10}b_{10}$$
$$35000 \le y_9 \le 38500$$
$$b_1 = b_2 + b_3 + b_4 + b_5 + b_6 + b_7 + b_8 + b_9 + b_{10}$$
$$b_2 + b_3 + b_4 + b_5 + b_6 + b_7 + b_8 + b_9 + b_{10} = 1$$
$$Z_1 = 3x_1 + 2x_2 + 6x_3$$
$$Z_2 = 0.05x_1 + 0.03x_2 + 0.01x_3$$
$$Z_3 = 0.04x_1 + 0.02x_2 + 0.08x_3$$
$$Z_4 = 1.92x_1 + 1.04x_2 + 3.94x_3$$
$$x_1 + x_2 + x_3 = 20000$$
$$0.88x_1 + 0.91x_2 + 0.97x_3 \ge 18400$$
$$0.02x_1 + 0.01x_2 + 0.06x_3 \le 600$$
$$x_i \ge 0 \text{ and } i = 1,\ 2,\ 3.$$
$$(i = 1,\ 2,\ \ldots,\ 10)$$
$$b_i \ge 0\ (i = 1,\ 2,\ \ldots,\ 10) \text{ are binary variables.}$$

The numerical example of the SCPLCC problem was solved using LINGO software [50]:

$x_1 = 0$ $x_2 = 15{,}750$ $x_3 = 4250$ $y_1 = 57{,}000$ $y_2 = 410$ $y_3 = 604$
$y_4 = 10{,}000$ $y_5 = 5000$ $y_6 = 15{,}000$ $y_7 = 6000$ $y_8 = 25{,}000$ $y_9 = 100{,}000$
$y_{10} = 35{,}000$ $b_1 = 1$ $b_2 = 0$ $b_3 = 0$ $b_4 = 0$ $b_5 = 0$
$b_6 = 1$ $b_8 = 0$ $b_9 = 0$ $b_{10} = 0$
$z_1 = 57{,}000$ $z_2 = 515$ $z_3 = 655$ $z_4 = 33{,}125$.

The results demonstrated that RMCGP with geometric mean weighting and penalty weighting was a suitable approach for solving the SCPLCC problem.

#### 4.1.5. RMCGP with Mean Weighting, Penalty Weighting, and WLGP

To verify the RMCGP SCPLCC model problem, we modified the model through mean-weighted–penalty-weighted linear goal programming. The objective function was expressed as an equation presented in a study [13]:

$$\text{MIN} = \frac{\theta_1}{T_1}dn_1^- + \frac{\theta_2}{T_2}dp_2^+ + \frac{\theta_3}{T_3}dn_3^- + \frac{\theta_4}{T_4}dn_4^- + \frac{\theta_5}{T_5}dp_5^+ + \frac{\theta_6}{T_6}dp_6^+ + \frac{\theta_7}{T_7}dn_7^- + \frac{\theta_8}{T_8}dn_8^+ + \frac{\theta_9}{T_9}dn_9^- + \frac{\theta_{10}}{T_{10}}dn_{10}^-$$

where

$\theta_i = i^{th}$ the $i$th weighted geometric mean; $i = 1, 2, 3, \ldots, 10$
$\theta_1 = 0.2958$; $\theta_2 = 0.0579$; $\theta_3 = 0.0863$; $\theta_4 = 0.0365$; $\theta_5 = 0.1291$; $\theta_6 = 0.1254$; $\theta_7 = 0.0392$; $\theta_8 = 0.0199$; $\theta_9 = 0.0151$; and $\theta_{10} = 0.1949$.
$T_i$ = normalization constant of the $i$th goal; $i = 1, 2, 3, \ldots, 10$
$T_1$ = 57,000 (total net cost goal)
$T_2$ = 515 (total rejection goal)
$T_3$ = 655 (total late delivery goal)
$T_4$ = 33,125 (total PLCC goal)
$T_5$ = 5000 (total capacity constraint goal of supplier 1)
$T_6$ = 15,000 (total capacity constraint goal of supplier 2)

$T_7$ = 6000 (total capacity constraint goal of supplier 3)
$T_8$ = 25,000 (total budget constraint goal of supplier 1)
$T_9$ = 100,000 (total budget constraint goal of supplier 2)
$T_{10}$ = 35,000 (total budget constraint goal of supplier 3)

Subsequently, the numerical example of the SCPLCC problem was solved using LINGO software [50]:

$x_1$ = 1034, $x_2$ = 15,000, $x_3$ = 3965, $y_1$ = 57,000, $y_2$ = 410, $y_3$ = 604, $y_4$ = 10,000, $y_5$ = 5000, $y_6$ = 15,000, $y_7$ = 6000, $y_8$ = 25,000, $y_9$ = 100,000, $y_{10}$ = 35,000, $z_1$ = 56,896.55, $z_2$ = 541.37, $z_3$ = 658.62, $z_4$ = 332,107.

An RMCGP SCPLCC model to which WLGP was applied should yield similar results to an RMCGP SCPLCC model to which WLGP was applied. The results indicated that supplier 2 is the most suitable.

## 4.2. Summary of Results Obtained under All Approaches

Table 5 presents a summary of the results obtained under all goal programming approaches.

**Table 5.** Results obtained through all goal programming approaches.

| Approach | Zimmermann's Additive Weighted (FMOLP) | Lin's Weighted Max-Min (FMOLP) | RMCGP with Mean Weighting NO Penalty-Weighted | RMCGP with Geometric Mean Weighting Penalty-Weighted | RMCGP with Geometric Mean Weighting Penalty-Weighted and WLGP |
|---|---|---|---|---|---|
| Objective | | | | | |
| Net cost z1 | 57,000 | 57,000 | 68,383 | 57,000 | 56,896 |
| Rejection z2 | 521 | 515 | 583 | 515 | 541 |
| Late deliveries z3 | 656 | 655 | 850 | 655 | 658 |
| Product Life cycle cost z4 | 33,162 | 33,125 | 42,116 | 33,125 | 33,210 |
| Order quantity x1 | 240 | 0 | 15,000 | 0 | 1034 |
| Order quantity x2 | 15,570 | 15,750 | 9166 | 15,750 | 15,000 |
| Order quantity x3 | 4190 | 4250 | 5833 | 4250 | 3965 |
| Capacity restrictions | | | | | |
| Supplier 1 | 5500 | 5500 | 5000 ($y_5$) | 5000 ($y_5$) | 5000 ($y_5$) |
| Supplier 2 | 16,500 | 16,500 | 15,000 ($y_6$) | 15,000 ($y_6$) | 15,000 ($y_6$) |
| Supplier 3 | 6600 | 6600 | 6000 ($y_7$) | 6000 ($y_7$) | 6000 ($y_7$) |
| Budget restrictions | | | | | |
| Supplier 1 | 27,500 | 27,500 | 25,000 ($y_8$) | 25,000 ($y_8$) | 25,000 ($y_8$) |
| Supplier 2 | 110,000 | 110,000 | 100,000 ($y_9$) | 100,000 ($y_9$) | 100,000 ($y_9$) |
| Supplier 3 | 38,500 | 38,500 | 35,000 ($y_{10}$) | 35,000 ($y_{10}$) | 35,000 ($y_{10}$) |

## 4.3. Discussion of Results Obtained under the Two Approaches

Notably, the weighted max–min method developed by Lin [39] 44 yielded the same solution to the SCPLCC problem as did the RMCGP method involving geometric mean weighting and penalty weighting.

RMCGP involving geometric mean weighting and no penalty weighting revealed that the lower-bound order quantity of supplier 1 was 5000. This is attributable to the lack of penalty-weighted constraints. Moreover, $b_1$ = 1 and $b_5$ = 1 (Table 5).

RMCGP involving mean weighting and penalty weighting yielded $b_1$ = 1 and $b_6$ = 1, and the upper bound of the order quantity of supplier 1 exceeded 15,000; specifically, $x_2$ = 15,750. To ensure that the net cost rejection or late delivery goal is met, zero should be obtained (e.g., $b_1$ = 1 and $b_6$ = 1), and the auxiliary constraints of $b_i$ should be applied to the order quantity adjustment (Table 5).

To compare the results of the two verification approaches and increase the accuracy, we applied RMCGP with geometric mean weighting and penalty weighting. We also implemented a WLGP model. Figure 3 shows that the order quantity, net cost, rejections, late deliveries, PLCC, production capacity, and budget under the MCGP model approach were lower than those under the MOLP model approach. In sum, the RMCGP SCPLCC model yielded the same results as did the RMCGP with geometric mean weighting and penalty weighting and the WLGP model (refer to the results for the fifth model in Table 5).

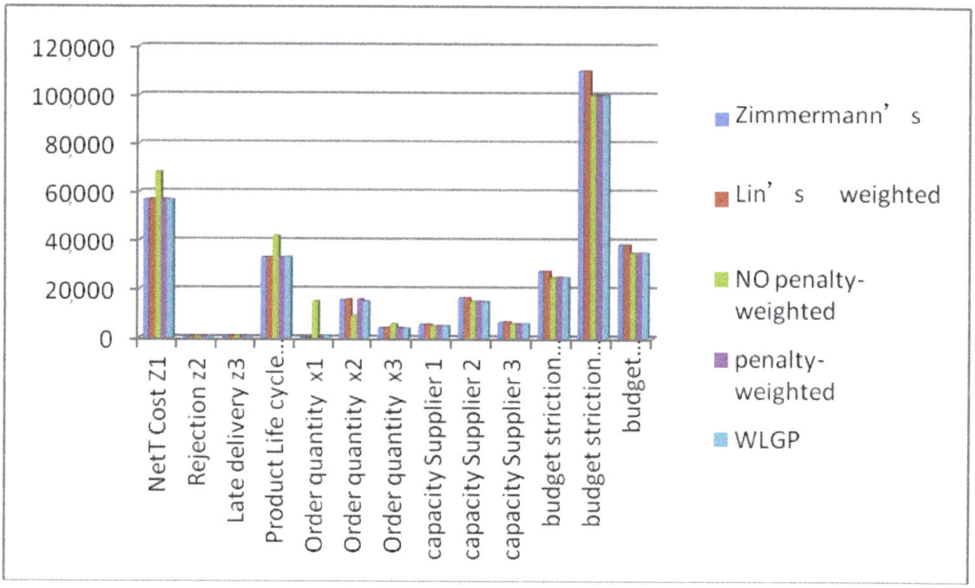


**Figure 3.** Results obtained using revised multichoice goal programming (RMCGP), multiobjective linear programming (MOLP) model approaches.

*4.4. Sensitivity Analysis*

To test the RMCGP and the WLGP models using geometric mean weighting and penalty weighting, we employed a method presented by Ho [51]. Table 6 and Figure 4 demonstrate that if managers wish to diversify the supplier selection risk, decision-makers can set λ to 0 or 0.2 to divide low order quantities among the three suppliers. Otherwise, they can set λ to 0.8 or 1 to obtain relatively higher achievement levels.

**Table 6.** Sensitivity analysis results of the mean-weighted–penalty-weighted linear goal programming model.

| | λ = 0 | λ = 0.2 | λ = 0.5 | λ = 0.8 | λ = 1 |
|---|---|---|---|---|---|
| Z1 | 56,897 | 56,897 | 56,897 | 56,898 | 57,000 |
| Z2 | 541 | 541 | 541 | 540 | 515 |
| Z3 | 658 | 658 | 658 | 658 | 655 |
| Z4 | 33,210 | 33,209 | 33,209 | 33,208 | 33,125 |
| X1 | 1030 | 1029 | 1026 | 1013 | 0 |
| X2 | 15,003 | 15,003 | 15,006 | 15,015 | 15,750 |
| X3 | 3966 | 3966 | 3967 | 3971 | 4250 |

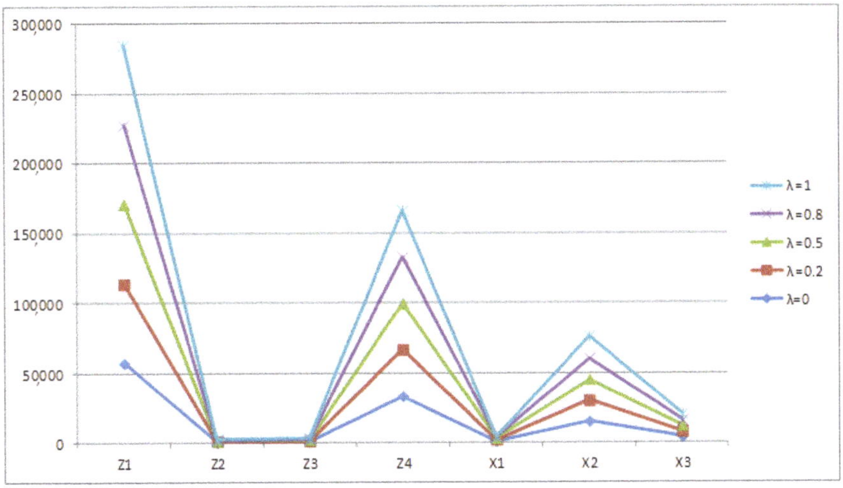


**Figure 4.** Sensitivity analysis results of the WLGP model.

### *4.5. Discussion*

A short review of the key results from the study is as follows. For the SCPLCC sustainable procurement challenge, the WLGP approach model and RMCGP with geometric mean weighting and penalty weighting are appropriate (from Table 6, we can discover $x_2$ order quantity value 15,750 is acceptable).

The two approaches to tackling the SCPLCC problem provided here can assist decision-makers in improving supplier selection and provide real-world consequences for the LED sector in terms of lowering PLCC costs, minimizing net costs, reducing rejections, and meeting PLCC objectives.

## **5. Conclusions and Managerial Implications**

### *5.1. Conclusions*

Because high-tech companies consider numerous factors (e.g., net cost, rejections, late deliveries, PLCC strategies), achieving only one objective is insufficient for PLCC reduction. SC models for solving this SCPLCC problem have seldom been investigated. In the present study, we solved the SCPLCC problem through two approaches, both of which employed geometric mean weighting. The results serve as a managerial reference for the examined company—specifically, for determining vendor quotas in SCM when the capacities and budget constraints of each vendor are uncertain (given a lower or upper bound). The first approach, which considered the uncertainty of the fuzzy model, used a linear membership function, and the entire formulation was solved through a fuzzy multiobjective programming approach. The SCPLCC problem was then transformed through a weighted max–min method, as well as through the application of an AW MOLP model and a corresponding crisp, single-objective linear programming approach. The second approach, which involved RMCGP with geometric mean weighting and penalty weighting, emphasized the supply of a high-quality product and PLCC reduction. Furthermore, this approach guided the relationships between goals in the multiobjective problem.

### *5.2. Managerial Implications*

Our study has the following managerial implications.

First, the SCPLCC problem can easily be solved using commercially available linear programming software such as LINDO and LINGO.

Second, our solution to the SCPLCC problem is more comprehensive than that achieved through linear programming, conventional goal programming, and other deterministic methods applied when information vagueness is a concern. We transformed the SCPLCC problem into a weighted max–min fuzzy programming model with lower computational complexity, thus simplifying the application of fuzzy methodology [20].

Third, in real-world SCM scenarios, the provision of deterministic values of system parameters (e.g., production capacity and constraints) is unnecessary.

Fourth, our findings can assist managers of LED manufacturers in identifying associations between SSCM and RSCM, thereby facilitating informed decision-making in terms of both sustainability and resilience. We assert that transparency, as well as the consideration of the PLCC model in SCM and stakeholder engagement, can help firms gain competitive advantage. Our findings revealed several complementary and conflicting relationships among the formative elements of SSCM and RSCM. For example, flexibility enables sustainability in SCs but is at least partially discordant with long-term firm–supplier relationships. Our examination of a real-world case adds to the empirical evidence on the implementation of SSCM and RSCM. However, the generalizability of these results to other industries or regions is questionable given potential differences in managerial practices and other features.

Fifth, under the RMCGP approach [47], auxiliary constraints can inform between-goal relationships in multiobjective problems. Regarding the example of the SCPLCC problem, auxiliary constraints $b_i$ facilitated goal realization (e.g., through the adjustment of order quota allocation). RMCGP is relevant to managerial decision-making problems. We modified an RMCGP approach proposed by Chang [47] involving weighted goal programming to solve the SCPLCC problem, accounting for multiple target levels by employing the multiplicative terms of binary variables. Regarding the mapping of one goal to numerous aspiration levels, under certain conditions, decision-makers may base their decisions on a goal that can be achieved on various aspiration level [52]. According to Little [53], operational models are only useful if they are simple, robust, easily controllable, adaptive, easily communicated, and address essential SCPLCC issues. In particular, our SCPLCC model satisfies the first four criteria [27].

We tested our SCPLCC model on a Taiwanese LED company. Because having precise knowledge of all parameters is not required, high-tech companies can easily use our two goal-programming approaches to select suppliers in a fuzzy environment. The two present methods for solving the SCPLCC problem can help decision-makers optimize supplier selection. They also enable managers (including purchasing managers) to manage SC performance and SSCM goals on the basis of factors such as the net cost, number of rejections, number of late deliveries, and overarching PLCC goals.

### 5.3. Limitations

To validate the SCPLCC problem, overcome the shortcomings of the two goal programming approaches, and ensure accurate outcomes, we compared the models with the RMCGP and weighted goal programming approaches. However, should decision-makers employ a hybrid RMCGP method with a multicriteria decision-making approach, they may obtain unexpected or confusing results.

### 5.4. Future Directions

To solve SCPLCC problems in scenarios involving multiple suppliers, our SCPLCC models can be integrated with other goal programming approaches, including multicriteria decision-making approaches such as de novo programming, the rough set approach [54,55], or the neutrosophic set approach [56].

**Author Contributions:** Conceptualization, Y.-J.T.; formal analysis, G.-Y.G.; C.-S.T. and Y.-J.T.; writing—original draft preparation, Y.-J.T. and C.-S.T.; writing—review and editing, C.-S.T. and Y.-J.T.; planning all works in the study, H.-S.L.; supervision, Y.-S.L. and H.-S.L. All authors have read and agreed to the published version of the manuscript.

**Funding:** This research received no external funding.

**Institutional Review Board Statement:** Not applicable.

**Informed Consent Statement:** Not applicable.

**Data Availability Statement:** The study did not report any data.

**Conflicts of Interest:** The authors declare no conflict of interest.

## References

1. Choi, J.; Bai, J.; Geunes, H. Manufacturing delivery performance for supply chain management. *Math. Compu. Model.* **2007**, *45*, 11–20. [CrossRef]
2. Li, G.; Yamaguchi, M.; Nagai, M. A grey-based decision-making approach to the supplier selection problem. *Math. Comp. Model.* **2007**, *46*, 573–581. [CrossRef]
3. Michaels, R.; Kumar, A.; Samu, S. Activity-specific role stress in purchasing. *Int. J. Purch. Mater. Manag.* **1995**, *31*, 11–19. [CrossRef]
4. Patton, W.E. Individual and joint decision-making in industrial vendor selection. *J. Bus. Res.* **1997**, *38*, 115–122. [CrossRef]
5. Wang, G.; Hung, S.H.; Dismukes, J.P. Product-driven supply chain selection using integrated multi-criteria decision-making methodology. *Int. J. Prod. Econ.* **2004**, *91*, 1–15. [CrossRef]
6. Vidrová, Z. Supply chain management in the aspect of globalization. *SHS Web Conferences* **2020**, *74*, 4031. [CrossRef]
7. Suryaningtyas, D.; Sudiro, A.; Eka, T.A.; Irawanto, D.W. Organizational resilience and organizational performance: Examining the mediating roles of resilient leadership and organizational culture. *Acad. Strateg. Manag. J.* **2019**, *18*, 1–7.
8. Fiedler, K.; Lord, S.; Czarnezki, J.J. Life cycle costing and food systems: Concepts, trends, and challenges of impact valuation. *Mich. J. Envtl. Admin. L.* **2018**, *8*, 1. [CrossRef]
9. Shashi, S.R.; Centobelli, P.; Cerchione, R. Evaluating partnerships in sustainability-oriented food supply chain: A five-stage performance measurement model. *Energies* **2018**, *11*, 3473. [CrossRef]
10. Wang, C.T.; Su, S.J. Strategic capacity planning for light emitting diode (LED) supply chains across Taiwan and China. *J. Oper. Res. Soc.* **2015**, *66*, 1989–2003. [CrossRef]
11. Negri, M.; Cagno, E.; Colicchia, C.; Sarkis, J. Integrating sustainability and resilience in the supply chain: A systematic literature review and a research agenda. *Bus. Strat. Environ.* **2021**, *30*, 1–29. [CrossRef]
12. Zhu, Q.; Sarkis, J.; Lai, K.H. Choosing the right approach to green your supply chains. *Mod. Supply Chain Res. Appls.* **2019**, *1*, 54–67. [CrossRef]
13. Karutz, R.; Riedner, L.; Vega, L.R.; Stumpf, L.; Damert, M. Compromise or complement? exploring the interactions between sustainable and resilient supply chain management. *J. Supply Chain Oper. Resil.* **2018**, *3*, 117–142. [CrossRef]
14. Alonso-Muñoz, S.; González-Sánchez, R.; Siligardi, C.; García-Muiña, F.E. New circular networks in resilient supply chains: An external capital perspective. *Sustainability* **2021**, *13*, 6130. [CrossRef]
15. Zavala-Alcívar, A.; Verdecho, M.-J.; Alfaro-Saiz, J.-J. A conceptual framework to manage resilience and increase sustainability in the supply chain. *Sustainability* **2020**, *12*, 6300. [CrossRef]
16. Backes, J.G.; Traverso, M. Application of life cycle sustainability assessment in the construction sector: A systematic literature review. *Processes* **2021**, *9*, 1248. [CrossRef]
17. Kaviani, M.A.; Peykam, A.; Khan, S.A.; Brahimi, N.; Niknam, R. A new weighted fuzzy programming model for supplier selection and order allocation in the food industry. *J. Model. Manag.* **2020**, *15*, 381–406. [CrossRef]
18. Noppasorn, S.; Navee, C. Integrated possibilistic linear programming with beta-skewness degree for a fuzzy multi-objective aggregate production panning problem under uncertain environments. *Fuzzy Inf. Eng.* **2020**, *12*, 355–380.
19. Amid, A.; Ghodsypour, S.H.; O'Brien, C.A. Weighted max-min model for fuzzy multi-objective supplier selection in a supply chain. *Int. J. Prod. Econ.* **2010**, *131*, 139–145. [CrossRef]
20. Kumar, M.; Vrat, P.; Shankar, R. A fuzzy goal programming approach for vendor selection problem in a supply chain. *Int. J. Prod. Econo.* **2006**, *101*, 273–285. [CrossRef]
21. Shen, C.W.; Peng, Y.T.; Tu, C.S. Considering Product Life Cycle Cost Purchasing Strategy for Solving Vendor Selection Problems. *Sustainability* **2019**, *11*, 3739. [CrossRef]
22. Kagnicioglu, C.H. A fuzzy multi-objective programming approach for supplier selection in a supply chain. *Bus. Rev. Camb.* **2006**, *6*, 107–115.
23. Nasr, A.K.; Tavana, M.; Alavi, B.; Mina, H. A novel fuzzy multi-objective circular supplier selection and order allocation model for sustainable closed-loop supply chains. *J. Clean Prod.* **2021**, *287*, 124994. [CrossRef]
24. Beiki, H.; Seyedhosseini, S.M.; Ponkratov, V.V.; Zekiy, A.O.; Ivanov, S.A. Addressing a sustainable supplier selection and order allocation problem by an integrated approach: A case of automobile manufacturing. *J. Ind. Prod. Eng.* **2021**, *38*, 239–253. [CrossRef]
25. Dickson, G.W. An analysis of vendor selection systems and decisions. *J. Purch.* **1966**, *2*, 5–17. [CrossRef]
26. Wind, Y.; Robinson, P.J. The determinants of vendor selection: The evaluation function approach. *J. Pur. Mater. Manag.* **1968**, *4*, 29–41. [CrossRef]

27. Mo'Ath, A.L.; Ahmad, K.J.; Mohd, H.Z.; Afifi, M.D. Weighted linear goal programming approach for solving budgetary manufacturing process. *Far East J. Math. Sci.* **2017**, *101*, 1993–2021.
28. Umarusman, N. Solution proposal for supplier selection: Problem an application in agricultural machinery sector with global criterion method. *Dokuz Eylul Univ. Iktis. Ve Idari Bilimler Dergisi* **2018**, *33*, 353–368. [CrossRef]
29. Budzinski, M.; Sisca, M.; Thrän, D. Consequential LCA and LCC using linear programming: An illustrative example of biorefineries. *Int. J. Life Cycle Assess.* **2019**, *24*, 2191–2205. [CrossRef]
30. Ojo, O.O.; Farayibi, P.K.; Akinnuli, B.O. Modified goal programming model for limited available budget alocation for equipment procurement under inflation condition. *Adv. Res.* **2020**, *21*, 25–35. [CrossRef]
31. Hocine, A.; Zhuang, Z.Y.; Kouaissah, N.; Li, D.C. Weighted-additive fuzzy multi-choice goal programming (WA-FMCGP) for Supporting Renewable Energy Site Selection Decisions. *Eur. J. Oper. Res.* **2020**, *285*, 642–652. [CrossRef]
32. Hardy, C.; Bhakoo, V.; Maguire, S. A new methodology for supply chain management: Discourse analysis and its potential for theoretical advancement. *J. Supply Chain Manag.* **2020**, *56*, 19–35. [CrossRef]
33. Al-Husain, R.; Khorramshahgol, R. Incorporating analytical hierarchy process and goal programming to design responsive and efficient supply chains. *Oper. Res. Pers.* **2020**, *7*, 100149. [CrossRef]
34. Giri, B.C.; Dey, S. Game theoretic models for a closed-loop supply chain with stochastic demand and backup sup plier under dual channel recycling. *Decis. Mak. Appl. Manag. Eng.* **2020**, *3*, 108–125.
35. Samanta, B.; Giri, B.C. A two-echelon supply chain model with price and warranty dependent demand and pro-rata warranty policy under cost sharing contract. *Decis. Mak. Appl. Manag. Eng.* **2021**, *4*, 47–75. [CrossRef]
36. Bahareh, S.O.; Kaveh, K.D.; Peiman, G. Solving a supply chain problem using two approaches of fuzzy goal programming based on TOPSIS and fuzzy preference relations. *Int. J. Ind. Syst. Eng.* **2021**, *13*, 27–48.
37. Tavan, M.; Kian, H.; Nasr, A.K.; Govindan, K.; Mina, H. A comprehensive framework for sustainable closed-loop supply chain network design. *J. Clean. Prod.* **2022**, *332*, 129777. [CrossRef]
38. Mabrouk, N.B. Green supplier selection using fuzzy delphi method for developing sustainable supply chain. *Dec. Sci. Lett.* **2021**, *10*, 63–70. [CrossRef]
39. Kumar, L.; Jain, P.K.; Sharma, A.K. A fuzzy goal programme–based sustainable Greenfield supply network design for tyre retreading industry. *Int. J. Adv. Manuf. Technol.* **2020**, *108*, 1–26. [CrossRef]
40. Zimmermann, H.J. Fuzzy programming and linear programming with several objective functions. *Fuzzy Sets Syst.* **1987**, *1*, 45–56. [CrossRef]
41. Bellman, R.E.; Zadeh, L.A. Decision making in a fuzzy environment. *Manag. Sci.* **1970**, *17*, 141–164. [CrossRef]
42. Lee, H.I.; Kang, H.Y.; Chang, C.T. Fuzzy multiple goal programming applied to TFT-LCD supplier selection by downstream manufacturers. *Exp. Syst. Appl.* **2009**, *36*, 6318–6325. [CrossRef]
43. Liao, C.N.; Kao, H.P. Supplier selection model using Taguchi loss fuction, analytical hierarchy process and multi-choice goal programming. *Comp. Indu. Eng.* **2010**, *58*, 571–577. [CrossRef]
44. Lin, C.C. A weighted max-min model for fuzzy goal programming. *Fuzzy Sets Syst.* **2004**, *142*, 407–420. [CrossRef]
45. Tiwari, R.N.; Dharmar, S.; Rao, J.R. Fuzzy goal programming—An additive model. *Fuzzy Sets Syst.* **1987**, *24*, 27–34. [CrossRef]
46. Soleimani, H.; Mohammadi, M.; Fadaki, M. Carbon-efficient closed-loop supply chain network: An integrated modeling approach under uncertainty. *Environ Sci Pollut Res.* **2021**. [CrossRef]
47. Chang, C.T. Revised multi-choice goal programming. *Appl. Math. Model.* **2008**, *32*, 2587–2595. [CrossRef]
48. Narsimhan, R. An analytical approach to supplier selection. *J. Purch. Mater. Manag.* **1983**, *19*, 27–32. [CrossRef]
49. Zimmermann, H.J. Fuzzy Mathematical Programming. In *Advances in Sensitivity Analysis and Parametic Programming*; Gal, T., Greenberg, H.J., Eds.; International Series in Operations Research & Management Science; Springer: Boston, MA, USA, 1997; Volume 6. [CrossRef]
50. Schrage, L. *LINGO Release 8.0*; LINGO System Inc.: Chicago, IL, USA, 2002.
51. Ho, H. The supplier selection problem of a manufacturing company using the weighted multi-choice goal programming and MINMAX multi-choice goal programming. *Appl. Math. Model.* **2019**, *75*, 819–836. [CrossRef]
52. Chang, C.T. Multi-choice goal programming. *Omega* **2007**, *35*, 389–396. [CrossRef]
53. Little, D.C. Models and manager: The concept of a decision calculus. *Manag. Sci.* **1970**, *16*, 466–485. [CrossRef]
54. Stević, Ž.; Tanackov, I.; Vasiljević, M.; Rikalović, A. Supplier evaluation criteria: AHP Rough approach. In Proceedings of the October Conference: XVII International Scientific Conference on Industrial Systems (IS'17), Novi Sad, Serbia, 4–6 October 2017.
55. Stanković, M.; Gladović, P.; Popović, V. Determining the importance of the criteria of traffic accessibility using fuzzy AHP and rough AHP method. *Decis. Mak. Appl. Manag. Eng.* **2019**, *2*, 86–104. [CrossRef]
56. Bausys, R.; Zavadskas, E.K.; Kaklauskas, A. Application of neutrosophic set to multicriteria decision making by COPRAS. *Econ. Comput. Econ. Cybern Stud. Res.* **2015**, *49*, 9–106.

MDPI
St. Alban-Anlage 66
4052 Basel
Switzerland
Tel. +41 61 683 77 34
Fax +41 61 302 89 18
www.mdpi.com

*Sustainability* Editorial Office
E-mail: sustainability@mdpi.com
www.mdpi.com/journal/sustainability

www.ingramcontent.com/pod-product-compliance
Lightning Source LLC
Chambersburg PA
CBHW061558220926
51850CB00005B/367
* 9 7 8 3 0 3 6 5 5 1 8 9 0 *